Learning and Analytics in Intelligent Systems

Volume 31

The main aim of the series is to make available a publication of books in hard copy form and soft copy form on all aspects of learning, analytics and advanced intelligent systems and related technologies. The mentioned disciplines are strongly related and complement one another significantly. Thus, the series encourages cross-fertilization highlighting research and knowledge of common interest. The series allows a unified/integrated approach to themes and topics in these scientific disciplines which will result in significant cross-fertilization and research dissemination. To maximize dissemination of research results and knowledge in these disciplines, the series publishes edited books, monographs, handbooks, textbooks and conference proceedings.

Indexed by EI Compendex.

Srikanta Patnaik · Fred Paas
Editors

Recent Trends in Educational Technology and Administration

Proceedings of the 2nd International Conference on Educational Technology and Administration

Springer

Editors
Srikanta Patnaik
Interscience Institute of Management and Technology
Bhubaneswar, Odisha, India

Fred Paas
Erasmus University Rotterdam
Rotterdam, The Netherlands

ISSN 2662-3447 ISSN 2662-3455 (electronic)
Learning and Analytics in Intelligent Systems
ISBN 978-3-031-29015-2 ISBN 978-3-031-29016-9 (eBook)
https://doi.org/10.1007/978-3-031-29016-9

This Springer imprint is published by the registered company Springer Nature Switzerland AG
The registered company address is: Gewerbestrasse 11, 6330 Cham, Switzerland

Editorial

On the behalf of the organizing committee of the International Conference of Educational Technology and Administration (EduTA-2022), held during December 18–20, 2022, I would like to present you all this proceeding. This conference has given a variety of researchers, educators, academicians, and scholars a forum to exchange ideas, theories, research findings, and experimental observations. This conference has also helped to offer a thorough picture of the state of the educational systems, taking into account current challenges and best practices while locating various research gaps and suggesting new regulations. Many attendees who still needed to submit research papers to the conference did so to learn more about their respective fields in the context of the fast-changing environment. With recent breakthroughs in information and communication technologies (ICT), practically all aspects of the world have been influenced by the digital era. Moreover, ICT presents a complex connection in the education sector; it has the potential to have the most significant influence. In today's world, educators play a critical role in utilizing ICT in the classroom.

Again, with new technologies such as IoT, Big Data, Cloud Computing, and Artificial Intelligence, the education sector is more focusing on student-centered model. Also, to capture the benefits of rapid technological advances, educational institutions must embrace ICT-enabled reforms while also modeling new teaching pedagogies and learning aids. Additionally, Pandemic-induced education reformation entails enhancing instructors' talents and attitudes about e-content preparation and disseminating outstanding knowledge in a variety of critical sectors. However, due to improvements in ICT, education is no longer restricted to physical participation in the digital era. Moreover, Virtual classrooms have emerged as the education system's savior in this global pandemic due to the extended benefits that come with removing geographical barriers, such as the ability to record and retrieve sessions, as well as speedier organizing and delivery of resources. Advanced interactive lecture sessions can be held over the internet and attended from any remote location with internet access. While virtual classrooms were formerly limited to MOOCs, conferences, and workshops, they have nearly surpassed traditional teaching methods in the current global pandemic scenario. Moreover, innovative policies and different practices must be revised to ensure the quality of education and learning materials distributed in virtual classes while encouraging strategic goals and experimentation.

With all of these urgent demands in mind, EduTA-2022 has concentrated on not only identifying deficiencies in existing educational policies, but also on recommending future directions for reforming new policies while providing a clear image of the current situation. The conference had a positive reaction from diverse contributors and received over 100 papers from around the world. Out of which 49 selective papers have been shortlisted by the organizing committee on the basis of various metrics including content quality and organization, illustration style, relevancy, and scope of the research, innovation, etc. These papers have been broadly categorized into five major sections as follows: (i) Big Data in Education, (ii) Reform and Policy in Education, (iii) Integrated-Learning, and (iv) Technology for Education.

The first section in this proceeding has been categorized as the Big Data in Education that presents the papers that are directly or indirectly related to adoption of Big Data to solve various problems in the administration of the education system. Around 10 papers have been categorized to this section that focuses on addressing issues such as construction of PE teaching evaluation system using Big Data, art and design education resources based on Big Data, Online Learning Evaluation Analysis based on Big Data, evaluation and decision-making based on Big Data warehouse technology, prediction of aggressive behavior of Adults in educational institutions, OBE based on big data management, intelligent classroom to improve students' oral English ability and literacy, etc.

The next section assembles around 10 papers that deal with the proposals presented by contributors suggesting potential various reforms and policies in Education. This section exhibits various research works that involve reformation of the education system, its underlying principles, pedagogies, tools, and technologies. It focuses on identifying the gaps and integration of new strategies into traditional teaching practices.

Adoption of situational teaching methods, reform and practice for traditional engineering courses, environment evaluation of kindergarten, reforms of talent training mode, outcome-based education as paradigm shift, ideological and political reforms, for CEC educational model, etc.

The third section categorized as Integrated-Learning lists 10 papers that explores and proposes novel approaches that involve intelligent education systems, mobile learning, e-learning, etc. The contributors not only present proposals but also present future directions for wide adoption of the proposed approaches. Some of these works include application of OLAP and web technology for evaluating the quality of higher education, construction of industry-based integrated education mechanism, translation teaching platform construction, campus information, aggregation platform, blended teaching model, intelligent labor education-based system, collaborative education platform, and integration of mathematical model into teaching platforms.

The last section is a collection of 18 papers into Technology for Education on the basis of the adoption of advanced technologies in education system by the contributors in the papers. These papers address problems like using data mining technology for managing college curriculum, adoption of block-chain technology to education arena, application of Bayesian network to vocational English ability examination, evaluation of cloud-based platforms, Industry-Education Integrated Ecosystem for Vocational Education, network Psychological Education Model Based on Cloud Computing, voice answer system for virtual volunteer teachers, English interpretation learning based on artificial intelligence, and Quantitative analysis of student emotion based on face recognition technology. With this the proceeding of EduTA-2022 presents an extremely informative compilation of papers addressing diverse issues in the current education.

We thank a lot to all the conference committee members and participants. We are also thankful to the organizers, support from the Education Department of Hunan Province Science Research Project: 21C1632, and conference service from IRNet. We want to put it on record the support, guidance, and timely help from the professional production team of Springer and I am sure the readers shall find lots of insights in this domain.

Prof. Srikanta Patnaik
Prof. Fred Paas

Contents

Research on the Construction of PE Teaching Evaluation System Under the Background of Big Data

Aimin Zhang and Feifei Li

Abstract In the reform and development of modern education, the application of big data technology to build a perfect PE teaching evaluation system provides perfect data support for practical teaching evaluation, which can not only help professional teachers to scientifically evaluate education and teaching results, but also directly feed back the final evaluation results. In college education and teaching, the rational application of big data technology can ensure that the evaluation indicators obtained in practice are reasonable and scientific, and continuously optimize the teaching level of professional teachers. Therefore, after understanding the research and application of big data technology, this paper deeply discusses how to build the evaluation system of physical education under the background of big data according to the accumulated experience of physical education evaluation in colleges and universities in recent years. The final research results prove that big data technology plays an important role in PE teaching evaluation.

Keywords Big data · College physical education · Teaching evaluation system · Index elements · The evaluation process

1 Introduction

Evaluating the effect of physical education teaching in colleges and universities is the basic condition of constructing the teaching evaluation system, which has a strong guiding effect on the educational content. By studying the current situation of physical education teaching in colleges and universities, we can see that the evaluation should start from the two aspects of teachers and students. On the one hand, when

A. Zhang (✉)
Yi Li Normal University, Yi Li, China
e-mail: zam19790315@163.com

F. Li
Physical Education College, Yili Normal University, Yi Li, China

S. Patnaik and F. Paas (eds.), *Recent Trends in Educational Technology and Administration*, Learning and Analytics in Intelligent Systems 31,
https://doi.org/10.1007/978-3-031-29016-9_1

evaluating teachers, we should focus on their innovation ability, teaching achievements, sports ability, application methods, scientific research ability, etc. On the other hand, in the evaluation of students, the main collection of theoretical knowledge, application skills, physical quality, sports interest, classroom performance, innovation ability and so on. According to the accumulated teaching experience in recent years, teachers become more and more active in the classroom, and the teaching effect can reach the expected goal within time. Students can participate in a number of activities organized by teachers independently and complete the learning tasks in strict accordance with the requirements of teachers, which can also improve the comprehensive teaching ability of teachers. But at present, the scientific research ability and innovation ability of physical education teachers in colleges and universities are insufficient, and the students' sports interest and innovation consciousness also have some problems. Therefore, in the innovation and development of big data technology, the evaluation system of college physical education teaching should be comprehensively evaluated from these aspects, in order to provide effective basis for practical education innovation.

According to the construction of the current physical education evaluation system in colleges and universities, it usually has to abide by the following principles: first, comprehensiveness. After defining the evaluation framework, the evaluation contents should be refined to ensure the perfection of the selected evaluation indicators and avoid duplication or similarity problems. Secondly, implementation. All indicators contained in the evaluation system should conform to quantitative standards, the actual content should be clear and perfect, and the influence of subjective factors should be avoided as far as possible. Again, the corresponding. The quantitative results of all indicators should meet the requirements of the big data analysis system to ensure that more effective data can be mastered in the research process, and the data information and analysis system can be corresponded one by one; Finally, personality traits. To ensure that all evaluation data can show students' personality characteristics in the analysis results, after the construction of a perfect physical education evaluation system, it can help teachers to understand students' learning ability and individual differences as soon as possible, so as to provide effective basis for practical teaching reform.

Integration under the background in the era of big data, to construct evaluation system of college PE teaching, let the online and offline together during class, clear the whole process of college PE teaching activities, to ensure that the university teachers can improve the understanding of the students' physical quality, targeted design to optimize the teaching content, to strengthen the students' interest in learning, improve the quality of classroom teaching. At the same time, under the guidance of the evaluation system, college teachers can make use of diversified teaching forms, organize and design rich and changeable teaching activities, and cultivate students' awareness of lifelong exercise actively on the basis of satisfying the lifelong physical education teaching concept.

On the basis of understanding the current research status of physical education evaluation in colleges and universities in the era of big data, and according to the basic requirements of physical education in colleges and universities, this paper deeply discusses how to build a physical education evaluation system with big data technology structure as the core, so as to provide effective basis for modern college education reform [4–6].

2 Method

2.1 Index System

The physical education teaching evaluation system is constructed based on big data technology. The specific contents are shown in Tables 1 and 2. Table 1 represents the teaching evaluation index system of physical education teachers, and Table 2 represents the teaching evaluation index system of students. Combining the two together, the final teaching evaluation results can be obtained:

2.2 Weight Distribution

In this paper, component analysis is used to clarify the weight coefficient of evaluation indicators. It is necessary to conduct comparative analysis on the correlation between indicators, and obtain the weight coefficient of each indicator after quantitative processing of the final results. First of all, the evaluation system should be constructed, focusing on the analysis of the obtained index content; Secondly, the hierarchical structure model should be built to divide all kinds of indicators according to different attributes and importance. Thirdly, the corresponding judgment matrix should be constructed according to the comparative analysis of all evaluation indicators at the same level. The specific matrix form is as follows [7–9]:

$$A = \begin{bmatrix} a_1 a_2 \ldots a_n \\ a_{11} a_{12} l, \ldots a_{1n} \\ a_{21} a_{22} \ldots a_{2n} \\ \ldots\ldots\ldots\ldots\ldots \\ a_{n1} a_{n2} \ldots a_{nm} \end{bmatrix}$$

Finally, the weight vector is calculated and the consistency test is performed. The maximum characteristic value of the matrix is, and the specific formula is:

CR = CI/RI.

In the above formula, CR stands for consistency ratio and CI stands for consistency index. The corresponding calculation formula is as follows:

Table 1 Teaching evaluation index system of PE teachers

Level indicators	The secondary indicators	Indicators show
Teaching preparation A1	Prepare before class B1	Arrive at the class place 5 min in advance, the venue equipment is fully prepared, safety measures in place, fully understand the situation of the students
	Teaching plan B2	The teaching documents are complete, the teaching plan is standardized and orderly, the current task is clear, the teaching content is arranged reasonably, the teaching method is applied appropriately, and the time distribution is reasonable
	Teaching Etiquette B3	Dress standard, good appearance, full of spirit, enthusiasm, greet students
	Class routine B4	Arrange novices reasonably and strictly abide by the time of arrival and departure
	Teaching attitude B5	Students as the main body, teaching seriously responsible, seriously guide students to participate in sports activities, and students actively interact
Teaching process A2	Teaching organization B6	Orderly organization, good adaptability, outstanding key and difficult points, reasonable allocation of learning and training time, continuity of teaching process
	Teaching Method B7	The explanation and demonstration are concise, accurate, standardized and tailored to students' abilities
	Teaching content B8	The teaching content arrangement is reasonable, the PE course progress is reasonable, the teaching content is novel
	Exercise load B9	Students should not practice at less than 40%, and practice at an appropriate intensity
	Classroom atmosphere B10	The classroom atmosphere is active, the interaction with students is good, the students participate actively
Teaching effect A3	Exercise consciousness to develop B11	Students have the consciousness of participating in sports activities and exercising themselves

(continued)

Table 1 (continued)

Level indicators	The secondary indicators	Indicators show
	Motor skills B12	Students have a basic grasp of the knowledge, technology and skills, students' progress in sports skills
	Physical quality B13	Student physical condition and improvement degree
	Basic knowledge of sports theory B14	Students' mastery of basic physical education theories

Table 2 Teaching evaluation index system of students

Level indicators	The secondary indicators	Indicators show
Study Preparation C1	Preparation before Class D1	Arrive 5 min early to prepare for class
	Teaching Etiquette D2	Dress code, good appearance, full of spirit, enthusiasm, and teachers say hello
Learning process C2	Class routine D3	Attendance, not being late and leaving early
	Learning attitude D4	Take the initiative to participate in sports activities, take the initiative to think about repeated practice, devote oneself to, respect teachers, seriously cooperate with teachers to practice, accept guidance, solidarity and mutual assistance
	Friendship shows cooperation D5	Obey the rules, respect the opinions of others, have the spirit of cooperation, strive to shoulder the responsibility of learning, overcome difficulties with confidence
	Classroom atmosphere D6	Active classroom atmosphere, good interaction with teachers, active participation in sports activities
Learning Effect C3	Exercise consciousness to develop D7	Active participation in sports activities, self-training awareness
	Motor skills D8	Basic grasp of the knowledge, technology and skills, motor skills progress
	Physical quality D9	Physical condition and improvement degree
	Basic knowledge of sports theory D10	Master the basic knowledge of sports theory

Table 3 Value range of RI

Order number	1	2	3	4	5	6	7	8	9	10	11	12
RI	0.00	0.00	0.58	0.90	1.12	1.24	1.32	1.41	1.45	1.49	1.52	1.54

$$\lambda \max = \frac{1}{n}\sum_{i}\frac{(AW)i}{wi} : CI = \lambda \max - n/n - 1$$

AW represents the result of judgment matrix and feature vector, and R I represents the average random consistency index. The specific calculation results are shown in Table 3.

When CR is less than 0.1, it can be considered that the evaluation judgment matrix has complete consistency, and the obtained weight coefficient can intuitively show the importance of the index; otherwise, the matrix needs to be readjust to ensure its complete consistency.

2.3 *Evaluation System*

According to subject classification, big data under the background of physical education teaching evaluation system is divided into two kinds, one kind is only self assessment, another is to point to others evaluation, in this paper according to the main role, combining the evaluation index and index weight for the analysis of the similarities between main body and not gay, specific structure as shown in Fig. 1.

On the one hand, the student-subject-based evaluation includes three contents: first, it refers to students' self-evaluation, which belongs to students' self-knowledge in the learning process; Secondly, it refers to students' evaluation of teaching, which

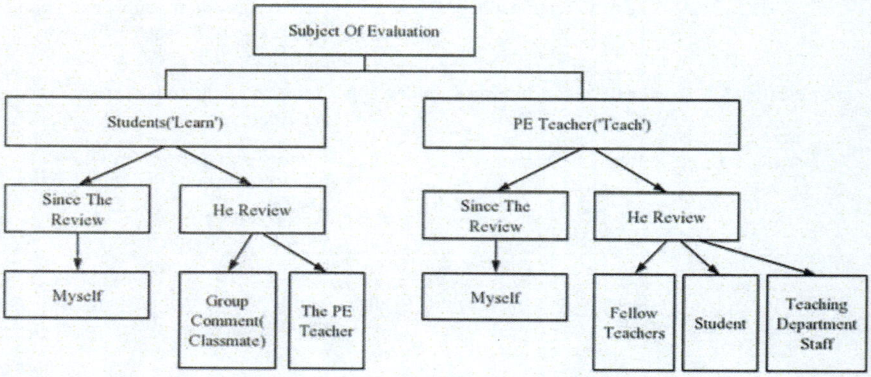

Fig. 1 Sports teaching evaluation architecture diagram...

needs to be conducted in a one-to-one manner. Finally, it refers to the evaluation of physical education teachers, who need to log in the teacher system to evaluate the learning status and final effect of students in the class. The framework diagram of the specific evaluation index system is as follows (Fig. 2).

On the other hand, the teacher-subject-based evaluation includes four contents: firstly, it refers to teachers' self-evaluation, which is a kind of recognition of self-teaching quality; Secondly, it refers to students' evaluation of teaching activities. Students can evaluate physical education teachers' class status after logging in their own evaluation account. Third, it refers to the peer evaluation activities, mainly using the way of audit evaluation; Finally, it refers to the evaluation and teaching activities of the competent department, which can be evaluated by sampling or auditing. The framework diagram of the specific evaluation index system is as follows (Fig. 3).

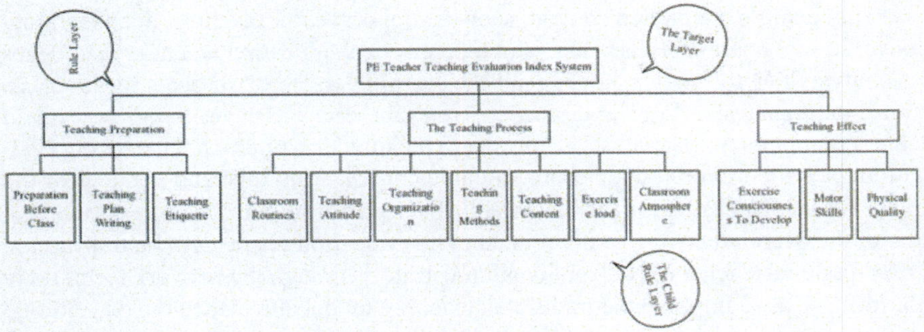

Fig. 2 Students physical education teaching evaluation index system of figure

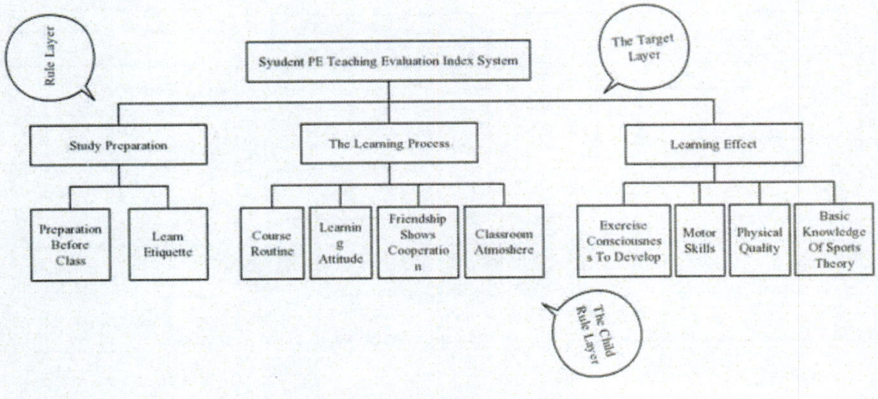

Fig. 3 Sports teachers' teaching evaluation index system of figure...

2.4 Evaluation Process

In the context of big data, the construction of college PHYSICAL education teaching evaluation system should also clarify the practical operation process, as shown in Fig. 4.

According to the above analysis, the actual work involves the following points: First, data collection. After the new project, the relevant evaluation data should be accurately input according to the project type and evaluation method. Therefore, during the operation of the evaluation system, it is necessary to systematically collect the basic information of the PE teachers, the basic information of the students, the evaluation information of college leaders and teachers and students, and finally choose the same format to store in the database; Secondly, data processing. On the basis of the acquisition of teaching data, the use of big data technology algorithms to deeply explore the hidden content, such as support vector machine, deep learning, convolutional neural network, etc., can help physical education teachers in colleges and universities to quickly understand the learning status of students in the class. Third, data analysis. After obtaining the relevant data of PE teaching, we should make a comprehensive analysis according to the direction of practical teaching evaluation, pay attention to obtain more valuable content from this, and make clear the final PE teaching evaluation results; Finally, the result is output. This stage is the final link of the overall teaching evaluation, and the evaluation results will be transmitted to professional teachers or university administrators through the network technology platform, so as to improve the traditional teaching model and clarify the educational objectives in the new era.

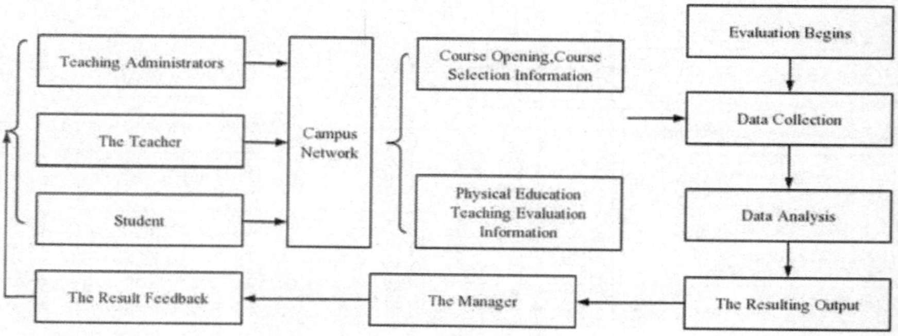

Fig. 4 Evaluation flow chart According to the above analysis

3 The Results

3.1 The Conclusion

First of all, the evaluation index should be clearly defined according to the knowledge of big data, to fully show the various elements during the physical education period, to ensure the distribution of the weight coefficient of the index is more reasonable, so as to present a scientific and objective evaluation. Secondly, we should master all the subjects of practical teaching evaluation, pay attention to the combination of this evaluation and other people's evaluation, and finally get two index systems of PE teachers and students. Finally, in the process of selecting the index system, we should fully consider the common and different characteristics of college PE teachers and students. For example, in teacher evaluation, the traditional teaching practice-oriented evaluation mode should be changed, and more attention should be paid to the teaching ability, professional quality and innovative consciousness of professional teachers. In student evaluation, the research focus should be placed on the cultivation of exercise consciousness and self-habit, to ensure that students can develop interest in physical exercise under the guidance of professional teachers, and gradually form a lifelong exercise consciousness.

3.2 Strategy

First of all, the evaluation content is comprehensive. In the context of the era of big data, the indicators of college physical education evaluation become more specific. The improved evaluation content is helpful for teachers to identify their own and students' strengths and weaknesses, so as to formulate more explicit teaching countermeasures. Secondly, the rational use of new technology. Big data as the basis of physical education teaching evaluation on the basis of technical architecture, with the continuous development of science and technology, the existing function more and more big data technology, can provide service more complex, so the university sports teaching evaluation work to also want to will shift, full use of big data technology platform to provide the service function, collect more sports teaching evaluation index system, Thus accelerate the pace of college education innovation. At the same time, the technology of data can help managers in colleges and universities and professional teachers forecast analysis the relationship between the various things and development patterns, for practice were used to predict the development trend of the education work, therefore the colleges and universities should strengthen the cultivation of professional and technical personnel, pay attention to in the process of education management, show with big data technology as the core advantage of teaching evaluation system, In this way, the prediction model is established to ensure that the data analysis results meet the needs of teaching activities. Finally, when obtaining the teaching evaluation results, the final quantitative results and corresponding relationship should be deeply analyzed to ensure that teachers and students can correctly understand the evaluation content, and effectively solve their own problems according to the results, so as to promote the orderly progress of college physical education teaching.

4 Conclusion

Under the background of comprehensive application of big data technology, physical education teaching activities in colleges and universities show diversified characteristics, gradually changing the limitation of traditional teaching concept, and build a physical education teaching evaluation system with big data technology as the core. From the perspective of practical application, the new evaluation system can better help universities and teachers and students to understand the practical teaching problems, rectify the practical teaching mode, and build a correct learning consciousness. On the one hand, colleges and universities can regard big data platform as the basic of development, and scientifically determine the evaluation index combined with the result of data analysis; on the other hand, college teachers and students should skillfully use big data technology service, integrate and analyze the final effect of practical teaching, change their teaching attitude, create a new teaching environment, fully demonstrate the advantages of physical education evaluation with big data technology as the core, and lay the foundation for the innovation and development of physical education in universities and China.

References

1. Z. Zhang. Construction and exploration of middle school physical education teaching evaluation system under the background of big data application. Education **8**, 1 (2021)
2. B. Ma, T. Wu, H. Sun, Research on PE teaching reform in colleges and universities under the background of big Data Era. J Heihe University, **11**(7), 2 (2021)
3. S. Li, Data analysis of special strength training in track and field jumping event from the perspective of big Data – Comment on "Research on Informatization Teaching Theory and Practice of PHYSICAL Education. Sci Technol Papers of China, **15**(8), 1 (2021)
4. Z. Zheng, Discussion on the construction of college PE teaching evaluation system under the background of big data application. Contemp. Sports Sci. Technol. **11**(28), 3
5. Y. Chi, Research on the Construction of college PE teaching evaluation system under the background of Big Data application. Commun World, **27**(2), 2 (2020)
6. W. Liu, W. Li, On the reform of physical education in colleges and universities under the background of big data era. New Curriculum Teaching: Electronic edition, Vol 4, pp 15–16 (2021)
7. L. Wang, Research on teaching behavior reform of college physical education teachers in the new era under the background of Big Data. Research on Innovation of Ice and Snow Sports **4**, 102–103 (2021)

8. X. Zhang, Physical education teaching and evaluation in primary schools – physical education teaching effect based on Big Data analysis. Education **36**, 1 (2021)
9. K Li, Research on the diversification of junior middle school PHYSICAL education teaching methods in the era of big data. Love, Marriage and family: Education Observation **8**, 1 (2021)
10. L. Wu, On the reform of PE teaching in higher Vocational Colleges in the era of big data. J. Jiangxi Electric Power Vocat. Technical College, **8**, 3 (2020)

Research on Ideological and Political Education in Colleges and Universities Under the Background of Big Data

Jing Zhou

Abstract Big data information technology offers new development chances for ideological and political education, enriches the content of education and innovates the methods of teaching in college. At present, the colleges' slow progress in building data platforms, the lack of data awareness of teachers, and the difficulties which is currently faced by the educators have seriously implication for ideological and political education. The institution ought to provide a platform for exchanging data, create a visual analysis platform for education, value the significance of cultivating the concept of teachers' data governance, and innovate the way of education in order to lay the groundwork for realizing big data technology and enhancing the efficacy of ideological and political education.

Keywords Dig data · Ideological and political education in Colleges · Promote

1 Introduction

The age of big data has arrived in human society, and all facets of social life are now represented as data. The way individuals study has substantially changed as a result of the usage of big data technologies has significantly altered how people study, work, and live. Big data information technology enables a thorough and methodical investigation of educational subject and object data and opens up new possibilities for education. The efficacy of education in the context of big data faces obstacles at the same time. Big data is not well understood by the majority of educators, and the shortcomings of the conventional educational approach are increasingly becoming more apparent, and the independent awareness of the object of education is not high. Political and ideological education is a crucial tool for assisting college students in

J. Zhou (✉)
School of Information Engineering, Wuhan Business University, Hubei Province, Wuhan City, P.R. China
e-mail: zhoujinglucy@163.com

S. Patnaik and F. Paas (eds.), *Recent Trends in Educational Technology and Administration*, Learning and Analytics in Intelligent Systems 31,
https://doi.org/10.1007/978-3-031-29016-9_2

developing the right morals. It has become a critical job to figure out how to use big data techniques to enhance educational efficacy.

2 Issues and Difficulties with Ideological and Political Education in Face of Big Data

Through big data conversion and other technologies, high-quality educational resources in the form of text, images, audio, and other formats can be presented in the form of data in the information age, and the sharing of resources through the platform creates favorable conditions for higher education. The growth law of students' ideas, the development trend of students' thoughts, and the provision of a data foundation for comprehending educational objects may all be determined through the summary and analysis of data information with enormous capacity and rich data kinds. Analyze the learning state and impact of students, as well as potential tendentious problems, to further anticipate them and to stop negative problems from happening.

The efficacy of education is impacted by students' understanding of instructional materials. Big data digitally exposes the textual material made available by educational items as well as the communication and engagement processes, which is also helpful for effect feedback and outcome evaluation. As a frontier of ideological education, colleges and universities should adapt to the development of the times [1]. It is easier for teachers to grasp students' learning progress, understand students' ideological trends, and effectively guarantee the effectiveness of ideological and political education when data is disseminated in a timely and authentic manner, such as the various learning data collected from students during class. There are several restrictions with traditional educational approaches. Using quantitative local analysis method is easy to ignore the ideological details of college students, which is not conducive to grasp the true ideological state of students. Big data technology uses advanced platforms to collect, mine and analyze various data related to students' thoughts and behaviors, which are presented in a visual form. In addition to enhancing the structure and environment of traditional educational providers, a thorough and scientific investigation and understanding of students' ideological dynamics may be used to create individualized education plans for students that are in line with their developmental needs.

It can not be ignored that college education is also facing challenges in the context of big data, which are mainly reflected in the following three aspects.

2.1 The Development of Ideological and Political Education Data Platform Falls Behind in Colleges And Institutions

The construction of data platform directly determines the possibility and reliability of data collection. Constructing a university data platform is essential to enhancing the impact of education. At present, the construction of educational platform in universities lags behind. Even if an online and offline education platform is established, the platform data is still scattered, and the weak data mining technology and integrated analysis technology also weaken the function of effective data information. Traditional channels are still the main body of the dissemination of education in universities. It might be challenging to get students enthusiastic in taking part in political education through courses, theory curriculum, reports, academic lectures and other forms. Due to excessive dependence on traditional channels, new data platform channels are not valued, and the energy and funds invested in the construction of data platforms are limited. Data platforms in college are currently separate, and the platform solely collects data that is used internally by the Department. In the study of realistic ideological and political education, the researchers actually exist outside the consciousness of the cognitive subjects [2]. The lack of liquidity in the data and the choppy sharing channels make it difficult to effectively integrate various types of information and data. In addition, the existing educational platform is lack of interaction with educational objects, less personal learning data mining for educational objects, and lack of high-level personalized technology settings.

The current use of big data in colleges results in an insufficiency of useful data sources for teachers to delve into. Colleges can obtain basic data information such as students' online speech and academic performance through the campus network, but it is difficult to obtain information outside the campus network that can reflect the dynamic changes of students' thoughts. Data resources available for mining are scarce, and data collection is lack of scientificity and standardization. Data collection is mostly carried out in the form of questionnaire. Before the scheduling of education resources big data, the network distance education n resources are fused and clustered [3]. The rationality of questionnaire design directly affects the authenticity of data. Due to the lack of data resources available for mining, teachers cannot obtain education related resources in time, cannot effectively grasp various data related to students, and it is difficult to integrate big data into innovative education, resulting in educational content or education appearance that does not conform to the current learning habits of college students. Due to a lack of more robust data mining and analysis technology, teachers are unable to swiftly and effectively integrate and deal with resources relevant to education, resulting in a waste of data resources.

2.2 *The Main Body of College Students' Ideological and Political Education Lacks Data Awareness*

Teacher lack the concept of statics, and they are still used to the traditional way when carrying out education, and their subjective consciousness has not changed into data consciousness. First, it is reflected in the compilation process of theory teaching materials. The textbook compilers are generally researchers of Ideological and political related majors. Their knowledge background leads to a weak awareness of big data, a high degree of dependence on the original knowledge system, and a low degree of keeping pace with the times in textbook compilation, which leads to difficulties in the docking of textbook content system and big data in the integration of big data thought into theory teaching. Second, the process of collecting and integrating data is not reflected in the teaching process, thus it is unable to adjust to the data era's empirical character. Combining classroom material with data thinking and data technology is difficult. Unlike the ideological and political education method of venture capital, using standardized and patterned methods to analyze the thoughts and behaviors of college students, big data technology can mine the real state of mind of students from multiple perspectives from diverse information. Using the traditional way of thinking to carry out teaching, there is path dependence on the traditional teaching method. Subjectivity and one-sidelines in the examination of the topic of education, as well as a lack of excitement for altering the teaching style and implementing new teaching techniques, are factors that contribute to the ineffectiveness of education.

Teachers make up the majority of the educational system and are crucial to the ideological and political indoctrination of students. Ideological and political educators need to be skilled at using big data technologies. Big data information comes from a variety of uncontrollable sources. False data coexists with true data during transmission. A technological difficulty encountered by educators is how to assess the useful data acquired throughout the educational process from the various data information, combine effective data, and then fully exploit the value of data information. The creative growth of higher education is now constrained by the lack of talent and weak data processing skills of ideological and political educators, as well as by the lack of big data technology and ideological and political theory expertise.

2.3 *Challenges Faced by the Object of College Students' Ideological And Political Education*

In the current era of big data, the Internet provides a better channel for massive information dissemination. However, due to the constraints of youth and experience, college students lack the capacity for autonomous judgment, and the complex and diverse information is easy to impact the thoughts of college students. College students' development needs and ideological knowledge are both diversifying at the same time. College students' ideological conduct is easily undermined by negative and even misleading information, which creates significant difficulties for education.

Clear understanding of political theory system is an important prerequisite for the organic combination of big data technology and theory courses. Teaching activities with the classroom as the main carrier, students lack interaction in the process of education, and the effect of absorbing and mastering theoretical knowledge is not very good. Students' study of political theory courses is utilitarian. Some students are unable to properly learn and understand theoretical information due to the mechanical recitation of knowledge points and the short-term memory chase of high scores, which prevents ideological and political education and instruction from integrating with big data.

Meanwhile, there are differences in data integration ability and technology among college students. Students in the humanities and social sciences face challenges when using big data technology to learn political theory. These challenges include unreliable big data systems and inadequate data integration tools, which to some extent reduce the effectiveness of education in the context of big data. Under the influence of big data, interactive teaching, autonomous learning, and cooperative learning are significant teaching approaches for education. At present, students are more inclined to traditional learning methods and lack the ability to master and explore in depth. At the same time, they lack basic training in special data processing and have difficulties in the collection, collation and analysis of massive data.

3 Countermeasures for Using Big Data to Improve the Effectiveness of Ideological and Political Education

3.1 Building Big Data Platform to Create Data Flow and Circulation

Big data is becoming an important scientific and technological means to promote educational innovation. With the wide application of big data and 5G related technologies, how to use these technologies to implement precise ideological and political education in classroom teaching, and how to make teaching activities more intelligent and convenient, has been concerned by many scholars.

Nowadays, college students use the internet frequently, and the internet is become the primary way for them to obtain knowledge and information. When college students obtain information on the internet, they also generate corresponding data on internet behavior. Mining and integrating these related data can analyze students' living conditions and ideological and psychological conditions.

Although the development of big data is the trend in today's society, however, the development of relevant technologies is not mature and perfect, and its application in ideological and political education needs to further research. First, we should establish and improve a standardized unified information platform to realize the efficient flow of data resources among various departments. Big data is often used to describe lots of unstructured and semi-structured data that takes too much time and money when downloaded to a relational databases[4]. At the same time, the unified information platform is used to convert digital information into data information available on the ideological and political education big data platform, improve the availability of data and openly share information and data, so that educators can analyze and observe students' ideological and political education status data in real time. Educators carry out relevant educational activities in a targeted manner, and implement personalized educational programs precisely. The second is to use data as the driving force, educators obtain big data of students through various channels, such as personal basic information data, classroom attendance data, classroom performance data, course selection and performance data, canteen consumption data, dormitory access card swiping data, book borrowing data, student association data, yiban data, students' second class data, etc. These students' daily study, life, social interaction and consumption behavior data are highly related to students' ideological dynamics. At the same time, through the collection and pretreatment, cleaning, processing, storage, analysis, visualization and other links of the whole process data of students in universities, it has converged into a big data resource library required for accurate ideological and political education. The structure type of big data is shown in Fig. 1.

While extracting structured data, educators should pay attention to the integration of semi-structured and unstructured data and the diversity of data types. By sampling and analyzing the big data of students' education, educators can form important index data that can quantitatively analyze the characteristics of students' ideological and political education. When storing and analyzing data, educators should desensitize sensitive data involving students' personal privacy, so as to ensure students' personal privacy and data security in colleges.

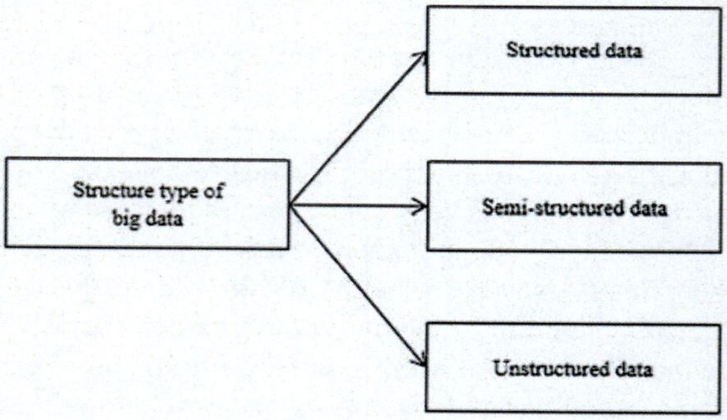

Fig. 1 Structure type of big data

3.2 Realize the "Data Portrait" of Students and Build a Visual Analysis Platform

Using technology to "data portrait", so as to achieve accurate ideological and political education for students. The system of big data technology is huge and complex. The commonly used big data technologies include Hadoop, Hive, Spark, etc., which can process massive data in real time. Using Hadoop can achieve distributed computing of massive data through clusters composed of computers. Through such parallel processing, the scale, efficiency and security of data storage can be improved step by step. Hadoop is also a core of data processing and analysis, it brings together structured and unstructured data.

In order to reasonably using "data portrait" technology among college students, the decision tree algorithm in data mining algorithm can be used to make an index quantitative analysis. In machine learning, the decision tree is a prediction model. It is first to construct a mutually exclusive and complete decision tree, and then classify the samples to be processed. In addition, the decision tree contains the physical concept of entropy. By data mining the sampled data of some students, constantly optimizing and adjusting the parameter weight value of the decision tree algorithm, and preventing the sample imbalance from causing over fitting, the accurate "data portrait" of all students can be can realized, and finally build a visual analysis and evaluation platform for students.

By establishing a quantitative evaluation index system of students' "data portrait", the abstract and complex data are indexed and quantifiable, so as to realize a comprehensive, three-dimensional, accurate, quantifiable and visual dynamic strategy mode for students, and realize the dynamic evaluation and closed-loop management mechanism for college students. Through a period of data precipitation and the continuous updating and iteration of big data platform data, the goal of all-round and multi angle ideological and political education can be achieved step by step. The implementation steps of data portrait is shown in Fig. 2.

3.3 Cultivate the Awareness of Using Data Resources and Strengthen the Construction of Ideological and Political Education Team

In the context of using data resources, building a professional ideological and political education team is an important guarantee and support force, and current related educators need to improve their abilities in this regard. The ability of educators to use data thinking and data analysis is the premise of ideological and political education in the context of data resources.

In the actual education process, due to the low quality of data resources, ideological and political educators are difficult to deal with massive data information in a timely and effective manner, resulting in inaccurate data results. At the same time, due to

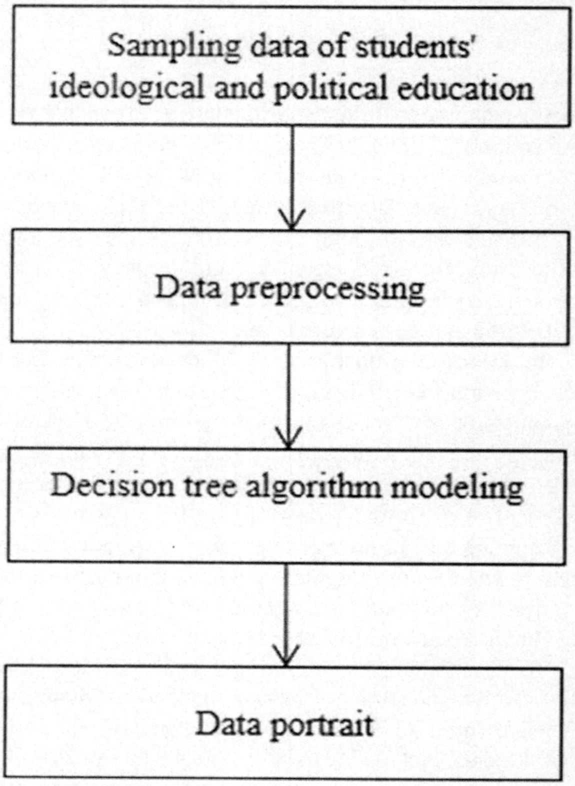

Fig. 2 The flow chart of data portrait

the dual pressure of teaching and scientific research, educators have limited time and energy to participate in the training big data technology. The study of knowledge is a prerequisite for using big data. To cultivate educators' data governance ideas, educators must effectively understand and use the processing technology of relevant data resources to carry out efficient ideological and political education related work, such as learning to mine effective data and analyze data, and gradually master the skills of processing and analyzing data resources by participating in training, further education, lectures, seminars, etc., so as to guide more educators to form the concept of data governance. On this basis, educators should improve their theoretical level and professional ability, be good at using related technology to analyze learning situation, identify the learning needs of students with the help of data resources, reasonably allocate relevant resources, and improve the efficiency.

By establishing the concept of multi-subject co-governance and promote the construction of interdisciplinary academic teams. Ideological and political educators need to break the gap between disciplines with open thinking and strengthen the deep integration with data processing technology. By introducing experts and scholars in the data resources industry for guidance, using the advantages of computer expertise to guide educators to use the data processing platform reasonably, participating in the process of collecting and analyzing data related to education, and giving reasonable play to the data of platform. The advantages of data resources are striven to build a multi-subject pattern of ideological and political education that cultivate students in an all-round way.

3.4 Optimize Educational Resources and Strengthen the Autonomy of Educators

Choosing and improving the methods of Ideological and political work is by no means a simple subjective judgment. Whether it reflects the reality of social development and the ideological reality of educational objects is related to the vitality and practical utility of the methods.

According to the actual demand characteristics of educational objects, expand the supply space and supply mode. Educators need to actively promote the effective supply of ideological and political education content. Data resources store various teaching resources on the network in the form of data. All kinds of fresh educational materials, cases and works can be used by different subjects, which can eliminate the spatial constraints of education to a certain extent. First, Educators should use data resource technology to timely understand students' psychological situation and learning needs, and constantly enrich and improve the ideological and political teaching and education content system. Educators carry out data visualization analysis of social hot spots that students care about, improve the attention of classroom students, and focus on the reality of Ideological and political education.

Through the data-based learning platform, provide learning resources that are more in line with students' own needs, adopt the correct value orientation, improve the information identification ability of educational objects, and guide students to make correct value judgments. Second, educators should establish a comprehensive three-dimensional content system, excavate the education elements in different disciplines in teaching practice link, and strengthen the construction of the curriculum. Adhere to the system principle, integrate the teaching objectives and moral education objectives, establish a content system including ideological education, political education, moral education, etc., and promote the in-depth intersection of various disciplines.

Educators should further actively explore new ways of ideological education and political education. In the new era, college students' media usage habits have changed. The network platform is a convenient way to obtain data resources. Data resources provides the new way for enriching classroom and teaching forms.

Educators should integrate data resources into daily teaching activities to stimulate students' learning enthusiasm. Relying on new data-based technology to establish smart classrooms, effectively integrate online and offline resources, explore new Internet multimedia technologies, carry out innovative daily teaching work, and solve the problem that lack of interaction in ideological and political course. For example, by using VR technology based on data to build a virtual education environment, educators can carry out immersive experience and enhance the sense of classroom acquisition of students. Using the fast and timely feedback mechanism of big data, educators can timely understand learning progress and teaching effect of educational objects, timely adjust the education methods that fit students, enhance the autonomy of educates, and improve the effectiveness.

4 Summary

Overall, this topic has conducted in-depth research and analysis in the ideological education using big data. In daily teaching activities, educators should focus on the target characteristics of curriculum teaching, build a new model of interdisciplinary national education, adopt advanced big data related technological innovation methods, and strive to integrate ideological education into the whole process of curriculum teaching, and form an iterative and optimized closed-loop education model. Big data is a valuable resource. Data currently serves as the world's representation. Big data's age has sparked unparalleled innovation and growth across all spheres of society. Education's evolution is a dynamic process. The presence of big data cannot be divorced from the evolution of higher education in the modern era. In the modern day, colleges and universities must actively integrate education and big data technology, support educational innovation, and implement technological change. College education adapts to the times in the context of big data and investigates how to combine big data with college education, opening up a new potential for the advancement of education in China. Of course, there are also practical difficulties with ideological and political education in China. Future work will concentrate on closely monitoring changes in teaching and learning, coordinating data resources both inside and outside of schools, enhancing data literacy among educators, implementing and optimizing a big data platform for college students' ideological and political education, integrating data resources, and developing a visual analysis platform for education based on existing knowledge and experience. In addition to creating a multi-subject educational model for everyone, it is necessary to adopt cutting-edge big data technology to innovate educational practices, fully utilize the benefits of big data platform and big data technology in the process of ideological and political education, and establish new standards for ideological and political education in colleges and universities in the big data era.

Acknowledgements Foundation Project: University-level scientific research project of Wuhan Business University in 2020, Project Name: Research on Improving the Effectiveness of Ideological and Political Education in Universities Based on Big Data—A Case Study of Wuhan Business University,2020 KY020.The author thank the staff of the team for their research and support.

References

1. Y. Chunyang, Innovative methods of ideological and political education for college students based on ideological cognition science. Educat. Sci. Theor. Practice **18**(6), 2989–2998 (2018)
2. J. Wang, Network distance education platform control system based on big data. Int. J. Internet Technol. **12**(3), 173–180 (2019)
3. F. Li, Research Method Innovation of College Students' Ideological and Political Education Based on Cognitive Neuroscience. NeuroQuantology **16**(5), 296–302 (2018)
4. W., Xinyue, Research on the difficulties and paths of ideological and political education of Big Data technology. J. Jilin Radio Television University **11**, 131–132 (2019)

Construction and Practice of Art and Design Education Resources Based on Big Data

Zuyin Zhou

Abstract In the rapid development of information technology and the Internet, the way of social production and learning entertainment has undergone new changes. In the traditional sense of the artistic design education form as emerging technological innovation put forward more requirements, so also should follow the time development of colleges and universities to integrate the education resources construction, information technology such as the rational use of cloud computing and big data to build sharing framework, design and develop digital curriculum resources, realizing the objective of the modernization of education development in at the same time, To improve the art and design education level and core competitiveness of colleges and universities. Based on the understanding of the status quo of art and design education in the era of big data, this paper makes clear the importance and technical characteristics of resource sharing architecture, and deeply discusses how to build personalized art and design education resource recommendation system based on big data technology.

Keywords Big data · Art design · Educational resources · Personalized

1 Introduction

In the development of big data construction, art design education has a unique resource advantage during the development of colleges and universities, which is directly related to social application and the demand of education industry. Nowadays, according to the service society, highlight the characteristics, principles of co-construction and sharing of development, in the department of education in colleges and universities around the special policy support, put forward the art design education project repository construction work, on the one hand, to build more high quality

Z. Zhou (✉)
Wuhan Business University, Wuhan, China
e-mail: zhouzuyin@163.com

S. Patnaik and F. Paas (eds.), *Recent Trends in Educational Technology and Administration*, Learning and Analytics in Intelligent Systems 31,
https://doi.org/10.1007/978-3-031-29016-9_3

and scientific repository, serve the society, on the other hand to take advantage of innovative thinking and ways to build high-quality goods project. Big data, as a common network communication tool for the development of the new era, has been paid attention to by researchers and the educational field, but it has not played its expected application value. In the face of this development situation, colleges and universities should systematically study the basic characteristics and organizational structure of the theme culture, understand the personalized needs of users and the service functions of theme resources, so as to design an innovative service system for the internal education system and promote the orderly implementation of talent cultivation in colleges and universities [1–3]. Therefore, the construction of art design education resources based on big data should start from the following points: First, strive to build a structured system with traditional and modern art design content, methods and development path, and form an overall framework with massive information basis and continuous improvement; Second, the overall framework of resource library as the basis for development, in the scientific topic selection and precise positioning, targeted selection and presentation of all kinds of resources, and finally in the depth of digging materials, multi-level performance of resource information; Thirdly, organize and excavate representative art and design education resources, focusing on the design of thematic content with local culture as the core, so as to reflect the leading direction function of serving society and education; Fourthly, scientific arrangement of characteristic art and design education resources, to build sustainable and serialized high-quality resources; Fifthly, based on resource expansion and analysis, it provides effective basis for art and design education in the new era. At present, the art and design education resource base with regional characteristics has formed more than 40 categories in accordance with the four directions of traditional craft, folk craft, modern design art and design creativity. After the initial infrastructure construction, the overall resource system has begun to take shape, and began to enter the stage of resource in-depth mining based on unique content, and finally formed a communication form with the characteristics of mobile browsing resource technology in the construction and promotion. From the perspective of practical development, art and design education resources with big data technology as the core should not only provide high-quality information services for professional students and practitioners, but also strengthen residents' art and design cognition level for the whole society. This paper mainly focuses on the beauty of the educational resources of art and design teachers and students in colleges and universities, and deeply discusses how to build a personalized recommendation system, so as to help them quickly discover valuable information resources.

2 Method

2.1 Design Objectives

In college education of artistic design specialized design resource recommendation system, to meet the demand of the era of big data university education, on the basis of secure electronic education resources can be quick and easy application, meet the requirement of information sharing between teachers and students, and liaising with other departments to fulfill management function, enhance the level of art design education informationization. The goal of art design education resource system based on big data is divided into two aspects. On the one hand, it refers to the business goal, which is mainly to use modern information technology to build a normative, scientific and unified information sharing business platform. On the other hand, it refers to the technical objectives. It should show the modern requirements of art design education management, improve the flexibility of system operation by using layered design, comply with the planning and design principles of campus integration, effectively connect with other application platforms on campus, and build an integrated information system on campus to meet personalized recommendation requirements.

2.2 Requirement Analysis

This research system should be designed to realize personalized recommendation and meet the needs of art design education in colleges and universities. Therefore, the main function of the system involves the following points: First, the main function of the system. To provide a platform for users to automatically search and recommend, active search needs to support fuzzy search, and advanced search is not considered in the initial construction of the application. The recommendation part should include two parts: universal ranking recommendation and personalized recommendation. Second, authority control. The art design education resources of the whole system should be classified and stored, and the data permissions do not involve restrictions. The system functions should be oriented to all teachers and students, while the system maintenance personnel should design the functional permissions and control requirements separately. Third, the user role. For the control authority of some functions, users are usually divided into two types. On the one hand, administrators make decisions, mainly protecting data storage of the whole system, and on the other hand, system users, usually teachers and students of colleges and universities, provide services according to their application learning needs. Fourth, behavior record. All user behaviors in the system need to be recorded in flow, which involves login information, access time, IP address and other key information. Fifth, personalized recommendation. Recommend functions should be provided to all users, among which non-logged-in users can obtain interests and hobbies through real-time interface operation; Sixth, resource management. All art and design education resources should be well summarized and relational processing, this part of the data information to use the results of the integrated data processing platform to achieve. In addition, users should be supported to analyze resource access and hotspot resources.

2.3 Overall Architecture

As the core requirement of constructing art design education resource system, business objectives can meet users' needs only by meeting the expected business criteria. According to the functional requirements of personalized recommendation system, this paper studies and constructs the following business framework:

Combining responsibilities and business planning, the art design education resource system is divided into five parts, among which the recommendation algorithm depends on other parts for implementation and has a dependency relationship from top to bottom.

In order to ensure that the system meets the unified design requirements and the art design education resource system can operate safely and stably, the boundary design of the system is shown in Fig. 2 [4–6].

The overall logical relationship of the system is shown in Fig. 3. The two subsystems built internally work in collaboration with the two external systems. For example, the personalized recommendation Web subsystem should provide system users with recommendation interface programs, facilitate users to search for applied art and design education resources, and provide rapid filtering function based on resource search, so as to further provide users with personalized and universal related

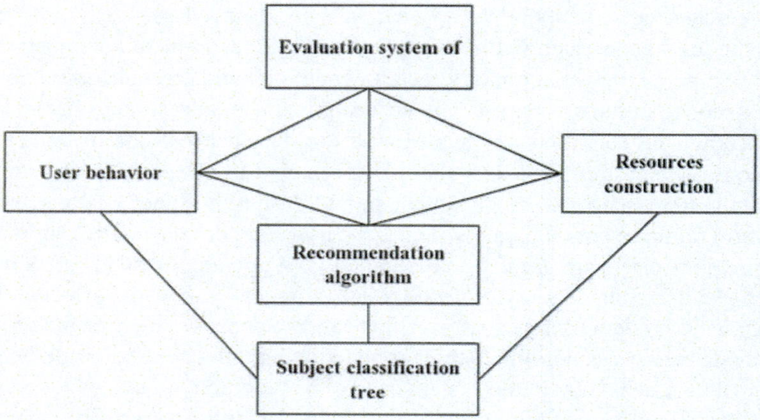

Fig. 1 System business architecture diagram

Fig. 2 Structure diagram of system boundary

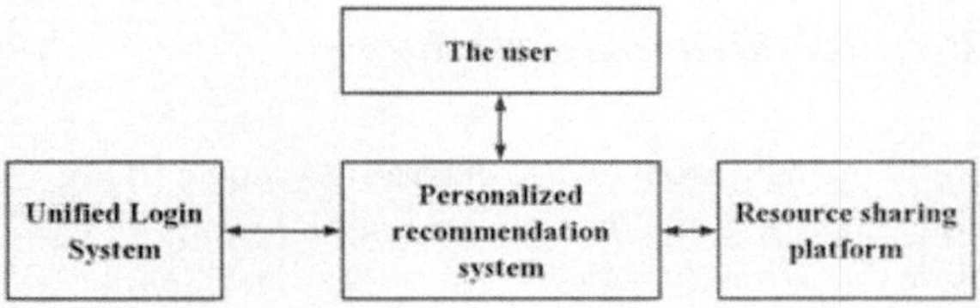

resources. This part is an important basis for the realization of personalized recommendation algorithm, and can also provide auxiliary functions for feedback user ratings, research behavior log and so on.

On the basis of the system mastering the functional requirements and system constraints, according to the planned system boundary and following the definition and implementation in the logical structure, the art design education resource architecture diagram as shown in Fig. 4 can be constructed:

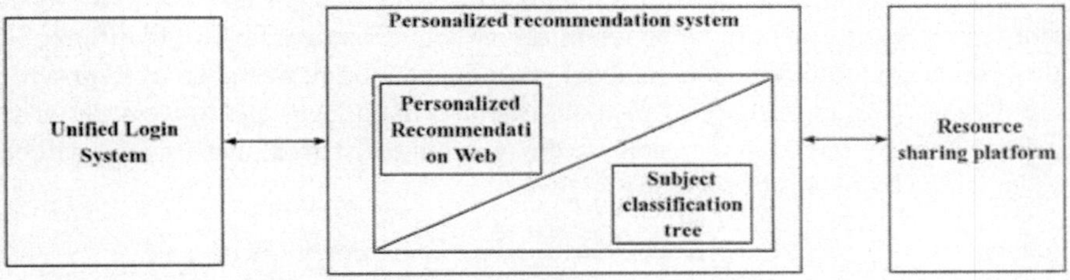

Fig. 3 Structure diagram of the overall logical relationship of the system

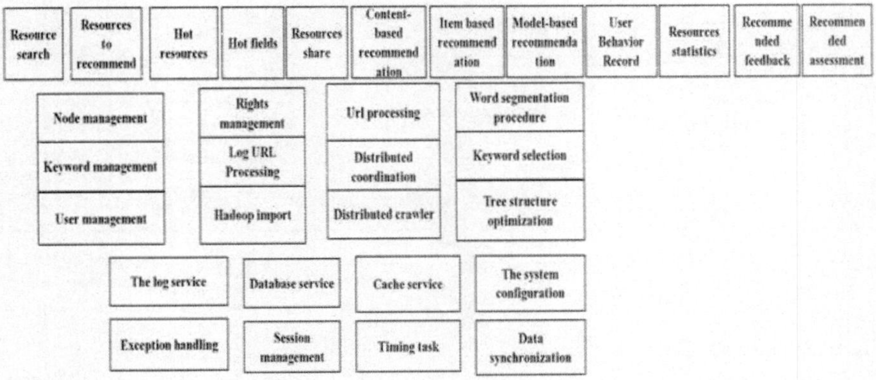

Fig. 4 Overall architecture of the system

2.4 Technical Architecture

Build the data architecture diagram as shown in Fig. 5, which contains the following contents: First, the data network monitoring library. This content is mainly used to store the log information in the education resource system. Sqlite database should be used to store it and provide behavior information for the application of art and design education resources. Classification tree can read URL information from the library and import the classification number content library, so as to facilitate the analysis and use. Second, the classification tree content library. In this content, HDHS should be used to store the intermediate results of text classification tree URL, user behavior content information, word frequency statistics operation of classification tree, etc., to provide a basis for the classification application of art and design education resources. Third, configure the information base. In this content, the dynamic configuration information of each node of distributed web crawler and distributed word frequency statistics application is coordinated by the application of Zookeeper classification tree. Fourth, recommendation information base. This content uses Mongo DB to store resource information, behavior information, recommendation algorithm and other contents required by personalized recommendation network, so as to provide core database for art design professional teaching in colleges and universities. The resource data contained in the database are all from the shared platform and will be implemented by means of regular import.

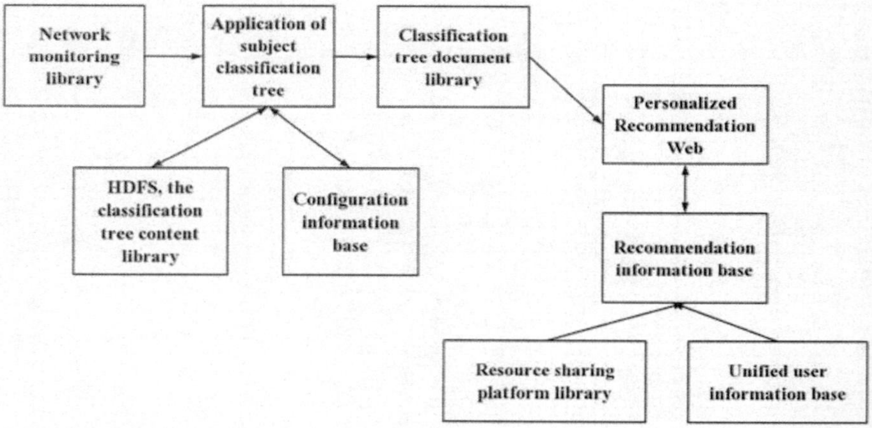

Fig. 5 System data architecture diagram

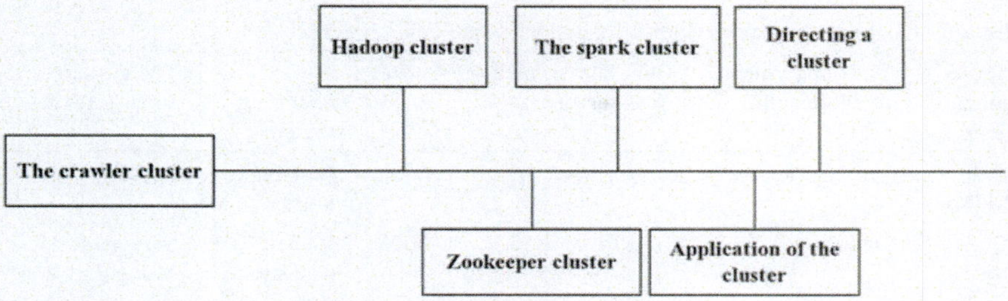

Fig. 6 Deployment diagram of the system operating environment

3 Result Analysis

3.1 Operating Environment

After the completion of the system development and design, a joint test should be carried out to systematically analyze the application performance of the construction of art and design education resource system with big data technology as the core. The operating environment of the system studied in this paper is the campus network, which includes HADOOP, SPARK, MONGODB, ZOOKEEPER and application cluster, as shown in Fig. 6.

Table 1 shows the corresponding software and hardware configurations:

The performance test of the construction system of art design education resource system in colleges and universities is carried out to discuss the application of personalized recommendation module. The actual performance indicators are as follows:

On the one hand, system response time. According to the analysis of the personalized recommended network concurrent test results as shown in Fig. 7, when 500 and 800 concurrent users are selected for the pressure test, the system response time of the former is less than 0.1 s, while the average response time of the latter is less than 0.15 s, which meets the requirements of performance requirements. It is proved that the increase of the number of users has a linear effect on the response time of the system, in other words, the limitation conditions of the system on the application load level in the hardware platform are linear expansion.

On the other hand, recommendation algorithms. The accuracy of the recommendation algorithm was verified by off-line experiments using data sets, and the common calculation method of root mean square error was used to test and analyze the algorithm. On the basis of clear prediction accuracy and operation efficiency, this paper mainly conducted 5 iterative tests on 100 K, 1 M and 10 M data, and the specific results are shown in Table 2:

Table 1 Configuration table of software and hardware environments

Server model	The deployment of application	CPU	The server	memory	storage	The operating system	The system software
Dell M910	The crawler node	2	Three	16	600G	64-Ubuntu Server 14. 04	JDK1.7
Dell M910	Hadoop	4	Five	16	600G	Mahout-0. 9 64-Ubuntu	Server 14. 04 MongoDB-2. 6. 4
Dell M910	mongoDB	4	Three	16	600G	64-Ubuntu	Server 14. 04
Dell M910	Zookeeper	2	Three	16	600G	Zookeeper-3. 3.6	Server 14. 04
Lenovo Wanquan R680	Spark	2	three	256	1.2 T	Ubuntu Server 14.04	spark-1.1.0-bin-hadoop2.4
Dell M910	Subject tree maintenance and log import	2	one	16	600G	Ubuntu Server 14. 04	Tomcat 7. 0.57
Lenovo Wanquan R680	Personalized Recommendation Web	4	one	32	1.2 T	Ubuntu Server 14. 04	Tomcat 7. 0.57
Dell M910	Keywords extraction node	2	three	16	600G	Ubuntu Server 14. 04	JDK1.7

Combined with the above table analysis, it is found that the change of data volume does not have a significant impact on the recommendation algorithm, and the continuous increase of data volume has a positive effect on the convergence effect of the algorithm [7–9].

To sum up, this paper studies the application system to understand the modern art design education resources, on the basis of, from the perspective of the special education college of art design demand, reasonable use of big data technology and the concept of personalized recommendation, constructing systematic art design education resource system, can provide users with a variety of service functions, not only can guarantee the art design education more reasonable utilization of resources. By testing and analyzing the application performance and liability level of the constructed system, it is found that the recommendation algorithm of the whole system has a high accuracy, and can provide users with a service platform with standard performance and effective functions during the operation of the system. Therefore, in the context of the development of the new era, researchers should strengthen the construction of art and design education resources with big data as the core, and only in this way can the development needs of modern art education be met.

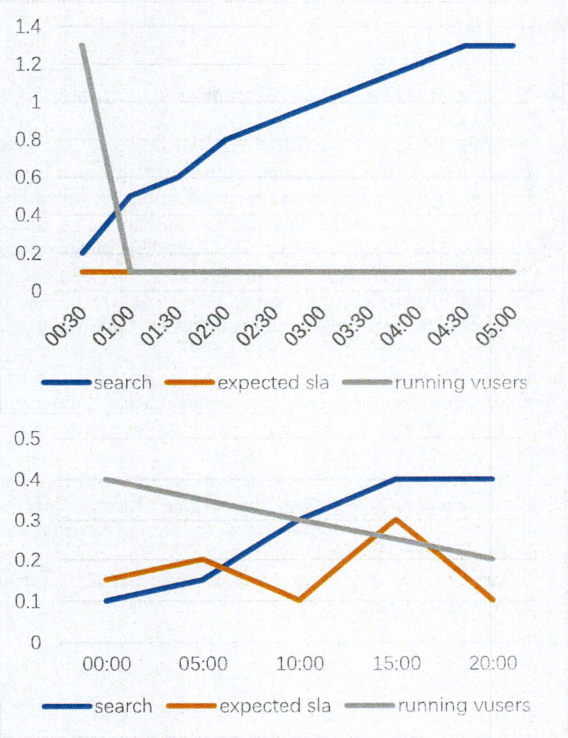

Fig. 7 Results of personalized recommendation network concurrent test

Table 2 Data results of 5 tests

The data set	1	2	3	4	5	Number of iterations
100 k	0.929670	0.921314	0.922895	0.93933	0.926531	5
100 k	0.920357	0.925393	0.926518	0.916293	0. 923,273	10
100 k	0.922105	0.923140	0. 914,839	0.915522	0.917939	20
1 M	0.891624	0.888054	0.891542	0.888635	0.889191	5
1 M	0.865903	0. 867,122	0.867379	0.867383	0.865579	10
1 M	0.859041	0.861062	0.858722	0.859944	0. 858,687	20
10 M	0.840759	0.838288	0.844987	0.840225	0.832224	5
10 M	0.818288	0.818212	0.818416	0.816341	0.815392	10
10 M	0.81 1033	0.810917	0.810562	0.810482	0.811102	20

References

1. Y. Chen, *Research on Brand Management of Art and Design Education.* Central China Normal University (2012)
2. Y. Wang, *Research on Talent Training Mode of Visual Communication Design under the Background of Informatization.* Northeast Petroleum University
3. F. Hou, Research on training methods of higher vocational clothing design talents under the interaction of art and technology. China Academy of Art (2016)
4. H. Wang, B. Yang, B. Wang, R. Chen, Competency-based, University-enterprise Linkage, Mixed Teaching, Multi-Party Sharing – The development and application of teaching Resource database for Art design Specialty. Design **32**(01), 89–91 (2019)
5. H. Zheng, On the necessity of resource construction for art and design education in the digital age. Art Appreciat. **12**, 116–117 (2016)
6. H. Li , Exploration from arts and crafts education to art design education – Comment on exploration of contemporary art design education. Educat. Develop. Res. **39**(21), 87 (2019)
7. T. Tian, *Research on teaching mode of Art design specialty in Distance and Open Education.* Hunan Normal University (2007)
8. R. Li, *Exploration and Practice of Improving Image Literacy in High School Art Visual Communication Design Teaching.* Shaanxi Normal University (2017)
9. R. Zhang, *Research on higher graphic Design Education and Teaching in Hainan in the New Economic Era.* Hainan University (2010)
10. S. Yang, *Research on Library Management and Service Innovation in Art and Design Colleges.* Tsinghua University (2005)

Research on the Application of Educational Big Data Analysis in Online Learning Behavior of Computer Basic Teaching

Xin Sui and Yi Sui

Abstract Big data technology is usually used to study and analyze big data and extract useful information. Big data of education has become a new research direction. This paper expounds the education data, analyzes the influence of big data on education, in order to improve education and teaching. We should further in-depth research, and then make role of big data in education fully play, in order to better improve the level of education and teaching.

Keywords Study · Analyze · Education · Influence · Advantage

1 Introduction

As early as the 1990s, the concept of big data was first proposed. Different fields and scholars have different understanding of big data. Big data generally refers to the collection of data that cannot be acquired, analyzed and utilized by traditional hardware platform or software technology. Data warehouse and data security technology, data analysis have become the focus of research in various industries [1]. Big data has far-reaching significance. Gradually penetrate into the core of education, change traditional education, teaching mode, and build new evaluation methods [2].

As an important strategic resource of the country, big data has become a new engine that drives society-driven economic and social development and intelligent transformation. 2015 is known as the "first year of China's education big data", and the domestic education sector has set off a boom in education big data research.

X. Sui
Changchun Humanities and Sciences College, Changchun, Jilin, China

Y. Sui (✉)
China Telecom Beijing Branch, Beijing, China
e-mail: 955932@qq.com

S. Patnaik and F. Paas (eds.), *Recent Trends in Educational Technology and Administration*, Learning and Analytics in Intelligent Systems 31,
https://doi.org/10.1007/978-3-031-29016-9_4

The coverage of the informatization teaching system of colleges and universities has been very high, and big data combined with cloud platform provides high-quality teaching resources for colleges and universities, and students' access to course resources is no longer limited by time and space. The use of data mining and analysis technology can perform real-time and timely analysis of learners' information, making personalized education possible. Through the analysis and mining of education big data, teachers can make teachers' teaching and students' learning more effective, teachers can timely understand students' learning progress, mastery of knowledge points, understanding of key and difficult points, can accurately classify, personalized recommendation, and make the whole teaching process more intelligent.

Conduct efficient analysis of students' academic performance and learning behavior, grasp students' learning situation, and further find the factors affecting students' performance. Use big data technology for data analysis and processing, manage and optimize a large amount of teaching data, timely understand the learning status of students, discover more hidden information to predict students' learning, timely discover possible problems in the learning process, predict the development of students' future learning, and determine whether to intervene in students' learning according to the prediction results. According to the collected education big data, predict the learning effect of students, and adopt different teaching strategies for different students according to the predicted results, so as to truly realize teaching according to aptitude and personalized teaching. Make full use of big data to improve the quality of education and promote equitable education, from teaching using experience to teaching based on educational data analysis.

2 The Impact of Big Data on Education

Big data includes massive data analysis, data processing and so on [3]. Education big data not only has an impact on higher education, some scholars have long believed that introducing big data into education will promote the process of education informatization, meet the individualized needs of students' education, and not be restricted by standardized traditional examinations, transforming traditional education into information-based education [4]. Making full use of big data technology, timely recording of relevant teaching process helps to discover new knowledge, optimize teaching methods and create maximum value [5].

Applying the majority to the field of education, teachers can objectively obtain the feedback of students' teaching effects, students can have a more equitable access to education, and students can obtain individualized teaching resources, educational administrators can better specify the talent-training program, teaching plan and Related Teaching Management System. Big Data is growing in the context of Internet, the rise of online education, all provide a favorable support for education big data. In the current Internet, the decisions made by educators, teachers, etc. Will be transformed into data-supported behaviors rather than subjective assumptions.

Data is the core of big data research. Big Data analysis technology and statistical research have both connections and differences, both from the traditional statistical theory, and higher than the traditional statistical analysis. Big Data analysis technology and traditional statistical theory research methods are not the same, but the two can complement each other.

2.1 The Rise of Online Education Has Brought Great Influence to Traditional Education

Big data, also known as big data, is a collection of massive data. It is generally believed that big data is a massive amount of data accumulated and grown by people's interaction on the network. Wikipedia defines big data as a collection of data whose content cannot be crawled, managed, and processed by conventional software tools over a certain period of time.

Online education makes learners receive education no longer limited by time and space, and they can learn autonomously anywhere and anytime. Learners can control the progress of learning independently and choose the content of learning individually. With the support of Internet technology, online offline hybrid teaching constantly optimize the teaching content, improve methods, and enhance learning effect of learners.

2.2 Big Data Supports Personalized Education

Big data promotes personalized learning. By acquiring education big data, we can more accurately understand education behavior, dynamically optimize modeling and make decisions, and realize efficient personalized learning [6].

Learners can learn online anytime and anywhere. Online learning platform may record students' online learning behaviors, such as the frequency of students logging in to the learning platform, which part of online video is watched repeatedly, the accuracy of answering questions raised by teachers, the situation of participating in interaction, the completion of homework and stage test, etc. Using the data recorded by these learning platforms, the study behavior is analyzed. Based on analysis the students' learning behavior, it helps the teacher guide the students to further study. It is more efficient than the traditional classroom-teaching teachers to guide students. Big data technology pays attention to learners' learning situation, progress and learning needs in real time. Students as the basis, pay attention to training students' interest, and guide students to learn scientifically according to the needs of big data age [7].

2.3 Pay More Attention to Process Evaluation

When testing students online, the online platform server can record the students' answers, the time of each question, the answers given by each question, the order of answers, etc. These are the behavior data that can't be recorded by ordinary traditional examination methods. Use behavior analysis technology to analyze the difficulty of test questions, compare the accuracy of students' answers, find the knowledge points corresponding to the test questions for the questions with high error rate, and then strengthen the explanation of knowledge points in the future teaching.

In the teaching process, we can pay more attention to the process evaluation from the original final evaluation. Using the online platform, in addition to recording teachers' teaching behavior and students' learning behavior, big data can record students' daily learning behavior, more records of the whole learning process, conduct diversified evaluation on students, analyze education big data, find students' daily learning state, and analyze students' performance and their usual learning, For effectively warn the students' achievements, effectively prompt the students who may not complete their studies, and urge them to complete their studies smoothly. Teachers can export students' scores from the online teaching platform. For example, if a class in a college chooses a course and the test score is lower than 20, the students' scores can be exported in order to give an effective warning to these students and promote them to complete the course better.

Everything has a life cycle, and so does data. The data collection, storage, analysis and utilization is the activity cycle of data. Among them, the data collection is the basis of big data analysis, and the collection of big data is the foundation of big data analysis and mining.

The traditional test shows the final answer of students. The online platform can master the more important "dynamic" answer process. For example, the students in which question is the longest time, which problem is repeatedly modified the most, etc. It can reflect the students' knowledge and strengthen the guidance.

Student-centered, collect the data learning behavior as fully as possible, analyze the students' learning behavior and learning achievement by using analysis and mining technology, find the relationship between learning results and learning behaviors, predict the students' learning performance, and send out early warning in time, so as to better complete the study of this course.

3 Design of Platform Architecture

According to the basic framework of big data architecture, the education big data platform is generally composed of infrastructure layer, intermediate support platform layer and user interface layer.

The deployment of shared network, the application of different terminals, and has the functions of network communication, storage, general computing and so on.

The intermediate support platform mainly provides various corresponding support services, which are In charge of the collection, storage, classification and summary the basic data. The collected and stored data includes unstructured, structured and semi-structured data. This layer includes all kinds of data which generated in the whole education and teaching activities, the basic data of students and teachers, various teaching resources used in teaching activities, as well as teachers' teaching behavior, students' learning behavior of participating in learning activities, etc.

Realize transparent and efficient application the big data. Because of storage of massive data, we need to use cloud storage technology to manage and store educational resources in the big data platform, and use the mainstream data management system to manage and store the data. The supporting layer also includes data mining and other systems. The data resource layer mainly constitutes the core data resource of the platform of education big data.

The user interface layer is the interface layer between the big data platform and users, providing various services. For example, information browsing, query and download, learners can use the online platform to learn and share high-quality education and resources. The platform of the evaluation can also predict and analyze the learning and find the problems in learning process, in order to improve and successfully complete the learning.

4 Data Mining Technology

The goal of data mining mainly has two aspects. One aspect is the summary and induction of laws, which classifies and describes the disordered and irregular data, such as classifying students' learning behavior habits and discovering valuable behavior laws. On the other hand, predict the future trend according to the existing analysis results, such as predicting students' academic performance, and then provide valuable guidance to help students successfully complete the whole learning, improve efficiency and achieve better teaching results.

The commonly used algorithms of data mining technology include clustering analysis algorithm, decision tree algorithm, association classification analysis algorithm and so on. Clustering analysis algorithm is to classify data according to specific requirements and laws, and the data of the same class have similarity. Cluster analysis of data samples according to different attributes and characteristics. If we study the relationship between different data sets, we should conduct correlation analysis on different types of data samples, and then find the correlation degree between different types of data and different attributes.

Decision tree algorithm classifies by the disordered data samples, and then form a tree like structure composed of root nodes, branches and leaf nodes. The root node of the whole tree represents the data set, and the branch of the tree represents the problem of splitting.

The integration of big data and education has promoted the reform and innovation of education, innovated teaching models, improved education standards, further promoted "Internet + education", and improved evaluation methods. Education big data generally refers to the data collection generated in the entire process of education and teaching activities or collected according to the needs of education and teaching, for educational development, education big data as a subset of big data, in the teaching process should make full use of the value of education big data, mining learners' daily learning behavior characteristics, and then realize personalized teaching according to aptitude. The analysis of education big data aims to mine valuable information from multi-dimensional massive data, discover teaching rules, improve the level of education and teaching management, make better decisions about education, change and innovate the original teaching mode, and improve teaching quality.

The main way big data technology solves problems adopts the divide-and-conquer thinking, that is, multiple different services need to be processed collaboratively at the same time, and these multiplications need to be arranged and installed on different servers. In a cluster, the number of master servers is not large, but it is important, they are the core of the entire cluster server. In order to facilitate maintenance and management, a part of the server can be set aside as the master node in the cluster design, which is specially used to install and deploy the master server. The architecture of big data analysis system mainly includes data warehouse, log processing system, etc. The main analysis methods used are indexing, parallel computing, etc.

Big data analysis tools can be used to analyze big data, and the results analyzed by the tools can provide a good data source for data visualization. The framework of the big data overall analysis system stores the preprocessed data into the data warehouse through data extraction, sorting, filtering, transformation, loading, etc. for structured, unstructured and mobile data. Data warehouse is a data central storage system built based on a specific data structure, and data management services can establish links between data to achieve integration services, and can access and query future data.

Big data for education is growing, and data is managed, controlled, and secured. Capture, store, transmit, integrate it into a data warehouse. Data warehouses implement standardized data formats. Data has a standard data format, and data warehouses can store large amounts of historical data, analyze its trends, and make future predictions more accurate.

5 Conclusion

In our country, the field of education has been developing, innovating and reforming. Education and big data technology are trends that are sure to continue.

Big data technology can also be applied in other aspects, such as the evaluation of students' family economic status, accurate assistance and funding for financial difficulties, and help to achieve targeted poverty alleviation through education. [8] Big data helps to promote the formulation of strategies for the accuracy of university administration, mainly including positioning accuracy, information accuracy, management goal accuracy and other aspects. [9] If we can find data, we can win the survival in the future; If we can mine data, we can adapt to the future development;

If we can make rational use of the mining data, we can provide personalized services and win the competition better. From data discovery, data analysis and mining to rational use of data, step by step. [10]

In recent years, our country has only carried out systematic research on education big data, and the relevant systematic research, the depth of relevant the practical application of technology are not very deep, which cannot effectively complete the construction of education data model.

The integration of big data and education big data has become the trend of future education development, and the further widespread application of education big data will also have a more far-reaching impact on education.

Acknowledgements "Analysis, research and practice of education big data in Colleges and universities", the key subject of the "13th five year plan" of Educational Science in Jilin Province in 2020, Project No.: ZD20057. In 2021, the special project of industry university cooperation and collaborative education in Jilin Province was "business data analysis teacher training industry university research project". Achievements of the 2022 project "Application Research of educational big data analysis in basic computer teaching" of the National Committee for Basic Computer Education, Project No.:AFCEC202200019. Achievements of the 2022 "teaching reform practice of deep integration of online and offline education and teaching" project of the National Committee for Basic Computer Education, Project No.: AFCEC202200020. Achievements of 2022 school science and technology project "design and research of personalized learning analysis platform based on educational big data". 2022 Jilin Province Higher Education Research Project "The reform and practice research of online-online-offline-deep integration education teaching mode".2022 Jilin Province Higher Education Teaching Reform Research Project "Research on Online Learning Behavior and Evaluation Analysis Based on Data Mining under the Background of Educational Big Data", project number: JLJY202253508131.

References

1. H., Long, Z. Xiaomei, T. Linhai, Research on the construction of smart education cloud platform under the background of big data. Comput. Knowl. Technol. **11**(20), 109–111 (2015)
2. Y. Changzhi, The possible turn of education in the era of big data. Jianghuai Forum, **04**, 188–192 (2020)
3. Y. Juan, Analysis of personalized application of big data in higher vocational education. Netw. Security Technol. Appl. **03**, 91–92 (2014)
4. W. Zhen, How far is education from informatization. It education in primary and secondary schools, 2012(12), 25–26
5. Z. Peng, Application of big data and its key technologies in education. Zhifu times, (08), 225–227 (2016)
6. Z. Jin, Z. Jianjun, W. Yijun, Rethinking the opportunities and challenges of education development under the big data thinking. Audio Visual Educat. Res **39** (06): 21–26 (2018)
7. F. Xin, Exploration of computer teaching in the era of big data. Appl. Microcomput. **30**(11): 32–34+3 (2014)
8. C. Ying, Research on accurate judgment and countermeasures of college students with difficulties under big data analysis. Hubei Agricultural Mechanizat. **23**, 170–171 (2019)

9. W. Biyu, Precision analysis of University Administration. Quality education in Western China **5** (05), 105–106 (2019)
10. W. Xiaobo, Big data promotes educational reform and innovation. Informat. Technol. Educat. Primary Secondary Schools **10**, 10–11 (2013)

Evaluation and Decision Making Research on Application Transformation of University Based on Big Data Warehouse Technology

Yuli Cui and Xinchun Wang

Abstract In order to process the big data of application transformation of higher education, the big data warehouse technology is applied to construct the evaluation and decision making system. Firstly, characteristics of application-orient university are discussed. Secondly, big data characteristics of application transformation effect of higher education is also summarized. Thirdly, big data warehouse of application transformation of higher education is designed. And then the multi join query optimization algorithm of big data warehouse is designed. Finally, the simulation analysis is carried out, and results show that the big data warehouse technology can deal with the complex data of application transformation of higher education, and provide theoretical guidance for evaluation and decision making of application transformation of higher education.

Keywords Big data · Data warehouse · Improved ant colony algorithm · Application transformation of higher education

1 Introduction

The higher education aims to culture senior specialized personnel with creative spirit and practical ability. The trained talents should adapt to the requirement of economic, technological and social development. The application oriented universities mainly develop the practical personnel. Many universities are unwilling to accept the "application type" positioning. Many universities think that the "application type" university has no hierarchy and no technical content. Currently "School summary" in many university websites has changed, the universities hope to construct the multiversity

Y. Cui (✉)
Information Engineering Department, Yantai Vocational College, Yantai, China
e-mail: yantaicyl@126.com

X. Wang
Scientific Research Division, Jinan Vocational College, Jinan, China

S. Patnaik and F. Paas (eds.), *Recent Trends in Educational Technology and Administration*, Learning and Analytics in Intelligent Systems 31,
https://doi.org/10.1007/978-3-031-29016-9_5

and application-oriented university, so many university administrators have accepted this reality. How, they do not clearly understand the application type undergraduate, the development goals of many teaching oriented universities are still positioned as the "teaching-research type" university. Many universities develop in research-based direction, and then the universities can survive and develop [1].

With rapid development of information technology, the chains of data acquisition and storage has been broken by big data torrent. The mass data has pervaded into every aspect of social life. The volume scale of data expands constantly, and the data types are various. The data dependency is analyzed, and the deep value of data mining is the advantage of big data. The big data can expand the channel of collect and acquire the information, and then the comprehensive analysis ability of information can be improved. Therefore the big data can improve precision of evaluation and decision correctness of application transformation of university. The application transformation effect of university can be monitored in real time. The big data is the theme of studying the application transformation of university, which can promote the effect of application transformation of university. The big data requirement mainly focuses on analysis, which can manage and process the data with huge scale and complex structure to achieve the evaluation of application transformation of universities. The data warehouse technology has framework, methods and commercial products. Based on further integration between big data analysis requirement, the column database, database internal analysis, in-memory, data compression and other techniques are studied. The large scale data should be analyzed and processed in real time. The big data warehouse technology can be applied to evaluate the application transformation effect of university [2, 3].

1.1 Characteristics of Application-Orient University

The application-orient university is different from researching and teaching oriented universities, which relies on vocational skills training. The application-orient university cultures the professional and vocational talents for locality. The application-orient university occupy an important position in higher education system of application orient university. The popularization of higher education ask the university to bear the training task of advanced applied talents. The higher vocational education idea has been extended in connotation. In developing country, the imbalance of technical personnel demand due to unreasonable educational structure has blocked the economic development of technical talents. The type and level of higher education has changed from single form to diversity, which is the irreversible global trends. The higher education has regarded the higher education as application education. The higher vocational education is the popular skills education, which can meet the self existence requirement of students, and can adapt to the social industrialization and urbanization demands. In traditional industrial society, the skills and cultural level requirement for labor is not high, and the repeatability of labor is strong. However the modernization, industrialization and urbanization require the labor adapt the changing social positions. The application transformation of university should be strengthened to adapt the modern development laws, and then the citizen can grasp the basic labor technologies suiting to modern development. The higher education should strengthen the wide vocational training. The university should culture the practical personnel through using courses and teaching contents adapting to social needs. The application oriented university can be formed [4].

The application oriented university is the local university connecting with market, industry and post. The application oriented university absorbs the school running experiences of regular undergraduate colleges. The discipline construction is used as basic logic, and the guiding ideology of major setting of higher vocational colleges is used for reference. The major is set according to the requirement of industry, post and skills. The application oriented university has specialty, undergraduate, graduate levels, which can construct the lifer long career learning system through establishing and perfecting vocational and technical qualification education.

2 Big Data Characteristics of Application Transformation Effect of Higher Education

The big data is data collection managed and processed, which is not captured based on conventional software tools in bearable time range. The application transformation effect data sample of higher education can be used to describe the development situation of higher education. The data collection can be used to carry our effective evaluation of application transformation of higher education. The data sample should be comprehensively analyzed. Under big data era, the modern information technology is used to application transformation effect of higher education. The mutual capacitance between information is strengthened, the independent information has close connection in deep level, the big data of application transformation effect of higher education has the following characteristics [5]:

(1) The volume of big data is massive, the content of big data has strong comprehensiveness. The data of application transformation of higher education is more open, and the coverage is more wide. One side, the information of application transformation of higher education involves a full rang of content of higher education. On the other hand, the information of it is affected by multilevel factor, such as social factors, economical factors and psychological factors.
(2) The big data permits a certain degree of confusion and fragmentation of data information. The big data of application transformation of higher education concludes massive irrelevant information, and is relatively scattered and vague.
(3) The core of big data is to mine the correlation between data. The university can find out development of application transformation of higher education based on big data. And the effective measures can be taken based on big data analysis results. The mutual relation can be constructed for independent data information based on big data.

The prediction is another core of big data, the missing connection can be mined and the possible connection can be predicted based on node of network, structure of network and other outer information. The traditional operational database has limitations in data analysis, which can not meet the real requirement of big data. The data warehouse can provide system analysis data with full view and high integration, and then the correlation between data can be mined. The big data warehouse technology is an effective tool of application transformation effect evaluation of higher education [6].

The big data of application transformation of higher education mainly concludes four characteristics, which are multi variety, high velocity, large volume and high value. The big data of application transformation of higher education has many types concluding logs, videos and pictures. Big data has quick data processing speed, and the analysis results should be given in second time range. The value density of these data is low, while it has higher application value. The big data technology has gradually tended to mature. The rich big data of application transformation of higher education should be integrated. The big data of application transformation of higher education not only concludes the structured data, but also concludes semi structured data. The non structured data is stored in local system, which is not benefit for being searched. The semi structured data is generally stored in file form. With increasing of data amount, the complex data amount generated can cause the data storing and processing pressures. The data warehouse can not be extended, and then the management difficulty increases. The efficiency reduces accordingly, Therefore the big data should be combined with data warehouse to carry out evaluation and decision of application transformation of higher education. The big data characteristics of application transformation of higher education is shown in Fig. 1.

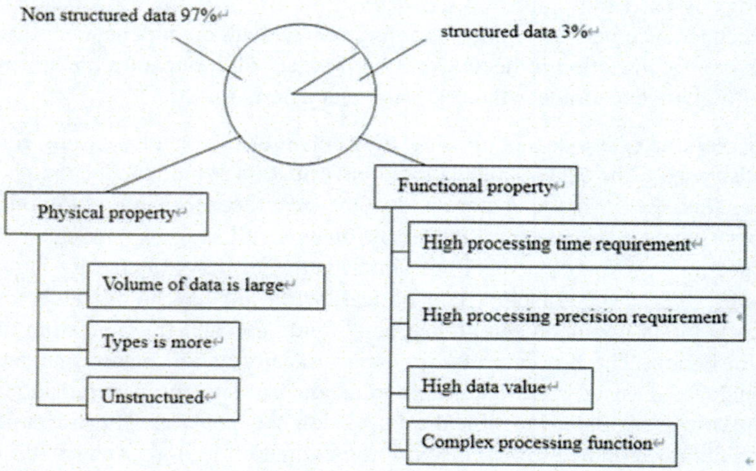

Fig. 1 .Property of big data of application transformation of higher education

3 Design of Big Data Warehouse of Application Transformation of Higher Education

The non structured characteristics of big data of application transformation of higher education brings out the difficulty for data warehouse design. The data warehouse design of application transformation of higher education has clear theme, which faces the research theme higher education administrator. The data warehouse has complete content. The time characteristics of big data covers history process of application transformation of university. The time synchronization characteristic of data has high precision requirement. The proper data granularity of data is benefit for storage and procession of data [7]. The basic framework of data warehouse of application transformation of higher education is shown in Fig. 2.

The data standard and preparation is to construct and maintain the dimensional model of data warehouse, which can manage the information and no structured big data of application transformation of higher education. The metadata is the relational tie between the data warehouse and procession system, which is the describing element of data warehouse [8].

The data source of data warehouse concludes two classifications, which are structured data and non structured data. Based on design standard, the data extraction, data transformation and data transformation should be carried out for data loading. The special logic relation of data is constructed, which is input into data warehouse to be managed. The data standard flow is shown in Fig. 3.

The metadata of structured data is defined by Wed service description language transmitted by SOAP protocol, HTTP protocol based on standard in procession of designing data warehouse. The metadata information in data management and procession is analyzed based on XML. The data base of binary based on binary file is designed autonomously is designed. The management of non structured massive data file and structured complex information can be achieved, and then the unified and efficient data application service expected by user can be satisfied.

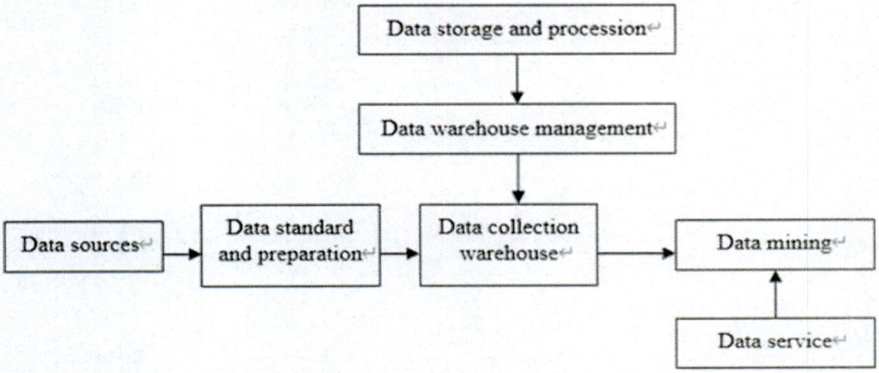

Fig. 2 Basic framework diagram of big data warehouse of application transform of higher education

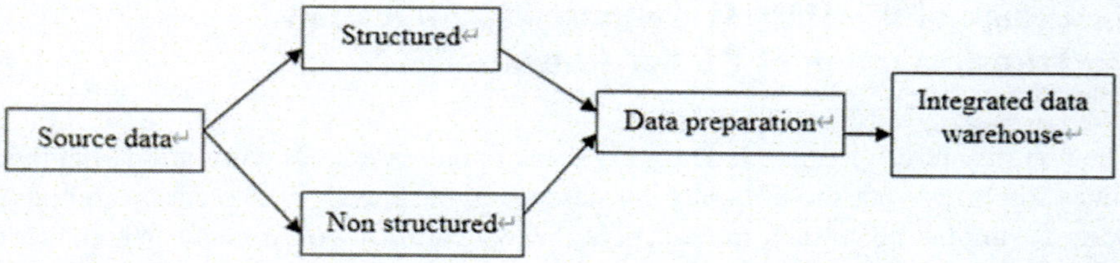

Fig. 3 Data preparation of source data

4 Multi Join Query Optimization Algorithm of Big Data Warehouse

For online transaction processing and business intelligence and data mining requirement, every query language involves one or several fact table and dimension table, the conjunctive ordering of different table decide the diversity of query execution plan [9]. Different query execution plan can constitute huge query policy space. In multi query, the join query graph is defined by $G(V, E)$, the connection to be connected is the node $R \in V$, and a connection operation is defined by edge $R_1, R_2 \in E$. The every query plan of query graph is denoted by connection tree, and relationship in database is denoted by leaf node. The result set of every connection of left and right nodes is corresponds to the intermediate node. Four query execution plans are shown in Fig. 4.

For the connection tree concludes *n* relationships, $(j_1, j_2 \ldots j_i)$ is the internal node of connection tree, and the cost evaluation model is expressed as follows:

$$\mathrm{COST} = \sum_{i=1}^{n-1} \cos \mathrm{t}(\mathrm{j_i}) \tag{1}$$

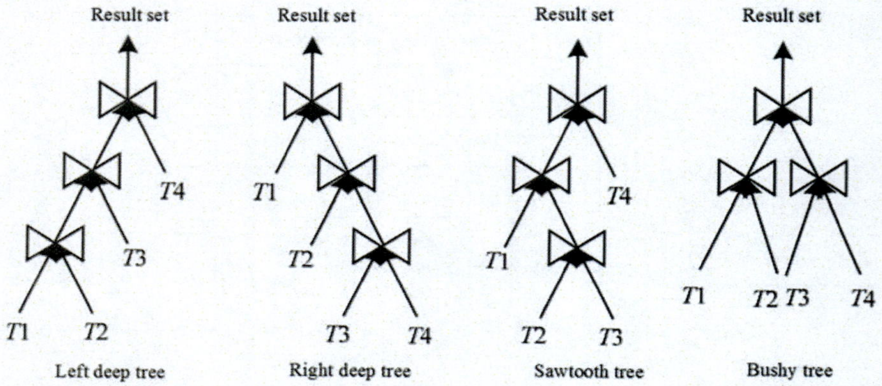

Fig. 4 Four kinds of connection trees

For a connection $J = RjoinS$, the cost of it is calculated based on the following expression [10]:

$$\cos t(J) = \frac{|R| \times |S|}{\prod_{c_i \in C} \max(V(c_i, R), V(c_i, S))} \tag{2}$$

where $V(c_i, R)$ is calculated the following expression:

$$V(c, J) = \begin{cases} V(c, R), c \in R - S \\ V(c, S), c \in S - R \\ \min(V(c, R), V(c, S)), c \in R \bigcap S \end{cases} \tag{3}$$

where $|R|$ is the number of tuples, C is the common attribute set between R and S, $V(c, R)$ is the number of property c with different value in relation R.

In data warehouse environment, the extraction of related data sets and multi-dimensional array in relation query technology should be achieved by connecting dimension table and fact table. The uncertainty of connection order constitutes the diversity of query execution plan. The optimizer can confirm a better connection order based on effective algorithm, and then the query efficiency can be improved. The improved ant colony algorithm is to optimize the query execution plan, and the corresponding algorithm flow is listed as follows:

Step 1: The maximum number of population is pre-defined by $N_{\max}$, and the pheromone concentration on all routes is obtained by ant of colony. The departure path is selected based on a certain probability, which is expressed by

$$p_i^j(t) = \begin{cases} \dfrac{\gamma_i^{\varsigma}(t) \cdot \eta_i^{\upsilon}(t)}{\sum_{i \in A_j} \gamma_i^{\alpha}(t) \cdot \eta_i^{\beta}(t)}, i \in A_j \\ 0, i \notin A^j \end{cases} \tag{4}$$

where, A_j denotes the selectable path of j th ant in colony, γ_i denotes the pheromone concentration of i th path; η_i denotes expected degree of path; ς denotes the inspire factor; υ denotes the expected factor.

The maximum evolutionary times of algorithm is defined by $I_{\max}$, and the parameters of chaos algorithm is initialized.

Step 2: The evolutionary times is initialized, $I = 0$, the population size of chaos algorithm is defined by H_g, and the model of chaos algorithm is defined by

$$\varphi_{i,j} = \lambda \varphi_{i,j}^0 (1 - \varphi_{i,j}^0) \tag{5}$$

where, $\varphi_{i,j}$ denotes the chaos vector, $\varphi_{i,j}^0$ denotes the original value, $i = 1, 2, \cdots N-1$, $j = 1, 2, \cdots, m$, m denotes the dimension of decision vector. The chaos variable is mapped onto the decision variable, the changing interval is chosen as $(x_{j\min}, x_{j\max})$, and the following equation is satisfied:

$$x_{i,j} = x_{j\min} + (x_{j\max} - x_{j\min})\varphi_{i,j} \tag{6}$$

Step 3: non dominant level is sorted for population H_g, which is equal to fitness degree. And then the genetic manipulation is used to generate the next generation population S_g with same scale.

Step 4: the intersect operation is carried out for previous and next generation populations, that is $C_g = H_g \cup S_g$. The leading edge of C_g is constructed based on non dominant level sort, which is defined by $U = (U_1, U_2, \cdots)$.

Step 5: The crowding distance of leading edge for non dominant level is calculated, and the intersect operation is carried out at same time, that is $H_{g+1} = H_{g+1} \cup U_i$, $i = i+1$. When the following condition is satisfied $|H_{i+1}|+|U_i| \leq N_{\max}$, the iteration operation is over. The sort is carried out based on crowding distance of U_i, the top $(N - |H_{g+1}|)$ solutions are chosen. 步

Step 6: The obtained population is judged to reach the optimization. When individual scale with non inferiority class 1 is equal to number of population, the chaotic refinement technique selects the top 10% individuals to construct the next population, and the corresponding searching interval is taken as $(x'_{j\min}, x'_{j\max})$. The calculation model is expressed by

$$\begin{cases} x'_{j\min} = x^*_{i,j} - \kappa(x_{j\max} - x_{j\min}) \\ x'_{j\max} = x^*_{i,j} + \kappa(x_{j\max} - x_{j\min}) \end{cases} \tag{7}$$

where, κ is optimization factor, $0 < \kappa < 0.5$.

When the condition $x'_{j\min} < x_{j\min}$ is satisfied, the following equality can be obtained $x'_{j\min} = x_{j\min}$. When the condition $x'_{j\max} > x_{j\max}$ is satisfied, the following equality can be acquired $x'_{j\max} = x_{j\max}$.

The chaotic variable $x'_{i,j}$ can be obtained based on mapping operation. And then the linear summation between $x'_{i,j}$ and $x_{i,j}$ is carried out based on the following expression:

$$x''_{i,j} = (1 - \gamma)x'_{i,j} + \gamma x_{i,j} \tag{8}$$

where, γ denotes the regulating factor.

Step 3 to step 5 is executed repeatedly. When the ending condition is satisfied, the step 7 is carried out.

Step 7: the algorithm is over, and the optimal solution is output.

5 Case Study

In order to verify the effectiveness of application transformation of higher education based on big data warehouse technology, the simulation is carried out. The ten universities in a city are used as researching objects, and the data of application

Table 1 Simulation results of query execution plan optimization

Number of relationship	Number of ants	Iteration times of optimal query execution plan		Execution time of optimal query execution plan/ms	
		Traditional ant colony algorithm	Improve ant colony algorithm	Traditional ant colony algorithm	Improve ant colony algorithm
8	5	43	21	76,543	68,545
10	8	48	23	134,563	123,793
12	8	45	28	256,438	23,658
14	10	46	29	325,672	32,186
16	12	50	26	473,278	465,248

transformation from 2006–2016 are used as sample. The optimization simulation analysis is carried out based on traditional ant colony algorithm and improved ant colony algorithm.

The iteration times of optimal query execution plan and execution time of optimal query execution plan are used as the index, and the simulation results are listed table 1.

As seen from Table 1, the improved ant colony algorithm has less iteration times and execution time than the traditional ant colony algorithm. Therefore the improved ant colony algorithm has quicker convergence speed, and it can obtain optimal solution with higher quality. The query performance of improved ant colony algorithm has been improved, and the premature can be avoided. And then the query efficiency can be improved.

6 Conclusions

The application transformation of higher education is a big trend, and the local universities should transfer the teaching method and teaching mode. The local universities should improve the school running characteristics under the joint efforts of university administrators and general teachers and students. In order to monitor the application transformation effect of higher education, the big data warehouse technology is applied to construct the evaluation and decision making decision. Simulation results show that the system constructed has an effect method of processing big data of application transformation of higher education.

Acknowledgements This work was supported by Horizontal Project of Yantai Vocational College (HX202240). In addition, the authors would particularly like to thank the anonymous reviewers for helpful suggestion.

References

1. L. Li, J. Lin, Ye., Ouyang, Xin (Robert) Luo, "Evaluating the impact of big data analytics usage on the decision-making quality of organizations." Technol. Forecast. Soc. Chang. **175**(2), 121355 (2021)
2. Li., Xuan, Big data-driven fuzzy large-scale group decision making (LSGDM) in circular economy environment. Technol. Forecast. Soc. Chang. **175**(2), 121285 (2022)
3. H. Zhang, Z. Zang, H. Zhu, M. Irfan Uddin, M. Asim Amin, Big data-assisted social media analytics for business model for business decision making system competitive analysis. Informat. Process. Manag. **59**(1), 102762 (2022)
4. Y. Kazancoglu, M. Sagnak, Sachin Kumar Mangla, Muruvvet Deniz Sezer, Melisa Ozbiltekin Pala, A fuzzy based hybrid decision framework to circularity in dairy supply chains through big data solutions. Technol. Forecast. Soc. Change, **17**(9), 120927 (2021)
5. Y. Niu, L. Ying, J. Yang, M. Bao, C.B. Sivaparthipan, Organizational business intelligence and decision making using big data analytics. Informat. Process. Manage. **58**(6), 102725 (2021)
6. K. Sun, D. Yang, Human health big data evaluation based on FPGA processor and big data decision algorithm. Microprocess. Microsyst. **81**(3), 103793 (2021)
7. L. Sfaxi, M.M.B. Aissa, DECIDE: An Agile event-and-data driven design methodology for decisional Big Data projects. Data Knowl. Eng. **130**(11), 101862 (2020)
8. P. Liu, Y. Long, H. Song, Y. He, Investment decision and coordination of green agri-food supply chain considering information service based on blockchain and big data. J. Cleaner Product. **277**(11), 123646 (2020)
9. F. Chiheb, F. Boumahdi, H. Bouarfa, A New Model for Integrating Big Data into Phases of Decision-Making Process. Procedia Comput. Sci. **151**(1), 636–642 (2019)
10. P. Martins, M. Abbasi, F. Sa, J. Celiclio, F. Morgado, F. Caldeira, Intelligent beacon location and fingerprinting. Procedia Comput. Sci. **151**(1), 9–16 (2019)

Management Culture of Art Colleges Based on Big Data

Sha Wu

Abstract With the rapid development of Internet information technology, big data has officially entered our work and life. Driven by big data, how to use big data and Internet information technology to innovate financial management in universities to improve the financial management and service level of universities, and promote the connotative development of school undertakings has become a key issue for college financial workers to think about and study. Facing the inadaptability in the construction of campus culture in the new situation, we must proceed from the characteristics of the times of data development, put the campus culture in the perspective of data refinement and task realization, and carry out classification and phased construction of cognition, recognition and practice, so as to better adapt to the changing trend of data culture. By analyzing and studying the management problems existing in the administrative management of art colleges, this paper puts forward a school management mode based on big data, and constructs the management culture of art colleges as a whole system, so as to further promote the update of teaching mode through the transformation of administrative mode, better improve the reform of teaching management mode, and provide new ideas for the development of art colleges.

Keywords Art school · Big data · Management culture

1 Introduction

The basic principles and laws of school management and enterprise management are common, and it is also an activity of employing people to manage affairs. School management is that managers adopt certain means and methods through certain systems, lead teachers and students, make full use of resources and conditions inside and outside the school, and effectively achieve school work objectives [1]. School

S. Wu (✉)
Feifanyilin, Chongqing, China
e-mail: 673854685@qq.com

S. Patnaik and F. Paas (eds.), *Recent Trends in Educational Technology and Administration*, Learning and Analytics in Intelligent Systems 31,
https://doi.org/10.1007/978-3-031-29016-9_6

management culture is a kind of management spirit and atmosphere, and it is the accumulation of humanistic spirit of leaders in management activities [2]. A school's unique management culture is the productive force of the school's connotation development and the supreme pursuit of school management. Art colleges, as an indispensable part of universities, have extensive connections and similarities with other universities in the construction of teaching management culture, but also have differences and personalities [3]. There is a natural and close relationship between school management and culture, and the influence of different cultural forms and psychology on school management behavior has long attracted people's attention. Especially since the beginning of the new century, with the continuous enrichment and development of modern educational management theory, people have realized more and more deeply that whether the school management objectives can be achieved efficiently will not only be directly influenced by management methods and means, but also be deeply influenced by culture [4]. It is of great significance to explore the connotation and characteristics, functions, ways and countermeasures of the cultural construction of teaching management in comprehensive art colleges for building high-level, rich connotation and individuality art colleges [5].

The construction and management of management culture in art colleges is a systematic project. To realize the vision of building advanced management culture in art colleges, we should make efforts in many ways, such as establishing the value orientation with service as the core, cultivating pioneering and innovative spirit, promoting the harmony between management culture and academic culture, and building a learning-oriented administrative organization [6]. Campus culture has strong interaction, which can not only absorb cultural elements from all walks of life, but also continuously publicize and spread its own cultural characteristics. From this point of view, the campus culture itself is open and progressive with the times [7]. At what level to adapt to the big data drive, especially while enjoying convenient and fast data services, the pursuit of updated technical means and the abandonment of traditional learning and living habits need to be guided and standardized in the campus culture [8]. By analyzing and studying the management problems existing in the administrative management of art colleges, this paper puts forward a school management mode based on big data, and constructs the management culture of art colleges as a whole system, so as to further promote the update of teaching mode through the transformation of management mode, better improve the reform of teaching management mode, and provide new ideas for the development of art colleges.

2 Innovation of Art School Management Culture Driven by Big Data

2.1 Rational Understanding of School Management Culture

The quality of teachers has always influenced the outside world's evaluation of a school. The headmaster should not only stick to the development goal of a school, but should try his best to improve the efficiency of each teacher to promote the progress of the school. Each teacher sends a daily work diary, which is sent to the direct leader and copied to the principal. It is not only a summary of daily work, but also a form in which every employee's voice can be seen by the principal [9]. Help to make an excellent plan in the next management strategy. Let teachers learn from each other through class competitions or regular sharing of excellent courses of senior talents in the school. Advocate common progress, not behind closed doors. Let teachers see the gap between each other and enhance their inner motivation for progress. The necessity of external training lies in broadening the horizons of faculty and staff, rather than just sticking to their own professional competition. Learning good methods and excellent styles outside the region and even outside the industry will help to open up the teacher's personal pattern.

For teachers with academic commitment, rewarding teaching achievements in time, such as the allocation of teaching and research funds, teaching assistants and teaching venues, can make teachers feel valued and produce products or processes that are more conducive to the development of schools. For teachers with perceptual thinking, care for birthdays and major holidays, care for family members, and consider family rewards such as children's study-related or parents' health-related when giving performance rewards [10]. Hold regular meetings of faculty and staff, and try to be presided over by the principal himself, because teachers usually don't want the meetings to be presided over by others. Although mobile devices such as online, principal's suggestion box and social software are convenient, face-to-face conversation with teachers can obtain the most direct communication clues from facial expressions and body language. Good suggestions from excellent teachers should be taken seriously. Excellent teachers hope that the school will prosper and succeed. If their enlightening suggestions can be heard, they will give practical support to the school.

Outstanding talents will be willing to work with outstanding talents. Employees' referral is a good channel to attract talents who are willing to settle in teaching and research through various channels. School reputation is an intangible and very important spiritual culture. Even if the school has a good spiritual culture, it needs to do a good job of propaganda. First of all, let people know the benefits of the company, and they will report back. In the information age, young people are easily influenced by external evaluation. Not every teacher wants to be a teacher, and not every teacher wants to be a manager. Combining employees' wishes with a competitive salary system, employees will develop and maintain a positive attitude in school.

Besides formal interviews, we can also use some informal methods to improve teachers' professional level. If you meet at the school gate or in the canteen, talk for a few minutes to collect information for oral or written evaluation. If you think the teacher has done a good job, you can praise it in public, or use notes, letters, etc. For the poor, you can use email, social mobile phone software or other multi-dimensional communication methods for personal and private communication. Pay special attention to methods. When collecting feedback information, don't destroy interpersonal relationships.

2.2 *Campus Management Culture Innovation Based on Data Mining*

The management culture of art colleges is the internalization of management culture in art colleges, a special organization. It is gradually formed by the long-term practice of the party and government organs and their staff in art colleges in organizing and managing school teaching, scientific research and social affairs. In the specific management culture environment of art colleges, the internal administrators and management departments of the schools will have corresponding administrative concepts [11]. A healthy, scientific, democratic and positive campus management culture will surely encourage managers to establish democratic, scientific, equal, service-oriented and promising administrative concepts in administrative practice [12]. Data mining is a process from data to model, and then from model to result, and it is realized through a cyclic process. Under the condition of modern information technology, analyzing and mining the massive data accumulated by these universities for a long time can help school administrators make decisions, improve the school, and provide reliable data basis for improving the teaching quality and optimizing the comprehensive teaching strength resources. The architecture of knowledge mining system of art digital archives in universities is shown in Fig. 1.

With the expansion of the scale of universities, the economic and business activities of universities become more and more frequent and complex. However, with the emergence of big data and Internet technology, the daily financial supervision becomes possible. Schools can embed the information of various economic and business activities into the comprehensive financial information service platform, so as to know all economic and business activities in time and realize the whole process management of business activities. The educational data mining process of art education management in universities is shown in Fig. 2.

Through continuous study and training, ANN can discover its regularity from a large number of archives data of teachers and students in unknown patterns. It usually needs to meet:

$$m = \sqrt{x + y} + R(10) \tag{1}$$

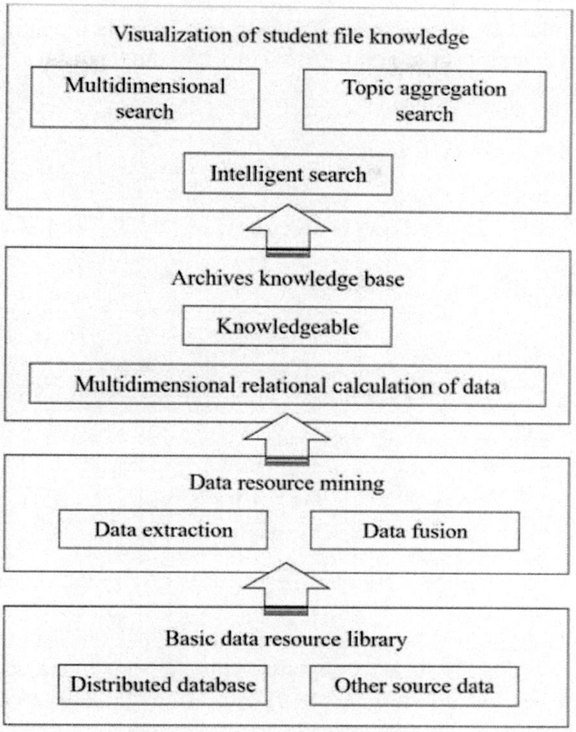

Fig. 1 Architecture of digital archives knowledge mining system in art universities

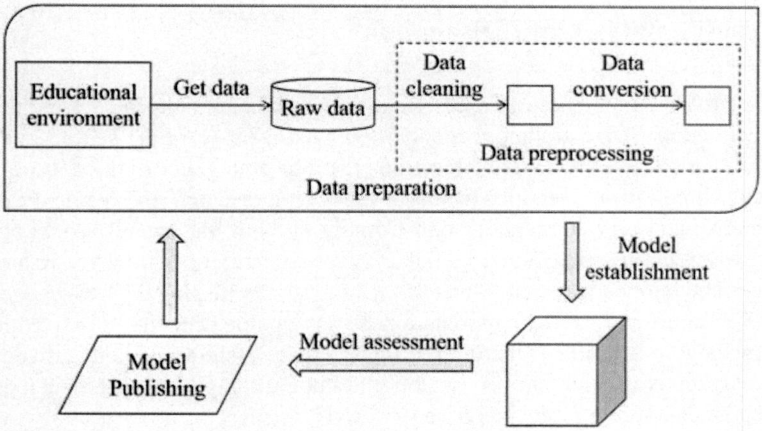

Fig. 2 Educational data mining process

where m is the number of neurons in the hidden layer, x is the number of neurons in the output layer and y is the number of neurons in the input layer. The roughness calculation process of X is:

$$R^{-}(X) = \{x \in U | R(x) \subseteq X\} = \{U_2, U_3, U_4, U_5, U_7\} \tag{2}$$

$$R_{-}(X) = \{x \in U | R(x) \cap X \neq 0\} = \{U_2, U_4, U_5\} \neq \emptyset \tag{3}$$

From this, we can get:

$$\rho(X) = 1 - \frac{|POS_C(X)|}{|R^{-}(X)|} = 0.6 \tag{4}$$

If $X = \{U_2, U_3\}$, it is undefined because:

$$R^{-}(X) = \{x \in U | R(x) \subseteq X\} = \{U_2, U_3, U_5, U_7\} \tag{5}$$

$$R_{-}(X) = \{x \in U | R(x) \cap X \neq \emptyset\} \neq \emptyset \tag{6}$$

It is necessary to reduce the attributes of the original file data of teachers and students to improve the efficiency of the algorithm. If we can set up a platform for archives digitization, this platform is also of extraordinary value, and can contribute to other types of work in art colleges and even in the society.

3 Result Analysis and Discussion

With the progress of the times, the people-oriented management concept and personalized service mode are gradually deeply rooted in the hearts of the people, and have a profound impact on the field of education [13]. The emergence of big data provides technical support for universities to realize the people-oriented management concept and personalized service management. Combining with the organizational characteristics to strengthen the comparison of archives content, it can mine the related file resources and knowledge in art universities, and visually display the knowledge relationship according to the topic association, so as to guide users to quickly complete the knowledge retrieval by clicking the links. Figure 3 shows the subjective rating results of campus administrators on traditional student campus management methods and archive data mining methods based on ANN.

Most campus administrators said that the data mining technology of college teachers and students files based on ANN can help campus administrators quickly locate the information they need in the massive resources. The application of data mining technology in the student campus management system greatly improves the

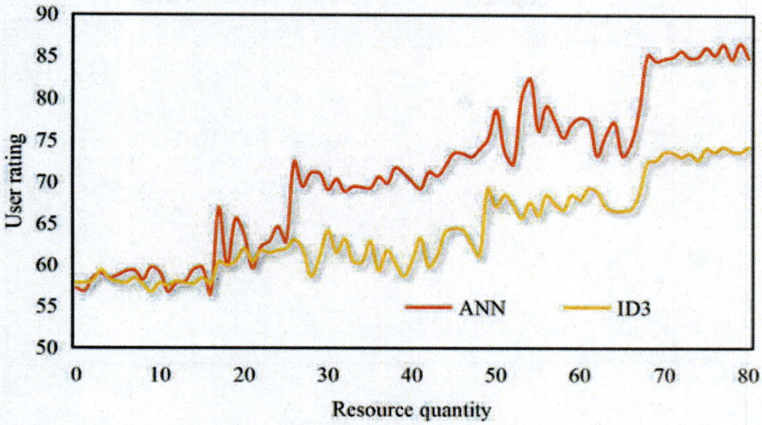

Fig. 3 Subjective rating of campus administrators

speed of processing archives information, thus saving the time spent by administrators in campus management and reducing the work intensity.

Making full use of the files of teachers and students stored in the cloud platform can provide cloud data to every campus administrator more conveniently. Table 1 shows the change of recommendation accuracy of data mining technology for college teachers and students files based on ANN when different iterations are selected. Select 9 groups of data with iteration times between 1500 and 9000, and observe the change of model accuracy.

It can be seen that the algorithm in this paper can achieve a high recommendation accuracy rate of 94.6% after continuous iteration. Therefore, it is of positive significance to apply this model to the management culture of digital campus in universities. The application of data mining technology to construct the campus management system of art school students should be combined with the actual situation of the school, proceed from reality, focus on meeting the application needs of the school, and make technology serve the actual needs.

The characteristics of big data require universities to establish a data-centric service decision-making idea, which will integrate a new perspective into the campus

Table 1 Accuracy of model with different iteration times

Iterations	Accuracy (%)
10	31.4–65.7
20	44.2–72.4
30	61.4–72.4
40	73.6–85.1
50	80.4–90.8
60	81.2–94.6

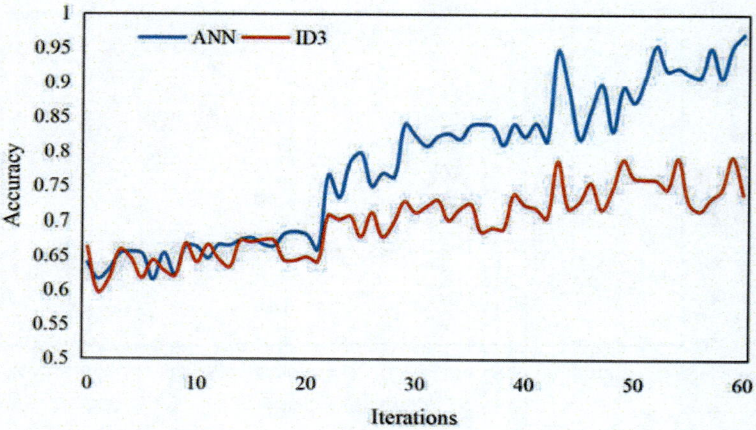

Fig. 4 Accuracy comparison results

information construction. Regardless of the macro level or micro level, relying on big data to promote the culture of decision-making based on educational data and empirical evidence in the education sector, and realize the transition of school decision-making from elite strategic decision-making to social decision-making. Samples in data analysis are represented by different characteristic attributes, and the diversity of big data can provide more characteristic dimension description data than before. The accuracy comparison results of ID algorithm and the algorithm in this paper are shown in Fig. 4.

The results show that the ANN algorithm is more reasonable, feasible and scientific than ID3 algorithm in the management culture of digital campus in universities. The application of data mining technology in the construction of students' management culture can improve data classification, improve the efficiency of file retrieval and realize the mining of file data information.

Art universities should adhere to people-oriented, firmly establish the concept of student-centered, so that all the work of the universities can be carried out around the center of "educating people", so as to truly achieve teaching and educating people, managing and serving them, and promote the healthy growth and all-round development of young students. Big data enables students, teachers and management departments to build a smooth information flow, realize effective management of students by teachers and management departments, and build an automatic feedback mechanism between teachers and students, departments and teachers and students. Intervene in the whole process of students' learning, understand and predict students' individual learning behaviors, procedures and attitudes, and the school provides targeted and personalized teaching contents, services and interventions. Teachers are also clients. They can dynamically pay attention to teachers' scientific research, educational affairs and other information, and provide personalized suggestions.

4 Conclusions

The basic principles and laws of school management and enterprise management are common, and it is also an activity of employing people to manage affairs. Art colleges, as an indispensable part of universities, have extensive connections and similarities with other universities in the construction of teaching management culture, but also have differences and personalities. Whether the school management objectives can be achieved efficiently will not only be directly influenced by management methods and means, but also be deeply influenced by culture. It is of great significance to explore the connotation, characteristics, functions, ways and countermeasures of the cultural construction of teaching management in comprehensive art colleges for building high-level, rich connotation and individuality art colleges. By analyzing and studying the management problems existing in the administration of art colleges, this paper puts forward a school management model based on big data, and constructs the management culture of art colleges as a whole system, so as to further promote the updating of teaching mode through the transformation of management model. The results show that the ANN algorithm is more reasonable, feasible and scientific than ID3 algorithm in the management culture of digital campus in universities. Campus culture has strong interaction, which can not only absorb cultural elements from all walks of life, but also continuously publicize and spread its own cultural characteristics. The application of data mining technology in the construction of students' management culture can improve data classification, improve the efficiency of file retrieval and realize the mining of file data information.

References

1. D. Yang, Analysis of the development and innovation path of college counselors in the era of big data. Sch. Party Build. Ideol. Educ. **2020**(20), 3 (2020)
2. C. Qiang, Enriching the school physical arts curriculum with folk culture. Prim. Second. Sch. Manag. **2019**(3), 1 (2019)
3. T. Zou, The possible direction and realization path of higher education management in the era of big data. Explor. High. Educ. **2017**(11), 7 (2017)
4. A. Xiang, Innovative thinking on college students' education management in the era of big data. Chem. Eng. Prog. **39**(1), 1 (2020)
5. J. Ma, X. Hu, Some thoughts on the integration of socialist core values into the construction of university campus culture. Res. Ideol. Educ. **2017**(1), 4 (2017)
6. A. Rao, K. Wan, W. Zou, Teaching leadership construction of university teachers in the era of educational big data. Mod. Educ. Manag. **2019**(1), 5 (2019)
7. X. Cao, Construction of university engineering audit management information system under big data. Economics **4**(6), 40–41 (2022)
8. Z. Jiang, J. Zhou, The path of campus culture construction in higher vocational colleges based on the integration of school-enterprise culture. Vocat. Educ. Forum **2017**(26), 4 (2017)
9. Z. Shi, C. Guo, Research on decision-making of university science and technology management driven by big data. Sci. Technol. Manag. Res. **41**(21), 6 (2021)
10. Y. Wang, The reform and development of college archives management mode under the background of big data. Arch. Manag. **2019**(6), 2 (2019)

11. K. Zou, P. Xu, Y. Guo et al., Construction of competency model of information management talents in universities under the background of big data. Inf. Theory Pract. **44**(12), 11 (2021)
12. Y. Wang, R. Zhao, L. Yan, Research on the operation mechanism and management countermeasures of university office information management under the background of big data. Inf. Sci. **38**(12), 6 (2020)
13. P. Yu, Y. Li, C. Wan, Research on the design of university scientific research service system integrating multi-source and heterogeneous education big data. Libr. Inf. Knowl. **2019**(1), 12 (2019)

An Adaptive Approach of Natural Language Processing (NLP) to Predict Aggressive Behavior of Adults in Educational Institution

Bin Hu, Qurat ul Ain, Muhammad Irshad, Ifrah Malik, Sohail M. Noman, Srikanta Patnaik, and Liying Hu

Abstract The paper examines why college students are aggressive. Data for the study were obtained from university professors through a questionnaire. Digital education has been transformed by natural language processing technology. Online literacy programs are increasingly incorporating NLP (Natural Language Process). In the modern age, NLP is being used for various educational tasks, including essay grading and feedback, student behavior, and example generation. The results show that boys use foul language and are involved in destructive activities compared to girls who were rarely observed in bad language or physical fights. The major causes of this include negative families or homes (parent and divorced and family issues), gross behavior of parents, dictatorial or mistaken teaching behavior, poor teacher/student contact, study pressure, unfriendly interpersonal relationships, and social injustices. The finding shows that there is no freedom of choice in the choice of subjects that makes boys aggressive and the gender-based society makes girls aggressive. Aggressive behavior of students in college should be addressed with the highest priority by organizing counseling services for students at the college level

B. Hu · L. Hu
Changsha Normal University, Changsha, China

Q. Ain (✉)
Department of Education, Gomal University, Dera Ismail Khan, Pakistan
e-mail: quratulaindgkhan@gmail.com

M. Irshad (✉)
Department of Electronic and Information Engineering, The Hong Kong Polytechnic University, Hong Kong, China
e-mail: ibrahim@stumail.ysu.edu.cn

I. Malik
Department of Management Sciences, International Islamic University, Islamabad, Pakistan

S. M. Noman
Yanshan University, Qinhuangdao, China

S. Patnaik
INTERSCIENCE Institute of Management and Technology (IIMT), Bhubaneswar, India

S. Patnaik and F. Paas (eds.), *Recent Trends in Educational Technology and Administration*, Learning and Analytics in Intelligent Systems 31,
https://doi.org/10.1007/978-3-031-29016-9_7

to provide psychotherapy for the relaxation of aggressive behavior. Parents' roles at home and interactions with teachers, strong teacher-student interactions, and cooperative learning activities for students will be a significant strength towards better educational future.

Keywords Educational behavior · Gender discrimination · University education system · Bad interactions · Descriptive research · Data analysis

1 Introduction

The research intended to study the impact of aggressive behavior on student academic success at university level in the rural areas of DG Khan (Dera Ghazi Khan). There is growing and widespread concern about the academic success of students in university. It is one of the most serious hurdles in effective teaching process [1]. It has also been reported that the students at university level feel uncomfortable because of lack of effective disciplinary measures and unpleasant situation. Teachers are not well prepared to control classroom discipline [2]. Student attitudes in example of aggressive talking, chronic avoidance of work, interfering with teaching activities, abusing classmates, verbal insult, physical aggression, and anger are some serious behavior problems in students of university [3].

The purpose of this study was to identify the forms and causes of aggressive conduct in university classrooms in the DG Khan area. The population of the study was made up of teachers who worked at DG Khan University. Using a multi-stage random sampling process, 60 teachers were chosen as a sample. The sampled instructors were given a questionnaire with seven scores on which they were asked to respond to the types and reasons of aggressive behavior in the classroom. Teachers devote a lot of time, effort, and energy to dealing with students' violent behavior in the classroom. It is of primary importance to investigate what actually are these behaviors inside classroom. For this purpose, many conditions have been used to know the effect of aggressive behavior in students [4].

When there are strict rules and regulations in university and classroom, then misbehavior and misconduct should be produced in the classroom. Angriness with teachers not complete the classroom needs. On the other hand, day dreaming in the classroom not fulfill the students work. Bad sitting arrangement produced disturbing during the studies [5]. This behavior is considered continuous feedback to students. Some of the aggressive behaviors were reported unanimously by most teachers. However, a significant deviation has been reported by most teachers. However, a large detachment has been reported. It is endorsed that aggressive behavior of university students towards DG Khan in Pakistan needs a lot of attention from educators, policy makers and government following the focus area due to the multiplication of learning in the country [6].

The development of a nation is above the characteristic of the lesson that is taught to students between schools [7]. If we look at the current instructional system, we see that it is low when evaluating values about life. In the absence of essential values about life, simply treat students in an undisciplined manner. In the current generation as your group about fears, struggles and frustrations. This is the youth of mental, sociable and personality disintegration. Aggression has always been a fundamental problem of humanity [8].

In [9], it is well-acquainted with the use of genetic, biological, temperamental, family, social, and cultural components in creating invasive behavior in children and adolescents under the age of 18. Television and other forms of media are also taken into account. It has asserted exclusive studies claiming that males are more violent than females in adolescence and childhood, or that physical violence is displayed by males more than females. Women's aggression, on the other hand, is rapidly increasing.

2 Literature Review

This section will review the literature on the topic of aggressive behavior in student academic success. A large body of literature provides much evidence that the strong relationship between behavior and positive behavior between teachers and students is an important part of the healthy academic development of all students. This type of literature involves various types of studies that have been conducted in recent decades to investigate aggressive behaviors among teachers and students and the impact of these aggressive behaviors on learning. Discuss the points of view of various disciplines from historical points of view to current thinking on the subject.

2.1 Aggression

Aggression can be described as an emotion that is often consistent with hurting, and destroying something and someone. In the case of humans, excessively harmful volition is performed to maintain the body and then the mind. Assault is an obvious act against a character with the intention of harming any other person physically and psychologically, and then damaging or absorbing that person's property [10]. The existence of ethnic groups is the sum of extraordinary noble or vile emotions. Aggression is something about these emotions, which manifest themselves at a certain stage of life, but which manifest themselves in a special way at a certain stage [11].

Aggression is defined as a spontaneous, impulsive act of anger in [12] and also the potential to degrade, threaten, or harm another person or object. It is unintentional and usually happens when you're stressed. It has been viewed as a loss of self-control or a failure to control one's impulses and also can manifest itself in both helpful and destructive acts, according to [13]. When employed for individual and collective well-being, it is constructive; when used for individual and collective distortion, it is destructive. Aggression is shown as an antisocial behavior by [14], whereas [15] identifies aggression as a disruptive behavior. Because of its growing importance, the notion of aggression has achieved international recognition [16], and it is well-known in the educational area, since extant literature describes its widespread application between middle childhood and adolescence [17]. Aggression has the potential to lead to violence and criminal activity, according to [18], and extreme forms of aggression may be linked to psychopathy [19]. Aggression is distinct in that it is linked to the expression of pain or harm [20].

2.2 *Features of Aggression*

The sub-dimensions of student aggression can be summarized into four categories namely verbal attacks, anger and resentment, physical attacks, and doubts. Verbal attack: It is defined as "verbal harm or harm to others, which represents a tool or component of behavior movement" [21]. Anger and resentment: Anger represents the emotional component of behavior, emphasizing the speed with which individuals adopt aggressive behavior. In addition, [22] implies "physical arousal and preparation for attack, which represents the emotional component of the emotion or behavior". Physical aggression: Physical aggression involves directly harming others in different ways (for example, instrumental or athletic behavior). Doubt: This includes "feelings of malice and injustice that represent the cognitive component of behavior".

2.3 *NLP Role*

The number of advanced NLP technologies is growing, and many of them are open source (like the latest Transformer NLP technology, which has several open-source models). In addition, neural technologies have become more accurate at performing human-like tasks on natural language. Here are some interesting potential applications of the NLP technology in education.

2.3.1 Personalized Plans

By using NLP technology, we can analyse students behavior that enhanced their aggression. Open text data from the internet can be analysed by NLP algorithms to compute the reading difficulty, and software can be built quickly to analyze student behavior in general.

2.3.2 Feedback

NLP systems that give human-like feedback on student assessment. As a result, the software can give immediate formative feedback to the teacher through the software. The training and use of these systems will require some ingenuity with data collection, but if done the right way, they can potentially lead to very innovative applications that will reduce the burden on teachers by reducing their work load.

2.3.3 Question Generation

We can generate various types of assessment questions from passages of materials. There are a number of companies that are attempting to develop commercially viable Question Generation Models (see Mostow et al.). Even though this is a very new area, scholars have established norms for evaluating Question Generation Models (see Mostow et al.).

2.3.4 Example Generation

NLP enables us to generate content, given some input data. With this power of the technology, we can utilize this to generate examples of self-confidence, control anger that show different examples of the same behavior, show examples of how good mental health is maintained, and so on, on how to use this type of technology effectively to achieve positive outcomes.

3 Methodology

3.1 Population

The point of the study was to examine in to different types of aggressive behavior and their causes in DG Khan, Pakistan. The population of the study was made up of teachers who worked in DG Khan's universities. Total of 60 teachers were taken as sample using multi stage random sampling method. One questionnaire was developed on seven scales and was distributed among the sampled teachers and collected their response.

3.2 Descriptive Research

Descriptive research focuses on the incidents of past. We find comprehensive detail of all the happenings around the world, use descriptive research. Administrators sort out solutions of different issues with the help of descriptive research [23]. The importance of the descriptive research in the field of education is because of its realistic approach. Although it is useful yet it is not much progress in descriptive research. We use the research to assist us define our issues. We give example to elaborate the mentioned study. No action is being taken without the intention of the parents; no fruitful planning is established without diagnosing the real issues and what should be the duty of the society [24]. Further, survey research and significant method of research has been being used since many past years to up till now, it is considered availed and most reliable technique to collect the data for research [25].

4 Conducted Evaluation

The authors in [26] has described the descriptive research and mentioned three major types, including:

(1) Survey studies.
(2) Developmental studies.
(3) Inter-relational studies.

4.1 Data Analysis

The term "data analysis" refers to the process of interpreting data. Data analysis provides a quick summary of the raw data [27], in example, it consists of procedures for analyzing data, techniques for interpreting the outcomes of such procedures, and methods for planning the collection of data in order to make its analysis easier, more exact, or more accurate. Raw data is turned into tabular form once it has been analyzed, making it very straightforward to interpret as shown in Table 1. The primary goal of data analysis is to describe, convert, and summarize the information. Differences between factors are detected and results are forecasted by studying data linkages. In addition, by [28], it is the process by which a researcher advances from a description of what is the case to an explanation of and why it is the case. The questionnaire was first analyzed statement by statement in this study, and then a percentage was calculated. Further, it is about analyzing and interpreting data from the chosen sample.

The first step's dependent variable was current academic achievement (GPA), whereas the second step's dependent variable was overall academic achievement. Class size (0 = ideal classroom, 1 = overcrowded classroom) and gender (0 =

Table 1 Analysis of multiple regression

Step 1: present academic achievements (GPA)			
Variables	β	*t*	VIF
Age	−0.029	−1.090	1.070
Gender	−0.160	−6.020	1.080
Class-size	−0.040	−1.550	1.020
Aggression	−0.055	−2.150	1.030
Academic achievement in general (CGPA)			
Age	−0.007	−0.270	1.070
Gender	−0.139	−5.220	1.080
Class-size	−0.042	−1.610	1.020
Aggression	−0.053	−2.020	1.030

female, 1 = male) were classified as binary variables. Because all variance inflation factor (VIF) values were less than 5, multi-collinearity was not a concern. Significant t values are those with a value of 2.00 or higher.

4.2 *Word Clouds*

Word clouds (also known as tag clouds or word art) are visual representations of any kind of information used for different purposes. They are capable of automatically detecting collocations (words that often go together) in sentences, paragraphs, and documents in order to provide a deeper level of context than word clouds with single words alone. It extracts useful information from the data and provide overview of words intensity. From the topic/situation which is more vital about students you want to discuss. Find out what their views are on the topic. Listen to what they have to say. If used at the beginning and ending of an intervention/workshop, this tool is able to help you measure progress and change. The understanding of a topic can be measured by looking at what the audience knows about it. This approach was used to provide a deeper understanding of the topic by providing a better overview of the data which was extracted from the research and analyzed using this approach. The overall scenario depicted in Fig. 1 gives a general overview of the situation.

5 Conclusion

The current study looked into the impact of aggressive behavior on university boys' and girls' academic accomplishment. It was discovered that girls had a higher level of violence than boys, and that boys have a higher level of academic success. The emergence of university anti-social behavior is frequently attributed to a combination

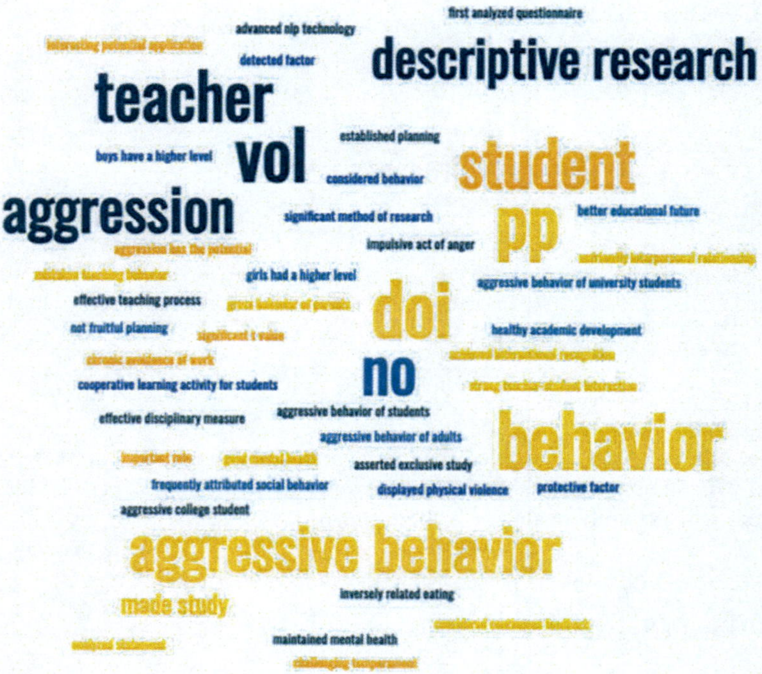

Fig. 1 Word cloud assessment method

of familial and personal causes. Children with challenging temperaments and early behavioral problems have been found to be at higher risk. Poor parenting is primarily to blame for later student hostility and behavior issues. As a result, parents must take a vital part in their children's social and emotional development; otherwise, they may engage in criminal behavior. The informal development of a person's personality occurs as a result of his or her family, neighborhood, and job environment, among other factors. The social aspect of his personality is developed in the school's community life, social customs, traditions, social contact, and cultural aspects, among other things, and plays an important role in grooming their personality.

Acknowledgements We would like to thank DG Khan University in Pakistan for allowing us to conduct this educational study, also this paper is supported by Ideological and Political Project of Changsha Normal University (CSSY202109).

Conflict of Interest The authors stated no conflict of interest.

References

1. S.L. Chew, W.J. Cerbin, The cognitive challenges of effective teaching. J. Econ. Educ. **52**(1), 17–40 (2021). https://doi.org/10.1080/00220485.2020.1845266
2. B. Hu et al., A pilot study of Global ICT strategy applications in sustainable continuing education. Procedia Comput. Sci. **183**, 849–855 (2021). https://doi.org/10.1016/j.procs.2021.03.009
3. I. Giannopoulou, M. Kotopoulea-Nikolaidi, S. Daskou, K. Martyn, A. Patel, Mindfulness in eating is inversely related to binge eating and mood disturbances in university students in health-related disciplines. Nutrients **12**(2), 396 (2020). https://doi.org/10.3390/nu12020396
4. C.C. Chuang, W.M. Reinke, K.C. Herman, Effects of a universal classroom management teacher training program on elementary children with aggressive behaviors. Sch. Psychol. (2020). https://doi.org/10.1037/spq0000351
5. M. Ainscow, K. Messiou, Engaging with the views of students to promote inclusion in education. J. Educ. Chang. **19**(1), 1–17 (2018). https://doi.org/10.1007/s10833-017-9312-1
6. H. Goren, M. Yemini, Citizenship education redefined—a systematic review of empirical studies on global citizenship education. Int. J. Educ. Res. **82**, 170–183 (2017). https://doi.org/10.1016/j.ijer.2017.02.004
7. K. Vuorinen, A. Erikivi, L. Uusitalo-Malmivaara, A character strength intervention in 11 inclusive Finnish classrooms to promote social participation of students with special educational needs. J. Res. Spec. Educ. Needs **19**(1), 45–57 (2019). https://doi.org/10.1111/1471-3802.12423
8. E. Hopper, A. Weyman, A sociological view of large groups, in *The Large Group: Dynamics and Therapy* (Taylor and Francis, 2019), pp. 159–189
9. A. Karinen, J. Tybur, R. de Vries, The disgust traits: self-other agreement in pathogen, sexual, and moral disgust sensitivity and their independence from HEXACO personality (2019). https://doi.org/10.31234/osf.io/dymqk
10. S.I. Benn, Privacy, freedom, and respect for persons, in *Privacy and Personality* (Taylor and Francis, 2017), pp. 1–26
11. A. Akram et al., A pilot study on survivability of networking based on the mobile communication agents. Int. J. Netw. Secur. **23**(2), 220–228 (2021). https://doi.org/10.6633/IJNS.202103_23(2).04
12. M.S. Stanford, R.J. Houston, C.W. Mathias, N.R. Villemarette-Pittman, L.E. Helfritz, S.M. Conklin, Characterizing aggressive behavior. Assessment **10**(2), 183–190 (2003). https://doi.org/10.1177/1073191103010002009
13. S.A. Haider et al., The inclusive analysis of ICT ethical issues on healthy society: a global digital divide approach. Procedia Comput. Sci. **183**, 801–806 (2021). https://doi.org/10.1016/j.procs.2021.03.001
14. S.M.A. Shah et al., A quantum spatial graph convolutional network for text classification. Comput. Syst. Sci. Eng. **36**(2), 369–382 (2021)
15. A. Murthy et al., Internet of Things, a vision of digital twins and case studies, in *IoT and Spacecraft Informatics* (Elsevier, 2022), pp. 101–127
16. F. Abasiubong, T. Abiola, O. Udofia, A comparative study of aggression amongst Nigerian University students in Niger delta region. Psychol. Heal. Med. **16**(1), 86–93 (2011). https://doi.org/10.1080/13548506.2010.507775
17. E. Kikas, K. Peets, K. Tropp, M. Hinn, Associations between verbal reasoning, normative beliefs about aggression, and different forms of aggression. J. Res. Adolesc. **19**(1), 137–149 (2009). https://doi.org/10.1111/j.1532-7795.2009.00586.x
18. B. Hu, S.M. Noman, M. Irshad, An essential vigilance analysis on security of web applications, in *International Conference on Wireless Communications and Applications* (Springer, Singapore, 2022), pp. 92–97
19. C.N. Gundry et al., Base-pair neutral homozygotes can be discriminated by calibrated high-resolution melting of small amplicons. Nucl. Acids Res. **36**(10), 3401–3408 (2008). https://doi.org/10.1093/nar/gkn204

20. M.N. Sohail et al., Group covariates assessment on real-life diabetes patients by fractional polynomials: a study based on logistic regression modeling. J. Biotech Res. **10**, 116–125 (2019)
21. A.H. Buss, M. Perry, The aggression questionnaire (1992), https://psycnet.apa.org/buy/1993-00039-001. Accessed 19 June 2021
22. B. Hu et al., Run-time prediction practices of multimedia web design in technology management, in *3D Imaging Technologies—Multidimensional Signal Processing and Deep Learning* (Springer, Singapore, 2021), pp. 179–186
23. W.R. Best, J.M. Becktel, J.W. Singleton, F. Kern, Development of a Crohn's disease activity index: National Cooperative Crohn's Disease Study. Gastroenterology **70**(3), 439–444 (1976). https://doi.org/10.1016/S0016-5085(76)80163-1
24. M. Sohail, M. Uba, M. Irshad, S. Abir, Why only data mining? A pilot study on inadequacy and domination of data mining technology (2018), https://www.researchgate.net/profile/Shake_Abir/publication/343809206_WHY_ONLY_DATA_MINING_A_PILOT_STUDY_ON_INADEQUACY_AND_DOMINATION_OF_DATA_MINING_TECHNOLOGY/links/5f40bb21299bf13404df6808/WHY-ONLY-DATA-MINING-A-PILOT-STUDY-ON-INADEQUACY-AND-DOMINATION-OF-DATA-MINING-TECHNOLOGY.pdf. Accessed 19 June 2021
25. S. Aziz et al., Protection of a smart grid with the detection of cyber-malware attacks using efficient and novel machine learning models. Front. Energy Res. 1102 (2022)
26. W. Balemans et al., Identification of a 52 kb deletion downstream of the SOST gene in patients with van Buchem disease. jmg.bmj.com. https://doi.org/10.1136/jmg.39.2.91
27. X. Contributions to the mathematical theory of evolution. —II. Skew variation in homogeneous material, Philos. Trans. R. Soc. London. **186**, 343–414 (1895). https://doi.org/10.1098/rsta.1895.0010
28. D. Hitchcock, G. Hitchcock, D. Hughes, *Research and the Teacher: A Qualitative Introduction to School-Based Research* (1995)

Exploration of Interdisciplinary Integration Mechanism Based on OBE Education Concept—Take Big Data Management and Application Major as an Example

Yu Sun and Liwei Tian

Abstract The interdisciplinary integration mechanism is one of the important paths for colleges and universities to cultivate innovative and compound talents. Promoting interdisciplinary integration not only conforms to the inevitable development trend of modern science, but also is an effective way to implement new development concepts and innovation driven national development strategies. As an advanced education and teaching concept, OBE has been recognized by the majority of higher education scholars and applied to the practice of education and teaching reform. This paper takes the interdisciplinary integration mechanism model as the breakthrough point, and based on the OBE education concept, ponders and explores the training mode of big data management and application professionals, aiming to promote the improvement of teaching quality and talent training quality, and has a certain reference role for the training of high-quality composite talents, and it can be used for reference to the cultivation of high-quality talents in China.

Keywords Big data management and application · Interdisciplinary · Integration mechanism · Composite · OBE education philosophy

1 Introduction

In January 2022, The Chinese Government issued Some Opinions on Deepening the Construction of World Class Universities and Disciplines, proposing to strengthen interdisciplinary talent training in key areas such as the digital economy, and comprehensively improve the ability to cultivate innovative, application-oriented and interdisciplinary talents. The Opinions on the Implementation of Deepening the Reform of Innovation and Entrepreneurship Education in Colleges and Universities issued by the General Office of the State Council clearly pointed out that interdisciplinary

Y. Sun · L. Tian (✉)
Guangdong University of Science and Technology, Dongguan, China
e-mail: 656453927@qq.com

S. Patnaik and F. Paas (eds.), *Recent Trends in Educational Technology and Administration*, Learning and Analytics in Intelligent Systems 31,
https://doi.org/10.1007/978-3-031-29016-9_8

courses should be set up to explore the establishment of a new mechanism for training innovative and entrepreneurial talents across departments, disciplines and specialties. Obviously, it is an important task for colleges and universities to cultivate interdisciplinary talents to improve national education level, enhance scientific and technological innovation ability, and develop economic society. With the arrival of the big data era, the demand for big data application-oriented talents from all walks of life has become increasingly strong, and China has also fully implemented the big data strategy. In 2018, the Ministry of Education approved universities to set up big data management and application majors, and encouraged universities to establish a cross-border compound talent mechanism to cultivate professional talents in the big data field. The student centered and results oriented OBE education concept is an advanced education concept today. This paper starts with the analysis of the connotation of OBE education concept, and analyzes and studies the training mode of big data management and application professionals based on OBE education concept, which has certain reference and reference value for Chinese universities to firmly establish OBE education concept, strengthen interdisciplinary integration mechanism, and improve the quality of talent training.

2 Research Status at Home and Abroad

The concept of "interdisciplinary" was first proposed by Professor Woodworth of Columbia University in 1926, emphasizing that interdisciplinary is the intersection, infiltration and integration of different basic disciplines. Since the 1960s, a group of top universities in the world, including MIT, University of Toronto, Manchester University, Princeton University, have been keenly aware of the increasingly prominent interdisciplinary characteristics in the fields of science, information, technology and production, and have integrated interdisciplinary education into the reform of talent training mode [1, 2]. After years of development, the current interdisciplinary integration mechanism emphasizes the continuous inheritance and innovation, intersection and integration of disciplinary development.

In contrast, China's modern discipline system and system mature late, and the construction of interdisciplinary integration mechanism lags behind. However, with the fast development of China's high education, the continuous exploration of experts and scholars, and the continuous practice of interdisciplinary construction in universities, China's interdisciplinary integration mechanism has entered a new stage of comprehensive development. Some colleges and universities have made remarkable achievements in interdisciplinary fields, setting up models for the development of teaching reform in colleges and universities in the new era, sharing experience and playing a leading role in demonstration [3]. Harbin University of Technology has continuously promoted the innovation of interdisciplinary system, and innovated the way to cultivate interdisciplinary innovation ability from the aspects of curriculum construction, teacher introduction, practical teaching content, innovative experimental research platform construction [4]. According to the industrial transformation and upgrading and innovation driven development strategy of Guangdong Province, Guangdong University of Technology has explored the cultivation of new engineering talents, and has established 10 new engineering talents cultivation models featuring interdisciplinary and multi professional integration of industry and education with the combination of science and engineering, the integration of commerce and industry, the infiltration of arts and engineering, and the penetration of arts and engineering [5].

3 Analysis on the Application of OBE Education Concept

The connotation of OBE education concept is an advanced education concept which is student-centered, goal oriented, reverse thinking based on curriculum system construction, and focuses on continuous improvement (see Fig. 1). There are lots of researches have confirmed that the OBE education concept has a good effect in the reform practice of talent training model construction for higher education [6], curriculum teaching reform research [7], integrated teaching system reform [8], integration of ideological and political education into professional courses [9], technical innovation education [10, 11], etc. This paper fully links the connotation of OBE education concept, and analyzes and summarizes the education innovation of big data management and application under the guidance of OBE education concept.

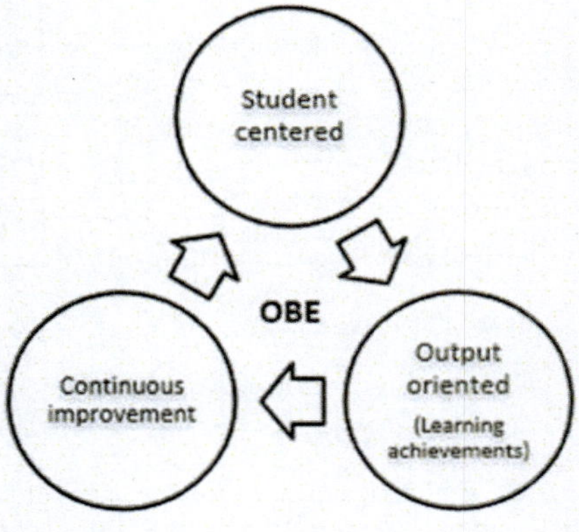

Fig. 1 Schematic diagram of OBE education concept

3.1 Analyze the Demand for Talents, Define the Learning Achievements, and Determine the Education Goals

To ensure the smooth implementation of a series of teaching activities based on results, the application of OBE education concept puts forward the first question for college educators to clarify the learning achievements. The demand of economic growth in China and social progress for application-oriented talents determines the principal direction and learning achievements of universities talented person raises. In the design process of the curriculum system for talent training of big data management and application professionals, the effective path to achieve the curriculum system is to follow the basic logic of determining training objectives based on talent needs, supporting training objectives to formulate graduation requirements, supporting graduation requirements to build the curriculum system and implement curriculum construction [12] (see Fig. 2). Determine the goal of education, realize the positive value guidance for students through the course teaching process, and cultivate students to face the complex problems in economic management, be able to comprehensively use management theory, economic theory, computer science and technology methods to establish models, analyze data, and provide decisions.

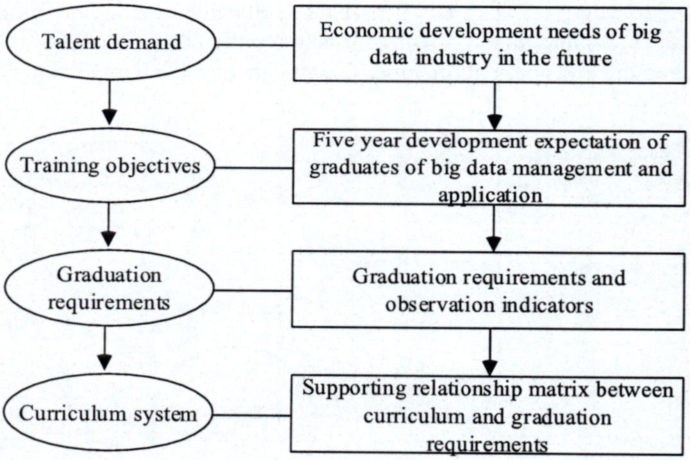

Fig. 2 The design process of the curriculum system for training big data management and application professionals under the OBE concept

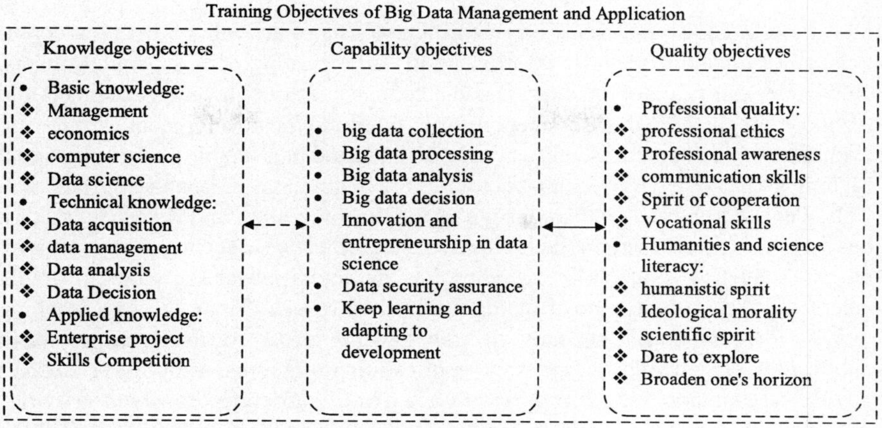

Fig. 3 Big data management and application major training target system

3.2 *Based on the Reverse Design Principle of OBE Concept, Set Appropriate Teaching Strategies*

As the big data specialty under the management discipline, the big data management and application specialty emphasizes the application of big data analysis theories and methods in economic management. The specialty has its own positioning consideration and talent training characteristics, focusing on how to use data analysis methods and technical means to solve problems in management, and using the perspective of big data to cultivate knowledge, ability and quality for comprehensive development of management decisions, and cultivate understanding of data A composite talent team that understands business and management, and can comprehensively control mathematics, data analysis, data visualization, statistics and other knowledge, and that's the only way that can also meet the requirements of current social employers. Therefore, big data management and application specialty integrate management, economics, computer science, data science and other basic disciplines into one, and determine their professional training knowledge based on OBE concept. The ability and accomplishment goals are shown in Fig. 3.

3.3 *Construct Evaluation Feedback Mechanism to Promote Continuous Innovation and Improvement of Professional Construction*

Under the guidance of OBE education concept, teachers need to carry out formative assessment and developmental assessment from the perspective of continuous innovation in education, based on scientific analysis of school and learning conditions, and evaluate students' performance in daily learning process, achievements, knowledge, ability, emotion, attitude and other aspects through various channels and methods with the help of assessment reform, Focus on the development of students' (evaluation object) potential, and carry out detailed and measurable phased or gradual teaching goals. When the teaching activities are behind, it is a conclusive evaluation of the degree to which students arrive at the teaching aims and demands, that is, the teaching effect, based on the pre-set teaching objectives. Technical support with big data statistics, students' comprehensive literacy is evaluated, so as to timely understand the students' grasp of the established results. Then, through teaching evaluation and feedback, we can judge the teaching effect, constantly improve and optimize the big data management and application professional training system, and provide data support for the progress of a series of subsequent teaching activities continuously. Achieve an effective closed loop of "teaching evaluation—teaching feedback—teaching improvement", and then will provide sustainable power for the innovative development of professional construction dimension.

4 Practice Strategy of Interdisciplinary Integration Mechanism Based on OBE Concept

On the basis of preliminary analysis, according to the training program for the major of big data management and application in China, this major construction and development should adhere to the educational concept of "moral education first, ability first, unity of knowledge and practice, courage to innovate, and comprehensive development". Guided by OBE concept, this paper determines the phased learning achievements of this major, then constructs an evaluation feedback mechanism. Taking "cross professional integration" as the starting point, the comprehensive development of ideological and political integration Industrial integration, research integration and innovation integration are the four dimensions of reform. We can see Fig. 4 for the specific implementation path of integration mode exploration.

Ideological and political integration: first of all, ideological and political education will be fully integrated with discipline education. In the integrated teaching mode, we will promote the sustainable development of ideological and political education through the selection and reconstruction of knowledge points, the overall design of core courses, and the construction of curriculum ideological and political element resource library, highlight the central position of "curriculum ideological and political" and "ideological and political curriculum", and comply with the professional ethics and norms of big data industry.

Industry integration: boost the comprehensive integration of big data industry and professional discipline education, through school enterprise cooperation, enhance the ability of social practice and ability to solve complex problems. It also integrates science and engineering courses, and can make use of China's MOOC platform,

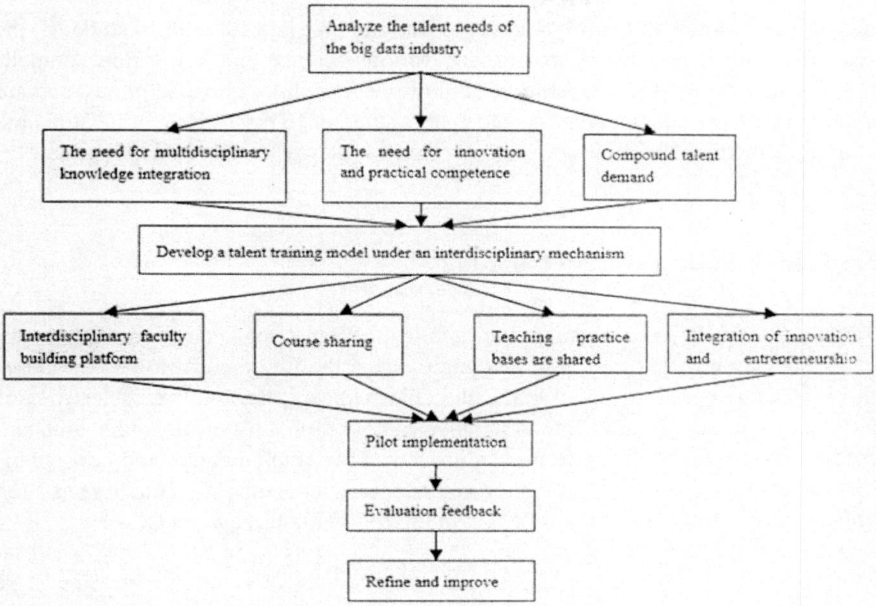

Fig. 4 Under the concept of OBE, the interdisciplinary integration mode of big data management and application is explored

micro courses and other platforms to enable students to obtain teaching resources across semesters and majors. The construction of teaching resources adopts the way of utility, multi-source, polymorphism and easy iteration to continuously provide diversified teaching guarantee services for students of this major. Cultivate students' multidisciplinary knowledge background, application ability and comprehensive ability.

Research integration: promote the comprehensive integration of scientific research and discipline practice education, that is, the integration of science and education, make the best of the teacher resources of other colleges (such as the Computer School, the School of Finance and Economics) and laboratory equipment resources, form a teaching resource pool of "teacher co construction, curriculum sharing, base co construction", and promote the cross integration of knowledge and technology of management, economics, computer science, data science and other professional courses, and ultimately serve scientific research.

Innovation integration: accelerate the education of creation and innovation and professional discipline education, and integrate the project driven method carried by comprehensive projects such as the "Internet plus" competition project for interdisciplinary teaching. Due to the complexity and implementation integrity of the introduced comprehensive project, it is difficult to complete the technical knowledge of a single discipline independently, and it needs to be completed jointly by interdisciplinary or even cross college teaching teams. Typical projects include national innovation and entrepreneurship training program for college students, the "Internet plus" competition project, and various science and innovation competitions, which require the cooperation of multiple disciplines and disciplines, and are completed in a complementary way to enable students to have strong innovation and entrepreneurship capabilities.

5 Construction Achievements

Taking the big data management and application major of Guangdong University of Science and Technology as an example, since the implementation of the interdisciplinary integration model under the OBE concept, the four characteristics of industrial demand-oriented, interdisciplinary integration, future-oriented layout, and comprehensive highlighting of innovation have been implemented, and new attainments have been made in teaching reform, and the talent cultivation quality has been greatly improved. It is mainly reflected in the following three aspects.

5.1 Students' Recognition of Learning Content and Training Mode Continues to Increase

The big data management and application major has three professional directions, namely logistics big data, financial big data and business big data. As the reform continued, satisfaction surveys were conducted on the learning results of each direction for three semesters. As can be seen from Fig. 5, the satisfaction of learning effect in the three directions is showing an increasing trend. From the survey data, it is shown that the student satisfaction of logistics big data increased from 66.8% in the first semester to 96.7% in the third semester, the student satisfaction of financial big data increased from 70.2% in the first semester to 95.6% in the third semester, and the student satisfaction of business big data increased from 68.9% in the first semester to 94.8% in the third semester.

5.2 The Industry's Satisfaction with the Quality of the Training Process Continues to Improve

Although there are no graduates of this major, in the three semesters of training, experts in the industry will be invited to evaluate the quality of talent training according to job requirements, and full-time teachers will be organized to investigate the satisfaction of students' learning effects. It can be clearly seen from Fig. 6 that the cultivation process of students in this major has been continuously promoted

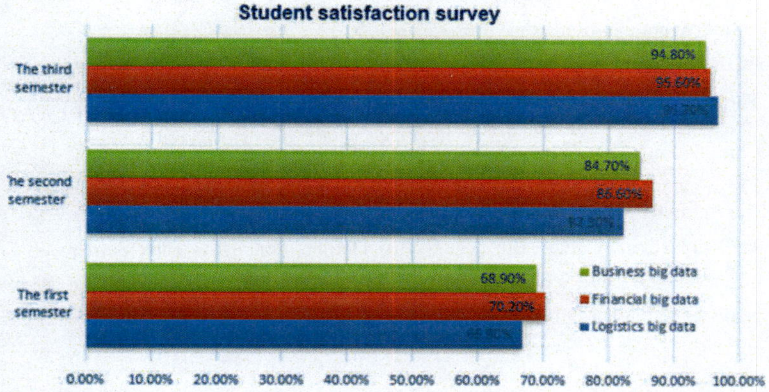

Fig. 5 Student satisfaction survey

by the industry. Taking the first semester and the third semester as an example, the industry's satisfaction with training quality has increased from 72 to 95.8%, with a growth rate of 23.8%; Full time teachers' satisfaction with students' learning effect has increased from 84.6 to 98%, with a growth rate of 13.4%.

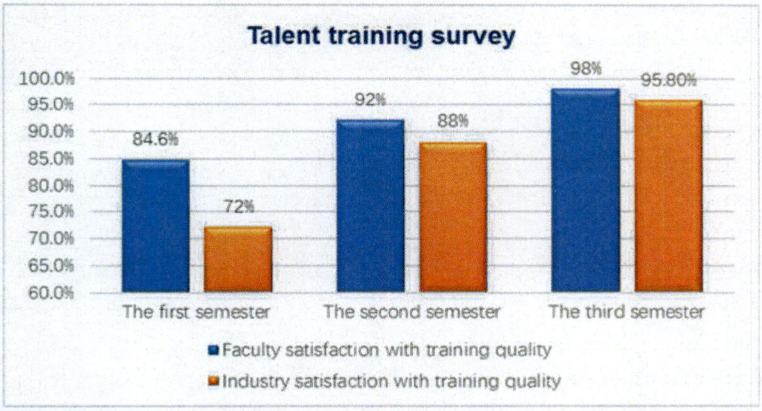

Fig. 6 Learning effects survey on the quality of cultivation

Table 1 Student competition award form table

No.	Prize	Grade
1	The 18th "New Way Digital Intelligence Talent Cup" National Undergraduate Numerical Enterprise Operation Sandtable Competition	The first prize
2	China International "Internet plus" Undergraduate Innovation and Entrepreneurship Competition	Bronze award
3	The 6th National College Business English Competition	The third award
4	2022 National College English Competition	The second award
5	National College English Writing Competition	Excellent works award

5.3 Increase in the Number of Students Enrolled in This Major and Improvement of Students' Comprehensive Ability

The first batch of enrollment of big data management and application specialty of the university will begin in 2021, with 153 students enrolled in the same year and 177 students enrolled in 2022, with an annual enrollment growth rate of 15.69%. With the improvement of the social attention of this major, it is estimated that the number of new students will exceed 200 in 2023. Although this major has only been set up for more than one year, compared with the training methods of other similar majors in this school, under the talent training reform of interdisciplinary integration mechanism based on OBE education concept, the comprehensive quality and ability of students in this major have been improved more significantly. Table 1 records part of students' competition results.

6 Conclusion

As an emerging major, big data management and application is an applied major that integrates multiple disciplines. The efficient implementation of interdisciplinary integrated education model in the existing training system is an important topic in higher education reform. The OBE concept reverse-designs the teaching process from the ultimate goal of education, determines the knowledge, ability and literacy goals from the talent needs, and constructs an evaluation feedback mechanism in the teaching process for continuous improvement. From the perspective of the realization effect of interdisciplinary integration, both student satisfaction and industry satisfaction have achieved good results.

Acknowledgements The paper is funded by 2022 "Quality Engineering" higher education teaching reform project of Guangdong University of Science and Technology: Research on interdisciplinary integration mechanism based on big data management and application (GKZLGC2022134).

References

1. W. Zhang, Interdisciplinary education: a case study of Princeton University undergraduate talent training. Res. High. Eng. Educ. **3**, 118–125 (2014)
2. D.R. Bie, T.Y. Qi, Trends of undergraduate teaching reform and construction in foreign first-class universities. China High. Educ. Res. **7**, 7–13 (2016)
3. Y. Chen, L. Xu, H. Li, Experience and reflection on interdisciplinary education practice in Chinese universities. Educ. Teach. Forum **17**, 269–270 (2020)
4. W. Zhao, X. Zhu, X.S. Ye, Research on the cultivation of interdisciplinary innovation ability in the context of new engineering—taking the landscape architecture discipline of Harbin University of Technology as an example. Heilongjiang Educ. (High. Educ. Res. Eval.) **07**, 13–15 (2022)
5. L.J. Li, W.B. Yang, M. Xiao, Y. Zhang, Exploration and practice of new engineering talents training mode of interdisciplinary and multidisciplinary integration. Res. High. Eng. Educ. **01**, 25–30 (2020)
6. Z.J. Yang, Q.M. Fan, P. Huang, M.H. Wan, Tan Hai construction and practice of talent training mode for computer science and technology major (computer audit) based on OBE concept. Sci. Educ. J. **22**, 101–104 (2022)
7. X.P. Xu, Y.R. Yang, Research on teaching reform of "engineering surveying" course based on "Internet plus" and OBE concept. Mod. Inf. Technol. **6**(22), 163–167 + 171 (2022)
8. M.L. Gong, F.F. Liu, F. Qu et al., A new innovative and integrated teaching system for engineering majors highlighting ideology and politics under the OBE concept—taking the material forming and control engineering major as an example. J. High. Educ. **8**(29), 27–30 (2022)
9. X. Li, G.H. Cheng, Y. Wang et al., Research on ideological and political education of logistics engineering major curriculum based on OBE concept—taking industrial engineering as an example. Logist. Technol. **44**(8), 168–171 (2021)
10. N. Sun, J.Y. Wang, J.D. Yang, Analysis and evaluation of the limiting factors of scientific and technological innovation education for college students based on OBE concept. Heilongjiang Educ. (Theory Pract.) **09**, 69–72 (2022)
11. T.D. Xu, N.W. Huang, X.L. Zhao, Innovation and entrepreneurship education system construction based on OBE concept: theoretical support, design ideas and implementation path. Innov. Entrep. Educ. **12**(6), 30–36 (2021)
12. X.Q. Shi, Thinking and practice of first class curriculum construction in the view of new needs and new ideas. Res. High. Eng. Educ. **04**, 52–58 (2022)

Research on the Construction of Higher Vocational Oral English Wisdom Classroom Under the Background of Internet Technology

Enming Du

Abstract The combination of Internet technology and education and teaching generates intelligent education methods. The establishment of a high-quality "wisdom classroom" can continuously enrich students' knowledge, improve students' individual personality, but also better develop students' personal ability, promote the wisdom of teachers and students to achieve symbiosis, is the progress of the new era, the new mission of education in the new era. Based on Internet technology, the three most important factors to build a smart classroom are material, teachers and students. The effective integration and coordination of the three is the key to the success of a smart classroom.

Keywords Internet technology · Higher vocational colleges · Spoken English · Intelligent classroom construction

1 Introduction

The state vigorously promotes the reform of education system and innovation of education methods, and actively integrates the products of the "Internet plus" era into contemporary education, which has become the mainstream trend of the development of the education industry. The rapid development of The Times, economy, science and technology level, the development of Internet technology has been greatly improved, based on the background of good development of Internet technology, social fields, all levels and all walks of life, are gradually popularize and use advanced information technology, including the field of education. The effective application of Internet technology promotes the gradual reform and improvement of the education system, and at the same time, it also promotes the further development of China's overall

E. Du (✉)
Baicheng Vocational and Technical College, Unit 1, Building 2, Baicheng Park East Road, Baicheng, Jilin, China
e-mail: enming217@163.com

S. Patnaik and F. Paas (eds.), *Recent Trends in Educational Technology and Administration*, Learning and Analytics in Intelligent Systems 31,
https://doi.org/10.1007/978-3-031-29016-9_9

education industry, and continuously advances and optimizes towards the goal direction of the education industry. From the English teaching course of English teaching, the application of Internet technology, establish wisdom classroom, strengthen its teaching effect, oral teaching mode, etc., has a very big role, can say, the more advanced means of education, for higher vocational colleges for oral English learning students brought the Gospel, can comprehensive complement of traditional teaching mode, make the oral teaching more convenient, intelligent and efficient. Under the background of the Internet, higher vocational colleges effectively build oral English wisdom classroom to explore and analysis.

2 The Necessity of Constructing Oral English Wisdom Classroom Under the Background of Internet Technology

First of all, the article will first analyze the necessity of establishing a smart classroom for spoken English in higher vocational colleges under the Internet background. From this perspective, it can be more clear why we should establish a smart classroom, and what the establishment of a smart classroom can do for spoken English teaching in higher vocational colleges, so as to promote the in-depth and comprehensive analysis of follow-up problems.

2.1 Promote the High Efficiency of Oral English Teaching Requirements

Today is a fast-paced new era, Every institution really needs very efficient and high-quality teaching, The Traditional teaching mode of Oral English in Colleges and Universities, Whether it is the teaching content or the teaching means, Have been unable to meet the development needs of the current era, With the relevant information surface, Traditional oral English teaching methods make students' oral English learning has a certain degree of error, This kind of traditional teaching mode with relatively low teaching effect [1]. It is not consistent with the current education and teaching purpose of higher vocational colleges, Cannot make the college teaching can be better satisfied, As a new type of teaching model, Smart classroom, both in terms of teaching content, teaching planning and teaching means, Compared with the traditional teaching model, They are more in line with and meet the development needs of the current era and education, To provide a good teaching form for intelligent talent training, As shown in Fig. 1, It can promote higher vocational colleges to gradually improve the teaching quality of their departments, Promote oral English teaching more efficient. Therefore, the establishment of intelligent classroom is a new era requirement to promote higher vocational colleges to continuously improve the efficiency of oral English teaching [2].

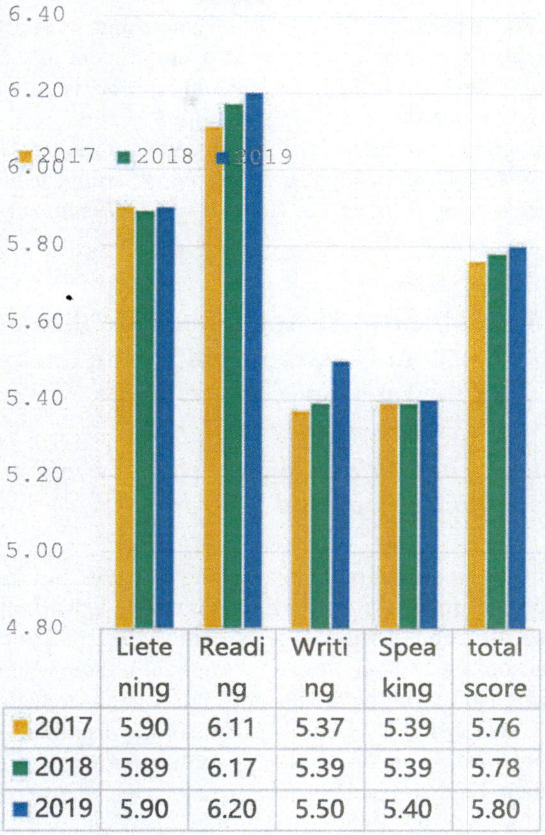

	Lietening	Reading	Writing	Speaking	total score
2017	5.90	6.11	5.37	5.39	5.76
2018	5.89	6.17	5.39	5.39	5.78
2019	5.90	6.20	5.50	5.40	5.80

Fig. 1 Average IELTS score of Chinese students from 2017 to 2019

2.2 *Promoting the Reform Requirements of the Education Model*

Will education industry, need to continuously in the teaching process into the new blood, only in this way can better promote the good development of college teaching goals, in higher vocational college teaching situation, the disadvantages of traditional teaching mode has been completely revealed, for example, cannot mobilize and arouse students' interest in learning, learning enthusiasm and independent learning ability [3]; Unable to fully grasp the specific learning situation of each student, it is difficult to develop targeted teaching methods for students with different levels. Therefore, it is easy to make the learning gap is large, and the establishment of intelligent classroom can help each student to better learn spoken English, and enable teachers to carry out more targeted and personalized teaching methods, so as to promote the reform and innovation of education mode [2]. Figure 1 shows a statistical chart of the average score of IELTS category A in higher vocational colleges in China from 2017 to 2019.

Can be concluded from Fig. 1, our country college students ielts test average score is rising year by year, 2019 an average score of 5.80 points, but in all test subjects, oral score and three years, the highest score is only 5.40 points, so you can conclude that our students oral English ability is poor, need to strengthen the reform of oral English teaching and the improvement of the teaching system. However, it can also prove that, combined with Internet teaching, it has a positive impact on the oral English performance of students in Chinese higher vocational schools [4].

3 Problems Existing in the Establishment of Oral English Wisdom Classroom Under the Background of the Internet

3.1 The Professional Quality of English Teachers Needs to Be Improved

In the process of building teaching activities and wisdom classroom, teachers' position is important, as an important knowledge transmission, teachers professional construction of wisdom classroom and teaching effect has a direct relationship, but according to some higher vocational colleges and related data, many vocational English teachers are still not high, mainly reflected in English professional level is not high, low teaching effect, some vocational English teachers are not English professional, many are not in further study after undergraduate graduation, so that their English teaching level cannot meet the needs of oral English teaching [5]. Moreover, the ability to use smart classroom is insufficient, and even some English teachers cannot deeply understand and master the way and role of smart classroom in oral teaching, and it is difficult to improve their own teaching ability and level. Finally, there is a lack of Internet technology. Some older English teachers with more traditional teaching concepts cannot fully master the Internet technology, so the demand for an intelligent classroom setting up cannot be effectively met [6].

3.2 The Intelligent Classroom Teaching Mode of Spoken English Needs to Be Strengthened

For the teaching of oral English in higher vocational colleges, the oral English wisdom classroom is still in its initial stage. Although it has achieved some results, the teaching mode is still relatively immature and needs to be strengthened. Through some higher vocational colleges to build the wisdom of the classroom, establish oral English wisdom classroom, the corresponding development mechanism needs to constantly establish and improve, for now, the development of wisdom classroom teaching mode has a long way to go, and this is immature teaching mode, is likely to make some higher vocational students when learning oral English resistance [7]. This is also not conducive to the effective construction of intelligent classroom teaching mode.

3.3 *Students Cannot Fully Accept the Smart Classroom Teaching Mode*

Higher vocational college students have some problems in oral English learning. For example, the poor cultural foundation requires a lot of time to learn oral English. Some students have poor learning ability, and it is difficult to grasp oral English quickly. What's more, some students have too low enthusiasm for learning, Most of the school time is given to electronic devices such as mobile phones and computers rather than learning spoken English. This situation affects the actual classroom effect whether in the traditional teaching mode or in the smart classroom. It is difficult for students in this state to fully accept the smart classroom teaching mode for a long time, which leads to the failure to form a good teaching effect.

3.4 *Optimizing the Teaching Environment*

Smart classroom has been practiced in the spoken English teaching system of higher vocational colleges for a period of time, and has accumulated a lot of excellent teaching experience. As far as it is concerned, one of the major problems in the practice of oral English wisdom class in higher vocational colleges is the improper handling of the relationship between modern technology and oral English teaching. For example, teachers integrate network technology, information technology, etc. into the oral English classroom for the construction of English language situations. However, it is difficult to introduce students into the English communication background by playing multimedia videos and audio only, so as to have deep thinking and understanding of oral skills and knowledge. The students' focus is on the data itself, and they have not formed enough awareness of active thinking, so the effect of smart classroom teaching is greatly reduced. When integrating big data technology and artificial intelligence technology into the oral English intelligent classroom in higher vocational colleges, first of all, it is necessary to clarify the function and applicable scenarios of different technologies in the intelligent classroom, reasonably arrange the time nodes when they appear, and assist in problem inquiry, so as to bring students' attention to the English communication scene and complete oral English teaching in a realistic English language environment.

3.5 Innovative Teaching Methods

The effective use of advanced teaching tools must be based on appropriate teaching methods. In order to enable big data and artificial intelligence to fully play the role of helping the smart classroom, it is necessary to innovate the teaching methods of oral English. In the English education system, the latest education concept requires that English learning should be placed in the corresponding cultural background to improve students' English cultural accumulation, master the significant differences between English and Chinese, and then achieve the goal of reducing learning difficulty and improving English literacy [3]. Oral English teaching in higher vocational colleges is highly practical and applicable, so it is necessary to take cultural education as the breakthrough point of teaching method innovation. For example, before formal oral teaching, use multimedia technology to introduce the corresponding English cultural background and Chinese cultural background, compare the differences between English system and Chinese system in terms of habits, and conduct rational oral skills teaching on the basis of perceptual cultural cognition; Recommend English culture learning websites, software, we media, etc. for students, so that they can learn English language and culture from multiple channels in their spare time; To truly integrate oral English learning into students' actual life, change their language learning thinking, and make them more interested in oral English learning.

4 Measures to Construct Oral English Wisdom Classroom in Higher College Under the Background of Internet

4.1 Improve the Comprehensive Professional Quality of English Teachers

Improve higher vocational English teachers overall professional quality can be from the following aspects, first of all, all the school English teachers regular training, not only to improve the level of English teacher teaching, but also make it through training correctly and comprehensively grasp the Internet technology and wisdom classroom mode, through training to optimize the English teachers. Second, timely hire professional, excellent English teaching talents, training teachers is always a long-term process, and establish wisdom classroom is imminent, so through the introduction of excellent professionals to further improve the overall English teachers, hired teachers in addition to the English level standard, also need to require a good grasp of the Internet technology and wisdom classroom [8]. Finally, higher vocational colleges need to eliminate the English teachers, completely eliminate the phenomenon of those people, eliminate those who muddle around and incompetent teachers, to ensure that the English teachers are more pure and professional [4].

4.2 Enhance the Main Body Status of Students

Under the intelligent classroom mode, students are the objects to pay attention to. When students' learning state is in passive acceptance, they cannot mobilize their learning interest and enthusiasm in time and effectively. Therefore, the subject position of students in the intelligent classroom must be enhanced [9]. First of all, when designing teaching content, teachers should focus on students to ensure that students' subject status and pay more attention to the learning of students and students. Secondly, for students' personality to establish targeted and personalized teaching form, effective application of Internet technology, combined with the actual teaching situation, fully understand and master the students' specific situation, in this technology is more suitable for oral English teaching, suitable for students' wisdom teaching goal, in English writing design, for example, as shown in Table 1, find better value export for teaching goals, and according to different students specify different solutions. Finally, improve the communication between students and teachers, promote the relationship between teachers and students and the classroom more harmonious, so as to better ensure the classroom teaching effect [10].

4.3 Improve the Oral English Wisdom Class

Set up relevant curriculum supervision system, effective supervision at the same time urge timely update of primary school means, correct some often unnecessary mistakes, innovation how far teaching evaluation system, so as to better promote the development of students' comprehensive, diversified, is not to take the test results as the only standard to evaluate students' ability. In addition, intelligent classroom pilot can be carried out in some higher vocational colleges, to test the teaching effect of practical oral English intelligent classroom through practical classroom experiment, so as to provide practical basis and experience to promote the establishment of intelligent classroom in higher vocational colleges.

4.4 Optimizing the Oral English Wisdom Classroom Model in Higher Vocational Colleges

In this regard, it is necessary to establish a development mechanism for the oral English wisdom class in higher vocational colleges, from the single higher vocational colleges and the overall higher vocational colleges are analyzed from two perspectives. First in terms of single higher vocational colleges, we can strengthen the monitoring of the use of smart classroom models on the basis of supervision, we will ensure the timely updating of teaching methods correct the mistakes that often

Table 1 3 D comparative writing methods and objectives

Target category	Target standard	Target value
Knowledge objective	1. Introduce and analyze the "comparative" English writing method, "draft the writing outline" and "three-step formation method" 2. Grasp the context of the article: use the "comparative" English writing method to quickly analyze and understand the structure of the article 3. Clear the writing points: select the subject, central words and central sentence	Just intake
Ability to target	1. Through the practice of "comparative" English writing method, to achieve the skilled use of comparative writing skills and improve the writing ability 2. Through the improvement of writing ability to promote the improvement of reading and communication ability, pay attention to the cultivation of students' output ability, so as to improve the comprehensive language application ability	Skills to improve
Emotional goals	1. Through learning. Compare the similarities and differences of the ways of making friends between different countries, enhance cross-cultural communication awareness and communication ability, and avoid cross-cultural communication conflicts 2. Based on the students' cognitive style. Classified teaching, develop students' independent learning ability. Cultivate the sense of cooperation and innovative spirit	Learning attitude and value cultivation

occur, and establish a diversified evaluation mechanism to promote enter the diversified development of students, rather than just taking the examination results as a test of students' ability standards; in terms of overall higher vocational colleges, one or several higher vocational colleges can the school conducts a pilot smart classroom to improve oral English the actual effect of the smart classroom is tested to establish the smart classroom in the overall higher vocational colleges provide development experience.

4.5 With the Help of Learning Platform Software, Teaching Oral English in Accordance with Students' Aptitude

In the process of oral English teaching, if English teachers only use hard homework assignments, it will go against students' oral English learning. This is because there is a learning gap between students, and hard learning assignments are often universal, which is not conducive to students with weak learning ability to practice and learn, but will polish their learning interest over time. Therefore, as an English teacher, we can use the smart classroom learning platform software to improve this situation, analyze the learning situation of different students through big data, and push the corresponding learning assignments from the system to achieve the teaching purpose of teaching students in accordance with their aptitude, so as to ensure that every student can get enough oral training and improvement.

Traditional English teaching assignments are usually uniformly arranged by English teachers according to the course content, while the assignments in smart classrooms are arranged to achieve the teaching purpose of teaching students in accordance with their aptitude through big data analysis and Internet technology. English teachers can ask students to open the intelligent learning software every day to sign in for learning, and the system will push oral practice that is consistent with students' learning situation and basic ability, and upload it to the software in the form of recording on time. When students upload oral assignments, English teachers should make timely corrections and give pertinent comments to build students' confidence and enthusiasm for learning. In this way, with the help of big data system push and students' online homework delivery, students' learning efficiency can be effectively improved, while teachers and students' learning burden can be reduced, which has a unique role in the development of students' oral ability.

Make full use of online teaching tools, for example, in oral English learning Upload oral test or oral training resources on the platform, and ask students to It is required to complete the test and training tasks, and use the data analysis function to Monitor students' learning. Organize diverse English in the platform Oral expansion activities. For example, English movie dubbing contest, English songs Singing competitions, etc. Attract students' autonomy through the atmosphere of games and competitions Carry out oral training activities, and comprehensively cultivate the oral English communication ability of vocational college students through various teaching methods to improve their English core elements Raise.

5 Conclusion

This paper first analyzes the necessity of constructing a smart classroom for spoken English in vocational colleges under the Internet background, then points out several problems it faces in its actual development, and finally makes targeted development measures based on a complete understanding of the problems. We must first improve the overall professional level of teachers and ensure the establishment of harmonious classroom relations, and the continuous optimization of the oral English smart classroom model in higher vocational colleges can ensure that the smart classroom can achieve better results in a wide range. Using Internet technology to build intelligent classroom teaching mode, broaden students' vision, and make students contact and understand more foreign local customs and English knowledge, effectively improve students' sense of English language. At the same time, the teacher should also in the teaching process of constantly update ideas, advancing with The Times, fully understand and master the students' ideas, combined with the student feedback and actual situation to improve the higher vocational oral English teaching mode, establish good communication with students, to establish more consistent with students' development of higher vocational oral English wisdom classroom, promote students to strengthen and improve their comprehensive ability. In the actual teaching work, teachers need to combine the professional characteristics and employment trends of vocational students, innovate the contents, forms and methods of oral English wisdom classes, enrich oral English teaching resources, and promote the overall development of vocational students.

References

1. C. Cheng, Research on the ath of oral English wisdom classroom based on SPOC. Campus Engl. (37), 2 (2020)
2. H. Cheng, Teaching mode of oral English in intelligent classroom environment. Campus Engl. (25), 2 (2020)
3. I. Calm, Practice analysis of higher vocational oral English teaching under the background of "Internet +" (2020)
4. C. Cheng, Research on the construction of higher vocational oral English under the background of "Internet +" (2020)
5. Y. Yuan, Q. Shen, Research on innovative strategies of English practice teaching in higher vocational colleges under the background of artificial intelligence. Campus Engl. (08), 50 (2020)
6. Y. Li, A preliminary discussion on the intelligent teaching mode of college English audio-visual theory based on the internet from the view of constructivism. North. Lit. (33), 169–170 (2019)
7. X. Jia, Exploration on the application of intelligence theory in higher vocational English teaching. J. Taiyuan City Vocat. Tech. Coll. (08), 144–145 (2018)
8. R. Qi, Student evaluation of higher vocational English teaching under the guidance of multiple intelligence theory. Crazy Engl. (Theor. Ed.) (03), 113–114 (2018)
9. Jinjing, The integration of multiple intelligence theory and English teaching in higher vocational colleges. J. Shaoxing Univ. Arts Sci. (Educ. Ed.) **36**(03), 57–60 (2016)
10. J. Lu, The application and research of MOOCs teaching mode in higher vocational English. New Campus (Read.) (07) (2016)

Intelligent Classroom Information Technology to Use English Reading to Improve Students' Oral English Ability and Literacy

Wei Zhou

Abstract Nowadays, the development of science and technology has promoted the innovation and optimization of China's education. In recent years, the development of smart classroom has been the attention of all walks of life. This teaching mode can create a good learning environment, but also improve the quality of education. Improving students' reading ability can help students open their thinking, improve their language comprehension ability, and enrich their learning world. Improving students' reading ability can further enhance their oral English ability, thus improving the quality of English teaching. However, at present, the teaching form of English teachers is still relatively rigid, leading to low students' oral English literacy. Therefore, intelligent classroom should be actively adopted to promote the development of English education.

Keywords Intelligent classroom · Read in English · Improve students' oral English ability

1 Introduction

English reading teaching is an important part of English teaching in junior middle school, which plays a key role in cultivating students' core English literacy. With the economic globalization, the world has entered the information age, and English plays an important role in the cooperation and communication between countries. In order to better communicate with other countries, China requires schools to focus on the training of students' oral English ability. At present, the related research on intelligent classroom is the hot spot in China's education circle. This article is on the basis of combining a large number of research data, fully analyze the specific characteristics

W. Zhou (✉)
Baicheng Vocational and Technical College, Unit 302, Building 3, 15-3, Minsheng West Road, Yihai New Village, Baicheng, Jilin, China
e-mail: 634085501@qq.com

S. Patnaik and F. Paas (eds.), *Recent Trends in Educational Technology and Administration*, Learning and Analytics in Intelligent Systems 31,
https://doi.org/10.1007/978-3-031-29016-9_10

of wisdom classroom, design in the English classroom wisdom classroom method, let the students to read effectively, hold before class, in class and after class these three periods, help students optimize their own learning methods, improve the mastery of English knowledge, further stimulate the interest in learning. The study has found that with the help of smart classroom in the Internet era, students' English level has also been significantly improved.

2 Junior High School English Teaching Status

2.1 Lack of Reading Awareness

In junior high school English class, teachers spend most of their time and energy on teaching English knowledge points. In the class, few students are organized to carry out English interest activities, and there is no good environment for students to learn English, which makes the classroom atmosphere dull, thus reducing students' expectations for learning English, attacking students' enthusiasm, and not conducive to cultivating students' interest in learning English. Learning in junior high school is very important. It is a bridge between primary school and high school, and also an important stage to improve students' comprehensive quality. In the process of English teaching in junior middle schools, due to the influence of traditional educational concepts, most teachers still focus on students' achievements and evaluate a student with their achievements. This kind of teaching mode which only pays attention to the results seriously neglects the development of students' comprehensive quality. In junior high school English teaching, most teachers only pay attention to the teaching of English words, grammar and sentence patterns. The boring teaching content and teaching mode make students lose interest in English, and the teaching efficiency cannot be effectively improved, making it difficult for English teachers to achieve their teaching goals. At present, many schools do not design relevant English reading teaching, which leads to English teachers do not realize the importance of English reading teaching in the teaching process, resulting in the waste of curriculum resources, thus restricting the improvement of students' reading ability. In addition, due to the deep-rooted teaching mode, many teachers attach great importance to English performance and lack the awareness of cultivating students' comprehensive ability [1]. In classroom teaching, they constantly reduce their English reading time and increase a lot of unnecessary textbook learning. At the present stage, education pays more attention to cultivating students' comprehensive ability, which requires teachers to fundamentally innovate their own teaching mode, establish people-oriented teaching consciousness, let students become the real subjects in the classroom, and provide students with opportunities to show themselves. However, in the actual teaching process, teachers will still encounter various problems, especially the lack of clear English reading goals, and the English knowledge conveyed to students is not clear enough, leading to the stagnation of reading teaching, resulting in meaningless English reading and learning.

Data show that 88% of the students' reading ability is less than 800L, which shows that the Chinese students' English reading and speaking ability is at a relatively low level [2].

2.2 Students Have a Rigid Sense of Thinking

Because of the influence of exam-oriented education, the thinking of students and teachers is more fixed. In teaching and learning, they can only care about the score problem, and can not correctly cultivate the comprehensive quality, resulting in students' thinking is not open and innovative enough [3]. In modern education, teachers should choose the most representative educational teaching materials, strict requirements and design of teaching objectives, and at the same time combined with teaching objectives to plan English reading practice, guide students to read articles carefully, to find the fun to learning English from the article, so that reading is no longer a task.

$$TotSco = (X - Mean)/SD \times 70 + 500$$

2.3 The Teaching Focus is Just the Scores

Many teachers do not focus on teaching on students' oral English, but blindly pursue the improvement of performance, and teachers in the key English vocabulary, generally will only focus on the content of the test, in the daily teaching of English listening teaching, so that students dare not and will not speak to speak English. In addition, the teaching methods used by teachers are relatively old, which cannot further excavate students' creativity, and affects the development of students' innovation ability [4].

2.4 Strengthen Students' Language Ability by Creating a Reading Environment

Language ability is an important part of English core literacy, and it is a necessary condition for junior high school students to learn and improve English. At present, from the effect of English reading teaching in junior middle school, students do not have effective language learning skills, and their English language reading ability is relatively poor. This is mainly due to the single teaching method and boring teaching content of teachers, which not only reduces students' interest in learning English and the efficiency of reading comprehension, but also prevents students from effectively cultivating their English language ability, thus hindering the sustainable development of students' English core literacy. For this vicious circle, teachers should study more effective teaching methods. For example, specific reading situations can be created according to the reading content to provide students with more interesting learning classes, which can not only improve students' interest in learning, but also better understand the reading content through reading situations to improve students' language ability [2].

2.5 Be Good at Using Question Guidance to Improve Students' English Thinking Quality

Thinking quality is another important part of junior high school English core literacy. Whether teachers can successfully cultivate students' English core literacy depends on whether they pay enough attention to students' reading thinking quality. In other words, the current junior middle school students are short of English thinking quality. Teachers should focus on cultivating thinking quality and analyze students' psychological activities and thinking patterns to improve their core quality. For example, junior high school students are slowly entering their adolescence. They are curious, eager to explore and active in their thinking mode, which makes them prone to problems in the learning of English reading. At the same time, they are willing to explore, which is conducive to the improvement of students' thinking quality and core literacy. Therefore, teachers can take advantage of these characteristics of students' adolescence to add a problem oriented teaching method in the process of English reading teaching, appropriately arouse students' desire to explore, guide students to think about problems, and help students solve problems, so that their English reading thinking quality can be improved.

2.6 Develop Students' English Learning Ability by Developing Cooperative Reading

The core quality of English also includes learning ability, which provides a guarantee for students to improve their English quality. In the previous history of English reading teaching in junior middle schools, most students have different levels of problems in English learning ability, which are mostly caused by inappropriate teaching models. In the current teaching of English reading, students are mostly passive recipients of knowledge. The teacher's unilateral explanation does not put students in the main position in the classroom, which seriously hinders the development of students' English language learning ability. Therefore, in order to better exercise students' English learning ability, teachers should reform the traditional teaching methods and increase the strength of cooperative reading activities. It is not only conducive to improving students' English learning ability, but also conducive to the sustainable development of students' comprehensive English ability to transform teachers' classroom dominant position into students' classroom dominant position and establish an interactive English reading classroom.

3 The Necessity of Using English Reading in English Teaching to Improve Students' Oral English Literacy

3.1 Master the Key and Difficult Knowledge of English Teaching

Because English learning is a relatively novel language for students, and students are in the development stage, their understanding ability is still relatively weak, coupled with the influence of the traditional teaching mode, leading to students' lack of enthusiasm to learn English. In the intelligent classroom, the reasonable use of English reading class to further innovate the teaching forms, help students to improve their understanding ability in reading, so as to quickly master English knowledge, but also can fully grasp the key and difficult points of English teaching content, and have a strong interest in English learning [5].

3.2 Improve Students' Understanding Ability

When carrying out intelligent teaching, teachers can provide students with opportunities to read English, help them understand English sentences, increase English memory and improve comprehension. Teachers can use the intelligent system effectively to understand students' learning situation, timely guide students to learn and ensure students' learning efficiency. Repeated reading and learning can not only inspire students active thinking, but also enrich students' imagination, interested in English; therefore, teachers can lead students in wisdom class, guide students to understand English vocabulary and reading rhythm, let students think in imagination, imagine in thinking, in order to fully understand English content, feel the artistic conception of the article, in a relaxed learning atmosphere to improve oral ability (Development of—VR/AR and Department classrooms, see Table 1).

Table 1 Smart classroom—VR/AR classrooms and department-wide classrooms

Metric	VR/AR classroom	Holographic class
Product configuration	VR glasses, AR terminals	Holographic screen, holographic 3D glasses, hologram glasses, etc.
State-of-the-art	Mainly used in early education, college education, the application is gradually increasing, but still face high cost, poor experience and other problems	In the initial stage, there are technological breakthroughs, incomplete industrial chain, high cost, and content resources need to be optimized
Superiority	Use VR/AR technology to achieve a more intuitive teaching experience, provide difficult to achieve courses and training, and integrate an immersive learning experience into education	More vivid, convenient and intuitive than VR/AR class, immersive experience and super virtual and real integration and virtual and real interaction
Prospects for development	Enhance and protect content and reduce costs with cloud high-performance computing rendering; The global VR education market will grow by more than 50% in 2018–2022	With cloud high-performance computing rendering, reduce cost input, enrich content and protect resources

3.3 Promote the Interaction Between the Students

In the smart classroom, group cooperative learning is needed. In such a learning environment, students can have heated discussions and exchanges, and fully express their views in the cooperation, so that each student can get the opportunity to show their strength [6]. Teachers guide students to answer questions in the classroom, with humorous evaluation methods to create an active atmosphere, so as to mobilize the enthusiasm of students, let students participate in the intelligent classroom, develop the habit of independent learning, can improve the efficiency of classroom teaching [7].

3.4 Use Information Technology to Create a Language Environment and Mobilize Students' Enthusiasm for "Speaking"

A considerable number of Chinese students believe that they have learned grammar knowledge and a certain vocabulary, even if they have learned English well, they do not pay attention to the training of spoken English at all. Spoken language ability refers to oral communication ability. It refers to a person's ability to communicate through listening and speaking. It is an external ability and the externalization of language ability. With the acceleration of China's internationalization process, good oral English ability is becoming more and more important. Multimedia teaching It is the hot spot of current school education and the general trend of teaching development. Teaching in English In learning, using modern information technology can create a language environment and mobilize students to "speak" To stimulate students' interest in learning and stimulate students to learn in the classroom Students' sense of ownership enables them to really participate in English learning activities is becoming the master of learning.

3.5 Use Information Technology to Tap Oral English Teaching Materials and Develop Themes Teaching

The Internet contains rich oral English teaching materials. Teachers should be good at digging. For example, teachers can divide the whole class into groups of six students and arrange different topics for each group, including birthday party, graduation party, classmate party, family party and other topics. Teachers use interactive whiteboards to search for relevant pictures, music, scenes and other materials according to the topics selected by students, Students choose the materials they need, and give students 15 min to discuss the selected topic. Teachers also join the students' discussion and give relevant guidance. After 15 min Each group performs in turn. After the performance, the teacher performs Line comments; In this way, create a good atmosphere for students to learn spoken English To effectively cultivate their language communication ability, and then greatly improve the junior high school Students' oral English ability.

4 The Practice of Using English Reading in the Intelligent Classroom to Improve Students' Oral English Ability

4.1 Build Interesting Reading Environments

A good learning atmosphere can affect students' learning mood, maintain a good learning mood can broaden the learning thinking, keep the thinking keen, if the mood is low will only make students' thinking blocked, the operation behavior becomes slow. The continuous development of science and technology make classroom teaching become more innovative, so in the English teaching teachers through wisdom, guide students to English reading, show related video, courseware, and pictures, etc., can let students improve English reading ability in an interesting environment, easily master English knowledge, to make oral learning more simple [8] (For the smart education industry chain, see Fig. 1) (Figs. 2 and 3).

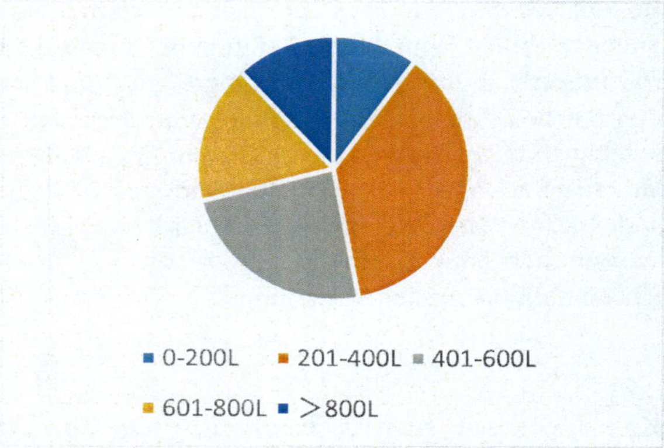

Fig. 1 Characteristics of English reading ability of students aged between 13 and 15 years

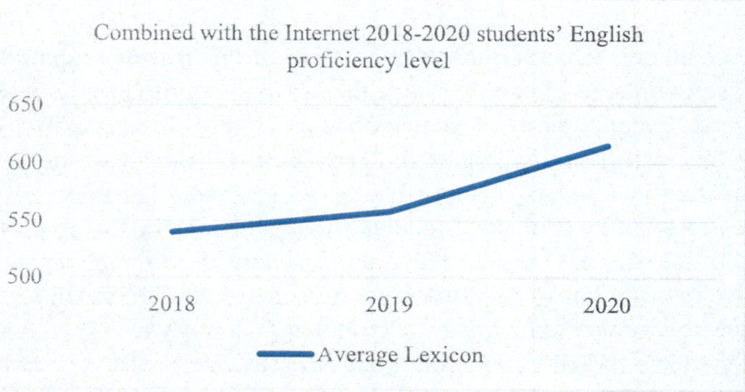

Fig. 2 Trend chart of students' oral English reading ability under Internet technology intervention from 2018 to 2020

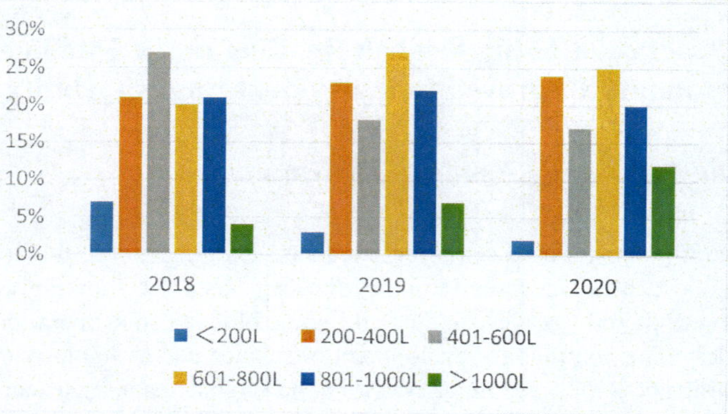

Fig. 3 Distribution chart of students' oral English reading level under Internet technology intervention from 2018 to 2020

It can be concluded from the data that, under the combination of information technology, students' English level is increasing year by year. The proportion of students with more than 1000L increased from 4 to 12%, and the proportion of students with less than 200L decreased from 7 to 2%, which is of great help to students to improve their oral English reading ability in the new era.

Such as teaching money related content, teachers can be in the classroom import stage or before the beginning of the classroom let students watch and teaching content related video, to stimulate the interest of students, attract the students' attention to the classroom, such as play "millionaire" highlights, let the student immersive feeling film content. After the main character in the film got a lot of money, the world changed his attitude towards him, which can let students understand human nature in advance, understand the impact of money on human nature, let students have doubts to read the article, so as to improve the reading effect and strengthen the learning memory. After the students complete the text preview, the teacher can carry out the group learning form, let the students read in sections, while maintaining curiosity, and carry out effective reading teaching. After completing the reading task, the teacher can guide the students to perform oral English in groups, and deepen their English memory in the interpretation environment, so as to improve the students' oral ability [9].

4.2 Develop Good Reading Habits

When carrying out intelligent classroom English reading teaching, teachers should guide students in an all-round way, guide the connection, English and reading connection after class, and carry out differentiated reading teaching according to the development characteristics of students. In addition, teachers can recommend interesting English reading articles for students, guide teachers to supervise and establish reading goals to accept English reading internally. Meanwhile, teachers should design appropriate reading tests according to students' reading ability, and add the oral expression link in the test, which can help students quickly understand the reading content and improve their oral level [10].

Such as in the teaching sports, can use spare time for students to play the relevant content of the Olympics, let the students to collect extracurricular books, reading their interest in English content, this can help students to further understand English, expand English learning thinking, and understand the sports of the Olympics, and can use oral English to express their like or understand sports knowledge, to active students thinking, enhance oral English ability.

4.3 *Broaden Information Channels and Enhance English Cultural Literacy*

Due to the huge differences between Chinese and Western cultures and different language thinking habits, students often feel "acclimatized" in the process of learning English. In addition, due to the lack of understanding of western cultural background knowledge, students sometimes can not correctly understand and grasp the content of the textbook, which requires teachers to use multimedia to introduce western cultural knowledge to students in teaching, such as religion, population, ethnic distribution, geographical location, history and humanities, to eliminate the psychological and reading brought by heterogeneous culture to students obstacles to reading.

4.4 *Pay Attention to the Role of Micro Lessons in English Preparation*

At the junior middle school stage, with the increase of English vocabulary, the difficulty of grammar knowledge has also increased a lot compared with the primary school stage. In addition, the length of the text is also relatively long, which has increased the difficulty of students' learning English. On the one hand, many students have not laid a good foundation in English in primary school, and their foundation is relatively weak. After junior high school, with the increase of difficulty, many students can not keep up with the progress of teaching, resulting in frustration; On the other hand, due to the differences between students and their acceptance of English knowledge, students with learning difficulties appear. On the whole, the junior high school students' attitude towards learning English is not active enough, and their performance in class is not very good. After analyzing the reasons, the author found that the reason is that the students did not preview before class, did not know the content of learning in advance, or did not master the method of preview, so that the students' learning objectives are not clear, and the classroom learning effect is not Poor results affect students' interest and enthusiasm in learning.

5 Conclusion

To sum up, the core quality of junior high school English includes students' language ability, thinking quality, cultural character, etc. At present, there are some problems in junior middle school English classroom teaching, such as dull classroom atmosphere, low students' interest in learning, and lack of effective interaction between teachers and students. The introduction of smart classroom into English classroom will help to build an interactive platform for teachers and students, improve classroom atmosphere, optimize students' learning feelings and experiences, stimulate students' learning potential, and improve students' core literacy. In short, teachers want to carry out wisdom classroom, effective use of English reading link to improve students' oral level, must pay attention to the cultivation of students' comprehensive ability, innovation teaching mode, create interesting teaching environment, guide students to read actively, cultivate students' thinking ability, thus converted into oral expression ability, to promote the students' all-round development.

References

1. S. Ma, Use English reading in intelligent classroom to improve students' oral English ability. Campus Engl. (36), 2 (2020)
2. J. Chen, X. Chen, Thinking on improving students' English literacy through reading in wisdom classroom. Knowl. Learn. Guid. (35), 2 (2021)
3. Y. Chen, Y. Chen, Exploration and practice of PBL teaching mode and wisdom classroom in graduate academic English writing teaching (2020)
4. C. Zhang, The application value of intelligent classroom English subject to improve students' language ability. Growth (5), 1 (2020)
5. Y. Yu, Research on teaching strategy of English majors based on multiple intelligence theory. J. Tianjin Sino-Ger. Univ. Appl. Technol. (05), 63–68 (2022)
6. L. Zhang, X. Cui, Takes the "theory and answer" artificial intelligence teaching auxiliary system as an example. Overseas Engl. (02), 96–97 (2022)
7. T. Su, Research on the innovative practice of college English mobile teaching mode under the background of intelligent education. J. Baise Coll. **34**(03), 133–138 (2021)
8. Q. Yu, Research on the application of intelligent teaching APP in English professional teaching in application-oriented undergraduate colleges. J. Hulunbuir Coll. **28**(06), 134–137 (2020)
9. X. Gao, On the impact of AI education products on English teaching in universities. Overseas Engl. (11), 111–112 (2021)
10. Y. Xue, Application of multiple intelligence theory in English teaching in preschool education majors. Campus Engl. **33**, 71–72 (2020)

Investigation and Research on the Deep Integration of New Information Technology and Situational Teaching Method—a Case Study of Statistics Teaching in Primary Schools

Shanmin Zhang

Abstract With the promotion and application of new information technology in subject teaching, traditional teaching methods are constantly being challenged. This study takes the teaching of mathematics statistics in primary schools as an example to investigate the integration of new information technology and situational teaching methods. Based on the analysis of the obtained data, the existing problems are discussed and suggestions are put forward.

Keywords New information technology · Big data · Situational teaching method · Statistics teaching

1 Introduction

At present, human society has entered a new information age with network, big data, artificial intelligence and other core technologies. Massive data has become an important information resource, and the development of an intelligent big data integrated management platform provides convenience for the collection, arrangement and analysis of massive data. Therefore, the huge data analysis, interpretation and prediction, which need a lot of manpower, can be done in an instant with the help of intelligent software. New Information Technology has brought convenience to all walks of life, but also brought the opportunity for the transformation of traditional teaching methods.

The main task of statistics teaching in Primary School is to explore the statistical problems in daily life and make students realize the application and value of statistics in daily life. Because the statistical content is more abstract and difficult to understand, the situation teaching method which turns the abstract into the image becomes the best teaching method. Situational teaching method is one of the most

S. Zhang (✉)
Liaocheng University Dongchang College, Liaocheng, China
e-mail: 2677009031@qq.com

S. Patnaik and F. Paas (eds.), *Recent Trends in Educational Technology and Administration*, Learning and Analytics in Intelligent Systems 31,
https://doi.org/10.1007/978-3-031-29016-9_11

widely used and important teaching methods in statistics teaching in primary schools [1]. However, due to the constraints of time, space, materials, equipment and other factors, it is difficult to give full play to the role of situational teaching methods under the traditional teaching conditions, it clears up the obstacles for the effective development of situational teaching method. The integration of new information technology and situational teaching method embodies the development trend of informationization and modernization of teaching methods. Therefore, this study is of practical significance to promote the development of informationization of teaching [2].

2 Research Design and Result Analysis

2.1 Research Design

Research Purposes: from the perspective of situational teaching method, this paper studies the deep integration of new information technology and subject teaching, discusses the problems existing in the process of teaching informatization, and puts forward countermeasures to further promote the informatization process of primary school teaching.

Research process: Starting from the statistical teaching of primary schools, the cognitive attitude, teaching ability, teaching effect and teaching evaluation of teachers using new information technology to set up teaching situation were studied. A total of 30 questions were designed and 70 mathematics teachers and 120 students from 7 primary schools were investigated.

2.2 Analysis of Research Results

2.2.1 Teachers' Cognitive Attitudes to New Information Technology Differ by Age

A total of about 64% of teachers think new information technology is important and important. In terms of age structure, most young and middle-aged teachers have a positive attitude towards new information technology, while teachers over the age of 50 have an indifferent attitude towards using new information technology (Table 1 and Fig. 1).

The above statistics reflect that there are age differences in teachers' cognitive attitudes towards adopting new information technology to set up teaching situations. Young teachers accept new information technology quickly, while old teachers are accustomed to using their existing experience and have low recognition of new information technology.

Table 1 Attitudes of teachers towards the use of new information technology to set up teaching situations

Degree of recognition	Very important	Important	General	Less important	Unimportant
Percentage	21	43	16	12	8

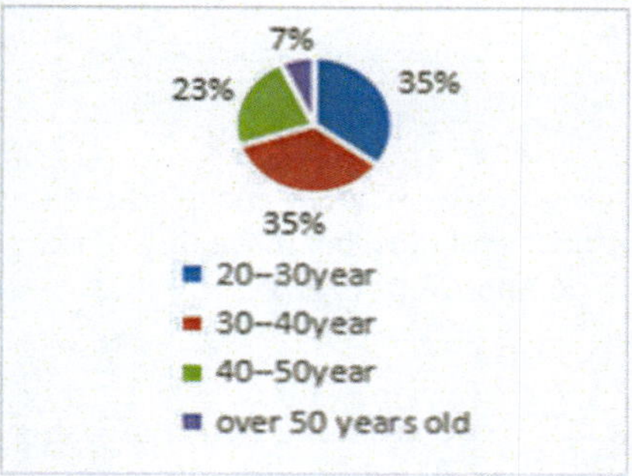

Fig. 1 Distribution of teachers who consider new information technology very important and important

2.2.2 There Are Differences in the Teaching Ability of Teachers Using New Information Technology to Set up Teaching Situations

In the method of creating a situation in teaching, most teachers are used to describing in language or borrowing pictures and video materials on the Internet, and only a few teachers use more advanced technology, such as VR simulation technology. This may be because language descriptions, pictures or video materials are easier to download. Although the new information technology represented by VR technology is becoming easier to operate, it is not widely used in primary schools due to certain technical requirements (Fig. 2).

The majority of teachers have a positive attitude towards the existing teaching resources when using the new information technology to set up the teaching situation, but the number of students using the new information technology to transform and construct the curriculum is not large. This shows that high-quality curriculum

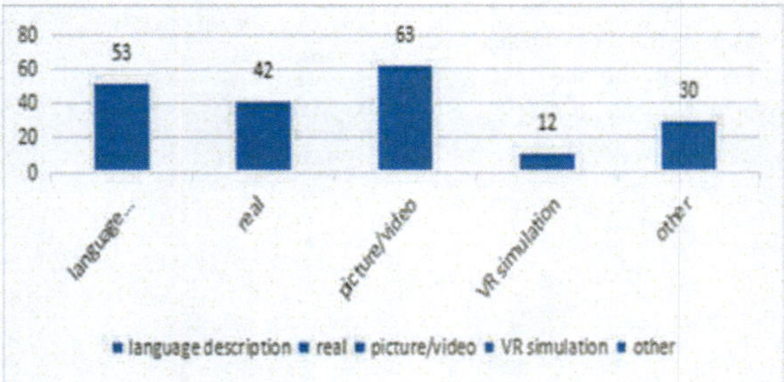

Fig. 2 Way to set up a situation

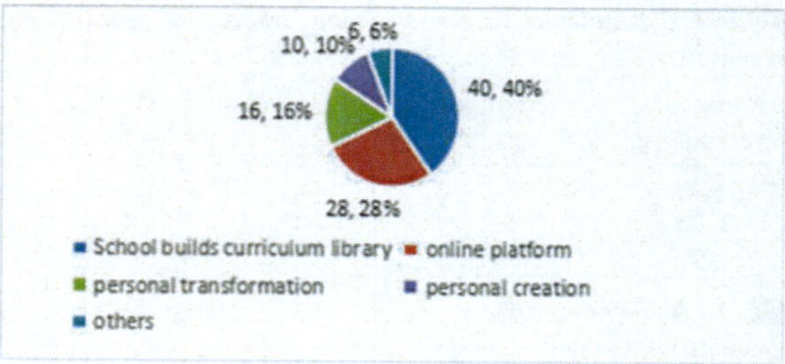

Fig. 3 Source of teaching materials

resources of schools and platforms are widely used in mathematics teaching in primary schools, but the ability of teachers to use new information technology needs to be improved (Fig. 3).

2.2.3 The Teaching Effect of Using New Information Technology to Set up Teaching Situation

By comparing the average scores of different classes in the statistical knowledge unit test, it is found that the average score of the classes taught by the new information technology is higher than that of the classes not taught by the new information technology, and the scores of each group conform to the normal distribution. Eight students were randomly selected. After calculation, the probability that the 8 students in the class taught by the new information technology was lower than the score of the 8 students in the class not taught by the new information technology was only 0.004, indicating that the use of the new information technology to teach the students (Table 2).

This shows that there are individual differences in the teaching effect of teachers using new information technology. Some teachers can stimulate students' enthusiasm for learning, but some teachers do not benefit students by using new information technology. This shows that there are still many problems in the practical use of new information technology in teaching.

Table 2 Comparison of the average grades of classes with and without the use of new information technology

Use of new information technologies	Average score	Standard deviation
Used class	655	20
Unused class	625	25

2.2.4 The Teaching Evaluation of Using New Information Technology to Set up Teaching Situation

About 70% of the students welcomed and appreciated the teaching situation set up by using new information technology, indicating that the application of new information is generally accepted by students, and primary school students have strong learning and acceptance of new information technology. When teachers self-assess the teaching effect, some teachers questioned the teaching effect of the new information technology, pointing out that the application process is difficult to operate, time-consuming, and the "overly gorgeous process" can easily cause students to be distracted and cannot cause students to think deeply.

3 Discussion: The Influence of New Information Technology on Situational Teaching Method

3.1 Higher-Order Thinking: A New Goal for Contextual Teaching

With the rapid development and wide application of new-generation information technologies such as artificial intelligence and big data, people's learning methods and learning needs have undergone tremendous changes, resulting in continuous updates of educational concepts, educational content, and educational methods. Basic education aims to lay a solid foundation for children to adapt to the future intelligent society, and help students acquire the ability to control new technologies and create a better life [3]. Chen Baosheng, Minister of Education, proposed that innovation, communication and collaboration, complex problem-solving, and human–machine collaboration will become essential abilities for human beings.

The implementation of teaching with the help of modern information technology is generally welcomed by teachers and students, but while based on vivid images, it is necessary to design situations around cultivating students' innovative ability, problem-solving ability, decision-making ability and critical ability, and focus on cultivating students' higher-order thinking ability. For example, in the fourth-grade learning folding statistics, the 3D scene setting is used to dynamically display the monthly average temperature changes in a certain place in 2010 and 2020, which not only allows students to visually perceive the increase or decrease in the monthly average temperature, experience Statistics are closely related to life, and at the same time, it is easy to cause students to think deeply about the temperature changes before and after 10 years through comparison.

3.2 *Combining Virtual and Real: A New Design of Situational Teaching Method*

The traditional statistics classroom adopts the situational teaching method, which uses objects or models to reproduce the life situation, and gives students an intuitive and vivid feeling and experience through practical operation [4]. However, some teaching contents are limited by time and space conditions, and it is impossible to achieve practical operation and personal experience. For example, for the characteristics of median and mode data, teachers only use words to describe, and it is difficult to achieve the ideal teaching effect. Nowadays, the application of new information technologies such as VR reality simulation makes the situational teaching method break through the limitations of time and space. Its advantages: First, teachers and students can freely design various scenes according to teaching needs, which is convenient for students to take initiative and enthusiasm, and stimulate their interest in learning; second, they can clearly display or observe content that cannot be perceived in real life, which is beneficial to students. Form the appearance and realize the transformation of knowledge from concrete to abstract; the third is to realize role interaction and scene switching in virtual scenes. It can be operated repeatedly to achieve breakthroughs in teaching difficulties. However, according to the survey, the ability of teachers to use new information technology needs to be further improved.

3.3 *Online Resources: The New Support of Situational Teaching Method*

3.3.1 Massive Online Course Resources and Materials Provide Convenience for the Use of Situational Teaching Methods in Primary School Statistics Teaching

It not only saves a lot of material resource costs, but also makes the situational teaching method simple and easy to implement. Teachers can make full use of courseware library, micro-teaching unit library, etc. to organize and reorganize materials such as MOOCs, micro-lectures, texts, graphics and images, animations, audios, etc. and creatively apply them to their own teaching to set the teaching situation.[5] This requires teachers to have certain curriculum reconstruction ability and good information literacy.

3.3.2 The Web-Based Learning Community has Greatly Expanded the Use of Situational Teaching Methods

Today, with the development of information technology, the learning community has gone far beyond the conceptual scope of the original constructivism theory, the network expands the learning community from the present teachers, experts, tutors, classmates and peers to the same learning interests and learning needs, but with different learning backgrounds, a Learner, facilitator, or helper who is in a different place and does not know each other. In the interactive learning activities based on the network environment, people discuss, negotiate, communicate information and communicate feelings through the use of words or sounds on the screen, which makes the situational teaching method break through the traditional scope of application, especially in less developed areas, the value of e-learning community is greater. Therefore, teachers and students' awareness of online learning needs to be further strengthened.

3.3.3 Relevancy: A New Evaluation of the Effectiveness of Situational Teaching Methods

The rapid development and wide application of information technology facilitate the implementation of situational teaching, but the only criterion for the application of situational teaching is the correspondence between teaching methods and teaching contents. Therefore, the effective implementation of situational teaching method in primary school statistics teaching needs to pay attention to the following points:

3.3.4 Suitability

Situational teaching method is characterized by intuitionistic and vivid, and its application value is mainly expressed by abstracting into intuitionistic and vivid, making the knowledge difficult for pupils to understand vivid and easy to understand by reappearing in the scene.

The thinking of primary school students is in the transition stage from concrete image thinking to abstract logic thinking. The logic and abstraction of thinking are gradually developing, and they have already had rich life experience [6]. Therefore, not all statistical knowledge is abstract and difficult to understand for them. Only those important knowledge points that are far from their life experience and have a high level of abstraction and logic need to set up scenes to help students learn. Therefore, it is the primary principle to measure the effectiveness of situational teaching method that the learning of a certain knowledge needs no situational setting. Setting up situational teaching first needs to effectively screen the knowledge points and select the necessary knowledge points, and for the general knowledge points which only need to use conventional teaching methods, we should put an end to the blind use of situational teaching methods just for the pursuit of intuitive effect and "going through the process" [7].

3.3.5 Moderation

Setting up situational teaching links in the teaching process should be moderate. It is advisable to use situational teaching to keep about 1/3 of the knowledge points for primary students. Situational teaching method should not be used too much due to the development of information technology, the convenience of drawing materials and the ease of setting up virtual scenes. In addition, long time situational teaching is easy to dilute the focus, so that students lose the freshness of the teaching content, have distraction, but reduce the learning effect, is not conducive to the cultivation of students' higher-order thinking ability.

In addition, the class size of situational teaching should be moderate. The number of students should not be too large. Even if the class is taught in network context, the number of students should not exceed 30 in order to avoid overcrowding and confusion, which will make it difficult for teachers to give effective guidance and maintain normal class order [8].

4 Suggestion

The information age is an era of resource sharing, with high-quality teaching resources and learning resources is to improve the quality of teaching and learning the premise. The massive amount of online course resources and materials has facilitated the development of situational teaching method. Teachers can borrow or reorganize materials according to their teaching needs, which not only saves time and energy, but also promotes high-quality resources, however, the effective application of new information technology in situational teaching also puts forward new requirements for teachers.

4.1 Effective Development of Situational Teaching Requires Teachers to Have Awareness of Network Course Resources

First, teachers should make full use of teaching materials to provide curriculum resources. The micro-lectures and micro-videos attached to the textbook are high-level resources carefully designed by relevant experts, which meet the requirements of the outline, and have a high station and a complete system between materials [9]. Therefore, teachers should dig deep into their connotations, guide students to perceive the situational process, and understand the conclusion. Second, teachers should learn to find and use high-quality resources on the Internet, and use these resources to optimize their own teaching. For example, websites such as the National Educational Resources Public Service Platform, the People's Education Publishing

House, and the China Teachers' Station all have abundant high-quality teaching resources. The third is that teachers should actively participate in the creation of high-quality courses and be the masters of the courses. Teachers should give full play to their own enthusiasm and initiative, conduct in-depth thinking and creation of courses, carefully design teaching, continuously accumulate course resources for situational teaching, share their own high-quality situational materials through the network, enrich online course resources, and be builders of high-quality courses. and promoters.

4.2 Effective Situational Teaching Requires Teachers to Have Strong Scientific Research Ability

Under the background of informatization, curriculum resources are extremely rich, but the effective use of these resources puts forward higher requirements for teachers' scientific research ability. Teachers need to clarify what kind of teaching purpose is to be achieved in the creation of the situation, how to realize the progressive and interlocking situation of the situation, how to use the situation to guide students to think and analyze in depth, to achieve the rise from perceptual to rational, and to constantly carry out teaching reflection and improvement. Teaching design, so as to create high-quality curriculum resources and achieve teaching goals.

4.3 Effective Situational Teaching Requires Teachers to Master Modern Teaching Techniques

Modern teaching technology changes static book knowledge into perceptive and dynamic content through audio and video images, which greatly reduces the difficulty of learning. Mastering modern teaching technology becomes an important condition for teachers to carry out situational teaching effectively. Each teacher not only to master the multimedia technology, more want to learn the network new media technology, not only can to collecting, sorting and application of teaching resources, can also according to the teaching needs of network teaching resources to carry on the design and development, so as to formulate reasonable teaching decisions, set the appropriate learning situations [10].

In short, under the background of new information technology, modern teaching technology and subject teaching are deeply integrated, and the application and development of situational teaching method in primary school statistics teaching is of certain practical significance to promote the modernization and informatization development of primary school teaching.

References

1. W. Guiying, Discussion on the application of situational teaching method in higher mathematics teaching. Educ. Teach. Forum **35**, 186–187 (2019)
2. W. Jinran, The application of situational teaching method in primary school mathematics teaching. Math. Learn. Res **24**, 78–79 (2021)
3. H. Li, Research on the strategies of life and situationalization of primary school mathematics classroom teaching. Chin. Extra-Sch. Educ. (Mid-School), **2020**(4), 109−111(2020)
4. W. Yuping, The application of situational teaching method in primary school mathematics teaching. Sci. Technol. Inf. **18**(9), 79–80 (2020)
5. Z. Anlin, Application of life situational teaching method in primary school mathematics teaching. Xue Wkly. **8**, 67–68 (2021)
6. Y. Zhao, The application of life situation teaching method in primary school mathematics teaching. Famous Teach. Online, **2020**(20), 72(2020)
7. Y. Ding,Analysis of the application of life situations in primary school mathematics teaching. Sci. Consult. (Education and Research), **2020**(7), 295(2020)
8. S. Yuanjun, X. Li, On the application of life situations in primary school mathematics teaching. Caizhi **12**, 90 (2020)
9. Z. Dequan, The application of life situation method in primary school mathematics teaching. China Rural. Educ **9**, 108 (2020)
10. C. Linqun, Research on the creation and utilization of effective scenarios in primary mathematics teaching. Charming China **2**, 77–78 (2020)

Reform and Practice of Traditional Engineering Courses in Application Oriented Local Universities Against the Background of Emerging Engineering Education

Take Engineering Drawing as An Example

Lijing Yan, Yanhua Shi, and Jiaojiao Wang

Abstract Under the background of emerging engineering education, the reform of traditional engineering courses in application oriented local universities is necessary and exemplary. The reform and practice of engineering drawing course are carried out from the ways of the implementation of curriculum ideological and political education, online and offline mixed teaching, 2-3D collaborative teaching, reform of evaluation methods, etc. Based on the analysis of teaching reform data, it discusses the effectiveness of engineering drawing reform in cultivating talents with emerging engineering qualities.

Keywords Emerging engineering education · Engineering drawing · Teaching reform · Application oriented local Universities · Cultivation of talents

1 Introduction

In respond to the new round of technological and industrial change, the Ministry of Education actively promotes the construction of emerging engineering education. In 2016, the concept of emerging engineering education was put forward. In 2017, the "Notice on the research and practice of emerging engineering education" was issued [1], and the 'Fudan Consensus', 'Tianda Action' and 'Beijing Guide' were formed, which laid the basic education governance pattern of emerging engineering [2]. In 2017 and 2020, two notices were issued to promote the research and practice of emerging engineering [3, 4]. In 2018, the Ministry of Education first identified 612 emerging engineering research and practice projects. In 2019, the Ministry of

L. Yan (✉) · Y. Shi · J. Wang
School of Electromechanical Engineering, Guangdong University of Science & Technology, Dongguan 523083, China
e-mail: lj-yan@qq.com

S. Patnaik and F. Paas (eds.), *Recent Trends in Educational Technology and Administration*, Learning and Analytics in Intelligent Systems 31,
https://doi.org/10.1007/978-3-031-29016-9_12

Education held several thematic exchanges to explore the formation of emerging engineering education system with Chinese characteristics. In 2020, the Ministry of Education passed the acceptance of 612 projects, leading the research and construction of emerging engineering to depth. Based on the orientation of talent cultivation in application-oriented universities, this paper selects the traditional engineering course-engineering drawing with a typical role, and carries out the teaching reform and practice on the basis of the concept of emerging engineering.

2 The Necessity of Traditional Engineering Curriculum Reform in Applied Local Universities Under the Background of Emerging Engineering

Professor Lin Jian proposed that the construction of emerging engineering should be carried out according to different types of schools with emphasis and distinction [5]. The orientation of application-oriented universities is to serve the local and regional economy. With the rise of the emerging engineering industry, application-oriented universities must focus on emerging engineering in personnel training, based on regional economic development, aim at the forefront of industrial development, focus on the new technological revolution, the new industrial revolution and the new economic revolution, start from the practical engineering application, take the results as the guidance, emphasize the learning of new knowledge and new technology, and cultivate graduates who meet the talent needs of new technology enterprises.

The final focus of any teaching reform must be curriculum reform. Curriculum is the core element of talent cultivation, and the quality of curriculum directly determines the quality of personnel training. Under the background of emerging engineering, the course must serve the regional economic and social development as the main battlefield, and reflect the integration of multidisciplinary thinking, industrial technology and discipline theory, cross-disciplinary ability and multidisciplinary project practice. It can serve the cultivation of innovative and compound talents [6]. Emerging engineering requires engineering drawing teaching not only to cultivate students' ability to make use of multidisciplinary knowledge for innovative design, but also to improve students' thinking and ability to apply the Internet through M-learning [7, 8], MOOC, SPOC, virtual simulation and other new teaching methods [9]. Obviously, new technologies make higher demands for the curriculum, but traditional engineering courses such as engineering drawing have been inherent in a set of teaching methods for many years, and the resistance to breaking the old and establishing the new is obviously greater. Therefore, it is of great significance to study how to make the traditional engineering drawing course serve the local emerging engineering talents effectively.

3 Contents of Traditional Engineering Curriculum Reform in Applied Local Universities Under the Background of Emerging Engineering

3.1 Guiding Ideology of Engineering Drawing Curriculum Reform

Engineering drawing is the first professional basic course that mechanical students contact after entering the university. Therefore, this course is not only the basis for subsequent professional knowledge learning, but also shoulders the burden of cultivating students' professional learning habits and thinking. In view of this, the guiding ideology of this curriculum reform is to lay a solid foundation, serve the profession, highlight innovation. Specific ideas are the following.

(1) Adhere to the moral education as the fundamental task, through strengthening the teaching content of "curriculum ideological and political education", cultivate students' dedication to lean, focus on innovation 'craftsman spirit' and tough character; shaping the professional spirit and scientific literacy of seeking truth and pragmatism and pursuing excellence; it emphasizes the national feelings and mission of loyalty, dedication and innovation [10].
(2) Guiding students to correctly use Internet resources, forming Internet thinking and 'Internet + learning' learning mode, guiding students to learn to think independently and cultivate self-study habits.
(3) Guide students to establish professional knowledge consciousness, and lay the foundation for the cultivation of professional core competence. According to the regional demand for talents, the core ability of mechanical related majors in our university is defined as the ability of computer aided design and manufacturing.
(4) Guide students to study hard, eliminate the opportunistic learning of cram for a test.

3.2 The Reform Scheme of Engineering Drawing Course

3.2.1 Implementation of Curriculum Ideological and Political Education to Cultivating New People of the Times

Emerging engineering endows universities with the burden of cultivating new people of the era. New people of the era should not only have the knowledge and skills required by the emerging engineering era, but also have the mission, sound humanistic values and professional quality of the new era. The important task of education is to cultivate students 'character. In addition to ideological and political courses, other courses should realize the educational function through curriculum ideological and political education [8]. The teaching object of engineering drawing course is the freshmen of Grade 1 in all engineering colleges, which belongs to the course with large demand and wide audience. Therefore, the teachers of this course should shoulder the responsibility of educating people, actively explore the educational connotation contained in the course content, integrate professional ethics, professional ethics, humanistic spirit, family and national feelings, and socialist core values into all aspects of drawing classroom teaching, and realize the organic the course's integration of imparting knowledge, cultivating ability and guiding value orientation, so as to realize the ideological and political education of the course in a silent way.

The concrete implementation ways of engineering drawing course curriculum ideological and political education are as follows: First, through the training of teachers to improve teachers' ideological and political ability and the implementation of curriculum ideological and political skills. The second, the curriculum group establishes the curriculum ideological and political resource database to provide materials and cases for teachers to carry out curriculum ideological and political work. The third, gradually integrate curriculum ideological and political education into the assessment project, causing students' attention to curriculum ideological and political education. The curriculum ideological and political education programs implemented in each part of this course are shown in Table 1.

3.2.2 Online and Offline Mixed Teaching to Cultivate Students' Autonomous Learning Habits

As a traditional basic course of engineering specialty, engineering drawing has far-reaching influence on cultivating students' habits of learning subsequent professional courses. The curriculum construction under the emerging engineering concept requires teachers to guide students to form Internet thinking, learn to actively explore and think independently. Therefore, engineering drawing teaching should make full use of the teaching mode of 'Internet + education', carry out online and offline mixed teaching based on MOOC and online open courses, stimulate students' desire to actively explore and develop the habit of autonomous learning. This course mainly adopts online and offline teaching mode based on self-built learning platform and supplemented by MOOC resources.

This course has a complete system of curriculum resources on the superstar learning platform, including task points, video, animation, documents, pictures, homework, etc. Before class, teachers first record the course learning video, upload relevant learning resources, and design self-test questions and offline tasks covering the corresponding knowledge points. Teachers publish preview guidance programs through superstar learning platform or WeChat, push relevant learning resources, and guide students to preview by using free and fragmentary time. Teachers understand students' preview situation through the statistical module of learning through. Students' doubts in preview are discussed and communicated between teachers and students or students through superstar learning platform WeChat. In classroom teaching, students first complete the online test, and then teachers adjust the teaching difficulties according to the test situation of learning feedback, so as to take students

Table 1 Ideological and political implementation plan of engineering drawing course

Engineering drawing course module	Implementation plan of curriculum ideological and political education
National standard module of mechanical drawing	By requiring students to standardize drawing, and introducing the spirit of dedication, lean, focus and innovation into teaching, students ' conscientious working attitude and meticulous working style are cultivated
Projection knowledge module	By explaining the position relationship of points, lines and planes on different surfaces and the projection of plane and three-dimensional, the viewpoint of universal connection, the viewpoint of development and the law of quality alternation of materialist dialectics are penetrated into the curriculum teaching, and the ability of students to analyze and solve problems by using the law of unity of opposites and the law of quality alternation of materialist dialectics is cultivated, so that students can master good thinking habits and learning methods, and extend to the use of dialectical thinking for innovation
Combinatorial knowledge module	By analyzing the relationship between the combination and the form of composition, the relationship between the whole and the individual, and the relationship between the state and the individual are introduced into the patriotic consciousness; secondly, scientific methodology is introduced in the introduction of the combined body analysis method; finally, by drawing the combination, cultivate students' craftsmanship spirit
Common expression methods of machine parts	Through the expression method of machine diversity in the course, students are guided to learn to think in transposition, to guide students to think from the perspective of others, to cultivate students' overall concept and global consciousness, and to guide students to form ideas that are convenient to others
Standard parts and common parts	Through the study of the basic knowledge of thread and thread fasteners in the course, the prescribed drawing method and labeling, the ' irregular, not square' of 'Mencius' is cut, so that students can develop the quality of complying with standards and laws, cultivate good professional ethics literacy and then extend to the rules of doing people, standardize doing things

(continued)

Table 1 (continued)

Engineering drawing course module	Implementation plan of curriculum ideological and political education
Part drawings and assembly drawings	When reading part drawings and assembly drawings, help students overcome the fear of difficulties, cultivate students' hard work, not afraid of difficulties, brave forward spirit; in the part drawing, assembly drawing expression scheme group discussion, cultivate students' team cooperation consciousness and helpful spirit; when drawing part drawings and assembly drawings, students can dialectically understand, analyze and solve problems, and cultivate dialectical thinking ability

as the center and improve the efficiency of classroom teaching. After class, on the one hand, according to the situation of students' learning before class and class, teachers will select the video to push students, as a supplement to classroom teaching. On the other hand, teachers will also recommend some professional resources to expand students' knowledge and.

extend the classroom to extracurricular activities. In order to stimulate students' enthusiasm for active use of online resources for learning, the data such as videos, chapters' learning duration, discussions, classroom interaction and online homework scores displayed on the learning passage will be used as the basis for assessment. The assessment results will be published regularly to enable students to understand their usual scores and rankings in real time, urge students to form a competitive learning atmosphere and help students develop the habit of autonomous use of Internet learning.

In order to expand students' ability to use Internet learning, teachers will also use MOOC to assist teaching. At the beginning of the school, on the basis of fully digesting the MOOC curriculum resources, teachers have targeted and differentiated selection of different teaching content resources regularly recommended to students through the WeChat group, and irregularly check the data of MOOC course students' learning as one of the assessment basis, in this way, students can completely get rid of the limitation of time and space, and can use fragmented time to learn whenever and wherever they want.

3.2.3 2-3D Collaborative Teaching, Education Industry Elite

With the maturity of CAD/CAM technology, enterprises have higher and higher requirements for students' computer-aided design ability. In the past three years, the team of this course has investigated the demand for talents of machinery-related enterprises in the region where the school is located. It is found that enterprises all need students to skillfully operate 2D and 3D drawing software. In view of this, it is necessary to organically integrate 2-3D software and engineering drawing teaching.

First, reconstruct the curriculum system. In order to support local service and cultivate applied talents with strong practical ability, the original separate parts mapping and computer graphics (Auto CAD) are integrated into the 'engineering drawing comprehensive training', and the three-dimensional modeling course of mechanical parts is moved forward. Second, revise the syllabus. It is required to introduce AutoCAD as one of the drawing tools in the basic knowledge and skills module of drawing, help students master the basic drawing interface and command, and learn modern design methods; three-dimensional modeling software is gradually introduced into the teaching of the follow-up module. With the help of three-dimensional display of graphics, it makes up for the defects of insufficient space imagination of some students, helps students develop the thinking ability of 2-3D conversion, and encourages students to learn drawing software. Third, focus on training student's 3D drawing software application ability. After a semester of engineering drawing knowledge, through the three-dimensional modeling course of mechanical parts, students' ability to use three-dimensional modeling software for medium-difficult parts modeling, component assembly and mechanism simulation is systematically cultivated, and students' practical ability of digital drawing and design is cultivated. Fourth, strengthen the drawing quality and ability training. After students have the basic knowledge of drawing and the ability to use computer software for drawing, they need to use practical methods to comprehensively train students' drawing skills and qualities. Engineering drawing comprehensive training comprehensively trains students' application ability of computer drawing and drawing standards by surveying and mapping parts, three-dimensional modeling of parts, drawing two-dimensional assembly drawings and parts drawings. At the same time, the team spirit, independent learning ability and innovative consciousness required by new technology enterprises are cultivated through project teaching, group teaching and task-driven teaching.

3.2.4 Adopt Formative Assessment to Encourage Students to Attach Importance to Learning Process and Improve Learning Enthusiasm

Assessment methods directly affect students' learning methods and efficiency. The traditional assessment method is based on the final examination results, and the application-oriented undergraduate colleges and universities to cultivate students' ability to flexibly use knowledge in practice as the goal, a test paper is difficult to assess the students' ability, also cannot fully feedback the teaching effect, often appear some students that cram at the eleventh hour can also obtain high scores, which often hit enthusiasm of students that usual efforts to learn. The teaching of emerging engineering is learning-centered and requires students to develop the ability of self-study and flexible use of knowledge to solve problems. Therefore, courses should adopt a variety of assessment methods, so that teachers and students can master the learning situation at any time, and take the assessment as the guide for students' daily learning. Figure 1 shows the process assessment mechanism of engineering drawing based on new engineering. The offline classroom performance mainly refers to students' answering questions and participating in discussions in class. Online learning performance mainly refers to students' online video and chapter learning time, participation in online discussion, etc. The project design mainly refers to the achievements of students' practice in class. In order to urge students to develop good professional quality and habits, this assessment set up 5% of the curriculum ideological and political assessment content, mainly from three aspects of professional quality, responsibility and teamwork. The proportion of usual assessments will be set in advance on the superstar learning platform before class starts. The offline classroom performance teachers will regularly update to the superstar learning platform. The online learning performance is automatically generated by the superstar learning platform, and the performance is completely open to students. Teachers will display the current performance ranking of students from time to time, so that students can grasp their current performance in real time, and improve students' enthusiasm for learning in ordinary times by urging backward students and praise actively promoting students, so that students can accumulate knowledge in ordinary times, learn each knowledge point solidly, and eliminate the disadvantages of mixing time in ordinary times and passing the customs in front of the exam.

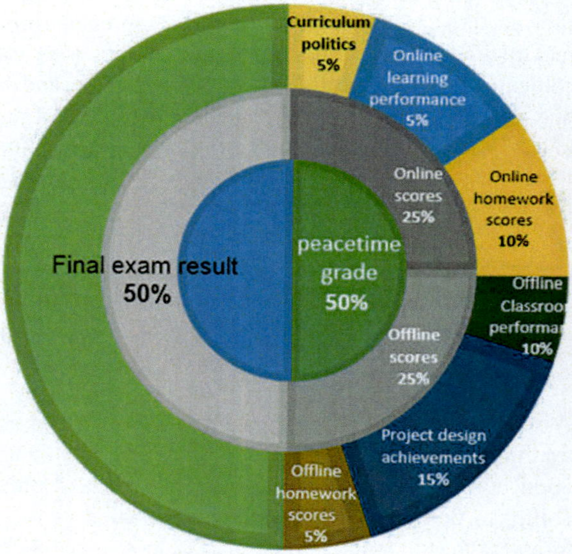

Fig. 1 Process assessment mechanism of engineering drawing course based on emerging engineering

3.3 *Teaching Reform Practice and Data Analysis*

The project team carried out a round of engineering drawing teaching reform in Mechanical and Electronic Engineering (ME), Mechanical Design, Manufacturing and Automation (MDMA), and Material Forming and Control Engineering (MFCE). The distribution of students 'assessment scores before and after the reform is shown in Table 2 below. After the curriculum reform, the average scores and pass rates of the three majors have improved, especially the pass rate has improved significantly. Through the return visit and the questionnaire survey of students, it can be seen that the rich online resources provide a powerful supplement for their classroom learning. The students with weak foundation can make up for the content they don't understand in the classroom through online resources in time, and can communicate with teachers and students online at any time to solve confusion, so as to greatly improve their learning enthusiasm. In addition, the reform of ME specialty has achieved the most remarkable results. After analyzing the online learning data of the superstar learning platform for one semester, it is found that the average task completion rate of ME specialty students is 87.24, and 70% of the students complete 90% of the task points issued by teachers, of which 51% complete all the task points issued by teachers. The average duration of watching video is 708.37 min per birth, and the longest duration is 1053.5 min. The average number of chapter learning is 216.98 times per student, and the maximum number of chapter learning is 522. It can be seen that online teaching can indeed promote students' self-study and receive good results. Through the questionnaire survey after class, it is found that most students approve the implementation effect of engineering drawing reform, and the specific survey results are shown in Table 3. In addition, the survey of teachers in subsequent professional courses also found that students after the curriculum reform had higher enthusiasm for autonomous learning, and they would actively require teachers to recommend online learning resources and carry out online teaching. In addition, students in subsequent curriculum design had stronger ability to use computers for modeling, design and simulation, and had a deeper understanding of the curriculum.

4 Conclusions

The emerging engineering has put forward higher requirements for the training of higher education talents. The application-oriented local colleges and universities must combine their own characteristics to carry out teaching reform. Engineering drawing, as a traditional engineering course, plays an important role in the teaching reform under the background of emerging engineering, and is also a good example of other professional basic courses. Engineering drawing course improves the teaching quality by focusing on emerging engineering, accurately docking local talent demand, paying attention to curriculum ideological and political education, and flexibly using

Table 2 Analysis of students 'examination results before and after curriculum reform

Score	The class of 2020				The class of 2021			
Speciality	Average	Maximum	Minimum	FailureRate (%)	Average	Maximum	Minimum	Failure rate (%)
ME	60.40	95	30	47.37	71.27	98	40	24.39
MDMA	64.16	95	34	48	64	96	44	37.21
MFCE	61.93	92	34	56.67	69.56	95	36	46

Table 3 Student feedback form for teaching reform model

Evaluation of project	Number of students scored Scores 5 to 1 represent a gradual decrease in student recognition				
	5	*4*	*3*	2	1
The necessity of teaching ideological and political content in professional courses	225	4	3	4	1
The Influence of teachers' ideal and faith, moral sensibility, solid knowledge and benevolence on your moral quality	215	6	8	3	5
The recognition of online and offline mixed teaching	231	1	1	4	0
The necessity of learning computer drawing software in advance	220	7	4	5	1
Acceptance of teacher evaluation	228	1	2	3	3

new teaching methods and evaluation system. It has been recognized by students and speeds up the process of cultivating emerging engineering talents serving local areas.

5 Fund Projects

1. Guangdong Higher Education Teaching Reform Projects in 2020: Research and practice of engineering drawing curriculum system reform based on the concept of emerging engineering education and the background of first-class curriculum construction.
2. First Class Curriculum Construction Projects of Guangdong University of Science and Technology in 2020: Engineering Drawing (CQ2020031).

References

1. Higher Education Division of Ministry of Education of the People's Republic of China. Notice on emerging engineering research and practice, http://www.moe.gov.cn/s78/A08/tongzhi/201702/t20170223_297158.html. 20 Feb 2017
2. K. Liu, Y. Dai, Z. Zhang, etc., Achievement and strategic analysis on the guideline of the first batch of emerging engineering education research and practice projects. Res. High. Educ. Eng. Issue. **1**, 31–38 (2021)
3. Higher Education Division of Ministry of Education of the People's Republic of China. Notice on recommendation of emerging engineering education research and practice Projects, http://www.moe.gov.cn/srcsite/A08/s7056/201707/t20170703_308464.html. 21 June 2017
4. Higher Education Division of Ministry of Education of the People's Republic of China. Notice on recommending the second batch of emerging engineering education research and practice

projects, http://www.moe.gov.cn/srcsite/A08/s7056/202003/t20200313_430668.html. 21 June 2017
5. J. Lin, The construction of China's new engineering disciplines for the future Tsinghua. J. Educ. **38**, 26–35 (2017)
6. Ministry of Education of the People's Republic of China. Opinions of the Ministry of Education on the Construction of First- class Undergraduate Courses, http://www.moe.gov.cn/srcsite/A08/s7056/201910/t20191031_406269.html. 2019/10/30
7. L. Chen, L. Wu, R. Zhang, etc., The teaching reform and practice of engineering drawing based on the system engineering. J. Graph. **39**, 1214–1219 (2018)
8. Y. Xiao, M. Li, J. Hu, etc, Exploration of the construction of emerging engineering Education in universities featuring of arts and science. Jiangsu High. Educ., Issue. **5**, 58–61 (2021)
9. X. Wang, Cultivation of innovative engineering talents in colleges and universities under the emerging engineering background. Xuexiao Dangjian Yu Sixiang Jiaoyu issue. **10**, 81–83 (2021)
10. The Ministry of Education. Guiding outline of ideological and Political Construction in college curriculum. http://www.moe.gov.cn/srcsite/A08/s7056/202006/t20200603_462437.html. 8 May 2020

Study on Outdoor Environment Evaluation of Kindergarten Based on Probabilistic Neural Network

Gao Ting and Jiangxi

Abstract In order to improve the outdoor environment evaluation correctness and efficiency of kindergarten, the probabilistic neural network is constructed. Firstly, the evaluation system for outdoor environmental design of kindergartens is established. Secondly, the probabilistic artificial neural network model is constructed. And then Outdoor environment evaluation procession of kindergarten is designed. Finally, thirty kindergartens are selected to carry out simulation analysis, and results show that the probabilistic neural network can efficiently and accurately evaluate the outdoor environment of kindergarten.

Keywords Outdoor environment · Kindergarten · Probabilistic neural network

1 Introduction

With the massive construction of kindergartens in China, the problem of outdoor environment of kindergartens has attracted more and more attention of educators and designers. As a supplement to the indoor environment, the traditional outdoor environment of kindergartens can no longer meet the needs of children's social development. At present, the research on children's social development in China has attracted the attention of scholars in various fields, especially in the field of psychology. However, the existing research results have no clear pertinence, especially in the field of design, there is a close relationship between children's social development and kindergarten outdoor environment design. The evaluation system of outdoor environment design of kindergartens to promote children's social development is even less [1].

At the same time, designers and managers of outdoor environment lack effective environmental safety assessment tools to guide the correction and improvement of

G. Ting (✉) · Jiangxi
Jiangxi Vocational and Technical College of Industry and Trade, Nanchang, China
e-mail: jxgmgaoting@163.com

S. Patnaik and F. Paas (eds.), *Recent Trends in Educational Technology and Administration*, Learning and Analytics in Intelligent Systems 31,
https://doi.org/10.1007/978-3-031-29016-9_13

space. The traditional interview method, observation method, questionnaire method and case method are often used to evaluate and analyze the injury cases that have occurred in the outdoor space environment, which has strong lag and subjectivity. The research results are lack of generalization, and it is difficult to form a universal principle to guide the design of similar space environment. To solve these problems, in recent years, some scholars have tried to introduce modern information technology into the research of outdoor space environmental safety assessment, such as computer simulation technology, ultra wide-band positioning technology, gravity sensor technology and so on. However, computer simulation technology is still difficult to fully simulate the actual situation, and its calculation results are difficult to be used for safety evaluation. In addition, ultra wide-band positioning technology and gravity sensor technology that require children to carry sensors will directly affect children's normal movement and weaken the credibility of experimental results. At the same time, due to the high cost investment of sensors, the experiment cannot be carried out on a large scale, resulting in insufficient samples for safety detection, Therefore, it is difficult to promote [2, 3].

This paper proposes a kindergarten outdoor environmental evaluation method based on artificial neural network. This method is based on the image data of space moving targets, obtains the data required for the evaluation of space environment through artificial neural network calculation, and carries out analysis of space environment.

2 Establishment of Evaluation System for Outdoor Environmental Design of Kindergartens

1. Design premise

Designers need to understand the overall idea and specific content of the kindergarten education plan, so that the kindergarten outdoor environment design can meet the curriculum plan of specific educational objectives, and achieve specific analysis of specific problems; Attention must be paid to children's main activity preferences, and consideration should be given to encouraging children's active and exploratory activities under the minimum amount of supervision and control of preschool education; In the design, the height difference, slope, spatial scale and other factors of the terrain in the outdoor field of the kindergarten need to be integrated into the design, and these factors should be suitable for the expected type of activities of children [4].

2. Site characteristics

All parts of the outdoor space of kindergartens should be able to be used by children to the greatest extent; In case of bad weather, there should be enough shelter space to provide children with a certain flexible place for activities; Outdoor activity areas should be divided into outdoor venues according to children's ages. At the same time, the division should be clear. The separators in the divided areas should not only effectively separate the potential conflicts in children's activities, but also make children pay attention to the activities in adjacent areas and allow mutual interaction; The outdoor environment shall be provided with an open space without any equipment and equipment for children to carry out activities such as chasing and fighting.

3. Device based environment

In the device based environment, traditional devices popular with children are essential. In order to adapt to the change of children's use mode in different periods, the devices should be composed of different components rather than a single huge integrated device. The selected devices must meet the relevant safety standards and be maintained regularly; In order to prevent children from being injured, the swing should be in the form of sling or tire; Elastic materials shall be laid under all instruments and around the whole falling area, which must comply with relevant standards [5].

4. Environment based on scene design

In the outdoor environment of kindergarten, we should provide sufficient environment based on scene design, and we must consider whether they are suitable for the growth needs of children of the corresponding age; If a sand area is provided for children, the sand area shall be provided with local shade and clear edges, which can also be used as seats, and the sand area shall be at a certain distance from the road and building entrance to prevent children from bringing the sand stuck to their bodies into other areas or classrooms; If a water source is provided for children, there should be both still and moving water if conditions permit.

5. Children's physical and psychological development

   (a) Physiological development.
   In the design of outdoor environment, corresponding activities should be provided for children's physiological development, such as a large number of opportunities to climb, slide and swing around; The design of stair steps, door handles, drinking fountains, isolation, etc. should consider the physiological scale of children.
   (b) Hands-on activities.
   It is necessary to provide children with hands-on elements, and children can promote their cognitive and emotional development by operating and controlling environmental elements by themselves [6].
   (c) Contact
   In the outdoor environment of Kindergarten Based on scene design, children should be provided with enough observation space, the distance between the observation point and most activity areas should be appropriate, and the field of vision should be good; Children should be provided with space for fantasy games, performance activities and social activities; The contact of each activity area should provide an opportunity for children to inadvertently join other activity groups; The equipment shall have a variety of choices and be equipped with enough moving parts, so that children with difficulty in

completing the equipment activities can exit the game with dignity and provide conditions for joining other activity groups at any time.

According to the above analysis, the evaluation index system of outdoor environment of Kindergarten is constructed, which is listed in Table 1.

Construction of Probabilistic Artificial Neural Network Model

The probability neural network is used to evaluate the outdoor environment of Kindergarten. The probabilistic neural network (PNN) is a branch of radial basis function network, which belongs to feed-forward neural network. It is a probabilistic neural network with four layer structure, including input layer, hidden layer, summation layer and output layer. The network structure of PNN model is shown in Fig. 1 [7].

The independent variable $X = (X_1, X_2, \cdots, X_n)$ composes n dimensional vector, the first layer of PNN is input layer, which can receive preprocessed sample data, and transmit it to hidden layer, and the number of elements is n.

The second layer of PNN is hidden layer, which is also named as mode layer, which connects with input layer through full connection, the Gaussian function is used as basic function between input layer and hidden layer, and the number of elements is equal to number of training samples. And the output of hidden layer is calculated by the following procedure:

Table 1 Evaluation index system of outdoor environment of Kindergarten

First grade evaluation index	Second grade evaluation index
Design premise (I1)	Meet course plan (I11)
	Concern course activity preferences of children (I12)
	Suitable for expected activity type of children (I13)
Site characteristics (I2)	Sufficient shelter space (I21)
	Divide outdoor venue according to age of children (I22)
	Open space that does not contain instruments and equipment (I23)
Device based environment (I3)	Adapt to changes of use pattern for children in different period (I31)
	Comply with relevant safety standards (I32)
	Regular maintenance (I33)
Environment based on scene design (I4)	Suitable for growth needs of children age (I41)
	Sand area with local shading and clear edge (I42)
	Water sources including still water and moving water (I43)
Children's physical and psychological development (I5)	Physiological development (I51)
	Hands-on activities (I52)
	Contact (I53)

Fig. 1 Diagram of PNN

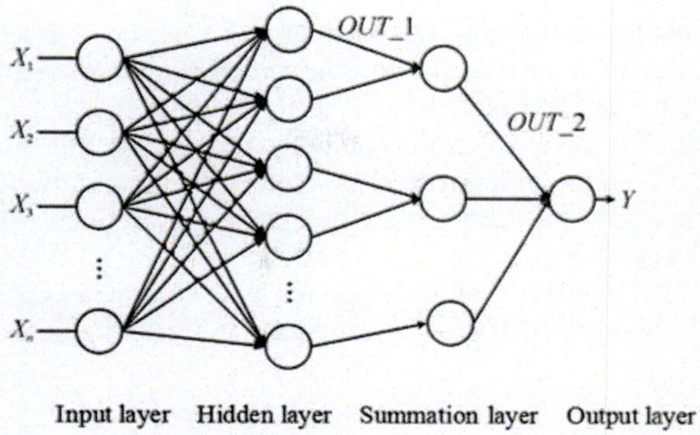

(1) Calculate the Euclidean distance between each input sample and the training sample center;
(2) The Gaussian function is used to calculate the similarity between the input sample and the sample center, as shown in Eq. (1) [8]

$$OUT\_1 = e^{-\frac{1}{2\sigma^2}\|X-c_i\|^2} \tag{1}$$

where X is the sample needing to be predicted, c_i is the center of training samples, σ is the smoothing factor, $0 < \sigma < 1$.

(3) The third layer is the summation layer. Different from BP, RBF and other neural networks, the PNN hidden layer is not fully connected to the summation layer, but only connected with neurons belonging to the same category. The number of neurons is the same as the number of categories of the sample data set. In this paper, that is, the outdoor environment levels of five kindergartens, and the output of the summation layer is the weighted average value of the output of the hidden layer, as shown in formula (2)

$$OUT\_2 = \frac{1}{m_i(2\pi)^{\frac{n}{2}}\sigma^n}\sum_{j=1}^{m_i} OUT\_1_j \tag{2}$$

where m_i is the number of the same category samples.

The fourth layer is the output layer, that is, the final category judgment of the output sample. Its judgment method is the maximum value of the output of the output summation layer as the final output category, as shown in Eq. (3)

$$Y = \max\{OUT\_2_i\} \tag{3}$$

Smoothing factor σ in probabilistic neural networks is one of its important adjustable parameters. Choosing an appropriate smoothing factor is the key to PNN model recognition. Based on this, this paper proposes to use the adaptive dynamic correction model to correct the smoothing factor in real time according to the error of the classification results, so as to establish an adaptive probabilistic neural network model and evaluate the outdoor environment of the kindergarten.

In order to improve evaluation efficiency and performance of probability neural network, the genetic algorithm is used to optimize parameters of probability neural network.

Number of samples is defined by N, y_i denotes output value of probability neural network, $\tilde{y}_i$ denotes real value, and the error is calculated by

$$E(y) = \frac{1}{N}\sum_{i=1}^{N}(y_i - \tilde{y}_i)^2 \tag{4}$$

The fitness function is calculated by

$$f(y) = \frac{1}{1 + E(y)} \tag{5}$$

When population of genetic algorithm is constructed, probability of individual i being selected for evolution is calculated by

$$P_i = \frac{f_i}{\sum_{i=1}^{N} f_i} \tag{6}$$

where f_i denotes individual fitness degree value.

Let two individuals x_A^t and x_B^t cross process at time t to get x_A^{t+1} and x_B^{t+1}, corresponding expression is listed as follows:

$$\begin{cases} x_A^{t+1} = \alpha x_B^t + (1 - \alpha)x_A^t \\ x_B^{t+1} = \alpha x_A^t + (1 - \alpha)x_B^t \end{cases} \tag{7}$$

Individual x_K is mutated to get x_k' that is expressed by

$$x_k' = W_{\min}^k + \gamma(W_{\max}^k - W_{\min}^k) \tag{8}$$

where γ is random number between 0 and 1.

Crossover and mutation probabilities are calculated by

$$P_c = \begin{cases} P_{c\max} - \dfrac{(P_{c\max} - P_{c\min})(f_{\max} - f')}{f_{\min} - f_{mean}} & f' \geq f_{mean} \\ P_{c\max} & f' < f_{mean} \end{cases} \tag{9}$$

$$P_m = \begin{cases} P_{m\max} - \dfrac{(P_{m\max} - P_{m\min})(f_{\max} - f')}{f_{\min} - f_{mean}} & f \geq f_{mean} \\ P_{m\max} & f < f_{mean} \end{cases} \tag{10}$$

where $P_{c\min} = 0$, $P_{c\max} = 0.9$, $P_{m\max} = 0.01$, $P_{m\max} = 0.1$. f_{mean} denotes fitness degree mean value of all individuals, f' denotes crossover fitness degree, f denotes mutation fitness degree.

3 Outdoor Environment Evaluation Procession of Kindergarten

The outdoor environment evaluation procession of kindergarten includes training and state prediction.

In procession of model training, the algorithm procedure is listed as follows [9.10]:

Step 1: Input the training sample data set, where the outdoor environment evaluation index of kindergarten is defined by $X = (X_1, X_2, \cdots, X_n)$, the health status level corresponding to the sample is represented by label y, and the sample size is m.

Step 2: Preprocess the sample data. The dimensionless process the index data to obtain new data based on expression (4). In the formula, max and min represent the maximum and minimum possible values of the evaluation index X_i respectively.

$$X_i' = \frac{X_i - \min}{\max - \min} \tag{11}$$

Step 3: Initialize the smoothing factor. For samples belonging to the same category of hidden layers, the same smoothing factor is used as the initial smoothing factor, i.e. Initializes the smoothing factor. For samples belonging to the same category of hidden layers, the same smoothing factor is used as the initial smoothing factor, i.e. $\sigma_1 = \sigma_2 = \cdots = \sigma_5 = \sigma$.

Step 4: Training PNN network model. Give the smoothing factor to the PNN network, start training, and calculate the hidden layer output, output layer output and error rate.

Step 5: Modify the smoothing factor and training threshold.

Step 6: Judge the error rate and the given threshold value. If the error rate is less than given the threshold, the PNN network parameters are output and enter the evaluation test stage. If the error rate is greater than given threshold, transfer to step 7.

Step 7: Compare this training error with the last training error. If the former is greater than the latter, turn to step 8; Otherwise, increase the smoothing factor.

Step 8: Calculate and compare the difference between this training error and the last training error. If it exceeds 1.03 times of the last training error, enter the training of the next sample; Otherwise, reduce the smoothing factor.

Step 9: Continue training the network until the conditions are met.

In the evaluation and prediction stage, the prediction process is described as follows:

Step 1: Enter the test sample index data.

Step 2: Carry out data preprocessing. The dimensionless operation of formula (4) is used to preprocess the index data to obtain new data.

Step 3: The trained PNN network is used to predict and evaluate the outdoor environment of kindergarten, and the evaluation value is obtained and output.

4 Case Study

In order to verify the effectiveness of the proposed method, thirty kindergartens are selected to carry out outdoor environment evaluation. The twenty kindergartens are used as training samples, and the other ten kindergartens are used as testing samples.

The outdoor environment level of kindergarten is divided into five grades, which are excellent (evaluation value ranges from 0.85 to 1), good (evaluation value ranges from 0.70 to 0.85), medium (evaluation value ranges from 0.65 to 0.70), poor (evaluation value ranges from 0.45 to 65), bad (evaluation value ranges from 0 to 0.45).

Parameters of PNN are set as follows: number of nodes in input layer is five, the number of nodes in hidden layer is 14, an the number of nodes in output layer is 5. The training times is 5000, and the excitation function uses Gaussian function, the initial smoothing factor is 0.04. The BP neural network (BPNN) and RFB neural network (RBFNN) are also used to evaluate the same training samples. The simulation results are listed in Table 2.

As seen from Table 2, the convergence time of PNN is less than that of other two models, and the mean square error of PNN is also lest among the three models. Therefore the proposed PNN has quicker convergence speed and less mean square error.

The correctness rate of different methods is listed in Table 3. As seen from Table 3, correctness of PNN is highest among three methods, therefore it has highest precision.

Table 2 Comparison results of different models

Method	Convergence time/s	Mean square error/%
BPNN	0.185	6.53
RBFNN	0.086	5.47
PNN	0.054	3.52

Table 3 Comparison results of evaluation correctness

Method	Evaluation correctness/%
BPNN	87.4
RBFNN	92.1
PNN	96.8

Table 4 Evaluation results of testing samples based on trained PNN

Number of sample	Evaluation score	Outdoor environment grade
1	0.55	Poor
2	0.76	Good
3	0.77	Good
4	0.66	Medium
5	0.84	Good
6	0.48	Poor
7	0.88	Excellent
8	0.69	Medium
9	0.64	Poor
10	0.57	Poor

The trained PNN is used to carry out outdoor environment evaluation for test samples, and the evaluation results are listed in Table 4.

As seen from Table 3, the outdoor environment grade of kindergarten can be effectively obtained, and the seventh kindergarten has best outdoor environment. The main reasons for that is the playground of this kindergarten place large children's toys, small children's climbing frames and children's educational toys. At the same time, it contains sand pits. In various forms of game activities, children's attention will be continuously improved, so as to make the brain think positively. This kindergarten also has swimming pool and children's maze, and the water storage capacity of the swimming pool does not exceed 0.3 m. This kindergarten has good greening, which is conducive to the growth and development of children.

5 Conclusions

The outdoor environment of the kindergarten not only represents the image of the kindergarten. Children are at an age when they like games, outdoor activities and are curious about new things. A good outdoor environment for kindergartens not only meets their physical and psychological needs, but also provides a good environment for their healthy development. The constructed PNN is complete and accurate, which can truly reflect the actual situation of the outdoor environment of the kindergarten, find the weaknesses of the outdoor environment of the kindergarten, and provide a new tool for the evaluation of the outdoor environment of the kindergarten, with accuracy, objectivity and economy.

References

1. S. Guak, K. Kim, W. Yang, S. Won, H. Lee, K. Lee, Prediction models using outdoor environmental data for real-time PM10 concentrations in daycare centers, kindergartens, and elementary schools. Build. Environ. **187**(1), 107371 (2021)
2. F. Mirkhond, Cheginia Abbas Norouzian, Baghania Mohammad Sadegh, Hassanvandaj Armin, Indoor and outdoor airborne bacterial and fungal air quality in kindergartens: Seasonal distribution, genera, levels, and factors influencing their concentration. Build. Environ. **175**(5), 106690 (2020)
3. V. E. Vitiello, T. Nguyen, E. Ruzek, R. C. Pianta, J. V. Whittaker, Differences between pre-k and kindergarten contexts and achievement across the kindergarten transition. J. Appl. Dev. Psychol., **80**(5), 101396 (2022)
4. L. Wang, W. Zaixing, M. Gong, X. Ying, Y. Zhang, Non-dietary exposure to phthalates for pre-school children in kindergarten in Beijing. China, Build. Environ. **167**(1), 106438 (2020)
5. I. Skalstad, E. Munkebye, How to support young children's interest development during exploratory natural science activities in outdoor environments. Teach. Teach. Educ. **114**(6), 103687 (2022)
6. K. L. Anderson, K. T. Nesbitt, N. A. Sheeks, A. Vrabec, K. Borise, W. M. Fuhse, Executive function mediates the relationship between Conscious Discipline fidelity and kindergarten readiness, J. Appl. Dev. Psychol., **79**(3), 101393 (2022)
7. Y. Yang, T. Wang, J.P. Woolar, W. Xiang, Guaranteed approximation error estimation of neural networks and model modification. Neural Netw. **151**(7), 61–69 (2022)
8. W. Jinlong, P. Wenjie, B. Yongjie, Y. Yuxing, C. Chen, VHCF evaluation with BP neural network for centrifugal impeller material affected by internal inclusion and GBF region. Neural Netw., Eng. Fail. Anal. **136**(6), 106193 (2022)
9. X. Zhao, X. Liu, Y. Xing, L. Wang, Y. Wang, Evaluation of water quality using a Takagi-Sugeno fuzzy neural network and determination of heavy metal pollution index in a typical site upstream of the Yellow River. Environ. Res. **211**(8), 113058 (2022)
10. Y. Hu, J. Su, Research on credit risk evaluation of commercial banks based on artificial neural network model. Procedia Comput. Sci., **199**(1), 1168-1176 (2022)

Exploration and Practice of First-class Specialty Construction of MDMA Under the Background of "Double 10000 Plan"

Xifeng Liang and Weihong Sun

Abstract The first-class specialty construction is an important way to improve the quality of talent training in colleges and universities. Under the background of "New Engineering" and the guidance of "double 10000 plan", The major of Mechanical Design, Manufacturing and Automation (MDMA) has conducted some explores and practices on the first-class specialty construction. We have conducted some researches from the aspects of training objectives, training programs, teaching staff, curriculum system, teaching methods and means, etc. We have made statistical analysis on students' achievement of graduation requirements through direct evaluation methods and indirect evaluation methods. The results showed that all students' achievement evaluation values are greater than 0.7, which met the graduation requirements of MDMA major. At the same time, 100% of the students in MDMA have innovative practice experience during their college years and won many awards in various professional competitions every year. After graduation, they can perform well in professional competitiveness and ability, etc. The research results can provide reference for other first-class specialty construction.

Keywords First-class specialty construction · Student training · OBE · High engineering education · Cultivation mode · Achievement evaluation analysis

1 Introduction

The construction of majors is the foundation of the ability cultivation for a student in colleges and universities. In order to improve the talent training ability of colleges and universities and realize the connotative development of higher education, the Ministry of Education of China implemented the "Double 10000 Plan" for the construction of first-class undergraduate majors in April 2019 in China [1]. The plan is to build 10

X. Liang (✉) · W. Sun
China Jiliang University, Hangzhou 310018, Zhejiang, China
e-mail: lxfcjlu@163.com

S. Patnaik and F. Paas (eds.), *Recent Trends in Educational Technology and Administration*, Learning and Analytics in Intelligent Systems 31,
https://doi.org/10.1007/978-3-031-29016-9_14

000 national first-class undergraduate majors and 10 000 provincial first-class undergraduate majors [2]. The plan would be implemented in "two steps". The first step is that the submitted specialty is determined as the national first-class undergraduate specialty construction point. Then the Ministry of Education of China will organize professional certification and re determines it as the national first-class undergraduate specialty in 3 years [3]. So the national first-class undergraduate specialty construction point is focus on construction [4]. The major of MDMA in our university has been determined as a national first-class undergraduate major construction point. At present, there are still some problems to be solved in all aspects of talent training. For example, the training objectives lack the guidance of modern advanced teaching concepts. The content of the training program lacks innovation and characteristics. The part of courses system is deviate from social needs. The teaching mode and method are relatively traditional. The teaching objectives of the course are not enough to support the graduation requirements, and the assessment methods of most courses are relatively simple. In addition, the university teachers have great pressure on scientific research and attach importance to scientific research rather than teaching, etc. Therefore, it is very important to strengthen the construction of first-class undergraduate majors and improve the quality of talent training in combination with the orientation and objectives of talent training of the university.

2 Research Status and Analysis

Major is the basic unit of talent training for high-level undergraduate education [5]. In recent years, Harvard University launched "general education reform", MIT launched "new engineering education transformation", Duke University implemented "interdisciplinary teaching strategy", Stanford University released "Stanford 2025 plan", etc. Under the guidance "Double 10000 Plan" specialty construction strategy of the Ministry of Education of China, many specialties in colleges and universities are actively exploring the strategies and paths of first-class specialty construction. Cheng et al. [6] designed teaching activities reversely based on the Guidance of OBE and determined the training objectives, clarified the graduation requirements, optimized the curriculum structure and revised the curriculum outline, etc. For example, Zhang et al. [7] practiced the training mode for transportation engineering majors by introducing the experimental teaching class system and the team tutor system. Zhang et al. Jiang [8] expounded that it should take the road of connotative development during building a first-class major from the aspects of building a high-level teacher team, strengthening the awareness of standards and quality etc. Liang [9] has studied on the cultivation mode for Chinese and American higher education of engineering talents from aspects of training objectives, curriculum, classroom teaching mode, etc. Ma et al. [10] have explored the cultivation of innovation ability of engineering college students in the construction of first-class Majors.

It can be seen that many colleges and universities have carried out relevant research and exploration on the construction strategies and plans of majors and have achieved certain research results. However, first-class major construction is long-term. There are still many problems as follows: (1) In terms of training programs, the design of the training objectives of undergraduate majors in colleges and universities lacks the guidance of modern educational concepts, which mainly emphasis on theoretical knowledge learning and exist differences with the current social development needs. (2) In the curriculum system of some majors, the proportion of compulsory courses is too high while the proportion of elective courses is low, the proportion of theoretical courses is too high while that of practical courses is low, etc. (3) The curriculum outline is not standardized and the teaching objectives of the curriculum are not specific. In addition, course assessment is still "emphasis on summative assessment and pay little attention on process assessment" and mainly based on examinations, which cannot support the graduation requirements effectively. (4) With the rapid development of high-tech information technology such as artificial intelligence and big data, traditional majors such as machinery are not attractive to excellent students. (5) The diversified classroom organization modes, such as "student-centered and teacher led" mixed teaching and flipped classroom, need to be further strengthened.

As a national first-class undergraduate major construction point, combining with the orientation and characteristics of our university, we will conduct professional construction research, discussion and practice from the aspects of training objectives, training programs, teaching team, curriculum system, teaching methods and means, and achievement cultivation, etc.

3 Basic Idea for First-Class Specialty Construction

Based on the idea of "demand oriented, goal oriented and capability oriented" of OBE, according to standard of engineering professional certification and new engineering construction, our major is dedicated to training professionals who focus on equipment manufacturing, integrate new technologies, new materials and new processes, and highlight the characteristics of standard and testing. In order to meet the talent training requirements of "double 10 000 plan", we will revise the training plan and curriculum system, innovate the talent-training mode firstly. Then we are to build an excellent teacher team and deepen the classroom teaching reform. At the same time, we will strengthen the application of Internet + and information technology during teaching. In addition, we will optimize the experimental training base and promote the integration of production and education. Our Basic ideas for first-class major construction is shown in Fig. 1.

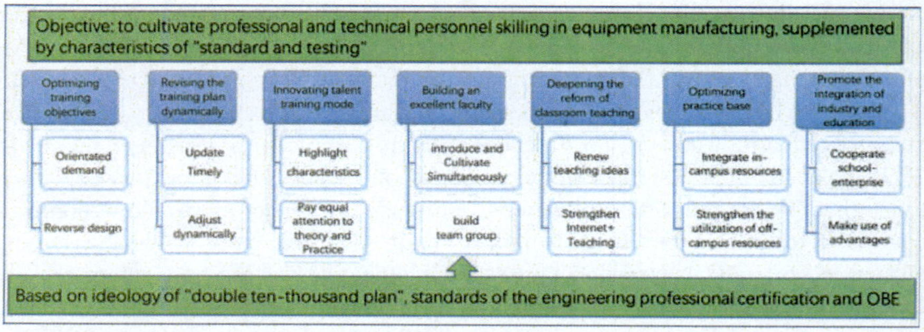

Fig. 1 Basic ideas for first-class major construction of our major

4 Exploration and Practice of MDMA Construction

4.1 Optimize the Training Objectives and Strengthen the Characteristics of Talent Training

In order to solve the positioning and characteristics of personnel training and increase talent competitiveness, we have conducted a large number of research and analysis on graduates, employers and industry experts, etc. We determined the orientation of talent training relying on the industrial characteristics of "measurement, standards and testing" of our university. The MDMA aims to train application-oriented engineering and technical personnel with the professional characteristics of "standard and inspection", and be able to engage in product design and manufacturing, research and development, detection and control and production management, etc. At the same time, we determined the ability goals of five years after graduation from knowledge and skills, professional quality, communication and cooperation, lifelong study as follows. In knowledge and skills, the graduates can effectively apply professional knowledge and technology principles, and have the ability to follow technological development and apply new knowledge to solve practical complex engineering problems in the field of mechanical engineering. In professional quality, graduates have the ability of communication and management, and can play the backbone or leadership role in the work team. In professional ethics, the graduates have humanities and social science literacy, professional ethics, social responsibility and innovation consciousness, and have the willingness and ability to serve the society. In lifelong study, they can competent for job responsibilities, able to increase knowledge and improve ability through continuing education or other means. In addition, we have establish a continuous improvement mechanism for training objectives to adapt to engineering certification.

4.2 *Optimize the Training Program and Curriculum System Dynamically*

Talent training program is one of the core contents of major construction. We adopt the dynamic adjustment mechanism to optimize the cultivation scheme every year combined with the needs of social development. We have built a three-dimensional training program of "ideological and political education + general education + professional education + innovation and entrepreneurship". The modular curriculum system fully embodies "Internet + manufacturing". At the same time, in view of the current social development trend, we introduced the course of artificial intelligence, intelligent manufacturing and intelligent detection and control, etc. into the training program to adapt the talent demand under the background of "intelligence + ".

4.3 *Innovate the Mode of Talents Training of MDMA*

Based on the OBE theory, MDMA refine the training characteristics of the major and have constructed the training mode of "basic course module + specialty characteristic module" relying on the background of the manufacturing industry. The Innovative talent training mode focuses on "One center, two directions, three characteristics, four practice platforms and five course groups", which is shown in Fig. 2. "One center" refers to "mechanics" as the center. The two professional directions is "digital design" and "intelligent manufacturing". Three professional characteristics means "manufacturing, standards and testing". Four practice platforms includes the laboratory in university, industry, enterprise and government practice basement. Five course groups consist of "modern design", "advanced manufacturing", "equipment and control", "product testing and quality control" and "engineering innovation practice". At the same time, we attach equal importance to theory and practice, guide innovation and entrepreneurship, and strengthen the ability of professional talents to serve the society and the industry. The MDMA major training mode is benefit for improving personnel quality, promoting teacher development, highlighting professional characteristics, increasing academic level and social recognition.

4.4 *Build an Excellent Teaching Team*

Teachers are the main body of undergraduate education in colleges and universities to building a first-class specialty. It is very important to improve the professional teaching level and teaching ability of teachers for the long-term development of the specialty. The teaching team of MDMA includes in-school teachers and out of

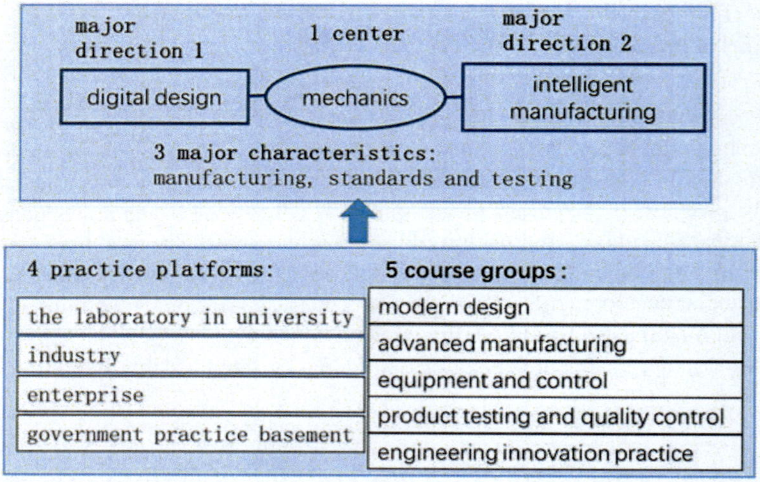

Fig. 2 The mode of talents training of MDMA

schoolteachers such as industrial enterprises. The scheme of teacher team construction of MDMA is shown in Fig. 3. Teachers in our university adopt both introduction and education. MDMA introduces leading talents and excellent doctors in the field of machinery from well-known universities at home and abroad every year. At the same time, to meet the needs of internationalization, characteristics and engineering of teachers, we encourage and subsidize teachers to visit universities at home and abroad, quality inspection departments and well-known enterprises for taking temporary training. For young teachers, we set up a teaching supervision group and a young teacher tutor system to improve the teaching level and ability of teachers. Teaching observation activities and young teachers' ability improvement activities are held every semester. Major construction working group in MDMA responsible for specialty development planning, decision-making, organization and management. We have built five course-teaching teams and carry out teaching seminars regularly. Teachers from well-known enterprises mainly provide practical course and graduation design guidance for students.

4.5 Deepen the Reform of Classroom Teaching

The realization of specialty training objectives and graduation requirements will finally be implemented in specific courses. Therefore, curriculum construction and teaching reform are the key and direct elements of first-class professional construction. Under the background of "Internet + ", we promote education and teaching reform from the aspects of teaching syllabus, teaching content, teaching methods and means as well as assessment and evaluation comprehensively. The specific teaching

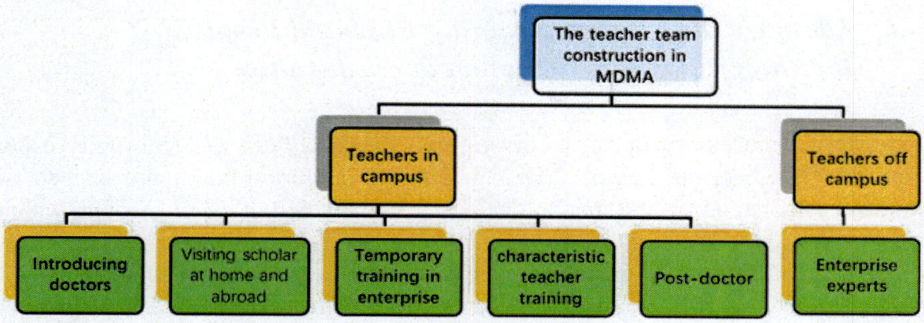

Fig. 3 The scheme of teacher team construction of MDMA

reform and construction ideas of MDMA are shown in Fig. 4. Firstly, it is important to guarantee all of teachers in the major MDMA can deeply understand the essence of OBE teaching theory and the requirements of professional certification standards through lectures and seminars held in every semester. According to the concept of OBE, we have revised the teaching syllabus of every course and optimized the course objectives to support the graduation requirements in the certification standards of the engineering education very well. We also discuss to optimize teaching content in the course group regularly. In teaching method and means, we make full use of internet resources and modern information technology to implement MOOC, Micro class, flipped class, rain class, etc. At the same time, we adopt the "ladder" first-class curriculum construction scheme to carry out the construction offline, online, and online and offline mixed courses. The specialty core courses are planned to construct two provincial-level courses and two college-level courses per year to realize the continuous construction of first-class courses.

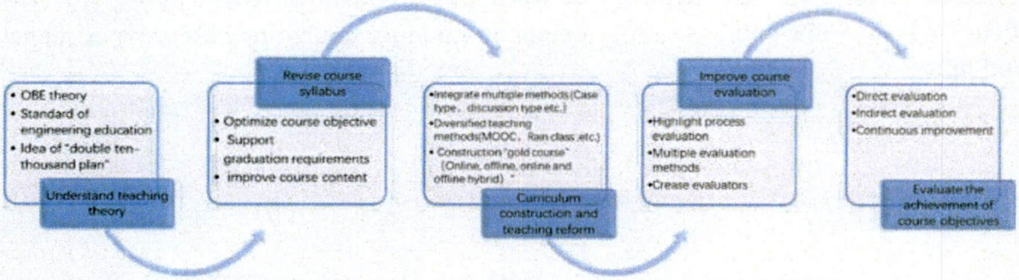

Fig. 4 The specific teaching reform and construction ideas of MDMA

4.6 *Optimize the Practice Training Base and Promote the Integration of Production and Education*

Practice is a necessary link in cultivating talents in college and university especially for engineering majors. Practical teaching can consolidate and deepen an intuitive understanding and theoretical knowledge learned in the classroom, which is an important way to cultivate students' practical and innovative abilities. As an important platform for practical teaching, the laboratory condition is necessary for supporting the construction of first-class major in colleges and universities. Therefore, we strengthen laboratory construction from the top-level planning, the renewal of experimental hardware equipment and software, and the construction high-level experimental teaching teams, etc.

To provide a multi-dimensional practice platform for students' social practice and enterprise practice, we have built four practice platforms and practice bases, namely campus laboratories, detection industry practice base, enterprise practice base and government public training platform. The internal experimental platform mainly strengthens the integration and rational utilization of resources of several majors to meet the requirements of students' basic experiments and some characteristic experiments. In recent years, we pay much attention to the construction of digital manufacturing and professional characteristic laboratories, such as 3D printing lab, small CNC machining lab, etc. The off campus industry laboratory mainly undertakes practical activities such as professional characteristic course experiment, cognitive practice and graduation design, etc. MDMA has established long-term cooperative relations with enterprise practice bases, such as Wahaha precision machinery laboratory, STAUBLI manufacturing laboratory, Geely Automobile Research Institute, etc. It has carried out enterprise practice for regular class and excellent engineer class, and promoted the integration of production and education. In addition, MDMA also make full use of the free experimental practice platform provided by the government of Hangzhou public training center to conduct course teaching experiments and practice.

5 Construction Achievements Analysis

In order to test the training effect of students, the major of MDMA adopts the direct evaluation method and the indirect evaluation method to carry out statistical analysis of the achievement of the graduation requirements of all the graduate students. According to the certification standards for engineering majors, 12 graduation requirements have been formulated for the majors of MDMA. Each graduation requirement is divided into 2–4 index points, and each index point is supported by 3–5 courses. The direct evaluation method, that is, the analysis of course achievement, is to calculate the achievement value of each index point through the achievement value of the course objectives, and then obtain the evaluation value of each graduation requirement. The calculation method is as follows:

$$GEi = \min[\Sigma(CE \times w)] \tag{1}$$

where, *GE* is the evaluation value of the ith graduation requirement; CE is Evaluation value of each course for supporting the index point; w is the weight of the course evaluation. If GE is greater than or equal to 0.7, the graduation requirements of the major students are met.

Indirect evaluation method, i.e. questionnaire survey method, is a qualitative analysis of whether the curriculum objectives have been achieved. After the completion of the course, the questionnaire on the achievement of the course objectives is distributed to students, and the survey results are statistically analyzed. The calculation method of the achieved value of the questionnaire method is as follows:

$$QVi = \Sigma(SV \times n)/Total \tag{2}$$

where, QVi is the Questionnaire evaluation value of the ith graduation requirement; SVj is one of student's questionnaire evaluation value; n is the number of questionnaires; Total is the total number of questionnaires.

Take the students graduated in 2021 as an example, the achievement evaluation value of graduation requirements are shown in Table 1.

It can be seen from Table 1 that the achievement value of students' graduation requirements is greater than 0.7 (the lowest achievement value), and the indirect evaluation value is relatively higher than the direct evaluation value.

MDMA passed the engineering education professional certification of the Ministry of Education of China, in 2021, which provide a "pass" with internationally recognized quality standards to the world for students. In recent years, it is reported from Zhejiang Provincial Education Evaluation Institute that graduates' satisfaction degree to the classroom teaching effect of professional courses and the job fitness both ranked third in 50 majors in 2019, while professional competitiveness of the students graduated in 2016 ranked second in 47 majors in our university. Even under the epidemic situation, the employment rate of students is over 95%, and the rate of postgraduate enrollment is over 30%. In the past 3 years, students have won 28 first prizes, 59 s prizes and 79 third prizes in national and provincial science and technology competition for college students. 100% of the students in MDMA have innovative practice experience during their college years. Parts of students' scientific and technological works in the major MDMA are shown in the Fig. 5. Graduates trained in MDMA are employed in Zhejiang manufacturing, industrial testing equipment, intelligent testing, equipment, and other industries.

Table 1 the achievement evaluation value of graduation requirements of students graduated in 2021

Graduation requirements	Statistics of achievement evaluation results(%)	
	the direct evaluation	the indirect evaluation
Graduation requirement 1	0.710	0.781
Graduation requirement 2	0.717	0.782
Graduation requirement 3	0.774	0.782
Graduation requirement 4	0.782	0.770
Graduation requirement 5	0.792	0.786
Graduation requirement 6	0.771	0.8
Graduation requirement 7	0.783	0.806
Graduation requirement 8	0.795	0.814
Graduation requirement 9	0.814	0.812
Graduation requirement 10	0.716	0.818
Graduation requirement 11	0.755	0.786
Graduation requirement 12	0.774	0.842

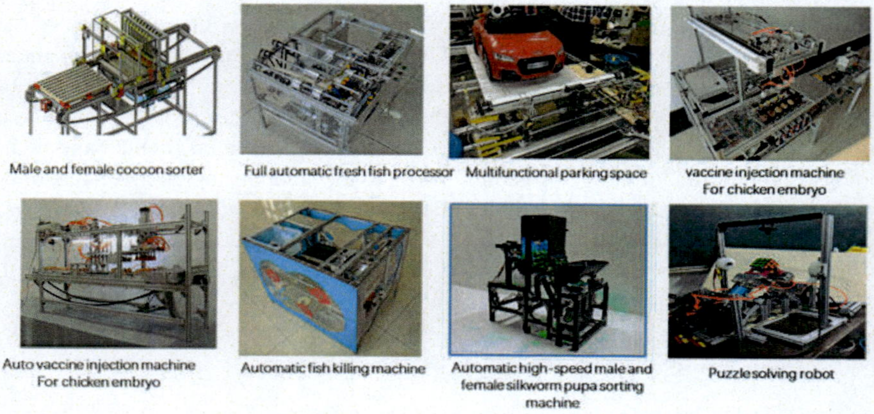

Fig. 5 Part of students' scientific and technological works of the major MDMA

6 Conclusion and Discussion

Under the background of the "intelligence + " era, facing the new situation of the current epidemic, the major of MDMA explores and practices the construction of first-class majors actively in recent years from training pattern, curriculum system, teaching reform, etc. Some achievements have been made in course construction and teaching reform, etc. However, the major construction involve all aspects of "teaching" and "learning", which is a long-term construction and accumulation process from the condensation of teaching ideas to the feedback of students' training effects. Therefore, it requires the unremitting efforts and investment of all departments and all teachers in universities. In the future, we will do further research on mobilizing all teachers' enthusiasm for teaching and promoting teaching through scientific research, etc. We also need to strengthen continuous improvement in training objectives and graduate requirements of the MDMA.

Acknowledgements This paper is supported by Research Project of Educational Science Planning in Zhejiang Province (2021SCG381) and Teaching Research Project of Teaching Guidance Subcommittee of Mechanical Basic Courses in Colleges and Universities of the Ministry of Education of People's Republic of China (2022JXJC-JXSJ-09).

References

1. S. Yikai, The core of first-class curriculum development and teaching is to improve instructors' teaching skills. J. Northwest Polytech. Univ. (Soc. Sci.) **1**, 50–57 (2020)
2. Z. Li, Build First-class undergraduate education and First-class specialty. Chem. Eng. High. Educ. (2), 12-18 (2020)
3. Ministry of Education of People's Republic of China, Implement the "Double 10000 Plan" for the construction of first-class undergraduate majors and vigorously promote the construction of the "Golden specialty." Window of Information **4**, 77–78 (2019)
4. B. Lyu, J. Yan,F. Xiong, Core elements and their combination of the first-class major constructions. Heilongjiang Res. High. Educ. (5), 149–153 (2022)
5. T. Zhang, M. Huang, Y. Sun, et al., Exploration and practice of the construction of first-class major in local comprehensive universities under the OBE Concept. Rev. High. Educ. (2), 19–24, 57 (2021)
6. C. Jiafu, L. Qing, P. Heping et al., Research on Innovative design of First-class Undergraduate Professional Training Program Based on OBE. J. Huaibei Norm. Univ. (Philos. Soc. Sci.) **41**(4), 98–102 (2020)
7. Z. Ye, C. Yanyan, C. Jing et al., Model of transportation engineering talents for the construction of "First-class" Major. Educ. Teach. Forum **39**, 261–262 (2020)
8. J. Zongli, Take the road of connotative development and build a first-class specialty. Chin. Univ. Teach. **8**, 7–13 (2020)
9. R. Liang, A comparative study on the cultivation mode for Chinese and American higher education of engineering talents under the background of double world-class construction—Take MIT and Xi'an Jiaotong University as an example. High Eng. Educ. (5), 1–12 (2022)
10. L. Ma, S. Li, H. Sai, et al., Exploration and practice on the cultivation of innovation ability of engineering college students in the construction of first-class majors. Chem. Eng. & Equip. (9), 294–295,282(2021)

Reform of Talents Training Mode in Computer Science and Technology

Wei Cong, HongYan Li, and Jing Liu

Abstract New engineering is the innovation of talent training which requires a reform of the talent training mode to be realized. According to the requirements of new engineering and the characteristics of the computer science and technology specialty, the ability and quality requirements of the students and the current situation of talent training are analyzed. However, there is a gap between student innovation ability and engineering professional quality and the requirements of industrial talent. According to the ideas of new engineering construction, an integrated talent training model of Learning-Training-Competition is produced. The special features of the new model are the progressive ability training route, a hierarchical and progressive course system, project-driven practical teaching system, and in-depth cooperation between schools and enterprises. The implementation results show that the new talent training model effectively promotes the coordinated development of student knowledge, ability and quality, and has achieved remarkable results in graduate employment and further education.

Keywords New engineering · Talent training mode · Integration of learning-practice-competition · Integrated cultivation of quality knowledge and ability · Application undergraduate colleges · School enterprise cooperation

1 Introduction

Xijing College is a provincial ordinary private undergraduate college. For many years, it has adhered to the mission of serving the regional economy and has cultivated a large number of applied engineering talents. The Computer Science and Technology specialty is one of the key construction majors of Xijing College. In recent years,

W. Cong (✉) · H. Li · J. Liu
School of Computer Science, Xijing University, Xi'an Shanxi 710023, China
e-mail: congweicw@126.com

S. Patnaik and F. Paas (eds.), *Recent Trends in Educational Technology and Administration*, Learning and Analytics in Intelligent Systems 31,
https://doi.org/10.1007/978-3-031-29016-9_15

with the vigorous development of the computer industry, the number of professional enrollments has been increasing year by year.

New engineering is a reform strategy for engineering education proposed in the context of a new technology revolution, the development of new industry and new economy [1]. In the construction of new engineering, different types of colleges and universities undertake different missions. The application undergraduate colleges actively explore new engineering construction models suitable for school characteristics to provide talent support for regional economic growth [2]. Therefore, it is an urgent task to explore the training mode of the Computer Science and Technology specialty.

The new talent training mode of the Computer Science and Technology specialty is designed with ability training as the core. This model realizes the simultaneous cultivation of knowledge, ability and quality through the integration of learning-practice-competition. The practical results show that the competitiveness of graduates in employment and further education has been improved.

2 Reform Background

The new engineering implements the concept of outcome-based education. The outcome means the maximum ability that students can achieve after a certain stage of learning, which represents a kind of ability and quality structure [3]. Therefore, the construction of competency structure is particularly important to achieve learning outcomes.

Numerous literatures put forward the capabilities and qualities of new engineering talents from multiple perspectives [4–9]. However, these views are talent evaluation standards rather than specific implementation plans. So the professional implementation faces the problem of turning concepts into realities [10]. Therefore, we combine the new engineering talent ability and quality standards with the actual practice of specialty, analyze the core competencies of specialty from the perspective of computer technology, and provide a basis for the implementation of the reform of the talent training mode.

2.1 Ability Requirements

When developing a computer system, it is the first work to convert application domain problems into computer domain problems. In the computer field, the work to be done includes architecture design, software and hardware co-design, hardware circuit production and software programming, system joint testing, etc. After the system is delivered, it needs to be optimized and upgraded to make it work normally.

Ability to solve cross-field problems. The application fields of computers are very wide, such as education, medical care, transportation, etc. The process of transforming problems in the application field into problems in the computer field is called requirements analysis, which is the key to the success of system development. The work requires students to be familiar with the work flow of the application field, and to have cross-field knowledge acquisition and analysis capabilities.

Computational thinking ability. Computational thinking is an abstract thinking mode that mainly refers to a series of thinking that covers the breadth of computer science, such as problem differentiation, step-by-step solution, system design, and understanding of human behavior based on the concepts of computer science [11]. Computational thinking has become a basic way to solve problems with computers, which is a necessary ability for students.

System thinking ability. System thinking ability is the ability to solve practical problems with computer software and hardware. It requires students to master the collaborative relationship between computing software and hardware, master the external characteristics and interaction methods, and optimize the overall system through collaborative design of software and hardware. The system thinking ability runs through the process of solving complex engineering problems and it is an ability that students must be cultivated.

Generalizability of knowledge. In the field of machine learning, generalization ability is the ability of the model to adapt to new samples, and the trained model can also give better results for new samples. The generalization ability can be understood as the ability to apply knowledge by drawing inferences from other facts and being able to solve new problems with learned knowledge.

Continuous learning ability. New technologies such as the Internet of Things, artificial intelligence, and big data have made amazing progress, and these technologies are continuing to evolve and lead the development of the industry. Students should have the concept of lifelong learning and strong continuous learning ability to quickly master these new technologies and apply them in practice.

2.2 *Quality Requirements*

The new engineering aims to cultivate engineering innovation talents with strong practical innovation ability, humanistic feelings, and a high sense of social responsibility. Therefore, a very important aspect of ability training under the background of new engineering is to shape the inner world of students and cultivate their humanistic quality through engineering education culture.

Quality of critical thinking. Critical thinking is the basic quality of engineering and technological innovation. To design and develop a new product, students need to use critical thinking to analyze the current problems and needs of the product, and only in this way can innovation be achieved.

Innovative Quality. Awareness of the pursuit of innovation and the courage to participate in the practice of innovation and entrepreneurship. To have creative thinking and skills, including the ability to express, write, and materialize innovation results, and more innovative enthusiasm, and good personality characteristics such as independence and diligence.

Quality of self-awareness. Students can self-reflection and self-transcendence, they must clarify their own abilities, constantly innovate themselves, and make continuous progress. Students dare to surpass their own abilities, dare to create and persevere.

2.3 Current Status of Talent Training Mode

In recent years, the Computer Science and Technology specialty has continuously strengthened the construction of new engineering, revised the talent training plan according to the OBE concept, and reconstructed the curriculum system. Some achievements have been made in talent training, but there is still a certain gap compared with the demand for new engineering talents.

First, the coordinated training of quality, knowledge and ability is insufficient, emphasis is placed on theoretical knowledge rather than practical ability training, and theory and practice are loose connection.

Second, the professional ability to practice is weak, the comprehensive innovation ability is slightly insufficient, the ability to innovate and entrepreneurship is insufficient, and the ability to learn lifelong has not been fully formed.

Third, the training method is single, lacking production, education and research promotion, lacking insufficient scientific research feedback and teaching, and weaker innovation and entrepreneurial training.

Therefore, we focus on strengthening ability training and propose an integrated talent training model of Learn-Practice-Competition to achieve the simultaneous growth of knowledge, ability and quality. Specific performance is integrating knowledge learning, practical training and innovative practice with project practice as a carrier. The learning link strengthens the knowledge understanding, and the practice-competition link promotes comprehensive quality training and innovation and entrepreneurship training. In-depth school-enterprise cooperation provides a guarantee for the reform of the talent training mode and improves the competitiveness of graduates in employment, entrepreneurship and further education.

3 Reform Implementation

3.1 Reform Principles

Update the training concept. The new engineering implements an outcome-based educational philosophy [2]. Student-centered, reverse design and continuous improvement run through the entire process of talent training. Student-centered means that teaching content, teaching methods and teaching evaluation revolve around what to learn, how to learn and how to learn. Reverse design means that the training objectives are determined by the needs, the graduation requirements are determined by the training objectives and the curriculum system is determined by the graduation requirements. Continuous improvement is finding problems and improving teaching by summarizing the previous teaching situation. At the same time, these summaries will become the starting point of the next stage of teaching [5].

Reconstruct training objectives. Faced with the needs of engineering development, pay attention to the comprehensive development of knowledge ability and quality [5, 6]. Quality training is necessary to establish a correct world outlook, outlook on life, and values represented by patriotism and dialectical materialism methodology; to cultivate enterprising spirit and truth-seeking spirit; to be diligent, practical, collaborative and sharing. Knowledge training should build a composite knowledge structure which includes the professional basic knowledge, professional technical theoretical knowledge and application knowledge of the discipline, and also have a broad and solid engineering knowledge and extensive mathematical knowledge. Capability training includes specific technical skills, the ability to further tap one's own potential and realize the ability to draw inferences from others through cognitive practice.

Broaden the training approach. Actively promote the synergy of internal and external environment of engineering talent training, and advance the supply of outstanding talents by promoting school-enterprise cooperation; actively promote the integration of teaching and scientific research in the internal environment of colleges and universities; in the professional system, promote the integration of disciplines and disciplines, majors and majors [9].

3.2 Reform General Ideas

The abilities and qualities of students can only be developed through practice. Under the theoretical teaching system is relatively complete, the focus of the reform of the talent training model is to design a practical teaching system that matches the theoretical teaching system and to reconstruct the practical content and

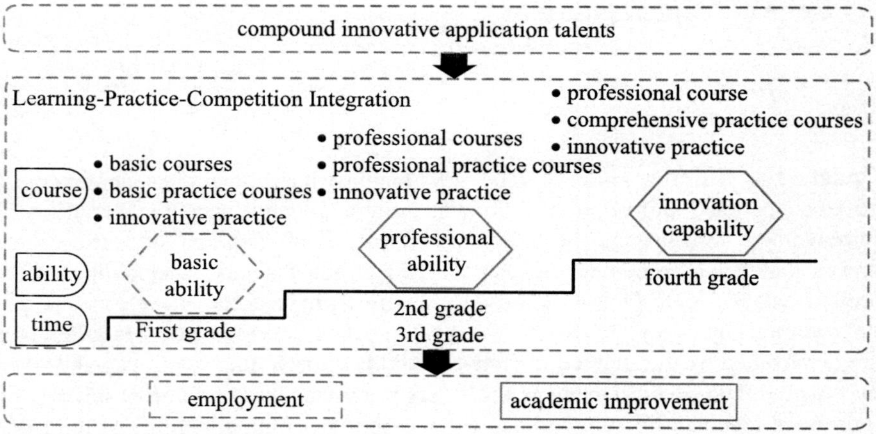

Fig. 1 Reform ideas of the talent training model

implementation methods. Based on the a new integrated talent training model of Learning-Practice-Competition is proposed, and the reform ideas are shown in Fig. 1.

This model takes ability training as the core and divides the ability into three progressive levels: basic ability, professional ability, and innovation ability. The three levels abilities are cultivated in three stages: freshman year, sophomore and junior year, and senior year. These abilities are supported by course teaching. The model provides graduates with two directions of employment and entrepreneurship.

Courses include theoretical courses, practical training courses, and innovative practice. Theoretical courses support the task of learning, while practical training courses and innovative practices focus on training for practice and competition.

Theoretical teaching, practical training teaching, and innovative practice adopt an integrated design to promote and coordinate each other. In the process of Learning-Practice-Competition, the coordinated development of knowledge, ability and quality is promoted.

3.3 Reform Process

Stage of basic ability training. This stage is mainly in the freshman year. Students study advanced mathematics, linear algebra, discrete mathematics, advanced language programming foundation and other theoretical courses, and complete the basic training tasks of programming. When first-year students enter the school, we provide professional education. Professional teachers introduce the professional situation, faculty, teaching and scientific research achievements, and publicize participation to stimulate students' awareness and interest in innovation and entrepreneurship. We organize students to participate in the subject competition team according to their interests and hobbies. Students do some extra work, such as word processing, graphic layout, improve their professional cognition, and cultivate their ability to participate. At the same time, students can take the computer grade examination to enhance their competitiveness in employment.

Stage of professional ability training. Professional ability training are developed in the second and third years of college. Students learn mathematics courses such as probability theory, mathematical statistics, and applied statistics, introductory engineering courses such as computer networks and computer composition principles, as well as data structures, database principles, object-oriented programming, network programming, Python language programming, machine learning, data mining, and other professional courses. It is also necessary to complete computer network practical training, computer composition principle practical training, network application practical training, scientific research training, and other practical training tasks. At the same time, we organize students to form teams to participate in various competitions, including but not limited to National Undergraduate Electronic Design Competition, National Undergraduate Computer Design Competition, Blue Bridge Cup National Programming Competition, China University Intelligent Robot Creative Competition, National Robcom Robot Competition, "Internet +" College Student Innovation and Entrepreneurship Competition, and guide students to apply for software copyrights, apply for utility invention patents, and write scientific papers. We use the design and development of competitions and small projects as the traction to cultivate professional abilities and professional qualities such as collaboration and communication.

Stage of innovation ability training. Students study some theoretical courses including cloud computing, and Java enterprise-level application development. Students attend practical training include intelligent application system practical training, computer system practical training, graduation practice, graduation design, etc. In terms of innovative practical training, students can inherit the achievements and experience of the previous stage, organize and participate in higher-level competitions, such as the ACM-ICPC International College Student Programming Competition, etc., and start from the design and development of medium-scale systems, through the restoration of damaged cultural relics. The training in medical image recognition, plant disease and insect pest identification, and other projects penetrate the three professional directions of "software + art", "software + medicine", and "software + agriculture" and cultivates students' preliminary ability to solve interdisciplinary problems.

3.4 Reform Implementation Method

The integrated talent training model of Learning-Practice-Competition is guided by the completion of specific projects to carry out teaching activities. The project links course experiments, practical training, and innovative practice.

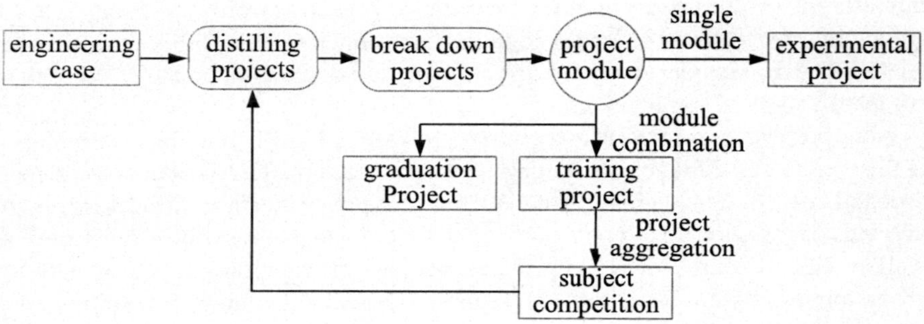

Fig. 2 Practical project setting ideas

Theoretical course experimental projects and practical training projects of the same grade can be used not only as subject competition projects, but also integrate into the curriculum teaching of senior grades, and innovative practice topics can be used as course teaching projects and experimental training projects. In the Learn-Practice-Competition cycle training from lower grades to upper grades, the knowledge chain and the spiral improvement of comprehensive ability are promoted. The fourth-year practical training practice, through wide project design, enables the in-depth synthesis of knowledge within the profession and the ability to solve interdisciplinary professional problems.

Practical projects are the key elements in the Learn-Practice-Competition. It is necessary to select famous engineering cases and to meet the requirements of course experiments, practical training practice and discipline competition in terms of type, function and scale. An idea of setting up a practice project is shown in Fig. 2.

First, we collect and organize engineering cases, extract teaching projects from them, and then decompose the teaching projects to obtain project modules. Second, the function, scale, and difficulty of the project module are reconstructed to be applied to teaching. Finally, the aggregation and optimization of the results of training projects can be used as an alternative for the discipline competition, and the projects in the discipline competition can be used for teaching cases. The practical project settings should pay attention to the following issues.

Engineering cases and teaching projects should meet the needs of different levels of ability training. Engineering cases can come from multiple channels, such as life, society, and enterprises. Typical cases include information management system, web application system, motion recognition, garbage classification, restoration of damaged cultural relics, etc.

Course experimental projects should focus on diversity and cohesion. The experimental project of the introductory course should reflect interest and guide its cognitive ability to solve practical problems. The experimental projects of professional courses should focus on typicality, cover comprehensive classic problems through knowledge points, and consolidate the application ability of professional basic knowledge and technology. Course experiments of the same grade should avoid duplication, and curriculum experiments of different grades should pay attention to the connection. The course experimental project is a part of the course teaching and is mainly based on individual projects.

The training program should focus on hotspots/openness and process standardization. The training program should focus on comprehensiveness, select hot issues/popular applications, cultivate the ability to flexibly apply professional knowledge to solve practical problems, and at the same time broaden students' horizons and understand industry development trends. The training project process is implemented through the analysis, design, development and testing process so that students are familiar with the IT products development process, including writing project documents, defense, etc., to promote the simultaneous growth of students' professional and nonprofessional capabilities [10].

Competitions should focus on the progressive cultivation of abilities. In the first grade, students can take the computer grade examination. Then they can also participate in the subject competition team according to their interests and hobbies, do some extra competition work, improve their professional cognition, and cultivate their ability to participate. In the senior grades, competitions can be organized at two levels (but not limited to). The first level is the school recognized B-type and A-type competitions, such as the Chinese College Student Computer Design Competition, the Blue Bridge Cup National Programming Competition and the Robcom National Robot Competition; The second floor is a national competition with high influence, such as the ACM International College Student Programming Competition and the 'Internet +' College Student Innovation and Entrepreneurship Competition.

3.5 Reform Guarantee

The implementation of practical teaching must give full participation to the key role of the enterprise. Following the collaborative education model of school-enterprise resource integration, industry-university, we jointly build practice bases inside and outside the school, allowing enterprise engineers to participate in all aspects of practical teaching, promoting the connection between professional talent training and job needs.

The company provides practice spaces, and the company engineers teach on-site and participate in practical assessments. Enterprise engineers participate in the preparation of experimental instructions, develop practical training projects, serve as practical training and graduation design instructors, and participate in practical training and graduation design assessments.

So far, we have established a school-enterprise partnership with 12 companies. The proportion of corporate mentors participating in graduation design guidance has increased year by year, reaching 35% in Class 2017 and 50% in Class 2018.

School-enterprise cooperation has been greatly expanded, students are placed in a real environment for training, engineering practice ability has been cultivated, and good training results have been achieved.

4 Reform Effect

The effect of the reform of the talent training model is evaluated from the three aspects of graduate employment, further education and entrepreneurship, and the 2016 students before the reform are selected as the baseline data.

Significantly improved employment competitiveness. The employment rate of graduates in 2017 and 2018 is 98%. The influence of the companies has greatly improved. Several graduates have joined large state-owned companies such as State Grid, China Post, and China Aerospace, and many students have signed contracts with IT companies such as Alibaba, JD.com, and Meituan.

Significantly increased awareness of innovation. Students participated in the National Undergraduate Computer Design Competition, Electronic Design Competition, China Engineering Robot Competition, International Open Competition, and other events and won more than 70 national and provincial awards.

5 Summary

Based on the background of new engineering construction, focus on strengthening ability training, the Learning-Practice-Competition integrated talent training model is proposed, which realizes the simultaneous training of knowledge ability and quality. After the reform of the talent training model, the competitiveness of graduates in terms of employment, entrepreneurship, and further education has been significantly improved.

In response to the current reform feedback results, improvements can be made in the following aspects: expand teaching projects in line with technological development and engineering applications; expand the scope of subject competitions and the number of participants, and strive for everyone to participate and win awards.

References

1. Z. Denghua, Connotations and actions for establishing the emerging engineering education. Res. High. Educ. Eng. **3**, 1–6 (2017)
2. Accreditation standards for Engineering Education, J. Electr. Electron. Educ. **41**(01), 1–4 (2019)

3. J. Shen, T. Li, W. Mingchuan, The new engineering education in China. Procedia Comput. Sci. **172**, 886–895 (2020)
4. J. Wang, Construction and practice of experimental teaching system of new engineering specialty based on CDIO mode. High. Educ. Res. **6**(1), 20–26 (2021)
5. X. Jiang, H. Zhu, Z. Li,A new model of talents training under the background of new engineering education. High. Educ. Dev. Eval., **34**(2), 17–24+103 (2018)
6. D. Wu, Y. Xu, X. Zhang, Z. Tao,Research on the cultivation of college students' innovation and entrepreneurship ability under the background of new engineering, J. High. Educ., **8**(23), 26-29 (2022)
7. Z. Kaifa, Z. Yuzhen, Exploration on the core competence and teaching mode of new engineering education. Chongqing High. Educ. Research. **5**(03), 22–35 (2017)
8. Y. Wang, Y. Ye, Study on quality education in sino-us engineering education accreditation standards. J. High. Educ., **7**(21), 1–4+10 (2021)
9. S. Jiguang, H. Shi, Exploration on the core competence and teaching mode of new engineering. J. High. Educ., **8**(23), 66–69(2022)
10. Y. Xu, Q. Zhang, Appeals,difficulties and choices of new engineering talents training mode reform in era of intelligent manufacturing. Heilongjiang Res. High. Educ., **40**(09),47-52 (2022)
11. F. Hu, Empirical analysis of the effectiveness of new engineering construction based on discipline competition. High. Educ. Forum, **12**, 33–39(2021)
12. L. Zhang, J. He, Y. Lin, J. Wang, Design and practice of project-centered talent training model of computer science and technology. Res. High. Educ. Eng., **5**, 76-81(2021)
13. L. Jinhui, The teaching reform of programming courses based on the output of computational thinking ability. Ind. Control. Comput. **35**(07), 167–168 (2022)

The Impact of Policies on Higher Education Based on Grey Prediction Model

A Case Study of China

SuYi Shi and JingSan Yang

Abstract By comparing the data of recent ten years, we found that compared with developed countries such as Britain and the United States, there is a big gap in the scale and quality of higher education in China, but the gap is gradually narrowing. In response to the extreme inequality in higher education in China, Chinese government has carried out reforms in four aspects. After the reform, while the gap between regions has been significantly narrowed, the overall level of higher education has also been steadily improved. Combining past data and using the Grey Prediction Model, we found steady increases in all indicators in the decade after the policy was implemented. Therefore, the policies proposed by the Chinese government in 2016 have certain sustainability.

Keywords Inequity · Higher education · Grey prediction model · Sustainability · Policy

Higher education, which determines the countries' future, is playing a more and more indispensable role in the contemporary education system and comprehensive development of a country.

Looking around the world from the United State to Japan to China, we can find a variety of national higher education systems that have both advantages and disadvantages. Thus, there are thousands of international students flocking to other countries, such as the United State and the UK, in search of better higher education.

In the wake of the pandemic, the situation varies from country to country and the education changes accordingly. How to measure the effectiveness of higher education policy is an unavoidable problem at present.

S. Shi · J. Yang (✉)
Nanjing University of Finance & Economics, Nanjing, China
e-mail: 9120051045@nufe.edu.cn

S. Patnaik and F. Paas (eds.), *Recent Trends in Educational Technology and Administration*, Learning and Analytics in Intelligent Systems 31,
https://doi.org/10.1007/978-3-031-29016-9_16

1 The General Situation of Higher Education in China

In order to understand Chinese current level of higher education more intuitively and more quickly, we collected a lot of data from the UN website for several typical countries and plotted the following charts from both horizontal and vertical dimension.

Figure 1 indicates the higher education enrollment of China, US, Japan, UK and India from 2004 to 2015. It can be seen that the USA has led the way in college enrollment rate, even more than double those of developing counties such as China and India. Compares with China and Japan, Britain and Japan also have relatively high higher education enrollment rate and Japan's rate is slightly higher than the UK's in recent years. Although China and India has relatively low higher education enrollment rate, enrollment rates in both counties are growing at a high rate. In particular, Chinese higher education enrollment went from being neck and neck with India in 2004 to far ahead in 2015.

We also collected proportion of foreign students of China, Japan, America, Brazil and India from 2000 to 2013. It can be seen intuitively from Fig. 2. that some developing countries, such as China, India and Brazil, almost have proportion of foreign students close to zero. For one thing, it may be because their large population. For another, it also indicates they may have a relatively little appeal to international students. Contrast with these developing counties, some developed counties, like America, Japan and Britain, have higher proportion of foreign students. In particularly, American international proportion has been rising to nearly 18% from 2000 to 2013 (Figs. 3 and 4).

Two figures above show that the percentage of students in tertiary ISCED level 5 and level 8 from different counties. We can easily find that in recent year America has an obvious advantage in the percentage of students in tertiary ISCED level 5 and its percentage of students in tertiary ISCED level 5 also relatively lead other countries. China has a good percentage of students in tertiary ISCED level 5, second only to America while the percentage of students in tertiary ISCED level 8 still remains to be improved.

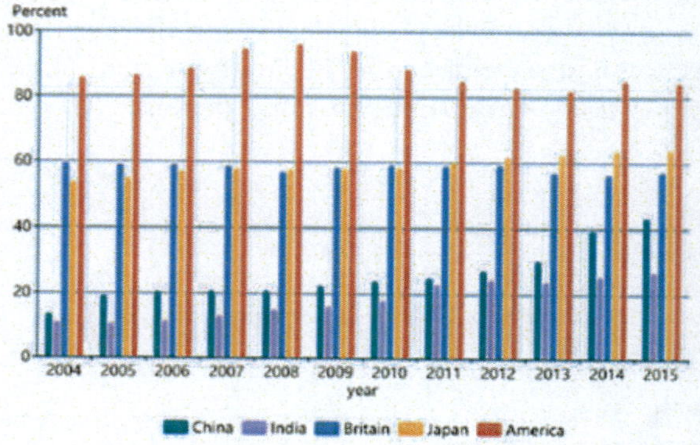

Fig. 1 Higher education enrollment rate

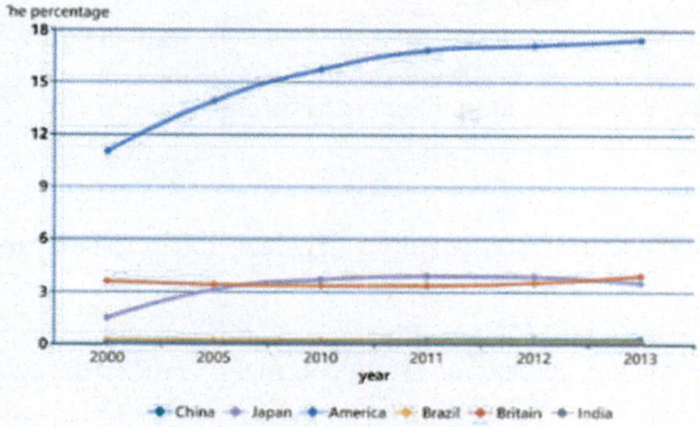

Fig. 2 Foreign students as a percentage of student size (higher education)

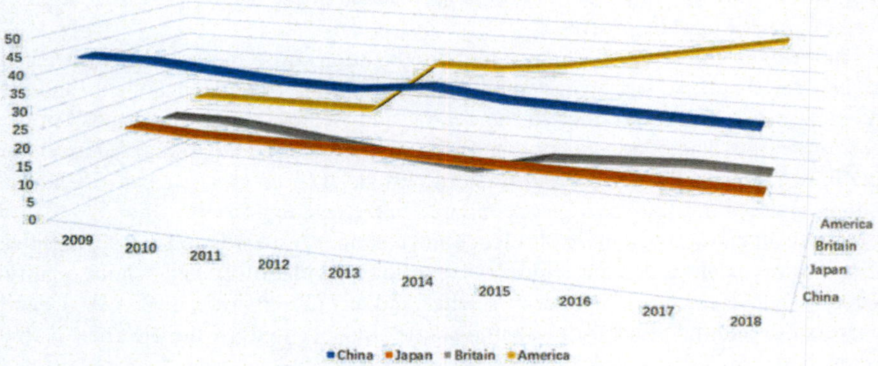

Fig. 3 Percentage of students in tertiary ISCED level 5

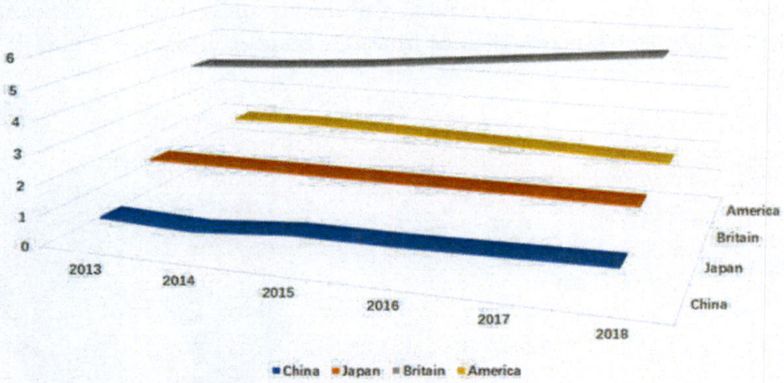

Fig. 4 Percentage of students in tertiary ISCED level 8

According to the above analysis, we can obviously see that compared with developed countries, China has a big gap in both the proportion of domestic higher education enrollment and its attractiveness to foreign countries.

2 The Distribution Situation of Higher Education in China

Compared with American, Chinese higher education enrollment rate, international students' proportion, the percentage of students in tertiary ISCED level 5–8 and other indicators are still relatively low but grow relatively quickly.

However, the geographic distribution of educational resources is really uneven [6]. We divide China into eastern, mid and western regions by geographical region. The chart below is drawn from the official website of the Ministry of Education of The People's Republic of China.

In terms of the scale of Chinese higher education, from the data above, we can clearly see that there is a great gap between the number of college students and the fixed assets of the universities in the central and western regions and those in the eastern region. The fixed assets in the eastern region exceed the total fixed assets of the universities in the central and western regions. That is to say, most of Chinese higher education resources are concentrated in the east region [9].

As we all know, the number of science and technology awards and key universities [5] to some extent reflects a region's scientific research ability and school quality. Therefore, I selected the number of science and technology awards of colleges and universities and the number of key universities in 2015 to draw the pie chart below (Figs. 5 and 6).

In terms of the quality of Chinese higher education [7], from the pie chart above, we can clearly see that in 2015 more than half of three-quarters of the awards were concentrated in the east and more than half of the key universities are concentrated in eastern China, which is nearly three times the number in western China. That is

Fig. 5 Number of awards for scientific and technological achievements in 2015

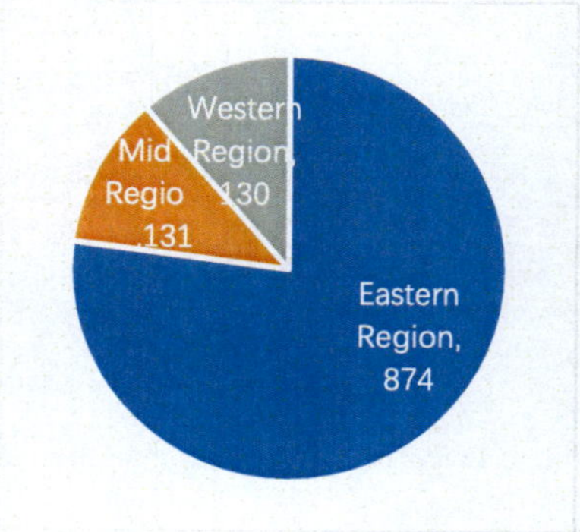

Fig. 6 Total number of key universities in 2015

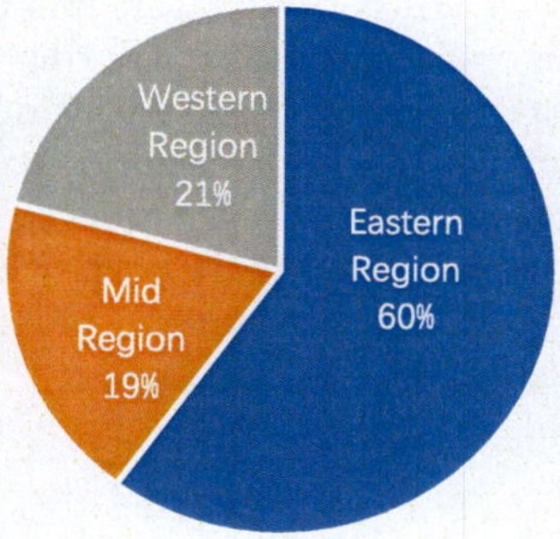

to say, most of China's high-quality higher education resources are concentrated in the east as well.

As we all known, an ideal higher education system should balance overall strength with regional balance. In this aspect, the statistics show that the enrollment rate of higher education, the proportion of overseas students, the proportion of students in tertiary ISCED level 8 and other indicators increase steadily, which means the overall trend of China's higher education is positive.

Therefore, how to further speed up the regional balance of higher education in China is an unavoidable problem at present. There are some policies for our reference.

3 Policies

In 2016, The General Office of the State Council issued a guideline on accelerating the development of education in central and western China in response to the uneven geographical distribution of educational resources [4].

The guideline reforms higher education from four aspects: discipline strength, infrastructure construction, form of aid and main body of aid.

In terms of **discipline strength,** the central government advocates greater emphasis on resource allocation and the introduction of high-level talents, and encourages local governments to support schools with foundations, characteristics and advantages based on actual conditions, so as to establish high-level universities through rational positioning and innovative development.

In terms of **infrastructure construction**, the central government advocates, the central government advocates that on the principle of "make up the blanks", the basic educational facilities and information construction such as basic teaching laboratories, specialized teaching laboratories, comprehensive experimental training centers and libraries should be strengthened, and teaching exp erimental rooms and necessary equipment should be built to improve the basic experimental ability of undergraduate education.

In terms of **aid's form and aid's main body**, the central government advocates diversification. In terms of **the form of aid**, it proposes to help recipient universities train in-service teachers in various ways, such as targeted doctoral training, preferential policy of government-sponsored overseas study, and encouragement of inter-university cooperative research projects. In terms of **main body of aid**, it proposes to jointly build r&d centers with government departments, industries, enterprises, unions and universities to promote the quality development of higher education in central and western China.

4 Impacts of the Policies

4.1 Impacts on Regional Fairness

To test the effectiveness of the policy, I selected post-2016 data and compared it with previous data. The results are as following (Tables 1 and 2).

In terms of the scale of Chinese higher education, from the data above, we can clearly find that regional differences still exist, but they are narrowing gradually, with the standard deviation falling from 2386871.469 to 2211236.98 (Table 3).

In terms of the quality of Chinese higher education [8], we also find that the standard deviation between regions fluctuates slightly, but the overall trend is down. That is to say, policies did promote the fairness of higher education among regions to some extent.

Table 1 Data about higher education in different areas in 2015

	Number of undergraduate and junior college students	Total Fixed Asset (in 10,000 yuan)
Eastern region	11,218,117	92,702,898.53
Mid region	8,581,435	46,516,697.25
Western region	6,453,416	401,011,49.93

Table 2 Number of undergraduate and junior college students from 2015 to 2018

	2015	2016	2017	2018
Eastern region	11218117	11414543	11520200	11733146
Mid region	8581435	8819973	9015490	9255351
Western region	6453416	6723917	7000179	7321851
SD	2386871.469	2349723.974	2264421.944	2211236.98

Table 3 Number of awards for scientific and technological achievements from 2015 to 2018

	2015	2016	2017	2018
Eastern region	874	1559	1667	1566
Mid region	131	1215	1185	1203
Western region	130	772	808	918
SD	429.2602	394.5364	430.5682	324.7815

4.2 *Impacts on Overall Development*

In addition, in order to explore the impact of the policy on the country's overall higher education, we also collected some data about Chinese higher education and plotted the following figures to show how the overall indicators changed before and after the policy (Figs. 7 and 8).

It can be seen obviously that after the implement of the higher education reform system, that is, after 2016 both the percentage of students in tertiary ISCED level 5 and the percentage of students in tertiary ISCED level 8 increased much more

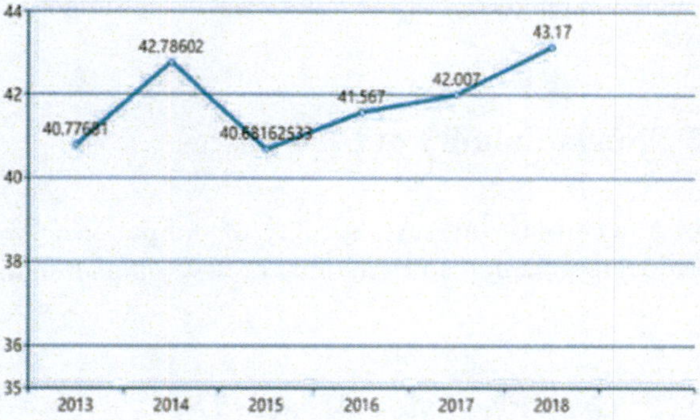

Fig. 7 Percentage of students in tertiary ISCED level 5 from 2013 to 2018

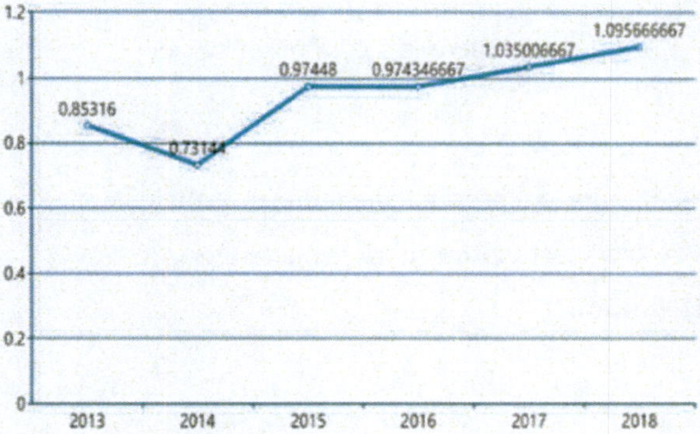

Fig. 8 Percentage of students in tertiary ISCED level 8 from 2013 to 2018

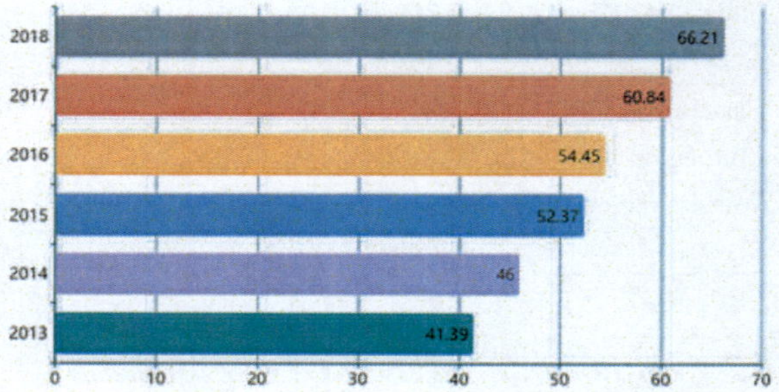

Fig. 9 The number of international students in higher education (unit: ten thousand people)

quickly, especially the percentage of students in tertiary ISCED level 8 [5]. More and more students enrolled in the colleges or universities across the country and the scale of the higher education became larger and larger after the more reasonable support policies in mid-west China (Fig. 9).

From the bar chart above, we can find that since the implement of the policy, the scale of international students became larger and larger and the growth rate also increased significantly. That is to say, China is becoming increasingly attractive to foreign students, which indicates the level of higher education of China is becoming higher and higher after the more reasonable support policies in higher education.

5 Sustainability of the Policies

In order to test the sustainability of the policies, we established Grey Prediction Model to predict the future trend of development of higher education in China.

5.1 Grey Prediction Model [1]

The specific steps are as follows.

Assume that there is a column of raw data:

$$x^{(0)} = \left(x^{(0)}(1), x^{(0)}(2), \ldots, x^{(0)}(n)\right) \tag{1}$$

where, n n n represents the number of the data.

Pre-test (test before model construction):

Definition :

$$\sigma^{(0)}(k) = \frac{x^{(0)}(k-1)}{x^{(0)}(k)} \tag{2}$$

If $\sigma^{(0)}(k) \in \left(e^{-\frac{2}{n+1}}, e^{\frac{2}{n+1}}\right)$, $x^{(0)}$ can be modeled as GM (1, 1).

Model Building

1. Preprocess data. The original data were to accumulated to weaken the volatility and randomness of random sequence. Then the new data sequence was obtained.

$$x^{(1)} = \left(x^{(1)}(1), x^{(1)}(2), ..., x^{(1)}(n)\right) \tag{3}$$

$$x^{(1)}(t) = \sum_{k=1}^{t} x^{(0)}(k), t = 1, 2, ...n \tag{4}$$

2. Set up a differential equation.

$$\frac{dx^{(1)}}{dt} + ax^{(1)} = u \tag{5}$$

where, a represents developing grey number [3]. Its feasible set is $(-2\nu 2)$; u represents endogenous control grey number.

3. Determine parameter.

Take the advantage value of the accumulated data and generate B:

$$B = \begin{bmatrix} -\frac{1}{2}\left(x^{(1)}(1) + x^{(1)}(2)\right) & 1 \\ -\frac{1}{2}\left(x^{(1)}(2) + x^{(1)}(3)\right) & 1 \\ \vdots & \vdots \\ -\frac{1}{2}\left(x^{(1)}(n-1) + x^{(1)}(n)\right) & 1 \end{bmatrix} \tag{6}$$

Take the constant term vector Y_n for the accumulated data. That is:

$$Y_n = \left(x^{(0)}(2), x^{(0)}(3), ..., x^{(0)}(n)\right)^T \tag{7}$$

Solove the parameter $\hat{a}$ by the least square method. That is:

$$\hat{a} = \begin{pmatrix} a \\ u \end{pmatrix} = (B^T B)^{-1} B^T Y_n \tag{8}$$

Put parameter $\hat{a}$ into $\frac{dx^{(1)}}{dt} + ax^{(1)} = u$ and solve. That is:

$$\hat{x}^{(1)}(t+1) = (x^{(1)}(1) - \frac{u}{a})e^{(-at)} + \frac{u}{a}, t = 1, 2, ..., n \tag{9}$$

Calculate $\hat{x}^{(0)}(t+1)$. That is:

$$\hat{x}^{(0)}(t+1)=\hat{x}^{(1)}(t+1)-\hat{x}^{(1)}(t) \tag{10}$$

Post-testing

1. Calculate the residuals $e^{(0)}(t)$ between $x^{(0)}$ and $x^{(0)}(t)$. That is:

$$e^{(0)}(t) = x^{(0)} - \hat{x}^{(0)}(t+1) \tag{11}$$

Calculate relative error. That is:

$$Q = \frac{\left|e^{(0)}(t)\right|}{x^{(0)}(t)} \tag{12}$$

2. Calculate $e^{(0)}(t)$ standard deviation. That is:

$$s_2=\sqrt{\frac{\sum\left(e^{(0)}(i) - \overline{e^{(0)}}\right)^2}{n-1}} \tag{13}$$

Calculate variance ratio. That is:

$$C = \frac{s_2}{s_1} \tag{14}$$

3. Calculate minimum error probability. That is:

$$P = P\left\{\left|e^{(0)}(i) - \overline{e^{(0)}}\right| < 0.6745 s_1\right\} \tag{15}$$

5.2 *The Results of the Prediction*

We used the exiting data to predict the future situation through the grey prediction model and plotted following figures to intuitively show how Chinese higher education's system trend is going Figs. 10, 11, 12, and 13.

It can be seen obviously that almost all the indexes listed are predicted to improve, including higher education enrollment rate, the number of postgraduate students and oversea students and number of Chinese universities in the actual QS top 100. That is to say, the policy will still be feasible in the future years. It is a really sustainable system.

Fig. 10 Higher Education Enrollment Rate Prediction

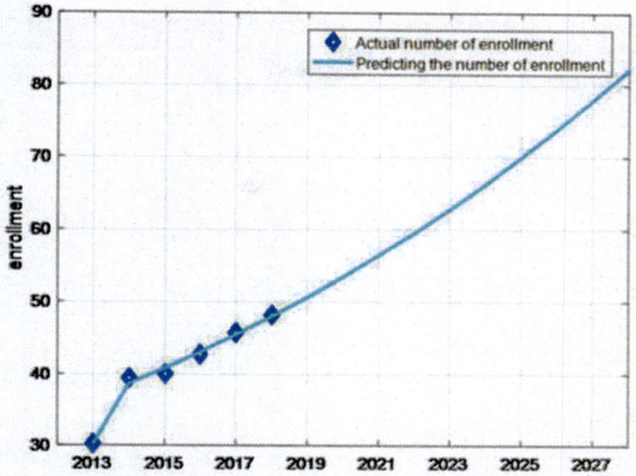

Fig. 11 Number of Postgraduate Students Prediction

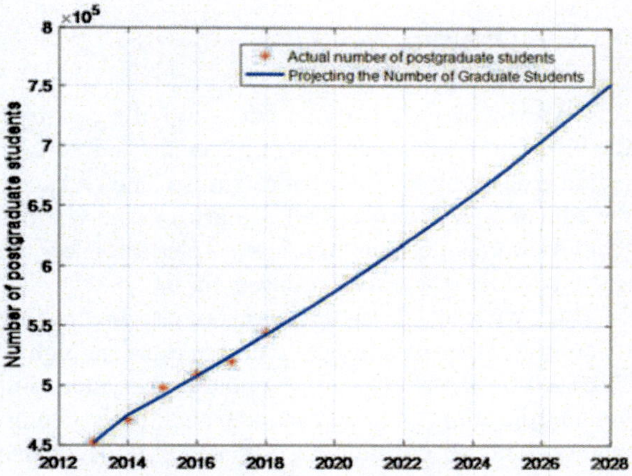

Fig. 12 Number of Oversea Students Prediction

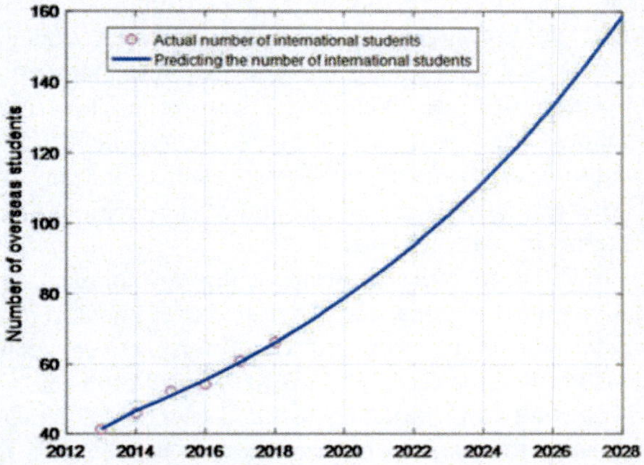

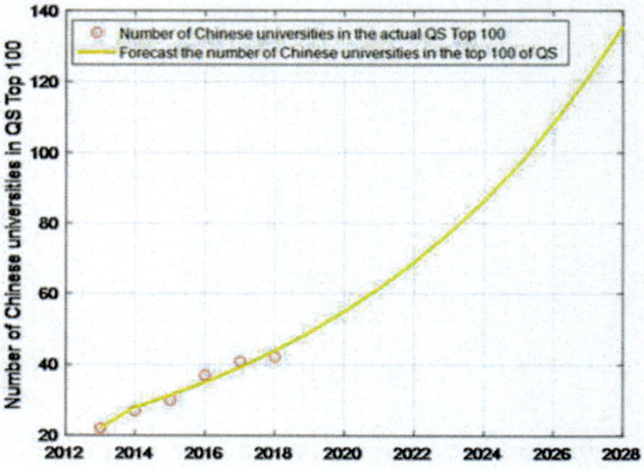

Fig. 13 Number of Chinese Universities in the Actual QS Top 100

6 Conclusion

As is known to all, the higher education of a country plays an indispensable role in the modern education system of a country and the comprehensive development of a country, deciding the future of a country. The global education system constituted by the educational systems of all countries in the world also plays an important role in the development of human society. Therefore, how to make extant higher education much sounder is a crucial problem for us.

Thus, we took China, the largest developing country in the world, as an example to study whether existing policies can promote higher education in the long run.

Based on our analysis of national higher education data, we found that the U.S. higher education system is sounder than other countries, both in terms of size and quality of higher education. Compared with developed countries such as the United States and the United Kingdom, there is still a big gap between China's higher education enrollment rate and its attractiveness to overseas students. But on the whole these gaps are narrowing gradually and Chinese higher education is developing well.

However, with the further study, we find that there is a serious imbalance in the allocation of higher education resources in China [10]. The eastern region, which accounts for less than a third of Chinese land, had nearly three-quarters of the country's science and technology awards and key universities. The resources of higher education in the middle and western regions are far behind the eastern regions in terms of scale and quality.

In 2016, in order to promote the development of higher education in central and western regions, the Chinese government reformed the current unbalanced higher education system from four perspectives: discipline strength, infrastructure construction, the forms and the main body of the aid.

By comparing the data before and after the implementation of the policy, we found that after the implementation, the gap between the number of students in higher education and the number of scientific and technological awards in the central and western regions has steadily narrowed, with the standard deviation of the number of scientific and technological awards dropping from 429 in 2015 to 324 in 2018 [2]. At the same time, with the improvement of the inequality of educational resources, Chinese overall higher education level has also been improved accordingly.

Through forecasting and analyzing previous data, we find that after the implementation of this policy, Chinese overall indicators will continue to improve in the following 10 years. It can be seen that this policy can indeed promote the perfection of the higher education system in the future, which has the advantage of sustainability.

Although changing is a long and arduous process, we believe that as long as we make unremitting efforts, a balanced, healthy and sustainable higher education is beckoning to us!

References

1. Y.M. Wang, J.L. Yuan, F.J. Yao, Technical prediction analysis of shandong province's high-end chemical technology industry based on grey prediction model//.The 34th China control and decision-making conference proceedings (4). [publisher unknown], 2022:71–76. The https://doi.org/10.26914/Arthur c. nkihy. 2022.020867
2. C. Hamnett, S. Hua, L. Bingjie, The reproduction of regional inequality through university access: the in China. Area Development and Policy, (2019)
3. Y Wang, Z.X. Wang, H. Zameer, Structural characteristics and evolution of the "international trade-carbon emissions" network in equipment manufacturing industry: international evidence in the perspective of global value chains. Environ. Sci. Pollut. Res., **28**(20), (2021)
4. Guideline of the general office of the state council on accelerating the development of education in Central and Western China. Bull. State Counc. People's Repub. China, (18), 22–29(2016)
5. X. Jiang, *Evaluation of the competitiveness of higher education in Anhui province*. Jiangxi University of Finance and Economics, 2020. https://doi.org/10.27175/d.cnki.gjxcu.2020.000744
6. S.H. Li, W.Y. Wang, Analysis of structural gap index of Midwest higher education. High. Educ. Res.,20, **41**(08), 42–51
7. J.Q. Liu, M. Yu, An analysis on ways of higher education evaluation in the New Era of our country. J. High. Contin. Educ., **34**(02), 22–25+72(2021)
8. Y. Zhao, H.Y. Wang, X. Tai, Accelerating the construction of higher education Power with quality improvement as the core//. Cultural inheritance and Innovation and the construction of Higher Education Power: Proceedings of 2012 International Forum on Higher Education. pp. 164–167(2012)
9. L.X. Ran, *Development of regional Higher education: Historical changes, realistic characteristics and future prospects*. East China Normal University, (2022)
10. X.X. Liu, Strong intervention: the western higher education development strategy selection. J. Chongqing High. Educ. Res., **10** (01), 80–91(2022). https://doi.org/10.15998/j.cnki.issn1673-8012.2022.01.009.

Outcome Based Education: A Paradigm Shift in Teaching and Learning Process

Bin Hu, Liying Hu, Nvdeng Chen, and Srikanta Patnaik

Abstract Our present surrounding which is divergence with many issues related to topological, sociological and economic transformations. So to face competitiveness in the business world, organizations must improve their human resources. This is achieved by outcome Based Education which assists to train students/graduates to certain level by joining superspecialized learning with energetic and representative competence through remodeling curriculums. The major aim of outcome-Based Education is to empowering students with intelligence, proficiency and exposures which are required for achievement after moving out the institution. Thus the prime perspective of OBE is to build skilled and self sufficient future generation. The objective of this paper is to provide and explore knowledge towards outcome-Based Education which is helpful to the learner to get sufficient idea about this emerging learning process. This paper compares the impact of outcome-based education (OBE) on student's educational accomplishment in connection with grade point average against traditional teaching learning method.

Keywords OBE · Traditional teaching learning method · Outcomes · Assessment · Operation

1 Introduction

May it in academic sector or in business sector, our globe is encountering with numerous changes at a fast speed. On such a changeable situation requirement of skilled persons is extremely high. In the employment process skills denotes to the

B. Hu · L. Hu · N. Chen
Changsha Normal University, Changsha 410000, China

S. Patnaik (✉)
INTERSCIENCE Institute of Management & Technology (IIMT), Bhubaneswar 751030, India
e-mail: patnaik_srikanta@yahoo.co.in

S. Patnaik and F. Paas (eds.), *Recent Trends in Educational Technology and Administration*, Learning and Analytics in Intelligent Systems 31,
https://doi.org/10.1007/978-3-031-29016-9_17

competency of an individual at the working place. Skill corresponds to the extent of acquaintance and proficiency of a person in a particular discipline as well as the capability to team up with multiple domains and utilize intelligence in similar fields efficiently.

For the pursuit of increasing human resources and skilled employees, teaching and learning performs a very crucial task. Majority of nations, prime focus is to give stress on to the tertiary pattern of education system because they require alumnae who are resourceful, inventive and well-informed in various fields. To build the nation and its financial system into a huge earning and powerful country, the above mentioned elements are needed. For these reasons, nations have financed a lot of money for the development of their education system by realizing this purpose.

The goal of education is to deliver teaching and learning insight, for improving the capabilities and skills of each learner and to bring proficiency as well as assist them with optimistic thoughts and values [1]. Javier [2] indicated that, the educational organizations have to bring provisions and facilities that are essential to accomplish the goals for which it plans to create.

To move in the direction of OBE system is equivalent to the overall quality shift in commerce and engineering. This contemplates a faith that the most excellent method for persons and associations is to find where they are decide to moving before that, they have to acknowledge where they are now and where they desire to be in future. All these are planned in reverse to obtain the best path to move from now to then.

Today's education system in institutions for advanced studies exceeds expectations of the conventional patterns where teaching, examinations and other kinds of appraisals which are related to educational achievement of a student executed through electronic based teaching and learning. Thus teaching and learning of learners concerning a subject not to have just classroom based but it can be conducted through online i.e. e-learning. All these are performed because of the increasing requirement to estimate the educational value of organizations for higher studies so as to keep pace with the international standard of teaching and learning strategies, competitiveness and other technological stands.

Therefore, an innovative teaching and learning approach comes out as well as a creative learning model namely outcome-based education (OBE) learning technique is preferred and accepted to improve and renovate the educational strategy globally. The teaching and learning process becomes more demanding due to this innovative strategy of learning. Outcome-based education has achieved distinct identification globally to encourage educational policy and reform.

OBE emerges in the form of skill based learning approach, helps to identify the necessities and outcome based excellent assurance controlling. It is considered as the utmost crucial instructive aspect of communities by means of informational based financial structure. OBE is planned in such a way that it attains the predetermined learning consequences. Accreditation of OBE generates specialization in achieving objectives and consequences in an engineering curriculum. The major role of OBE starts with the apparent image to measure the capability of a student to perform certain task, course structure planning, teaching learning skill and at the end assessment

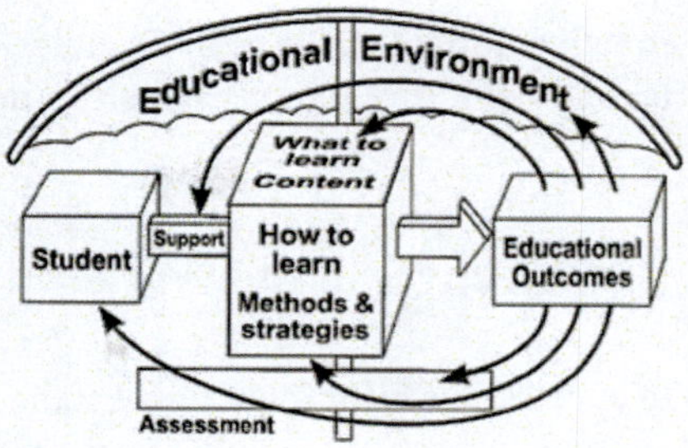

Fig. 1 A curriculum structure highlighting the significance of educational outcomes in planning

process to make sure that the targets are achieved. Adaptation of OBE facilitates ceaselessly brilliant development.

It is obvious that learning outcomes be supposed to hold a major place in designing and planning the curriculum as shown in Fig. 1. Learners/Students go through an educational structure by getting some support as per need. They learn the content that is previously set and use a suitable learning technique and during this period accomplish the educational outcomes as mentioned. Debating regarding the diverse mechanisms of the curriculum is worthless if they are not made in the framework of these learning outcomes. Reflection of outcomes ought to be the foundation for curriculum formation and assessment.

2 Literature Review

Many types of earnings/revenues are generated through education to record growth of economy over worldwide by exact referral to growing countries. Education and economic expansion are directly proportional to each other in a positive way which is demonstrated by universal statistics and they have also dependability with each other [3].

Many nations improve their academic structure and system due to the increasing need of excellence and competent graduates to accomplish their industry requirements. Further causes to reform the education system because of some political influences and other lobbies who are giving support to OBE [4, 5]. Although with a great deal of disapproval from contenders who are doubtful regarding OBE, consequently it is the most excellent choice which must be carried out. The nations who include OBE as part of their curriculum are New Zealand [6], South Africa [7], USA [8], Hong Kong [9], United Kingdom [10] and Western Australia [11].

Despite that, many people who are distressed by the OBE structure forced to eliminate it, like OBE unsuccessful in South Africa and Western Australia was discarded in 2010 and 2007 accordingly [12, 13]. However countries like Canada, Ireland, South Africa, Australia, United States, New Zealand, United Kingdom and Hong Kong along with Singapore, Malaysia and Germany effectively executed the OBE system of learning for undergraduate engineering students [14]. OBE system also effective in the United Arab Emirates Universities and they plan their Information Technology Curriculum based on learning outcomes [15].

3 Objectives of the Work

To encourage a clear focus on what the teacher wants learners to be acquainted with, comprehend and be capable to perform so by completion of the academic session.

To design the curriculum with a noticeable description of the planned outcomes which students should achieve at the finish of the course.

To set up challenging and distinguishing levels of achievement, motivate students to involve them totally in the learning process.

To provides extensive possibilities for all the students equally.

4 OBE Against Conventional Teaching and Learning

In this changing world, when considering about the change in the education pattern to adapt and update the new method of teaching and learning, trouble must be avoided generally. OBE is the type of instructive academic form which involves the reformation of syllabus, evaluation methods and pedagogy to contemplate the accomplishment of best kind of learning, rather than simple aggregation of credits for a particular course. The conventional pattern of education gives stress on what is taught while the OBE pattern focuses on what is learned and this differentiation is very crucial. Here the former is the teacher centric model while the later is the student based structure which includes practical problems.

At the finish of the academic session, the skills, talents and knowledge that the students get are more precious than what they taught. The conventional educational form depends a lot on uniform methods and in this form students are taught by a teacher in a classroom at a particular time. When the lecture is over, then students clear their doubts with the corresponding teacher; which shows that the success of learning process hugely rely on the efficiency of the lecturer. Contrastingly, the OBE pattern of education developed on specified consequences. OBE gives emphasis on the ability of students to gain after the end of their course. While designing the curricular activities, OBE focuses on the under mentioned skills.

- Basic skills,
- Life skills,
- Intellectual skills,
- Professional and vocational skills,
- Personal as well as interpersonal skills [16].

In OBE, the performance of students are evaluated through their outcomes while in conventional approach the performance of students based on the type of inputs or resources that are provided to them. Also implementation of OBE integrates a mass of numerous revolutionary models and techniques. It need not follow a specific pattern of teaching and learning rather than it demands that students must show their skills what they have learned.

5 Interpretation of OBE

5.1 Analysis of Outcomes

Major focus of outcome based education model is to evaluate the outcomes instead of measuring the inputs like the amount of time spend in a classroom, the types of textbooks that are delivered etc. Outcomes involve different talents and experience. Normally, outcomes are desired to be accurately computable. Subject outcomes usually contain all, from simple information to multifaceted examination and assessment. Interpreting exact and computable outcomes is a hard task and the selection of particular outcomes generates some disputes.

Every educational institution is accountable for fixing their respective outcomes. OBE structure recommends that an educational institution can mention any outcome.

Challenging issues

- Re-orienting the teaching staff members to adopt OBE in their classroom practice requires a considerate amount of effort.
- The entire curriculum needs to be restructured according to the OBE model.

5.2 OBE

OBE (outcome based education) is a learner centric educational structure which emphasis on evaluating the performance of a student based on outcomes. Knowledge, skills and outlooks are the main ingredients of outcomes. These are evaluated after the end of a course to judge student performance. In this structure the skill and knowledge that are needed designed for specific course is decided previously and the performance of students are calculated based on these specifications throughout the session.

5.3 Why Organizations Should Pursue OBE?

Adaptive and confidence building type of learning is OBE. The objective of it is to providing students with competence, orientations and knowledge that are required for attainment of goal after leaving the organization. Achievements at the educational organizations have little importance until the students are ready to shift this educational victory to challenging, advanced technology and complex future. OBE affects both input as well as output processes [17]. Main function of OBE denotes that all the actions of assessment and learning are moved in the direction of what is the consequences of teaching be supposed to be and what the student should perform and at which level but not what the teacher should going to teach.

5.4 *How to Evaluate OBE?*

The OBE system uses three parameters to calculate the improvement of graduate as shown in Fig. 2 and 3, they are.

- Program Educational Objectives (PEO)
- Program Outcomes (PO)
- Course Outcomes (CO)

Program Educational Objectives (PEO) portrays the professional and career activities which the program helps to train the graduates to accomplish. After 4–5 years of graduation, the PEO's are calculated.

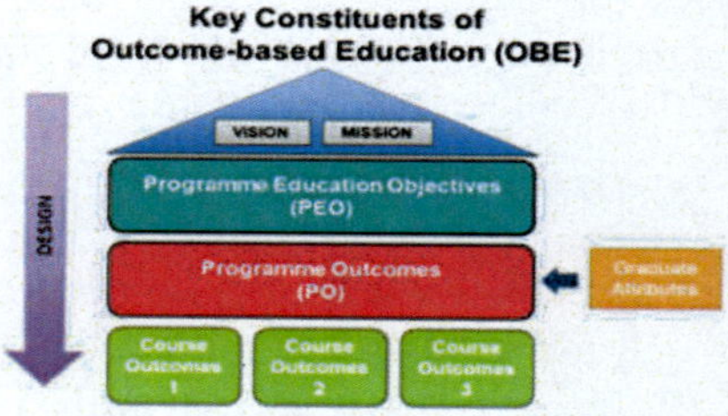

Fig. 2 Key features of OBE

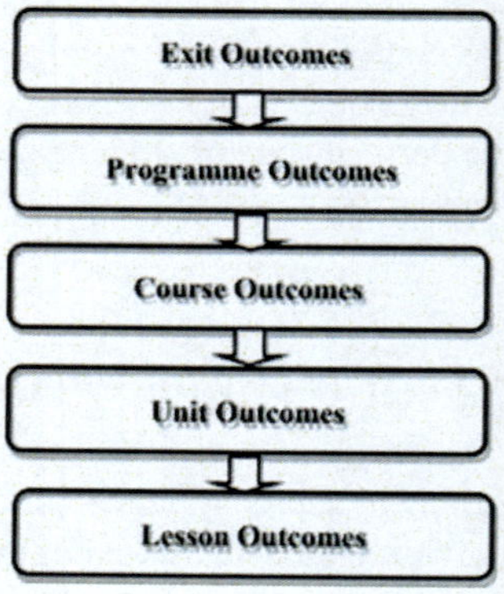

Fig. 3 Ordering of outcomes

Program outcomes explain what the students should know as well as capable to perform at the time of graduation. These are the computable parameters which estimate every student's results obtained from each subject that the student chooses in each semester.

Consequence of a course might be achieved at various stages of a program. Here the results of all the stages are not compulsory. Different stages are combined and the collective estimation is carried out. The measurement scale provides important values to consider the outcomes at different stages [18].

Significant features of the OBE

1. **Course** is specified as practical based or theory based, or theory cum practical based topic taught in a semester.
2. **Course Outcome** states the important and necessary knowledge that students must attain along with practically implement it when the program end. Usually many number of course outcomes are designated for a particular course depending on the priority.
3. **Programme** is considered as certain branch of a Degree course which is a correlated ordering of subjects, extracurricular and other curricular actions to achieve pre-decided goals which help in to achieve a degree.
4. **Programme Outcomes** explain what the graduates are predicted to be capable to perform at the time of their graduation level and this programme outcomes are associated strongly with the graduate attributes.
5. **Program Educational Objectives** illustrate the predictable performances of students while studying along with the accomplishment of learners during the session of their study.
6. **Programme Specific Outcomes** demonstrates what the graduates or learners must perform while continuing their study with indication to a particular branch. Generally a programme contains more than two programme Specific Outcomes (PSOs).
7. **Graduate Attributes** are those attributes possible for a graduate from an accreditation course.

Programme outcomes is the satisfactory pointer of quality of education as well as course outcomes also the efficient sign of brilliance achievement at the finish of graduation level.

5.5 *Assessment Methods*

This method includes the evaluation of learner's achievement in two different ways.

- Direct method includes learners should show their performance which is calculated through numerical techniques.
- The other method which is indirect evaluated through views.

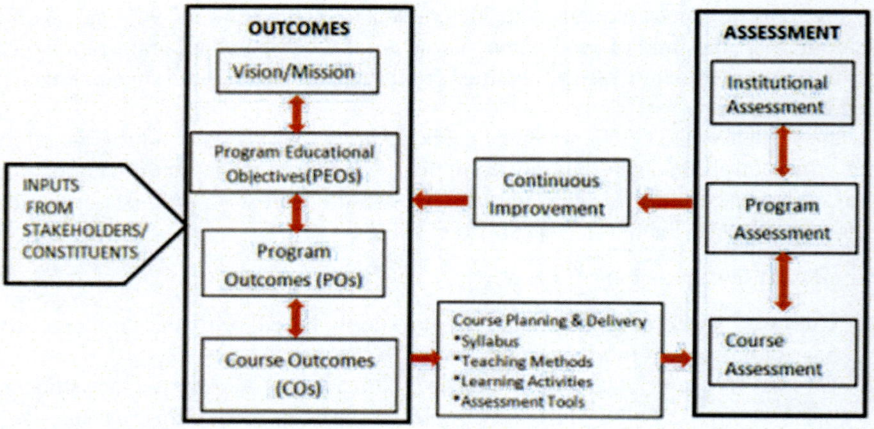

Fig. 4 OBE framework

The achievement of students in direct method based upon thorough assessment of their presentations which is given in Fig. 4. This technique helps to gather the proof of the student learning performance straightly from the students via various work assignments, practical exam, term end exam, seminars, projects etc. [19, 20]. On the other hand, indirect method includes the proof of student accomplishment through teacher assessment who gathers information about student capabilities, acquaintance, and standards. Indirect techniques give the perceptions of learners, teacher or other members who are bothered about the syllabus or curriculum or organization.

5.6 *Operational Procedure of OBE*

Attention for achievement—Educationalists should be aware and responsive for the consequences of learning system where every student should come out as per the program orientation which assist in the accomplishment of top rank outcomes. Therefore, during the beginning stage of planning the better curriculum, the improved results provide better path for educationalists in stating and explaining the central point of the program.

Extreme expectation—Previously mentioned that examining outcomes at the ground level consistently accessed superior level results. These links help educationalists to generate extreme level of achievement from learners. At this point, the whole accomplishment makes sure that the learners proficiently attain the expected results which is set for them and then let them to demonstrate outcomes at superior levels.

Improve possibilities—The above mentioned notions are too complex which are handled by the educationalists. Many learners consider it as complex in fulfilling the needs that are set for them. These findings demoralize the educationalists to establish the evidences from learners and believing that every student is absolutely a beginner. For this an augmented chance should be allowed to pupils in order to understanding and estimating their whole performances. This technique has multidimensional aspect.

6 Benefits and Drawbacks of OBE

The major benefit of outcome based education is it promotes the significance of transparency in education. Parents as well as students can choose any organization, syllabus and lessons that are clearly mentioned to them regarding the objectives of learning. This type of sharpness additionally found in their excellence in teaching and demonstration beyond sections and sectors where the teacher can regulate their subject more correctly. Further attraction and the most beneficiary one is the flexibility in education. It provides authorization to students to select what they want to learn and in which way they desire to learn it. OBE not just measures the potency and flaws of a student but offers enough time to achieve expertise and confidence in the corresponding subject. The OBE structure also permits the students to shift their credentials and move to other organizations that has the same accreditation. Based on this approval, institutions are identified, compared and also measured with each other. So from the OBE structure of learning each member can get benefit from it.

Similar to other academic frameworks, OBE goes along with few short of drawbacks. Form of evaluation is the first one. A prescribed set of informational plan is deficient in OBE model. Here the consequences are evidently established but there is a much gap in the analysis process. In addition to this, the quantity of terminologies present in the technical expressions it is very difficult to interpret all of them while studying. It is hard and takes time to design the OBE structure which is a great challenge. OBE, best suits for vocational education but not for other streams like arts, philosophy and literature subjects. The major drawbacks in the OBE system is the way of evaluation process because pen and paper exam do not provide best result in OBE format. It needs different types of evaluation methods like quizzes, seminars, projects etc. Finally by coping with the challenges, it gets to know what are really present and what is expected to be.

Our commercial world might be hindered with the advancement in technology along with economical and sociological modifications. In this context, OBE stands at the entry way to help students to find the way where they want to move in the globe and this is the demand of the time. A large group of students wants to learn the emerging skills to boost their future in high demanding courses like competency-based programs, flexible degrees and vocational training. Teaching also enhanced when teachers revive their positions to promote students for learning.

7 Teaching–Learning Outcomes

7.1 Skills and Knowledge

- **Academic session**

Finishing a particular approval course is characterized by more than four years of further education.

- **Information about technical courses**

Implement the concept of basics of engineering, majors of engineering, science and mathematics for the creation of technological structures.

- **Result creation**

Develop results for complicated technological problems which fulfill particular requirements by proper analysis of physical conditions like social, economical, cultural, and environmental along with health and security concern.

- **Inquiry**

Perform exploration of complicated issues like experimental work design, study and elucidation of facts and finally integration of findings to present legitimate outputs

- **Utilization of current device**

Develop, choose and utilize proper tools and techniques in complex engineering problems related to prediction, modeling, and measuring activities, with visualizations of boundaries.

- **Work single or in a group**

The purpose of an organization must achieved by working effectively as a single or as a part of a varied group in a multifaceted environment.

- **Communication**

In complicated engineering problems, communication plays an important role because if issues are not communicated or established properly to the technical people in the reports or the presentation of documents then it is difficult to resolve the problem

- **Technical people**

Technical people help to recognize issues related to society, fitness, security, and authorization and cultural and perform requisite duties that are important for them.

- **Principles**

Recognize and entrust to expert beliefs and tasks along with the standards of engineering methods.

- **Surroundings and survivability**

Know the influence of technological practices in a social background and show awareness towards effective growth.

- **Handling project Management and funding**

Express and admit different kind of administration and trade performances, with their boundaries.

- **All-time knowledge**

Identify the requirement and possess the skill to employ in self-regulating and lifetime education.

7.2 *Outcome Based Learning*

Students plan to attain the learning outcomes through competently finishing a particular course. Every part of a learning process aims to achieve more than one benefit of acquaintance, capability and expertise. Learning can be defined as an investigated, comprehensive and isolated form of teaching or training which contains a training process and other related subjects.

7.3 *Benefits of Outcome-Based Education for Students*

- Brings clarity among the teachers and students
- Every student has the flexibility and freedom of learning in their ways.
- There is more than one method of learning
- Reduces comparison among the students as everyone has a different target
- Completely involves students taking responsibility for their goals.

7.4 *Assessment for Teaching–Learning OBE*

Research demonstrates that the learners who pursue the OBE format of learning and technique are considered as more dynamic students in comparison with the students who are adopting the traditional approach. This approach is obvious by learner's individual assessments outcomes on their educational knowledge by utilizing entrance-exit outcome form which confirmed the average scores of all level outcomes as mentioned in Figs. 2, 3 and 4.

Application and execution of OBE format education system designates improved grade point average score with regard to educational achievement when compared with traditional academic system.

So the inclination towards future work is suggested to involve a vaster outlook with respect to additional potentials to involve the educationalists and the students' opinions as well as adoption of OBE of learning. This is quite motivating to write down the views of the end-users of the structure contrary to the controller's viewpoint. Learning through OBE productively developed the academic efficacy of students in spite of different branches and programmes.

Figure 5 depicts the theoretical diagram for the outcome based learning assessment structure, summarizing thorough ideas and concentrating on linking points. Assessment through learning outcomes performed in following steps.

Design time: learning evaluations are intended to calculate the accomplishment of definite planned learning outcomes.

Run time: learning process give details about concrete learning outcomes which is attained by the student.

Results obtained from the evaluation process are then normalized and maintain in the evidence records. Rarely, learning assessment procedures work separately, at the

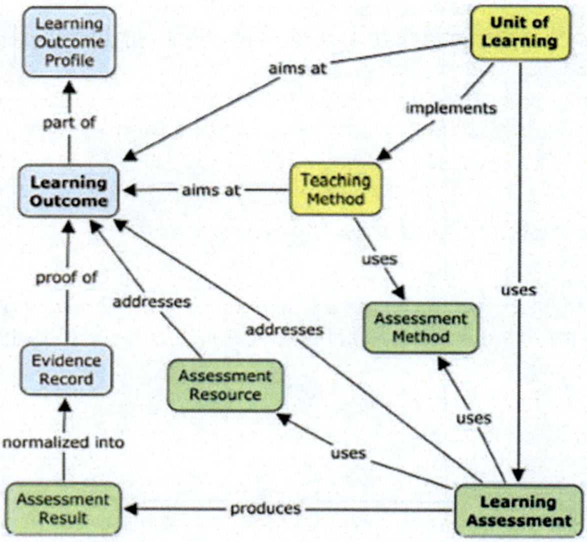

Fig. 5 Map for teaching–learning outcomes-based assessment

time when it just used for certifying the performance of learning outcomes without any allied learning method.

7.5 Applications of OBE

Area	Problem analysis	Year
Information technologies	Studies different types of performances of OBE and talks about its application in library teaching and orientation service, predominantly for the implementation of new technologies	1999 [21]
Medical education	Summarizes the competency model of the "Outcome-based Project" conducted by the Accreditation Council for Graduate Medical Education(ACGME) in the U.S., but also analyzes the successful experience of implementing this model at the School of Medicine and Public Health of the University of Wisconsin	2011 [22]
Accountancy	Planned to obtain the demographic report of the accountancy students as well as evaluate the influence of OBE approach	2014 [23]
Religious studies	Examining OBE application in madrasah studies and outlines their plans and purposes of studies to recognize the set of skills planned for graduates	2020 [24]
Economic management	Examine the correlation among the factors related to ability to detect and resolve the issues, team work, expert knowledge, communication skills, and work approach with students' supposed value along with practical implementation capability	2020 [25]
Engineering	Implements critical ideas for displaying the structure for the plan and work out of the appraisal and estimation of students achievement in engineering discipline	2021 [26]

8 Conclusion and Future Work

Academic methodologies are switching into a learning outcomes based technique for the dynamic learning framework. Skills, abilities and knowledge accomplished by the student do a progressively crucial role as the expert brings innovative influential challenges. This article offers an outline of the essential features of the outcomes Based Education on teaching and learning and its implementation in different domains.

This OBE method developed on valid academic ethics and also brings a healthy structure for learners to obtain the essential strength to perform. Degree of personnel dismissing OBE is very few. For effective application of OBE, the educationalists ought to know the OBE scheme. Unexpectedly, the conventional method of education must not be discarded, rather will be utilized as a way to implement OBE. Educationalists must revise and develop their methods of teaching and accessing the student's effort. Learners assessing system, curriculum structure, and teaching methods must be designed by the affiliating universities in such a manner that the learners be supposed to understand the significance of OBE approach.

OBE has lots of intrinsic benefits which necessarily build it as a striking model for curriculum designing and also helpful for curriculum planners, learners, teachers, employers and for the community. Even though OBE has noticeable demand, research on demonstrating its consequences are quite limited. Further research comprises the explanation of thorough classification of assessment techniques along with an investigation of the correctness of diverse assessment approaches based on the learning outcomes.

Acknowledgements This paper is funded by the Project of Educational Science Planning of Hunan Province (XJK22CGD060).

References

1. L.D. Borsoto, J.D. Lescano, N.I. Maquimot, M.J.N. Santorce, A.F. Simbulan, A.M. Pagcaliwagan, Status of implementation and usefulness of outcomes-based education in the engineering department of an Asian University. Int. J. Multidiscip. Acad. Res. **2**(4), 14–25 (2014)
2. F.V. Javier, Assessing an Asian university's organizational effectiveness using the Malcolm Baldridge model. Asian J. Bus. Gov.Ance, **2**, 37–55(2012)
3. N. Abd Rahman, S.R.S. Abdullah, Assessment tool of course learning outcomes for mechanical design of process equipment. Procedia - Soc. Behav. Sci. **102**, 116–121(2013)
4. S.P.T. Malan, The 'new paradigm' of outcomes-based education in perspective. J. Fam. Ecol. Consum. Sci.=Tydskrif vir Gesinsekologie en Verbruikerswetenskappe **28**(1), 22–28(2000)
5. M. Martin, A. Alderson, Outcomes Based Education: Where has it come from and where is it going? Issues Educ. Res. **17**(2), 161–182 (2007)
6. A.N. Sundar, Changed assessment, changed focus in curriculum delivery: what do teaching staff have to say. in *HERDSA Annual international conference* (Melbourne, 1992). 1999
7. R.J. Botha, Outcomes-based education and educational reform in South Africa. Int. J. Leadersh. Educ. **5**(4), 361–371 (2002)
8. R.M. Harden, J.R. Crosby, M.H. Davis, M. Friedman, R.M, AMEE Guide No. 14: Outcome-based education: Part 5-From competency to meta-competency: a model for the specification of learning outcomes. Med. Teach. **21**(6), 546–552 (1999)
9. P. Ewell, Applying student learning outcomes concepts and approaches at Hong Kong higher education institutions: Current status and future directions. Natl. Cent. High. Educ. Manag. Syst. (2006)
10. N. Ross, D. Davies, Outcome-based education: Part 4-Outcome-based learning and the electronic curriculum at Birmingham Medical School. Med. Teach. **21**(1), 26–30 (1999)

11. A. Alderson, M. Martin, Outcomes-based education: where has it come from and where it is going? Issues in Education **17**(2), 161–182 (2007)
12. O. Akir, T. H. Eng, S. Malie, Teaching and learning enhancement through outcome-based education structure and technology e-learning support. Procedia - Soc. Behav. Sci. **62**, 87–92 (2012)
13. K. McNaught, R.G. Berlach, Outcomes based education?: rethinking the provision of compulsory education in Western Australia. Issues Educ. Res. **17**(1), 1–14 (2007)
14. H. Basri, A. B. Che Man, W. H. W. Badaruzzaman, M. J. Mohd Nor, Malaysia and the Washington accord: What it takes for full membership. Int. J. Eng. Technol.**1**(1), 64–73((2004)
15. F. Bouslama, A. Lansari, A. M. Al-Rawi, A. A. Abonamah, A novel outcome-based educational model and its effect on student learning, curriculum development, and assessment. J. Inf. Technol. Educ.: Res. **2**(1), 203–214((2003)
16. S. M. Meyer, S. E. Van Niekerk, *Nurse educator in practice.* (Juta and Company Ltd., Cape Town, 2008). p. 25. ISBN 978–0–7021–7299–1
17. B. Bloom, Learning for mastery. Eval Com **1**(2), 1968
18. J. O'Neil, Aiming for new outcomes: The promise and the reality. Educ Leadersh **51**(6), (1994)
19. B.S. Bloom, *Taxonomy of educational objectives, handbook I: The cognitive domain* (David McKay Co Inc., New York, 1956)
20. L.W. Anderson, D.R. Krathwohl, *A taxonomy for learning, teaching, and assessing*, Abridged. (Allyn and Bacon, Boston, MA, 2001)
21. M. Lorenzen, Using outcome-based education in the planning and teaching of new information technologies. J. Libr. Adm. **26**(3–4), 141–152 (1999)
22. MIAO, Yun, Li Liu, Y. P. ZHANG, The application of outcome-based education in medical education. in *Fudan Education Forum*, (2011), p. 05
23. I. L. An, Impact of outcome-based education instruction to accountancy students in an Asian University. Asia Pac. J. Educ., Arts Sci. **1**(5), 48–52((2014)
24. M.I.M. Jazeel, Application of outcome-based curriculum in religious studies: the case of Madrasas in Sri Lanka. J. Pol. & L. **13**, 196 (2020)
25. P. Phuc., N. Vinh, Q. Do, The implementation of outcome-based education: Evidence from master program in economic management at Hanoi universities. Manag. Sci. Lett. **10**(14), 3299–33069((2020)
26. D. Pradhan, Effectiveness of outcome based education (OBE) toward empowering the students performance in an engineering course. J Adv Educ Philos **5**(2), 58–65 (2021)

Research on Landscape Types and College Students' Restorative Evaluation Based on the Evaluation of the Greening

Wanyue Suo

Abstract A campus is an important place for college students to live and study, which carries the function of college students to communicate and relax. Therefore, the reasonable arrangement of campus landscape is related to whether students can effectively relax their body and mind, relieve pressure, and reduce the probability of college students suffering from mental diseases. Existing studies mostly focus on the factors affecting the perception of recovery in a single landscape case, little knowledge exists regarding the diversity of landscape types. Based on the evaluation of greening level under the entropy weight calculation of Subjective Greening Evaluation(SGE), Vegetation Coverage Index(VCI), and Visible Green Index(VGI), four typical landscape types in the campus were selected: natural waterfront landscape, lawn landscape, square landscape, and avenue landscape, and their influence on college students' recovery evaluation was studied. The results showed that the restoration evaluation scores were as follows: natural waterfront landscape > lawn landscape > avenue landscape > square landscape. There was a significant positive correlation between the evaluation of greening and the score of recovery evaluation. Compared with the artificial landscape, the natural landscape has a higher recovery evaluation. The waterfront landscape has a higher recovery evaluation than the land landscape.

Keywords Subjective greening evaluation · Landscape type · University students · Visible green index · Vegetation cover index

1 Introduction

Due to the rapid development of society, the competition of society become increasingly intensified, and the pressure and anxiety of college students are increasing

W. Suo (✉)
School, Hainan University, Hainan, China
e-mail: wanyuesuo@163.com

S. Patnaik and F. Paas (eds.), *Recent Trends in Educational Technology and Administration*, Learning and Analytics in Intelligent Systems 31,
https://doi.org/10.1007/978-3-031-29016-9_18

sharply. Many studies show that the problems related to school are largely about students' pressure and mental health. The campus needs to improve students' mental health through greening and natural elements, creating a restorative and stress-relieving environment [1].

According to Kaplan's research conclusion, the experience that can make mental fatigue obtain restorative benefits or other benefits is called restorative experience, and the environment with such experience is called restorative environment [2]. There are two classic theories: Attention Restoration Theory (ART) [3] and Stress Recovery Theory (SRT) [4], which have been widely discussed. Although these two theories all have a common characteristic. SRT, the main focus on the characteristics of natural visual, can help people from sharp pressure or sad mood for emotional and physical recovery, while ART focused on the part of the people and the environment interaction, the interaction can promote the attention from the mental fatigue or cognitive resource depletion recovery [5].

Kaplan's ART is widely cited in the current study, which describes the composition of a restorative environment from four dimensions of fascination, being away, extent, and compatibility. Being away refers to the need to stay away from the ability to use directed attention in daily life; fascination refers to the ability of the environment to attract attention without mental effort, also known as "involuntary attention" [6]; The environment provides the ability to explore space and consistency as extent [7]. As well as the ability to organize and construct scenarios, a compatibility environment should be compatible with individual goals and tendencies [8].

In recent years, studies on the perception of recovery and mental health benefits have been a hot topic. Stigsdotter et al. explored the resiliency of green space in park cities from eight perception dimensions; and concluded that the two most important dimensions of perceived recovery for users are "social" and "serene" [6]. Berg et al. tested the mediating role of recovery in environmental preference and found that emotional recovery accounted for a large proportion of people's preference for the natural environment [9]. Liu Chang has found in his research that the visiting behavior of green space has a positive regulating effect on the emotional regulation of college students [10]. It can be seen that the landscape environment is closely related to residents' living standards, and its quality affects residents' satisfaction with life. Many studies have proved that the natural environment has more restorative experience than the artificial environment, but few studies have explored the discrepancy between different landscape types on the actual restoration of college students' restorative perception.

The research on the evaluation of college students' resilience by different landscape types can not only confirm the significance of the scale, but also discuss the recovery effect of different landscape types on college campuses, and put forward targeted adjustment and reference for future campus landscape construction planning.

2 Overview of the Study Area and Data Sources

2.1 *Overview of the Study Area*

The research site is located in Haidian Campus of Hainan University, Meilan District, Haikou city, Hainan Province, China. Combining with the basic landscape types of universities, four common landscape types, namely waterfront landscape, lawn landscape, square landscape, and avenue landscape, are selected for research and discussion. Dongpo Lake in the east of Hainan University, Starting Lawn in the middle, Tourism College Square in the north, and East Haiyun Road in the northeast are taken as waterfront landscape, lawn landscape, square landscape, and avenue landscape respectively.

2.2 *Data Sources*

2.2.1 Landscape Site Picture

Landscape site images are used to calculate the Visible Green Index (VGI) of landscape sites. The VGI was first proposed by Japanese scholar Yoji Aoki [11], which represents the percentage of green in the total visual field, and has become an important indicator of greening level evaluation. Taking 360° panoramic photos of the observation point from a fixed human perspective through the camera to adopt 360° panoramic images. This method can more accurately express the green area observed from the human viewpoint compared with the traditional 2D image calculation of the green visual ratio. In this paper, the convolutional neural network SegNet model proposed by scholars at Cambridge University is used to extract vegetation information by image semantic segmentation technology [12, 13]. This convolutional neural network segmentation method naturally includes the gaps between tree trunks into the green visual acuity, but just because of this feature, it can significantly reduce the impact of the density of plant branches and leaves in different seasons and phenological periods on the calculation results of green visual acuity [14].

2.2.2 Satellite Remote Sensing Images

VCI of the site was calculated using satellite remote sensing images. Hainan University was located by Google satellite map, and satellite remote sensing images of Dongpo Lake, starting point lawn, Tourism College Square, and East Haiyun Road were obtained. Import the satellite remote sensing image into Photoshop2020 to calculate by manual interpretation. Opening the histogram, selecting the whole image and green area respectively to get the pixel value of the whole image and green vegetation to calculate the vegetation coverage rate of the whole site through the ratio.

2.2.3 Evaluation of Greening Level

In this paper, the entropy weight method was used to determine the weight of VCR, VGI, and a Subjective Greening Evaluation(SGE) in the overall evaluation of greening level. The entropy value of each index in the entropy weight method reflected the degree of information disorder. In general, the lower the entropy weight, the lower the disorder degree of the system. The entropy weight method can reflect the difference of each index and make the evaluation result more scientific and reliable [15].

The calculation steps of the entropy weight method are as follows [16]:

$$p(x_{ij}) = \frac{x_{ij}}{\sum_{i=1}^{m} x_{ij}} (i = 1, 2, \cdots, m; j = 1, 2, \cdots, n) \tag{1}$$

$$E_j = -k \sum_{i-1}^{m} p(x_{ij}) 1n\, p(x_{ij}) \tag{2}$$

$$d_j = 1 - E_j \tag{3}$$

$$w_j = \frac{d_j}{\sum_{j=1}^{n} d_j} (j = 1, 2, \cdots, n) \tag{4}$$

Notes: In the formulas, $p(x_{ij})$ is the standardized value of each indicator; E_j is the information entropy; w_j is the entropy weight of each indicator.

The entropy weight proportion of VCR is 0.477, the entropy weight proportion of VGI is 0.235, and the entropy weight proportion of SGE is 0.289. Using the formula (5), the evaluation of the greening level of waterfront landscape, lawn landscape, square landscape, and avenue landscape is calculated as 1.273, 1.556, 1.023, and 1.453.

$$\text{Evaluation of afforestation level} = \text{VCR} * \text{w}_1 + \text{VGI} * w_2 + \text{SGE} * \text{w}_3 \tag{5}$$

3 Questionnaire Design and Data Collection

3.1 Questionnaire Design

The questionnaire was collected anonymously and divided into three parts. The basic information of students is the first part of the questionnaire, including grade and gender. The second part is the Perceptive Restoration Scale(PRS), which evaluates the recovery experience respectively of Dongpo Lake, Qidian Lawn, Tourism College

Square, and East Haiyun Road. The third part is subjective greening evaluation. Students make 1–5 points of evaluation on the campus landscape based on their personal subjective feelings. All measurement scales in the questionnaire used a 5-point Likert scale, which was expressed as 1 (strongly disagree) -5 (strongly agree).

Measures of recovery environment, Hartig et al. and Laumann et al. produced the Perceptive Restorative Scale (PRS) and Restoration Components Scale (RCS) [17] respectively. The main difference is that PRS measures four structures, including being away, fascination, extent, and compatibility, while PRS measures five structures, including fascination, novelty, escape, extent, and compatibility dimensions. Both tools are ultimately composed of 22 items. In this study, 17 revised projects of the Huangzhang Exhibition were used for reference, including five questions about being away [18], four questions about fascination, four questions about the extent, five questions about compatibility, and four dimensions to measure the restorative experience of the landscape environment.

3.2 Data Collection

Data will be collected by questionnaire survey on the campus of Hainan University from February 20 to February 25, 2022. The efficiency of data collection will be improved by offline code scanning and online questionnaire filling. Prizes will be offered to students who actively participate in the questionnaire to improve the rate of questionnaire collection. A total of 98 questionnaires were collected, including 13 invalid ones and 85 valid ones. The questionnaire recovery rate was 86.7%. The proportion of male students was 50.6% and that of female students 49.4%, which was almost the same. Import the questionnaire data into SPSS26.0 to process and calculate the data.

4 Data Processing and Analysis

4.1 Reliability and Validity Analysis

SPSS26.0 was used to analyze the reliability of the results of the PRS for Dongpo Lake, Starting Lawn, Tourism College Square, and Haiyun East Road respectively. The obtained reliability coefficient (Cronbach's α) ranged from 0.803–0.929, higher than the minimum reliability requirement of 0.600, indicating that the four dimensions of the four scales had high content reliability.

Convergence validity analysis is mainly used to judge the correlation between items in the same dimension, which is mainly represented by AVE value and combination reliability. In the analysis of the validity of the Greenland Landscape scale, it was found that AVE values of the two dimensions of distance and compatibility

Table 1 Analysis of Greenland landscape

	AVE	CR	Cronbach's α
Being away	0.436	0.792	0.929
Extent	0.629	0.871	0.913
Fascination	0.637	0.875	0.922
Compatibility	0.434	0.780	0.918

were 0.436 and 0.434, respectively, less than 0.500, but were still within the acceptable range given that their reliability combinations were both greater than 0.700 (see Table 1). The content validity of other dimensions meets the requirements.

4.2 Descriptive Statistics

Waterfront landscape, square landscape, Lawn landscape, Avenue landscape rehabilitation evaluation statistics as shown in Table 2, students can be found on the score of the campus landscape types generally low, in addition to the evaluation of waterfront landscape restorative scored an average of four dimensions are greater than three, avenue landscape, square landscape, lawn landscape dimensions are averaged less than three. According to Liu Qunyue et al. 's study on environmental preference and recovery evaluation, it was concluded that environmental preference had a significant positive impact on recovery evaluation [19]. It shows that many students on campus are not satisfied with the campus landscape. To sum up the reasons, it may be caused by the lack of landscape management on campus.

The score of square landscape in being away, fascination, and extent are not ideal, but the score in compatibility dimension is higher than lawn landscape and avenue landscape. Among the four landscape types selected, the waterfront landscape is most close to the natural landscape, while the other three landscape types are more inclined to the artificial landscape. By analyzing the chart, it can be concluded that natural landscape has more restorative benefits than artificial landscape. The restorative benefits of the four landscapes were as follows: waterfront landscape > lawn landscape > avenue landscape > square landscape.

Table 2 The average score of four landscape types

	Being away	Extent	Fascination	Compatibility	PRS
Square landscape	2.883	3.010	2.951	3.057	2.975
Lawn landscape	2.953	3.094	3.068	3.049	3.037
Waterfront landscape	3.548	3.643	3.851	3.501	3.623
Avenue landscape	2.925	3.082	3.042	3.044	3.023

Table 3 Correlation analysis

	Being away	Extent	Fascination	Compatibility	PRS
VCI	−0.486	−0.451	−0.458	−0.578	−0.487
VGI	0.008	0.049	0.040	−0.100	0.007
SGE	0.877	0.889	0.888	0.820	0.873
GE	−0.064	−0.025	−0.033	−0.172	−0.066

4.3 Correlation Analysis

Pearson correlation analysis was used to analyze the data, as shown in Table 3. It was found that VCI and greening evaluation levels were negatively correlated with the recovery evaluation, which was contrary to previous research conclusions. Considering the particularity of waterfront landscape, in the waterfront landscape of Dongpo Lake, the proportion of water bodies within the visual area reaches 23.6%. White and his colleague's study have found that aquatic components in the natural environment had the greatest influence on the recovery of scenes, and called this finding "the dose–effect of water" [20]. It can explain that the waterfront landscape Dongpo Lake has the highest recovery benefit.

Therefore, the relevant data of Dongpo Lake were removed and the correlation analysis of the data was conducted again. The obtained results are shown in Table 4. A significant positive correlation was observed between the evaluation level of greening and PRS. In the four dimensions of the evaluation of restoration, except for compatibility, PRC has a positive correlation with the evaluation level of greening and VCI. VCI and VGI were positively correlated with extent, and the evaluation level of greening was positively correlated with fascination. However, compatibility is negatively correlated with the evaluation level of greening. It can be seen from the statistical chart in Table 2 that the compatibility score of the square landscape is the highest. A compatible environment means that a person's purpose is consistent with the function that the environment can carry. Wang Yangyang et al. found in their research that residents' satisfaction can be achieved when the green visibility is about 30%. If the level of green visibility in a landscape is higher than 30%, other factors will become prominent influencing factors [21]. Therefore, after meeting the basic requirements of the green visual ratio, the square landscape provides students with a series of interesting projects such as class activities and club activities, so it has a high score in the compatibility dimension.

5 Conclusions and Discussion

The results showed that the restoration evaluation score was in the order of natural waterfront landscape > lawn landscape > avenue landscape > square landscape. The restoration evaluation of natural landscape was higher than that of artificial landscape,

Table 4 Corrected correlation analysis

	Being away	Extent	Fascination	Compatibility	PRS
VCI	0.957	1.000[b]	0.996	– 0.844	0.996
VGI	0.953	1.000[a]	0.994	– 0.851	0.995
SGE	0.999[a]	0.942	0.968	– 0.611	0.967
GE	0.973	0.998[a]	0.999[a]	– 0.811	1.000[a]

[b] Correlation is significant at the 0.01 level (2-tailed)
[a] Correlation is significant at the 0.05 level (2-tailed)

and the restoration evaluation of waterfront landscape was higher than that of land landscape, and the greening level was significantly positively correlated with the restoration evaluation score.

Waterfront landscape Dongpo Lake has the highest recovery benefit evaluation for students, indicating that the natural waterfront landscape can well restore students' ability of directional attention and relieve the mental fatigue caused by daily learning.

The VGI of the starting lawn reached 78.3%, which was the highest among the four landscape types. But the overall recovery evaluation score is not very ideal. Lawn landscape is an indispensable part of the campus landscape, and it is hoped that in the future campus lawn landscape design can be combined with the hierarchical design of plants, to stimulate students' desire to explore and enhance the fascination and extent. Building pavilions and corridors on the lawn to enhance the functionality of the lawn landscape and enhance the compatibility of the lawn landscape.

Campus square is an important venue for students' activities, while the square of tourism college is the worst in the above research and analysis, and the evaluation of greening level is not high as a whole. In the future, it can be combined with water features or plants to enhance the creation of plants inside the square to improve the fascination and extent of the square landscape.

In this study, the canopy of trees in the avenue landscape was larger, and the VCI reached 80%, indicating a higher overall greening level. In future campus avenue design, under the condition of satisfying the shading rate, the richness of plants should be appropriately increased, which should be combined with campus culture and local characteristics to improve the compatibility of the landscape. In addition, the campus landscape should be maintained regularly to ensure the beauty of the landscape and improve the overall recovery of the campus landscape.

This study using more innovative mathematical methods discussed the campus waterfront landscape, lawn landscape, square landscape, landscape avenue landscape restorative benefits for college students, respectively, and the four types of landscape greening level are the multi-dimensional research, using the entropy weight method to green rate, vegetation coverage, subjective evaluation respectively empowerment landscape, various types of overall green landscape level. Pearson correlation analysis was usually used to analyze the data, and it was found that VGI, VCI, and SGE were positively correlated with the evaluation of recovery.

Since quantification of environmental restorative benefits in the form of a questionnaire survey is subjective, objective methods such as electroencephalogram and electrocardiogram can be considered in future studies. In addition, heterogeneity exists in each landscape, which may affect the experimental results. In future studies, scholars can investigate multiple sites of the same landscape and use mathematical methods such as Moran index analysis to eliminate the interference of other factors and improve the reliability of research results.

References

1. Z. Kasim, M.F. Shahidan, Y. Yusof, Use of landscape environmental setting for pedestrian to enhance campus walkability and healthy lifestyle; proceedings of the WIT transactions on ecology and the environment, F, 2018. WIT Press
2. R. Kaplan, S. Kaplan, Cognition and environment. functioning in an uncertain world. Ulrichs bookstore. Ann Arbor, (1983)
3. R. Kaplan, S. Kaplan, *The experience of nature: A psychological perspective.* (Cambridge university press, 1989)
4. R.S. Ulrich, R.F. Simons, B.D. Losito et al., Stress recovery during exposure to natural and urban environments. J. Environ. Psychol., **11**(3), 201–230(1991)
5. A.E.V.D. Berg, A. Jorgensen, E.R. Wilson, Evaluating restoration in urban green spaces: Does setting type make a difference? Landsc. Urban Plan. **127**, 173–181 (2014)
6. K.K. Peschardt, U.K. Stigsdotter, Associations between park characteristics and perceived restorativeness of small public urban green spaces. Landsc. Urban Plan. **112**, 26–39 (2013)
7. T. Hartig, F.G. Kaiser, P.A. Bowler, Further development of a measure of perceived environmental restorativeness. Institutet för bostads-och urbanforskning, 1997
8. S. Kaplan, The restorative benefits of nature: Toward an integrative framework. J. Environ. Psychol. **15**(3), 169–182 (1995)
9. A.E.V.D. Berg, S.L. Koole, N.Y.V.D. Wulp, Environmental preference and restoration: (How) are they related? J. Environ. Psychol. **23**(2), 135–146 (2003)
10. Liu Chang, LI Shu-hua, Chen Song-yu. Research on the moderating effect of campus green space visiting behavior on emotion under the influence of multiple factors: A Case study of three universities in Beijing. Landsc. Arch., **25**(03), 46–52(2018)
11. Y. Aoki, The relation between the open field of vision and green quantity sense. J. Landsc. Arch. **51**(1), 1–10 (1987)
12. J. Lin, Y. Chen, X. Wang, Evaluation of road greening level in Gulou District of Fuzhou City based on green vision rate. Chin. Urban For., **19**(03), 73–7+84(2021)
13. V. Badrinarayanan, A. Kendall, R. Cipolla, Segnet: A deep convolutional encoder-decoder architecture for image segmentation. IEEE Trans. Pattern Anal. Mach. Intell. **39**(12), 2481–2495 (2017)
14. W. Zhang, Y. Zhou, M. Yang, Research on automatic recognition and Calculation of panoramic green view rate. Landsc. Arch., **26**(10), 89–94(2019)
15. J. Zhang, L. Shu, C. Zhang et al., Application of comprehensive index method based on entropy weight in groundwater quality evaluation. Hydropower Energy Sci., **28**(08), 30–32(2010)
16. F. Song, X.H. Yang, F.F. Wu et al., Development and evolution of human settlements ecosystem in the Lower Reaches of the Yangtze River based on information entropy. Soil Water Conserv. Res., **26**(01), 245–251(2019)
17. T.R. Herzog, P. Maguire, M.B. Nebel, Assessing the restorative components of environments. J. Environ. Psychol. **23**(2), 159–170 (2003)
18. H. Zhangzhan, H. Fangming, Z. Xianjie, The relationships between environmental preference and restorative perception of the environment: A case of mountainscape. Outdoor Recreation Study **21**, 1–25 (2008)

19. Q.Y. Liu, Y. Chen, W. Zhang et al., Study on the relationship between environmental preference, recovery evaluation and health benefit evaluation of tourists: A case study of Fuzhou National Forest Park. Resources Science **40**(02), 381–391 (2018)
20. M. White, A. Smith, K. Humphryes et al., Blue space: The importance of water for preference, affect, and restorativeness ratings of natural and built scenes. J. Environ. Psychol. **30**(4), 482–493 (2010)
21. W. Yangyang, H. Jinlou, Construction of urban ecological comfort evaluation model based on green visual ratio. Acta Ecol. Sin. **41**(06), 2170–2179 (2021)

Research on the Ideological and Political Reform of Mobile Communication Technology Courses Under the CEC Education Model

Tong Liu, Yu Liu, Haitao Jiang, and Rong Deng

Abstract The CEC education model, which emphasizes the participation of enterprises in international exchanges and cooperation, has created a new situation in international vocational education, and provided a new solution for improving the level of international vocational education. However, in addition to knowledge transfer at the professional level, colleges and universities that hold the CEC education cooperation model should also strengthen the worldview education of international students to improve the overall quality of students. Taking the major of mobile communication technology as an example, this paper discusses the ideological and political teaching of courses under the CEC education model. Subsequently, through the analysis of the survey data, we prove the effectiveness of the ideological and political construction of the curriculum in the international students' classroom under the CEC education model.

Keywords CEC education model · Communication technology · Course ideology and politics

1 Introduction

Recent years have witnessed the great progress in higher vocational education in China. The development of higher vocational education also made irreplaceable contributions to economic development, employment promotion and improvement of people's livelihood [1]. At the same time, with the deepening of economic globalization and the internationalization of higher education, the international cooperation of higher vocational education has become closer. China attaches great importance

T. Liu (✉) · Y. Liu · H. Jiang · R. Deng
Department of Big Data and Internet of Things, Chongqing Vocational Institute of Engineering, Chongqing, China
e-mail: liuyyshiwo@163.com

S. Patnaik and F. Paas (eds.), *Recent Trends in Educational Technology and Administration*, Learning and Analytics in Intelligent Systems 31,
https://doi.org/10.1007/978-3-031-29016-9_19

to the international development of higher vocational education. Some higher vocational colleges take the initiative to jointly carry out exchanges and mutual visits of teachers and students with foreign colleges and universities, forming a multi-level, multi-form and wide-ranging international exchange and cooperation pattern. With the continuous improvement of the quality of China's higher vocational education, some scholars even pointed out that China's higher vocational education is expected to provide experience to the world. Some higher vocational colleges have also taken "going to the world" as their development goal.

However, there are still some problems with the achievement [2]. For example, the overall level of international cooperation in higher vocational education is relatively low, the scale of international exchanges is not commensurate with the development scale of higher vocational education and the form of international cooperation and exchange is single. The current running mode of international cooperation is mainly based on college-to-college cooperation model, which lacks the in-depth participation of enterprises, resulting in the inability to cultivate talents that enterprises needed. The problem of the management system of both parties restricts the breakthrough development of international vocational education cooperation. The differences in the thinking mode of the two sides have led to differences in the curriculum setting, curriculum construction, teaching mode, teaching organization, assessment and evaluation and teaching staff construction. Another important reason is the lack of corporate participation in international cooperative education. The form and content of cooperation between colleges are relatively simple, which results in the lack of practical ability of the students. Even some colleges invite enterprises to participate, but the enterprises do not have a dominant position and lack the enthusiasm and initiative to participate in international cooperation [3]. In some cases, the role of enterprises is limited to participating in teaching and providing some practical training facilities, which leads to the weak competitiveness of the students.

2 CEC Education Model

In order to adapt to the economic globalization and the international development of higher vocational education, colleges need to cultivate international and local talents for enterprises. Chongqing Vocational Institute of Engineering (CQVIE) tries to change the current mode of international exchange and cooperation in higher vocational education, and explores a talent training model namely College-Enterprise-College (CEC) [4].

CE in CEC means that colleges and enterprises jointly fund the establishment of joint talent training bases, carry out mixed-ownership colleges, and jointly cultivate talents needed for the international development of enterprises. EC is the establishment of the "Belt and Road" international college jointly established by colleges in cooperative countries to cultivate localized talents for international enterprises. This education mode has been proved to be an effective way to promote the rapid development of international exchanges and cooperation in higher vocational education in these countries.

2.1 Ideological and Political Construction Under the CEC Education Model

When developing vocational and technical education in the context of CEC, it should also be noted that the success of students depends not only on the amount of professional knowledge they have, but also on their overall comprehensive quality. In order to further improve the success rate of students, it is also an important part of international vocational education cooperation to transmit and convey correct values and world views to the students.

2.2 The Classroom for International Students is an Important Place to Show China

President Xi Jinping once emphasized that "Running ideological and political courses well is to carry out Marxist theoretical education, and use socialism with Chinese characteristics in the new era to cast the soul and educate people". President Xi Jinping's words provide a fundamental method and point out the direction for us to further develop the ideological and political teaching of the curriculum. However, the understanding of the theoretical education of Marxism should not stop at the level of explaining the principles and methods of Marxism. It cannot be limited to explain Marxism-Leninism mechanically and dogmatically but should focus on the study and transformation of the application results of Marxism-Leninism. For example, the concepts of democracy, equality, justice, and the rule of law emphasized in the core socialist values, a typical local result of the of Marxist materialism, are not only the values advocated by the Chinese government, but also the spiritual beliefs that are vigorously promoted by many countries in the world. At this level, the pursuits of China and these countries are similar, and these concepts can be used as the core line in the process of curriculum ideological and political construction. At the same time, the good expressions of dedication, integrity, friendliness, etc. in the core socialist values are closely related to the growth and development of students throughout their lives. Whether it is a Chinese student or an international student, having these good qualities will play an important role in their lifelong growth.

In addition, the world today is undergoing profound changes unseen in a century. Countries around the world are intertwined and agitated with each other. Most countries follow the development trend of the times and take national construction and people's well-being as the goal of development, strengthen exchanges with other countries to promote mutual development. However, very few countries and political parties in the world are still going against the current, trying to smear China through improper speech and behavior. Such despicable methods prevent many foreigners from truly understanding China. With the gradual increase in the influence of China's international education, a large number of foreign students who have a strong pursuit of China's advanced culture have come to China for study. Their university classroom is the best place to show and promote China to the world. By promoting and publicizing Chinese culture, displaying Chinese image, and interpreting Chinese policies according to incorporating positive "course ideological and political" elements into the classroom, these students can become promoters of China's international image building, contribute to the friendly exchanges between Chinese and foreign cultures, and make the world more really understand China.

2.3 Ideological and Political Development of Courses is an Important Purpose of Teaching

Teacher's duty is to teach and educate people, and to cultivate people's virtues. People here are regardless of nationality, skin color, or race. No matter what country or ethnicity students come from, as long as they come to Chinese universities, they should receive the same education as Chinese students. The educational function of the university is not just the imparting of knowledge and skills. During the period of university study, students should also shape a correct outlook on life, values and world outlook under the influence of teachers, and develop good characters of self-confidence, optimism, tenacity and integrity.

During college, students should exercise their will to be active and hard-working, cultivate good professional quality, comprehensively improve their self-control ability and sense of social responsibility, and make adequate preparations for entering the society in the future, and for their long-term life lay a solid foundation for development. The ideological and political curriculum has the educational function of cultivating the above-mentioned excellent characters. Therefore, the development of curriculum ideology and politics in the classroom of international students is not only the essential requirement of education, but also the purpose of teaching and educating.

2.4 Ideological and Political Development of Courses is an Important Guarantee for Our China's Modernization Construction

As more and more foreign students come to China, their wonderful experiences in China, the wonderful things they see in China, and the friendship and enthusiasm of the Chinese people are deeply attracted to them. Many students choose to stay in China after their studies and jointly participate in China's modernization drive. President Xi Jinping also expressed China's welcome to these people with lofty ideals in his speech during a discussion with foreign experts. He mentioned that "the Chinese nation has always had a fine tradition of valuing and cherishing talents. We must implement a more open talent policy. We should not only introduce talents from different regions, but also make good use of talents without sticking to one pattern. While vigorously cultivating domestic innovative talents, we will introduce foreign talents more actively, and warmly welcome foreign experts and talents to participate in China's modernization drive in various ways."

3 Teaching Cases

During its more than 70 years of education, CQVIE has attached great importance to international exchanges and cooperation. Under the framework of CEC teaching model with ZTE Corporation and Beijing Huasheng Jingshi Information Technology Co., Ltd., the college has recruited students from Uzbekistan, Russia, Laos and other countries to carry out international vocational education training with the major of mobile communication technology as a pilot. In the teaching process, we not only attach importance to teaching students professional and technical knowledge, but more importantly, incorporate some Chinese cases, stories, and methods into the teaching process, and integrate professional teaching and value influence throughout the teaching. Below are a few cases from the teaching process.

3.1 Telecommunication Network Structure

The above figure is a typical communication network structure topology including the access layer, transport layer and core layer in the communication network. Each layer is composed of different devices connected to each other, such as PTNs, routers, switches, etc. The function of the transmission device is to send information from the source node to the destination node by choosing different routes. During this process, if a certain node has an error in encoding or decoding the information, the subsequent node will continue to transmit the wrong message, which will cause the destination node to receive the wrong message. Our usual social network also has the same structure as the transport network. If someone with ulterior motives is spreading rumors, and the next network node will pass on the information without any judgment and analysis, it will cause rumors diffuse and panic. Therefore, when we discuss this part with the international students in the classroom, we should tell them to keep a clear head and maintain independent problem analysis ability while not believing or spreading rumors.

3.2 FDMA/TDMA/CDMA

The emergence of multiplexing technology has greatly improved the resource utilization of communication systems. FDMA improves the utilization of wireless spectrum [5]. The emergence of TDMA has significantly increased the capacity of the system [6]. CDMA uses different coding methods to further improve the capacity [7]. And later OFDMA, NOMA is to further enhance the capacity of the system [8, 9]. Multiplexing is an important technology in mobile communication technology, and multiplexing can also be achieved in life, especially in time multiplexing. For example, while waiting for the bus, you can make full use of this time to read a book. While riding the bus, listening to a popular song is a good choice to relax the mood. In this context, we will further talk about the importance of time and guide students to cherish time.

3.3 Application of 5G and Mobile Edge Computing

The future mobile communication network will no longer be just a communication network, it will also be a computing network at the same time. The emergence of 5G provides users with fast data transmission channels, while edge computing sinks the center of computing to the edge of the network, further shortening the latency of user task execution [10]. The emergence of new information technologies such as 5G and edge computing will change the way of life of human beings to a great extent. For example, 5G technology has played an extremely important role in the fight against the epidemic. On this basis, review the process of China's fight against the epidemic, show the Chinese government's emphasis on people's life and health, and share the stories of China's help in Serbia, Italy and many other countries in fighting the epidemic will make the international students know more about China. According to these stories we can show them the importance the Chinese government attaches to people's lives, the institutional confidence of the socialist system with Chinese characteristics and China's responsibility as a major country.

4 Data Analysis

In the course of teaching, we tracked and investigated the impact of adding ideological and political content in the classroom to international students. The subjects of the survey are the students of the 2017 and 2018 grade of mobile communication technology in CQVIE. Among them, the 17 students of the 2017 grade are all from Uzbekistan while the students of the 2018 grade are from Belarus, Laos, Bangladesh and other countries. When testing this model of teaching reform, we selected some typical stories in traditional Chinese culture, such as "Iron pestle grinding needle", "repairing the prison after the sheep is gone" and "the idiot moving the mountain". The test is divided into two parts, the first is for international students to retell the story. The second is to share what they learned after reading the stories Figs. 1 and 2.

The test results are shown in Fig. 3. The abscissa in the figure is the time interval, in months. The ordinates are the students' completeness of retelling and the recognition of their values. As can be seen from the figure, with the longer the course ideology and politics are carried out in the classrooms of international students, the students' understanding and mastery of Chinese is also improving, and their understanding of traditional Chinese culture is also gradually deepening. For example, after one semester, that is, 5 months of study, students in the 2017 grade can retell the stories correctly by 42%, an increase of 13% compared with 29% in the first month.

From the data in Fig. 4, it can be seen that with the introduction of ideological and political courses in more and more depth, international students have a more thorough understanding of fables in traditional culture. There is also growing recognition of the value proposition contained in these stories. For example, after completing one year of study, the 2018 students' recognition of the values contained in the fables in traditional Chinese culture increased from 89 to 98%, an increase of 9%.

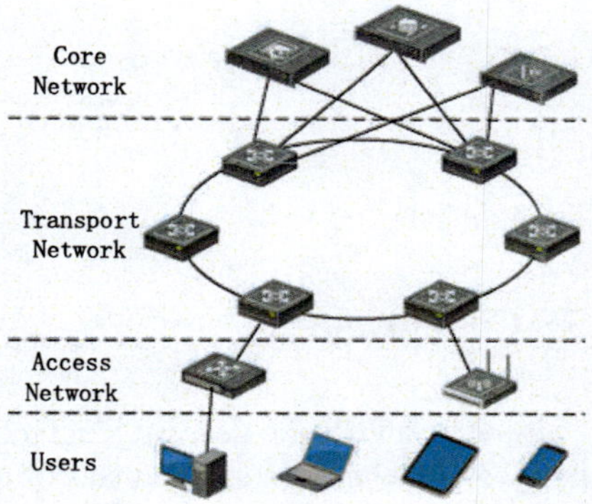

Fig. 1 Telecommunication network structure

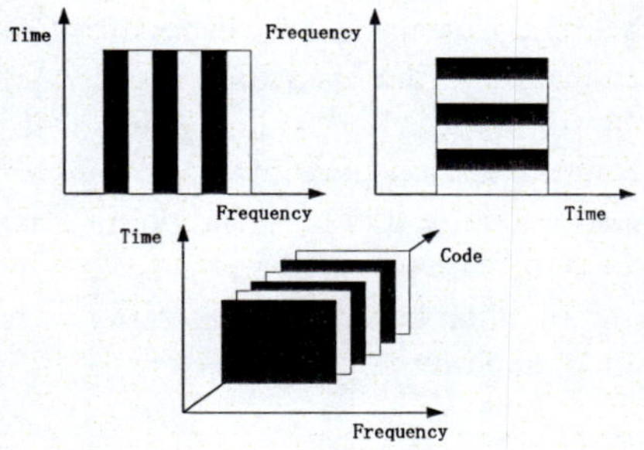

Fig. 2 Multiplexing technology

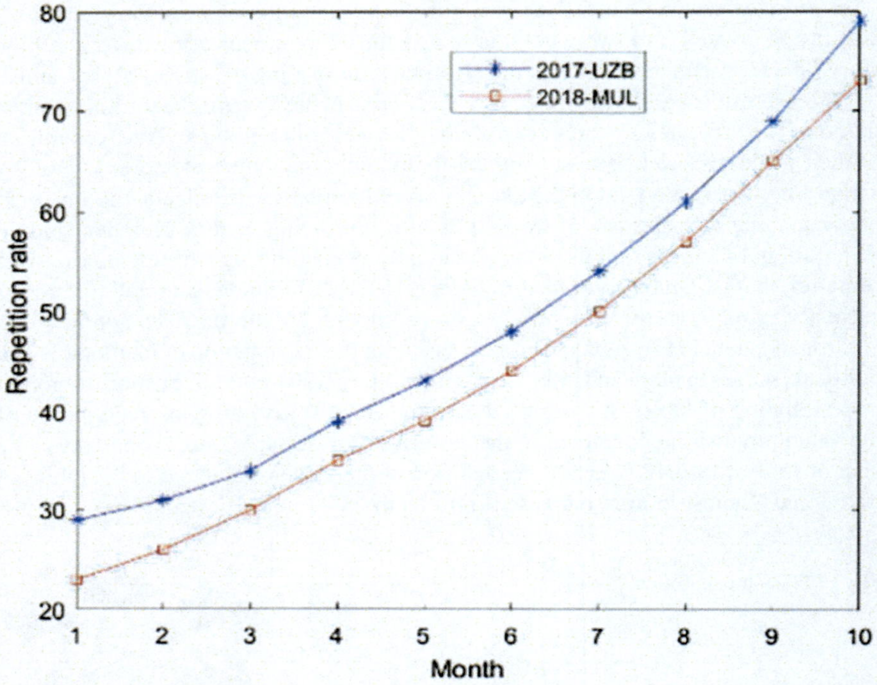

Fig. 3 Multiplexing technology

5 Conclusion

This paper studies the necessity and method of developing curriculum ideology and politics in the professional courses of mobile communication technology under the CEC education model. First, we introduce the CEC education model that can effectively improve the level of international higher vocational education cooperation. Secondly, we discuss the necessity of developing curriculum ideological and political education in the classroom of international students under CEC model. After that, taking the ideological and political teaching reform of the 2017 and 2018 mobile communication technology courses of CQVIE as an example, the case of introducing course ideology and politics into professional teaching is presented. Finally, the data shows the effectiveness of carrying out curriculum ideology and politics in international students' classrooms in helping international students to deepen their understanding of China.

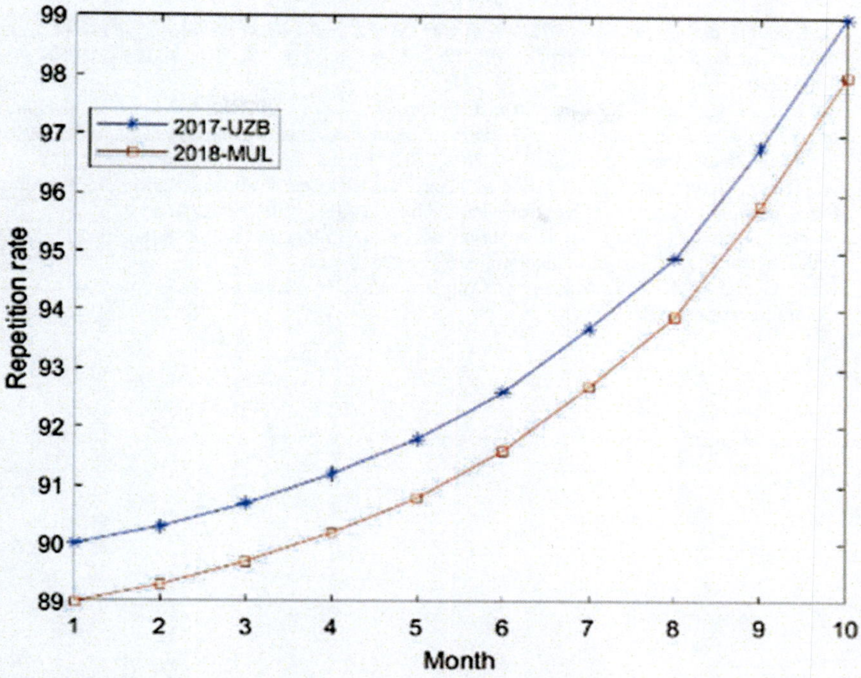

Fig. 4 Multiplexing technology

Acknowledgements This paper is supported by the Science and Technology Research Program of Chongqing Municipal Education Commission (KJQN202103402), the Key scientific research project of Chongqing Vocational Institute of Engineering (Grant No. KJA202135), Chongqing Higher Education Teaching Reform Research Project (Grant No. 203706 and No.202072S) and Chongqing Vocational Institute of Engineering Education and Teaching Reform Key Research Project (Grant No. JG191006 and Grant No. SDA202001).

References

1. J. Li, M. Pilz, International transfer of vocational education and training: a literature review. J. Vocat. Educ. Train. **2**, 1–34 (2021)
2. X. Zhang, Research on the path of agriculture-related higher vocational colleges in serving the Belt and Road Initiative. Mod. Econ. & Manag. Forum, **2**(6), (2021)
3. J. Li, *Study on the ways of creating the upgraded version of higher vocational education with Chinese characteristics*, (2019)
4. Y. Zhou, Z.Y. Yang, Exploration and practice of diversified school-running system of higher vocational education. Theory Pract. Educ. **39**(12), 27–29 (2019)
5. F. Ananasso, E. Saggese, Survey On the technology of multicarrier demodulators for FDMA/TDMuser-oriented satellite systems. 162–168 (1985)

6. A. Urie, *On the application of TDMA in UMTS, IEE colloquium on advanced TDMA techniques and applications* (Digest No: 1996/234, 1996), pp. 1/1–1/4, https://doi.org/10.1049/ic:19961230
7. L. Hanzo, L-L. Yang, E-L Kuan, K. Yen, CDMA *overview, in single and multi-carrier ds-cdma: multi-user detection, space-time spreading, synchronisation, networking and standards* (IEEE, 2004), pp.35–80. https://doi.org/10.1002/0470863110.ch2.
8. A.S. Khan, J. Abdullah, H. Lenando, J.M. Nazim, Green resource allocation for multiple ofdma based networks: a survey. J. Electron. Sci. Technol. **14**(02), 170–182 (2016)
9. Wang Z , Cheng Q , Fan Z , et al. A survey of resource allocation in non-orthogonal multiple access systems[J]. Telecommunications Science, 2018.
10. Liang C, Jiang J J . A Survey on 5G Communication Scenarios and Techniques[J]. Mobile Communications, 2015.

Exploration on the Integration of New Generation Information Technology and Curriculum Ideological and Political Under the Background of CEC Education Model

Tong Liu, Haitao Jiang, Yu Liu, and Rong Deng

Abstract On the basis of inheriting the traditional international higher vocational education cooperation and combining the characteristics of vocational education, the CEC education model introduces enterprises as an important part of the cooperation, which injects new development impetus into the international higher vocational education cooperation. However, except imparting knowledge and skills, education also has the function of shaping international students' character. Therefore, in the teaching process based on the CEC model, we combine the features of the new generation of information technology, carry out curriculum ideological and political education in the classrooms of international students. The survey data proves the effectiveness of curriculum ideological and political construction in international students' classrooms under the CEC education model.

Keywords New generation information technology · International student · CEC education model · Course ideology and politics

1 Introduction

Chongqing Vocational Institute of Engineering (CQVIE) has always attached great importance to international exchanges and cooperation in running schools. In more than 70 years of practice, through the joint efforts of teachers and students in the school, CQVIE have achieved good results in this regard. At present, CQVIE has established a long-term inter-school exchange mechanism with universities in the

T. Liu (✉) · H. Jiang · Y. Liu · R. Deng
Department of Big Data and Internet of Things, Chongqing Vocational Institute of Engineering, Chongqing, China
e-mail: liuyyshiwo@163.com

S. Patnaik and F. Paas (eds.), *Recent Trends in Educational Technology and Administration*, Learning and Analytics in Intelligent Systems 31,
https://doi.org/10.1007/978-3-031-29016-9_20

United States, Canada, the United Kingdom, Australia, South Korea and other countries, and has carried out activities in professional cooperation, curriculum certification, vocational qualification certification, teacher training, student exchanges, and education improvement.

In 2013, after China proposed the "Belt and Road" international cooperation initiative, many countries around the world responded positively, strengthened cooperation in politics, economy, trade, culture, etc., and joined the historical process of global integration. Based on its own school-running characteristics, CQVIE provides overseas student training and technical services to countries along the "Belt and Road" such as Malaysia, Laos, Uzbekistan, Belarus and other countries.

However, in the process of cooperation and exchange with foreign universities, we found that the students jointly trained to a certain extent cannot well adapt to the requirements of the current globalization development for vocational and technical personnel. After in-depth exploration, we believe that the core problem is the lack of corporate participation in international vocational education exchanges and cooperation.

2 CEC Education Model

The characteristics of vocational education determine that enterprises must be an important part of vocational education. Many years ago, the Chinese vocational education community has realized that the cost of human resources accounts for a huge proportion of all the costs of enterprises, and the expenditure on employee training is particularly huge. In order to reduce the expenditure on training, many companies hope that college graduates can be able to master the basic knowledge and skills required for the job when they graduate.

After the above problems appeared, many higher vocational colleges and enterprises actively carried out school-enterprise cooperation. Through joint school running, mutual assignment of teachers, joint development of teaching resources, etc., after years of trial and practice, school-enterprise cooperation has achieved certain results. CQVIE has improved its own school-running level, the company has recruited more outstanding talents, and the students have received good vocational education, forming a good situation of win–win for all parties.

For example, CQVIE signed a school-enterprise cooperation agreement with ZTE Corporation and Beijing Huasheng Jingshi Information Technology Co., Ltd. in 2016. Taking modern mobile communication technology and cloud computing technology as a pilot program, the two parties jointly funded the establishment of the Ministry of Education-ZTE The ICT industry innovation base, including the optical communication training room and the telecommunications engineering training room. They have carried out vocational skills training in the construction, design, construction and optimization of LTE and 5G networks, and has achieved good results. In recent years, graduates have been highly recognized by enterprises, and students have won national awards in skill competitions and innovation and entrepreneurship competitions [1].

Through the in-depth development of school-enterprise cooperation, we deeply realize that only by introducing enterprises into schools can we know what kind of talents the communication industry needs in a zero-distance way. This conclusion can also be applied to international vocational education. The unsatisfactory level of international vocational education cooperation is largely due to the absence of the participation of enterprises. The two sides of the cooperative education are only carrying out talent training from the perspective of education, but there is a lack of real understanding of the market's demand for international talents. As a result, the training objectives cannot be focused on the needs of the industry, and the trained students cannot meet the requirements of the market.

In this context, CQVIE has pioneered the creation of the CEC international vocational education cooperation model. The first and second C in the CEC model represent the two sides of the cooperative colleges while character E means the enterprises participating in the cooperative education. The CEC model is committed to cultivating international high-quality technical and technical talents for multinational communication companies and their affiliated enterprises, including dual-purpose talents who are familiar with the domestic environment as well as international standards and international rules. At the same time, it can actively respond to the national "Belt and Road" initiative and cultivate talents needed by communication companies in other countries.

3 Ideology and Politics in the New Generation of Information Technology

The classroom of colleges and universities has the unity of multiple goals of knowledge imparting, value shaping and ability training. In the past, the classrooms of colleges and universities in China focused more on the imparting of knowledge and the cultivation of skills, but neglected the shaping of students' price and quality. In order to give full play to the educational function of college classrooms, many teachers actively carry out curriculum ideological and political construction. It is worth noting that the development of curriculum ideology and politics cannot be limited to the classrooms of Chinese students, and the development of curriculum ideology and politics is also required in the classrooms of international students. The university classroom is the best place to show China and promote China to these international students who play the role of a bridge between Chinese and foreign cultures. By promoting Chinese culture, displaying Chinese image, and interpreting Chinese policies by incorporating positive "course ideological and political" elements to the international students, they can become promoters of China's international image building, contribute to the friendly exchanges between Chinese and foreign cultures, and make the world more really understand China. At the same time, international students are still in a critical period of formation and establishment of values during their studies. It is very important to convey and share correct values to them during this period, which will help them grow and grow in the future. Therefore, in the context of the CEC education model, it is particularly important to carry out curriculum ideology and politics in the classrooms of international students. Below we share some cases of curriculum ideology and politics in the international student classroom combined with the new generation of information technology.

Since human beings entered the third industrial revolution, information technology represented by computers and communications has greatly promoted the development of productivity and changed human life and production methods to a large extent. Taking mobile communication technology as an example, the 1G network allows people to freely make voice calls with another person at any time and any place, breaking through the mobility limitations of fixed phones. Then, with the large-scale construction and popularization of 2G networks, mobile communication is gradually becoming market-oriented, allowing more people to experience the network experience without distance restrictions, and bringing the distance between people. The emergence of mobile E-mail supported by 3G networks has improved people's work efficiency, and online video-on-demand has made people's entertainment more diversified. As the market's demand for data traffic services continues to increase, the 4G network based on LTE supports higher-speed data transmission. In the 4G network, services such as online car-hailing and vehicle navigation services have changed the life of human, while live webcasting and AR/VR experiences have changed the way people enjoy entertainment. The new generation of information technology represented by mobile communication networks will continue to impact on human life and production methods in the fourth industrial revolution, and the impact will be greater and greater.

In this context, many countries are actively planning the development layout of the new generation of information technology, such as the Industry4.0 plan proposed by Germany and the Made in China 2035 strategic goal formulated by the Chinese government. A large number of promising new information technologies have emerged, such as mobile edge computing, artificial intelligence, blockchain, etc. While helping students understand the principles of these advanced technologies, we will also add appropriate ideological and political elements to the curriculum to guide students in the establishment of values.

4 Mobile Edge Computing

With the continuous emergence of various new applications, especially in the fields of Industrial Internet and Internet of Vehicles, many services with low-response delay need puts forward extremely demanding requirements. Such as emergency braking and merging assistance in the Internet of Vehicles, excessive network delay will lead to serious consequences. However, the current network structure based on cloud computing is difficult to provide the guarantee of low latency [2]. The main reason is that cloud servers are generally deployed in distant places, and the increase in distance means an increase in transmission delay. In order to solve the above problems, the industry has proposed the concept of edge computing. By deploying distant computing and storage servers at the edge of the network, close to the distance between users and network resources, and providing users with reliable service guarantees with low latency [3].

The core idea of edge computing is to deploy computing resources and storage resources in the network at the edge of the network, thereby saving users' task solving delay and content reading delay. In an increasingly fast-paced modern life, time is an extremely valuable asset for anyone. However, there are still some students who can't cherish their time well and save time, thus wasting the opportunity to improve themselves. Therefore, at the end of explaining edge computing technology, we will summarize the importance of sharing time with students, improve students' time awareness, and share some useful methods and tools to help students make reasonable time arrangements.

4.1 Digital Twin

Digital twin (DT) is a multi-disciplinary, multi-physical, multi-scale simulation process that makes full use of physical model, operation history and other data to complete the mapping in virtual space, thereby reflecting the full life cycle of the corresponding physical equipment process [4]. DT is promising in creating a virtual world that is highly consistent with objects in the real world by collecting real-life data. It can be predicted that digital twins will have a very wide range of applications in future engineering construction, product design, medical analysis and other fields [5].

However, in the process of constructing a virtual world, massive amounts of real-world data need to be collected. For example, in order to have a more real-time and comprehensive understanding of a person's health status, it is necessary to collect basic data such as his blood type, height, and weight, and on this basis, use technologies such as three-dimensional reconstruction, combined with high-end data such as gene banks, to create a twin space. An avatar that looks exactly like this person. And by simulating the person's eating, living and other living habits, it analyzes the bad habits that may lead to diseases, and proposes targeted improvement measures. Although the future depicted by digital twin technology is extremely beautiful, the process of building a twin space for personal health involves the collection of users' private data. If these data are not properly stored, they will be stolen by network hackers and cause bad consequences. Therefore, in the classroom, we will use this as an introduction to publicize the importance of improving the protection of their own private data to international students. On this basis, privacy protection is extended to network security education, and active guidance is given in preventing network intrusion and network fraud.

4.2 Blockchain

In the past ten years, the value of a bitcoin has skyrocketed from a few dollars to hundreds of thousands of dollars. With the continuous improvement of the value of Bitcoin, the key technology behind it, the blockchain, has gradually entered people's vision. In essence, the blockchain is a shared database, and the data stored in has the characteristics of "unforgeable", "traceable", and "collective maintenance" [6]. Based on these characteristics, blockchain technology has laid a solid "trust" foundation and created a reliable "cooperation" mechanism. Blockchain has been widely used in product traceability, economic transaction vouchers, etc. [7]. Behind the technology is another very serious problem—the lack of integrity. Honesty is a fundamental moral principle and code of conduct shared by human society, and also an important principle for everyone to engage in social activities. If all the people were honest and trustworthy in the spirit of the contract, there might not be blockchain technology. Therefore, the honesty and spirit of contract in the introduction of blockchain technology is very consistent with the technical background of blockchain.

4.3 Artifical Intelligence

Artificial intelligence (AI) technology has developed rapidly in recent years, especially deep learning (as shown in Fig. 1) and enhanced learning represented by DDQN and DDPG have achieved unprecedented success in image recognition, face recognition and other fields [8]. Among them, the most famous is AlphaGo. From 2016 to 2017, AlphaGo defeated Li Shishi, Ke Jie and other top Go players. The emergence of AlphaGo even caused a big discussion of "whether machines will rule humans in the future".

However, what is unknown is that behind the success of AlphaGo, 1202 CPUs and 176 GPUs are continuously supporting computing power. The realization of AI needs to go through massive data screening, data training and data prediction steps, and each step means a lot of power consumption [9]. In the current environment

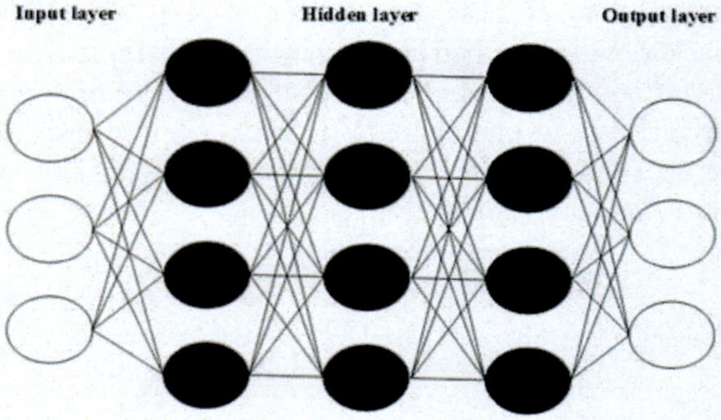

Fig. 1 Deep learning model

where new energy sources have not been developed on a large scale, whether it is necessary to consume such a high amount of energy to verify the superiority of AI is a topic worth discussion. Introducing a discussion on this topic can awaken students' awareness of environmental protection and use the classroom to promote environmental protection knowledge.

In addition, reinforcement learning in AI has many similarities between its learning process and human learning. For example, in the image recognition supported by Q-learning, by repeatedly showing the image to the agent, when the agent answers correctly, it can get a reward, and when the answer is wrong, the reward is 0. By describing the learning process of Q-learning, we can convey the concept to students that if we want to learn knowledge seriously and master it solidly, we need constant repetition and repeated drills. The scientific method of learning is to obtain knowledge through constant repetition and summarization.

4.4 5g

Mobile communication networks have undergone a gradual transition from 1 to 5G over the past few decades [10]. The original network can only support voice calls, and now the 5G network has supported the Internet of Things. 1G and 2G technologies originated in developed countries such as the United States, Finland and Japan. China can only choose to purchase equipment from these countries at an expensive price at this stage. But the Chinese have not given up the spirit of hard work. With the hard work of generations of communication practitioners, the Chinese have gradually mastered these advanced technologies. Among the four global 3G network standards announced by 3GPP, TD-SCDMA independently developed by China has become one of them. In the 4G era, LTE-TDD is also the network construction technology chosen by many operators around the world. In the 5G, Chinese companies represented by Huawei have made China's 5G technology the world's leading technology through their efforts. From being nothing in the past to the present, the technological leadership relies on the diligence and wisdom of the Chinese people, and on the spirit of the Chinese people's hard work. Through the review of the development history of mobile communication technology, we can show students the importance of hard work and hard work, and at the same time, we can show and promote China to international students.

5 Data Analysis

In 2017, CQVIE recruited 17 international students from Uzbekistan, and in 2018, the number of students expanded to 20, and the source countries covered Belarus, Laos, Bangladesh and other countries. We tracked and investigated the impact of adding ideological and political content in the classroom to them. We take "understanding

Fig. 2 Results

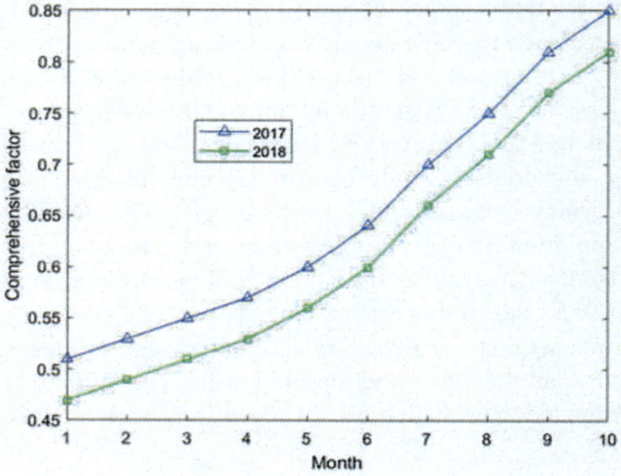

of traditional Chinese culture", "understanding of China's foreign policy", "professional ethics", "friendliness", "honesty and trustworthiness", "optimism", and "abiding by the law" as indicators. All metrics are finally multiplied by the corresponding weights and normalized. The duration of the investigation is 10 months. The results are shown in Fig. 2.

It can be clearly seen from the figure that the students in grade 2018 scored slightly lower on these indicators than the 2017 students, mainly because they came to China later and naturally had less knowledge of China. But no matter what grade international students came from, their performance on these metrics improved over time. The above data fully demonstrate the effectiveness of our curriculum ideology and politics in the classroom of international students.

6 Conclusion

The CEC model can effectively improve the education level of international vocational education cooperation and cultivate talents that enterprises really need. Cultivating talents in this mode not only needs to pay attention to the teaching of professional skills, but also pay attention to the cultivation of students' ideological character. Combined with the school-running example of Chongqing Engineering Vocational and Technical College, this article shares the case of developing curriculum ideology and politics in the new generation of information technology classrooms and cultivating international students to develop good ideological and moral character. Finally, the data analysis shows the effectiveness of carrying out curriculum ideology and politics in the classroom of international students.

Acknowledgements This paper is supported by the Science and Technology Research Program of Chongqing Municipal Education Commission (KJQN202103402), the Key scientific research project of Chongqing Vocational Institute of Engineering (Grant No. KJA202135), Chongqing Higher Education Teaching Reform Research Project (Grant No. 203706 and No. 202072S) and Chongqing Vocational Institute of Engineering Education and Teaching Reform Key Research Project (Grant No. JG191006 and Grant No. SDA202001).

References

1. Y. Zhou, Z.Y. Yang, Exploration and practice of diversified school-running system of higher vocational education. Theory Pract. Educ. **39**(12), 27–29 (2019)
2. M. Desertot, C. Escoffier, D. Donsez, T*owards an autonomic approach for edge computing: research articles.* (John Wiley and Sons Ltd, 2007)
3. Y. Mao, C. You, J. Zhang, K. Huang, K. B. Letaief, A survey on mobile edge computing: the communication perspective. in *IEEE Communications Surveys & Tutorials*, vol. 19, no. 4 (Fourthquarter, 2017), pp. 2322–2358. https://doi.org/10.1109/COMST.2017.2745201
4. Y. Wu, K. Zhang, Y. Zhang, Digital twin networks: a survey. IEEE Internet Things J., **8**(18), 13789–13804 (2021). https://doi.org/10.1109/JIOT.2021.3079510
5. D. Yang, H.R. Karimi, O. Kaynak et al., Developments of digital twin technologies in industrial, smart city and healthcare sectors: a survey. Complex Eng. Syst., **1**(1), (2021)
6. K. Yue et al., A survey of decentralizing applications via blockchain: The 5G and beyond perspective. in *IEEE Communications Surveys & Tutorials*, vol. 23, no. 4 (Fourthquarter 2021). pp. 2191–2217. https://doi.org/10.1109/COMST.2021.3115797
7. D. Yca, C.B. Hao, Z.B. Yang et al., A survey on blockchain systems: Attacks, defenses, and privacy preservation.High-Confid. Comput., (2021)
8. I. Ahmed, G. Jeon and F. Piccialli, From artificial intelligence to explainable artificial intelligence in industry 4.0: a survey on what, how, and where. in *IEEE Transactions on Industrial Informatics*. https://doi.org/10.1109/TII.2022.3146552.
9. B. Yu, S. Li, S. Xu et al., Deep learning: a key of stepping into the era of big data. J. Eng. Stud., (2014)
10. C. Liang, J.J. Jiang, A survey on 5g communication scenarios and techniques. Mob. Commun., 2015.

The Application of OLAP and Web Technology in the Evaluation of Higher Educational Quality and the Design of Management System

Yuru Liu

Abstract In recent years, with the rapid progress of computer science and the rapid development of higher education, the level of informatization of education management has improved significantly. The education quality evaluation system is a comprehensive management of education quality and an important tool for evaluating teacher education quality. Therefore, this article aims to study the application of OLAP and web technology in college education quality evaluation and the design of management system. On the basis of analyzing the main characteristics of OLAP, OLAP multi-dimensional data analysis methods and the performance requirements of the education quality evaluation system, the education quality assessment system was designed, and finally the performance of the system was tested. The test results show that when the system is operated by 500 people concurrently, it still maintains a system response time of less than 1 s and a CPU response rate of 17.5%, which is normal.

Keywords OLAP technology · College education · Quality evaluation · System design

1 Introduction

At present, China's higher education has entered the stage of mass education, and the quality of higher education has declined. Higher education, the government and the people of various disciplines pay more attention to guaranteeing the quality of higher education [1, 2]. The education system of higher education is affected by many factors and consists of many links. Therefore, the education system of higher education needs a good teaching mechanism. For this reason, education departments at all levels and schools of all kinds need to master and accurately grasp the progress,

Y. Liu (✉)
Xi'an Conservatory of Music, Xi'An 710061, China
e-mail: liuyuru@xacom.edu.cn

S. Patnaik and F. Paas (eds.), *Recent Trends in Educational Technology and Administration*, Learning and Analytics in Intelligent Systems 31,
https://doi.org/10.1007/978-3-031-29016-9_21

effectively guide, manage, organize, further improve and perfect the science degree, and need to adopt effective methods[3, 4].

Since the 1980s, China has introduced western modern education evaluation concepts, and domestic education quality evaluation research has developed rapidly. Especially in recent years, due to the continuous deepening of higher education reform, the reform of curriculum model, educational thinking, and education model has always been paid attention to, but the research on the evaluation of education quality has not received enough attention [5, 6]. In some Western countries with a higher education level, assessing the quality of education is a daily task, but in our country, there are still many universities that have not done this. As one of the evaluation tools, the content of the evaluation index is the same as that of classroom teaching 10 years ago. The model is basically the same [7, 8]; most of this teaching quality evaluation model is based on knowledge transfer, overemphasizing the selective function of teaching evaluation, and often neglecting the evaluation of the teaching process [9, 10]; the evaluation index system is lacking scientificity, technical unity and greater randomness, ignorance of quantitative analysis and qualitative evaluation, focusing on each part of the evaluation system, the overall study of the education quality evaluation system has not been reflected, the evaluation function has not been fully utilized, and the evaluation system is incomplete. The evaluation result is not very reasonable [11, 12].

On the basis of consulting a large number of relevant domestic and foreign references, this paper combines the main characteristics of OLAP, OLAP multidimensional data analysis methods and the performance requirements of the education quality evaluation system, and designs the university education quality evaluation system. The system includes four modules, They are login control module, data management module, online evaluation module and evaluation result management module. After implementing these four modules one by one, we tested whether the performance of the system meets the requirements of this article.

2 The Application of OLAP and Web Technology in the Evaluation of Higher Educational Quality and the Design of Management System

2.1 The Main Features of OLAP

(1) Multidimensionality is a key feature of OLAP. OLAP supports all end users through dynamic multi-dimensional analysis. Multi-dimensional, multi-channel, deeper big data integration view data analysis, so as to master the rules behind big data. In fact, multidimensional analysis is the most efficient way to analyze corporate financial data, and it is also the soul of OLAP.
(2) In linear performance, it refers to the response capabilities and interactive functions required by the OLAP system application. Although the application of

OLAP system is mainly aimed at online data query and problem analysis. However, because the operating system always has to transform and process according to the user's point of view in the process of search, analysis and processing, it can truly reflect some specific aspects of the application problem, and at the same time enable users to observe the data according to their needs.

(3) Informatization means that the system can receive information in a timely manner and manage a large amount of information. Here you need to check the data reproducibility, available disk space, OLAP product performance, and the degree of integration with the data warehouse, etc.

2.2 *OLAP Multidimensional Data Analysis Method*

2.2.1 Roll Up

Rolling up is to monitor more general data by performing aggregation tasks on the data cube to increase the dimension level or delete one or specific dimensions.

2.2.2 Drill Down

Drilling down is to take a closer look at the data by deleting dimension levels or introducing one or more dimensions. By collapsing the current member tree, the drill-up operation can enter a higher level of the hierarchy. If you drill down to the first quarter and observe the data based on the month, you can also drill up again to only view the entire first quarter members. When drilling up, the base shrinks and the visible cube part shrinks.

2.2.3 Slice

Slicing is a selection operation performed on the dimensions of a specific data cube. Two-dimensional data can be obtained through the slicing function.

2.2.4 Cut into Pieces

The dicing operation is a selection operation performed in two or more dimensions of a specific data cube. The cube cutting function can provide a sub-cube, and OLAP allows you to select a subset of the data presented to the user.

2.2.5 Rotating Shaft

The axis of rotation is to change the direction of the dimension. For example, you can move the dimensions from one row to one column. Conceptually, rotating around an axis can be thought of as rotating a cube to display analysis data from different angles. Demonstrate how to change the visualization of sales data by changing the dimensions of the market. You can also use rotation to move the size from the grid to the page. If the attribute is on the page, it will be used as a reference filter, and the filter data will be displayed on the current page of the current member.

2.3 System Performance Requirements

2.3.1 Practicality

Regardless of the system type, the focus should be on practicality. If a system is impractical, no matter how good it is in other aspects, it is not a good system. The system developed in this paper is used to manage the evaluation of university education quality. Therefore, the system mainly includes all the contents and procedures of university evaluation and quality control. Only when the content and process are perfect can we see all aspects of system performance. For example: How to maximize the response time of the system, provide the system's ability to handle incorrect input, and provide the maximum throughput of the system.

2.3.2 Ease of Use

After the system meets the practicability, the ease of use of the system must be considered, and the system interface should not be designed too finely. In addition, the interface should be as simple as possible for users to use the system. It is best to ensure that the system interface and the user's Windows operating system maintain a certain style compatibility. It also has some common office functions, such as creating Word documents and editing spreadsheets. You can use off-the-shelf Office software for integration without additional development, which can significantly reduce system development costs and improve system performance.

2.3.3 Compatibility

It should be noted that the university education quality evaluation management system we developed is not a separate system. It must share and transmit information with many existing information management systems in universities. This requires the system interface design to be carried out during system design and development.

Careful consideration makes the system compatible with the universities currently operating.

2.3.4 Scalability

When developing the system, you can not only check the current operation and usage of the system, but also consider the needs of future system upgrades and expansions as much as possible. For example: if the university needs to transplant the education of the university education quality evaluation management system course to the exam and other course management, if the system upgrades and expands the score system, how to realize from the perspective of network configuration, all this requires system scalability.

3 Experiment

3.1 Login Control Module

The login control module is implemented in the Web browser, which belongs to the first level of the B/S architecture. The browser's expression format is the login window. The function of this part is mainly for the user to log in to the system. All users must enter the correct user name and password. In order to enter the system, after logging in, different types of users use the corresponding permissions to perform some functions on the system. There are four types of users involved: Students, teachers, experts and department leaders.

3.2 Data Management Module

Using the system's OLAP data analysis tools can simplify complex data analysis functions, and realize direct mouse transmission and deposition data navigation, as well as more flexible OLAP classification, including slices, dicing, drilling up, drilling down, and interactive icons Etc., this data analysis tool creates a good graphical user interface and a direct data analysis and exploration environment for users, allowing them to present data analysis from different levels in the workflow. It is relatively easy to implement data management capabilities, because it can add, delete, modify and control basic information such as teachers, students, and courses. Therefore, when calling this information in the system, the method used is the same. These data do not require additional Processing.

3.3 *Online Evaluation Module*

The online evaluation module is the most basic functional module of the system. The module mainly lists the evaluation indicators in the system one by one, and provides the function of online evaluation of various indicators. When a student conducts an online assessment, he must first select the teacher currently being assessed, and then the system displays all the information of the course that the teacher is teaching the student in the form of selection. When the student confirms that the course information is selected for evaluation, the system will enter the course online evaluation page. Fig. 1 shows the process of online evaluation function implementation.

3.4 *Evaluation Result Management Module*

3.4.1 Query Statistics

After the online evaluation of the teaching evaluation work is completed, the evaluation results will be reviewed, passed, summarized and stored in the database. The person in charge of the teaching evaluation and the teacher can query the evaluation results through the query function, and can query with custom conditions. The statistics of teaching evaluation results can help teaching managers understand a teacher's teaching situation over a period of time. It can also measure the teaching performance of different teachers in a specific course. This is most helpful for teaching managers when planning courses. And make adjustments based on these statistics.

3.4.2 Evaluation Results

Evaluation of evaluation results is the main module designed in this paper. Through the use of data mining technology to analyze the evaluation results, the factors that influence each other between education and learning are obtained. For the analysis of the evaluation results, this paper applies the more classic association rule method in the data mining algorithm. $Support(X \Rightarrow Y)$ is the number of transactions that contain both X and Y in the entire database. The formula can be expressed as:

$$Support(X \Rightarrow Y) = Support(X \cup Y) = P(X \cup Y) \tag{1}$$

$Confidence(X \Rightarrow Y)$ is the proportion of transactions that contain both X and Y to the transactions that contain X in the entire database, expressed by the formula:

$$Confidence(X \Rightarrow Y) = \frac{Support(X \cup Y)}{SupportX} = P(X/Y) \tag{2}$$

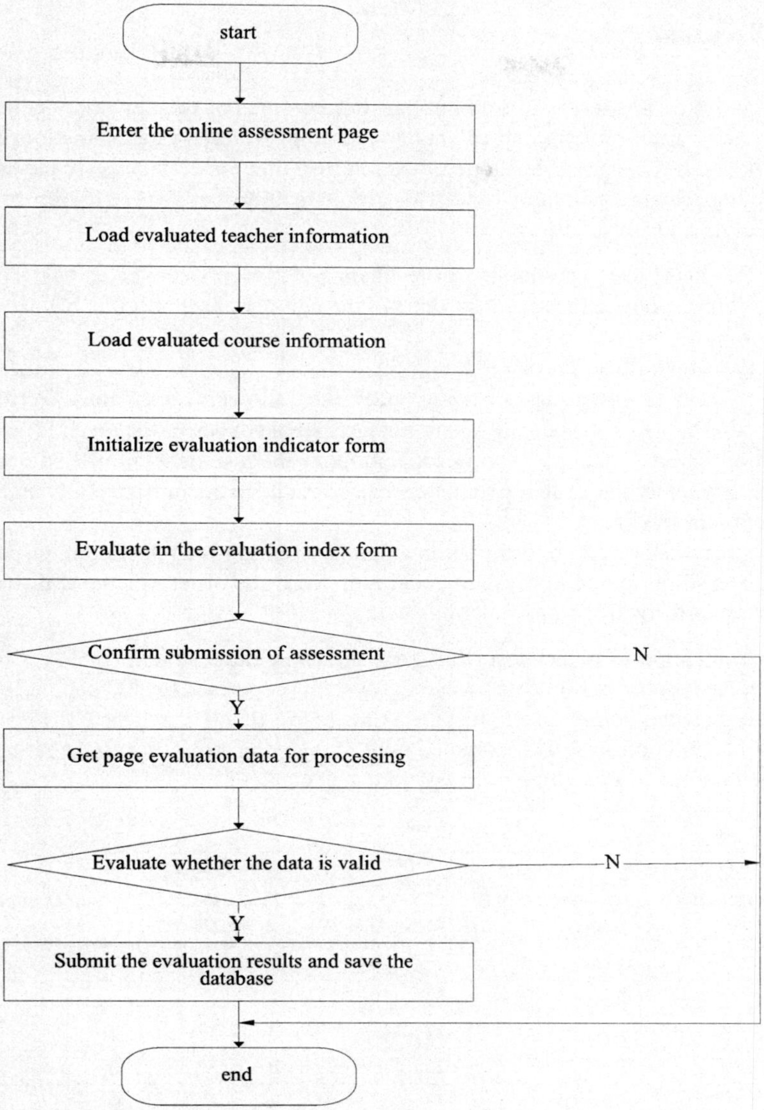

Fig. 1 Online evaluation function flow chart

Investigate the useful information hidden in the analysis of the evaluation results, create a reference knowledge base for teaching reform, and analyze the evaluation results based on the views of educators.

4 Discussion

The purpose of this test is to find out the shortcomings of the university education quality evaluation system, and to change and improve the system functions many times before the system is developed. After performing each operation in the evaluation system, run a performance test. This article mainly starts with response testing, load testing and stress testing.

(1) Response test: The number of teachers and students connected to the server must correspond to the response value of the number of schools that have reached saturation.
(2) Load test: Load test can evaluate performance characteristics. Check whether the system is operating normally under the following conditions. Perform a concurrent load test when the maximum number of concurrent accesses is reached, and observe the data processing capacity after the test lasts for at least one day. Check whether the system parameters can continue to operate normally under the maximum workload.
(3) Stress test: Refers to some technical means of high-load operation, inspecting the operating state of the system under high load, and observing the ability of the system to recover various functions in the event of a failure.

The system performance test results are shown in Table 1.

It can be seen from Fig. 2 that when the system is operated by 500 people concurrently, it still maintains a system response time of less than 1 s, and the CPU response rate of 17.5% is normal. The expected goal of ensuring that 500 people access and process data at the same time is completely acceptable.

Table 1 System performance test results

Number of clients	Average response time/s	CPU usage of server 1/% (%)	CPU usage of server 2/% (%)	Number of errors
300	0.15	7	7	0
350	0.18	8	8	0
400	0.24	11	10	0
450	0.27	15	16	0
500	0.3	18	18	0
550	0.5	23	23	0
600	0.57	35	37	0
650	0.63	41	41	0
700	0.68	43	45	0

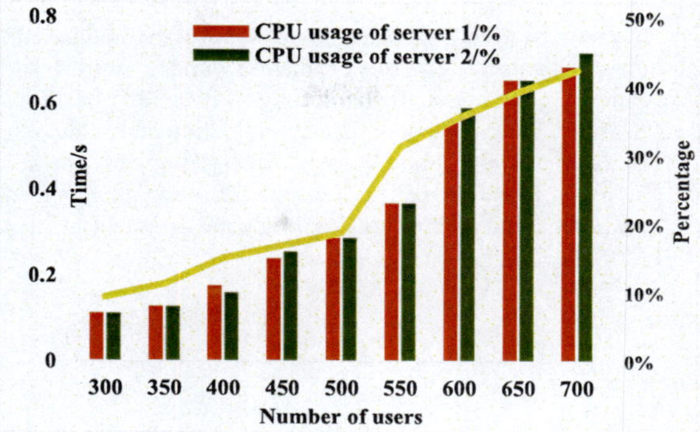

Fig. 2 System performance test results

5 Conclusion

The college education quality evaluation system based on OLAP and web technology provides an effective control and management method for the normal development of teaching work, can promote the education level of colleges, meet the needs of college informatization construction and transformation, and evaluate the education quality of colleges. It provides a highly scientific and effective way to formulate a targeted evaluation and formulation system based on the actual situation of colleges, which can accurately and objectively implement educational evaluation.

References

1. L. Jiang, X. Wang, Optimization of online teaching quality evaluation model based on hierarchical PSO-BP neural network[J]. Complex. **2020**(7), 1–12 (2020)
2. Y. Chen, College English teaching quality evaluation system based on information fusion and optimized RBF neural network decision algorithm[J]. J. Sens.S **2021**(5), 1–9 (2021)
3. W. Huang, Simulation of English teaching quality evaluation model based on Gaussian process machine learning[J]. J. Intell. Fuzzy Syst. **40**(2), 2373–2383 (2021)
4. F. Peng, Application of deep learning and cloud data platform in college teaching quality evaluation[J]. J. Intell. Fuzzy Syst. **39**(4), 5547–5558 (2020)
5. H. Tang, Research on teaching quality evaluation method of network course based on intelligent learning[J]. Int. J. Contin. Eng. Educ. Life-Long Learn. **30**(1), 1 (2020)
6. T. Yuan, Algorithm of classroom teaching quality evaluation based on Markov chain[J]. Complex. **2021**(21), 1–12 (2021)
7. Q. Sun, Evaluation model of classroom teaching quality based on improved RVM algorithm and knowledge recommendation[J]. J. Intell. Fuzzy Syst. **40**(2), 2457–2467 (2021)
8. C. Lu, B. He, R. Zhang, Evaluation of English interpretation teaching quality based on GA optimized RBF neural network[J]. J. Intell. Fuzzy Syst. **40**(2), 3185–3192 (2021)
9. Y. Luo, X. Zhao, Y. Qiu, Evaluation model of art internal auxiliary teaching quality based on artificial intelligence under the influence of COVID-19[J]. J. Intell. Fuzzy Syst. **39**(6), 8713–8721 (2020)

10. S. Zeng, N. Wang, C. Zhang et al., A novel method based on induced aggregation operator for classroom teaching quality evaluation with probabilistic and Pythagorean fuzzy information[J]. Eurasia J. Math. Sci. Technol. Educ. **14**(7), 3205–3212 (2018)
11. S. Jia, Y. Pang, Teaching quality evaluation and scheme prediction model based on improved decision tree algorithm[J]. Int. J. Emerg. Technol. Learn. (iJET) **13**(10),146 (2018).
12. T. Wang, T. Wu, A.H. Ashrafzadeh et al., Crowdsourcing-based framework for teaching quality evaluation and feedback using linguistic 2-tuple[J]. Comput., Mater. & Contin. **57**(1), 81–96 (2018)

Research on the Construction of Integrated Education Mechanism of Industry and Education in Applied Undergraduate Colleges

Hong Yuan Wang

Abstract Based on the analysis of the current situation of the inconsistency between the orientation of talents training and the orientation of colleges and universities, and the unreasonable match between curriculum and social demand, this paper puts forward the ideas of constructing the management mechanism of school-enterprise cooperation platform and strengthening the teaching quality guarantee system of school-enterprise cooperation for application-oriented talents construction. From the school-enterprise cooperation platform construction, teaching concept reform and security system and other aspects of the talent training model research, this paper takes the network engineering major as an example, puts forward the professional construction standards. The implementation of this topic will help guide the major to actively meet the needs of regional industry and industrial development, and help improve the quality of talent training.

Keywords Industry-education integration · School-enterprise collaboration · Network engineering · Talent training model

1 Introduction

In March 2016, the national policy emphasized the construction of a five-in-one collaborative education mode of government, industry, learning, research and application, so as to cultivate a large number of applied talents with innovative and entrepreneurial ability. Through the investigation and research on some undergraduate majors that have been withdrawn in recent years. These majors may have the following problems in the process of talent cultivation [1]:

H. Y. Wang (✉)
Guangdong University of Science and Technology, Guangdong 523083, China
e-mail: 563431669@qq.com

S. Patnaik and F. Paas (eds.), *Recent Trends in Educational Technology and Administration*, Learning and Analytics in Intelligent Systems 31,
https://doi.org/10.1007/978-3-031-29016-9_22

(1) Personnel training target positioning is not clear. In the process of application-oriented transformation, some undergraduate universities lack a careful analysis of their history and reality, deviating from the school-running characteristics of application-oriented universities that should serve regional economic development, leading to the lack of clear professional training objectives [2].
(2) The formulation of talent training program is single and lacks the effective participation of industry organizations. Lack of effective participation of industry organizations. Even if the participation, its participation for a variety of reasons, it is difficult to achieve a certain depth [3].
(3) Curriculum is not closely related to social needs. The curriculum system of some colleges and universities only draws lessons from or copies the curriculum system design of other colleges and universities, and the curriculum is set up because the research on social needs is not thorough enough. In the process of teaching, the teaching content is only according to the teaching material, the teaching material on the lack of flexible deletion, supplement mechanism.
(4) The application ability of the teaching staff needs to be improved. At present, most of the teachers in colleges and universities have a high academic level and good educational background, but the deficiency is that most of these teachers do not have rich practical experience, and many teachers just teach book knowledge, so it is difficult to cultivate high-quality applied talents [4].
(5) The experimental conditions were poor and students could not get effective exercise. At present, due to some expensive equipment and limited space, many experiments are carried out on simulators, and some experiments even cannot be carried out, resulting in students' practical ability cannot be comprehensively improved.

2 Methods and Materials

Schools and enterprises participate to optimize training programs. According to industry trends, technology development and talent demand, integrate existing majors, discuss with leading enterprises such as Huawei on specialty positioning and optimize talent training programs; Introduce Huawei industries, standards, technologies, products, courses and certifications into the college, and improve the teaching course support system of the Industrial College [5].

To build a high-level professional teaching team and encourage teachers to participate in enterprise projects, the school can also introduce senior engineering and technical personnel from enterprises to teach students specialized courses, undertake core curriculum teaching, and participate in the development and implementation of talent training programs. Improve the teaching quality of teachers, and improve the management level of industrial colleges.

Create a modern engineering training center. In order to serve the strategic needs of the country and the needs of enterprises in the new economy industry, it helps the industry to cultivate innovative and outstanding craftsmen, jointly builds a sharing engineering practice platform integrating lessons and research with enterprises, and explores a new engineering construction model. The engineering training Center provides a shared engineering practice platform integrating teaching, training and certification, and scientific research and development, covering the most cutting-edge technical direction of the profession [6].

We will strengthen our capacity for innovation and commercialization in scientific research. Through joint scientific research, teachers' development ability and industrial service ability are enhanced. The college and enterprises jointly apply for the relevant scientific research projects of the Department of Education and the Department of Science and Technology, establish a joint scientific research center to jointly develop product prototypes and apply for intellectual property rights, and apply the achievements in industrialization to promote the development of local economy [7].

To build a laboratory integrating new professional technology and create a practice teaching base. In order to meet the needs of practical teaching, the construction of network engineering major keeps up with the latest development direction of this major field, carefully builds Huawei enterprise culture atmosphere, builds Huawei enterprise experimental environment, leads the construction of smart campus, enables the construction of major, accelerates the transformation of results, and helps students improve their professional skills. Jointly build a teaching-research-integrated sharing engineering practice platform with enterprises to explore the construction mode of new engineering. The engineering training Center provides a shared engineering practice platform integrating teaching and training, training certification, scientific research and development.

Build a modern industrial college. Relying on the major and facing the industry, the College of modern Industry is established. As a model of the integration of industry and education, the operation service center of the college is firstly established, including the board of directors and the steering committee. The board is composed of senior leaders of the school and the enterprise, which is mainly responsible for the overall control and coordination of the operation of the whole industrial college. Implement all tasks decided by the Council. The second is the physical base carrier, including engineering training center, training and certification center, innovation and entrepreneurship center, scientific research application center and operation service center, to meet the four functions of teaching and training, training and certification, innovation and entrepreneurship, scientific research and development. The third is the cooperation service content. The industrial College is oriented towards the hot information technology directions such as network, big data and software, covering various links such as professional design, curriculum reform, teacher promotion, engineering training, training certification, innovation and entrepreneurship, competition support, joint scientific research, internship and employment, online learning and so on. School-enterprise co-construction management system, as shown in the Fig. 1 below.

Actively employ backbone technical personnel from industry associations and enterprises as part-time teachers, and increase the proportion of full-time teachers

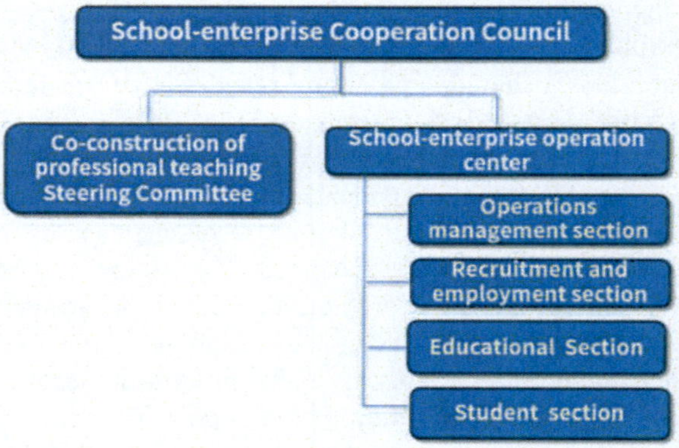

Fig. 1 School-enterprise co-construction management system

from enterprises as part-time teachers. The company has actively introduced front-line talents to participate in professional construction, curriculum reform, whole-week training and graduation design guidance, and gradually formed a professional practice system combining campus and off-campus practice bases. Take incentive measures to encourage teachers to get certificates and practice in enterprises. The school attaches great importance to the construction of dual-ability teachers, and takes measures to encourage teachers to take vocational qualification examinations and obtain relevant vocational qualification certificates [8]. At the same time, teachers are encouraged to go to school-enterprise cooperative enterprises for vacation practice. So that teachers can master advanced technology at any time, improve their practical ability. Taking network engineering major of Guangdong University of Science and Technology as an example as shown in the Fig. 2 below.

It can be seen from the Fig. 2 that in recent years, the proportion of teachers with high academic qualifications and professional titles in network engineering major has been increasing year by year, which will be conducive to the overall development of teachers and the integrated and innovative development of universities and enterprises.

3 Construct a Teaching Guarantee System Based on OBE Concept

Take the course of network engineering as an example. According to the OBE concept and the certification standard of engineering education, the network engineering major has revised the teaching content, improved the graduation requirements, added the relevant systems and documents such as the evaluation method of course achievement degree and the feedback of graduates, and completed the teaching quality assurance system based on the OBE concept [11].

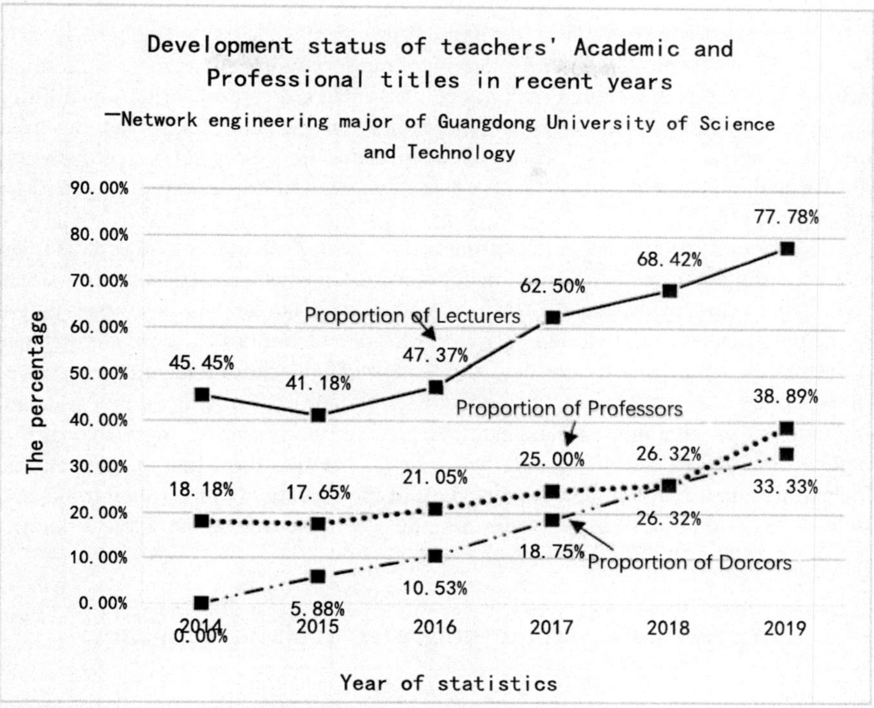

Fig. 2 High academic and professional title development status—Take network engineering major of Guangdong University of Science and Technology as an example

In each semester, teachers of each course directly evaluate students' core abilities according to curriculum evaluation standards through examination or homework assessment, so as to conduct teaching reflection on students' learning results and core abilities. From the evaluation of the teachers on the course to the achievement of the core competence of the graduates upon graduation, as well as the achievement of the school education goal of 3–5 years after graduation, it is necessary to evaluate and continuously improve the training goal on a regular basis according to the effect evaluation of the graduates and the evaluation of the society on professional training personnel.

The teaching quality of network engineering specialty is to monitor and evaluate the quality of teaching process and evaluate the influencing factors of teaching quality in time. According to the quality objectives and standards, the information of quality control points is collected, sorted and analyzed, quality effect evaluation, problem feedback and improvement tracking, so as to achieve the purpose of quality assurance. It provides powerful working status data for the strategy formulation, problem diagnosis, control and implementation of school teaching quality to ensure the continuous improvement of teaching quality. On the basis of school teaching quality monitoring, network engineering major has formulated a network engineering professional development plan. Professional principals, backbone teachers listen to the evaluation system; Graduate employment questionnaire, enterprise employment information feedback system. Finally, according to the collected information, training objectives, graduation requirements, and evaluation and analysis of teaching links are fine-tuned, revised and improved to ensure the achievement of core competence. Has formed a relatively perfect teaching quality standard implementation guarantee mechanism [12, 13].

It is necessary to evaluate and continuously improve the training objectives regularly according to the effect evaluation of graduates and the evaluation of professional training personnel by the society, from the evaluation of students in the courses taught by the teachers when they are in school to the achievement of the core competence of graduates at graduation and the achievement of the school education objectives for 3–5 years after graduation. Formed a relatively perfect teaching quality standard implementation guarantee mechanism. At present, there are five graduates of this major from 2017 to 2021. Through the survey, the results show that the educational goals of this major are in line with the needs of the industry and the social trend, and each goal is "important" and "very important" [9]. As shown in the Table 1 below.

4 Collaborative Education Integrated Training Program

The implementation of "person-job matching" practice education, the establishment of production and education integration, collaborative education of personnel training and evaluation system. Through the senior year, students will enter the cooperative enterprise for one year's practical study, participate in the first-line activities of the enterprise, learn the corporate culture and management system, study the development status and trend of the industry, consolidate professional knowledge, expand professional ability, and improve employment competitiveness.

The university and enterprises shall jointly develop talent training programs. With industry and market demand as the guidance, to cultivate high-quality comprehensive talents as the fundamental goal, to develop the curriculum standards of relevant professional courses. Timely investigate the responsibility requirements of corresponding positions of relevant majors in the industry, set up the professional curriculum system according to the job responsibility requirements, and timely reform the professional curriculum system. Centering on Huawei ecosystem, in the course system setting process, we strictly combine post ability with professional curriculum training objectives, mainly taking project-driven teaching as the theme, integrating OBE education and teaching concept, learning-centered, output-oriented teaching reform [14].

To build a team of high-level teachers, realize the mutual integration and communication of school-enterprise teachers, through engineering training, summer study, enterprise practice, etc., to strengthen the teaching and research level of the teacher team in a diversified way. Send teachers to practice, improve teachers' ability to

Table 1 Statistical Table of Questionnaire Survey on Employers of Education Objectives (2017–2020)

Level education target	Very important-5 (%)	Important-4 (%)	Middle-3 (%)	Unimportant-2 (%)	Very unimportant-1 (%)
(1) Qualified engineer in the field of Internet technology with basic theory and professional knowledge	61.33	25.33	13.33	0	0
(2) Expand the vision of science and technology, pay attention to scientific ethics, and have the spirit of communication, coordination and cooperation with the team	62.67	29.33	8	0	0
(3) Lifelong learning habit and innovation consciousness	64	30.67	5.33	0	0
(4) Have the ability to read foreign literature of the major and international vision	46.67	29.33	24	0	0

analyze and solve problems in the process of practical development, achieve the goal of training dual-teacher, and truly realize the integration of production and learning.

Establish a sound management system, relying on universities, enterprises, governments and other diverse entities, and form an organizational structure of joint construction and joint management. Establish a scientific, efficient and powerful system to optimize resource allocation and enhance self-hematopoietic ability of industrial college, so as to deepen the integration of production and education and realize the guarantee of deep integration of education chain, innovation chain and industrial chain [15].

We should build an industry-university-research system for coordinated and sustainable development, try to jointly build a number of new technology development and application platforms with local governments and enterprises, and take the innovation of technical skills required by posts and industries as the main research object. Relying on the real business scenario and real demand direction of enterprises, carry out scientific research cooperation, establish a scientific research center centered around professional construction, jointly apply for scientific research projects, strive to make new breakthroughs in the establishment of high-level scientific research projects and the output of high-level scientific research results, and actively promote the transformation of results.

We will promote the transformation of digital education. Build a new generation of digital-information-based industry-university-research alliance to promote the development of regional digital economy industry. The scientificity and precision of education management and decision-making achieved through data drive will become more and more strongly, which will promote the transformation of information technology in higher education. Information technology will be integrated into all levels of school education and teaching. Through data analysis and evaluation, it will help to better connect education and industry, and promote the fundamental reform of education governance decision-making. Network technology, artificial intelligence and sensor technology together create a networked virtual space and cloud environment. The virtual world seamlessly connects with the real society, making the learning environment more free, open and intelligent.

Enterprise experts on the platform, students practice into the enterprise, enterprise-project into the classroom, vocational certification into the course. The goal is to improve students' examination rate, practice ability and innovation ability, innovate the school-enterprise collaborative education mechanism, and build an integrated training program for senior application innovative talents with comprehensive cultivation of morality, intelligence, physical education, aesthetics and labor, innovative spirit and entrepreneurial consciousness, self-learning and improvement ability and good engineering practice ability [10].

The implementation of academic education and industry certification integration of the "3 + 1", the implementation of school education and enterprise training of the integration of the "dual track system" talent training, to promote the school education and enterprise employment seamless docking, to achieve students from graduation to employment seamless docking. As shown in the Fig. 3 below.

5 Achievements

Professional teaching quality standards have been formulated. In the integration of production and education under the guidance of network engineering applied talents training mode, compiled the network engineering professional teaching quality standard and the talent training scheme, the standard and the scheme from the training objectives, graduation requirements, quality assurance, etc., has established the demand oriented talent training plan, promote the development of education and industry linkage.

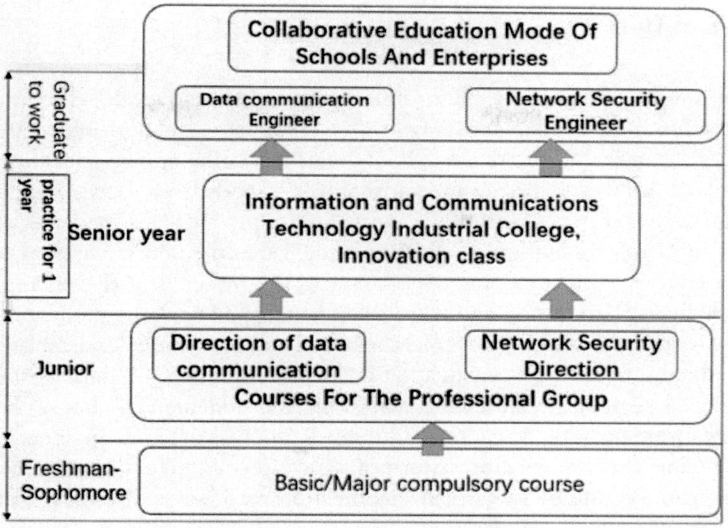

Fig. 3 Integrated talent training program

Passed the Institute of Education IEET engineering certification. Promote professional teaching form to certification, promote the adjustment of course, with certification to carry out the continuous improvement, to certification achievements in teaching self-management, with results oriented teaching quality manages view has a good guiding role, set up to ensure that the education of a target assessment mechanism, It includes the survey and evaluation of graduates' achievement degree of core competence, the survey and evaluation of students' satisfaction degree of educational goals, and the evaluation of professional education goal achievement degree conducted by professional advisory committee for continuous improvement to ensure the achievement of educational goals.

Network engineering major "take this as the foundation, four regression", implement the fundamental task of talent training, adhere to the integration of industry and education, deepen the integration of industry and education, promote the cooperation of industry, university and research, and improve the quality guarantee system. Network Engineering was successfully declared as "Provincial First-class Undergraduate Major Construction Point".

The network engineering major of Guangdong University of Science and Technology was evaluated as the provincial first-class professional construction site in 2019 and won the second prize of provincial teaching achievements. The employment rate of the graduates of the network engineering major has reached more than 97%. Through the questionnaire survey of employers and graduates, the students have a high degree of satisfaction in employment. To a certain extent, it shows that the mode of university-enterprise integration has a certain promoting effect on the cultivation of talents.

6 Innovation

Oriented to industrial demand, multi-dimensional collaborative talent training mode. With students as the center, universities as the main body, and market demand as the guidance, we actively explore the school-enterprise collaborative education mechanism, make full use of the resource advantages of both sides, effectively integrate and share the resources of both sides, jointly build a practice base, and cooperatively cultivate teachers, through professional practical training and graduation practice, and cooperatively train students. According to the research of the talent training mode integrating production and education, the stepped teaching module of "basic training—comprehensive improvement research and innovation" is established.

Vocational qualification certification in line with national (industry) standards. According to personal interests and market needs, students can choose their own certificates from different vocational skill levels, such as network engineer, Huawei Digital engineer and other professional qualifications, to take the exam for credit. The high passing rate of vocational qualification certificates also drives students' enthusiasm for professional learning.

7 Conclusion

Based on the education mechanism of the integration of industry and education, it is necessary to formulate a multi-dimensional collaborative talent training mode oriented to the needs of industry. We should highlight the education mechanism, make full use of the resource advantages of both schools and enterprises, effectively integrate resources and achieve a high degree of resource sharing improve the construction of teachers, and cooperatively train students.

The integration of production and education, under the guidance of applied talents training mode of the construction of the professional evaluation standards should take the initiative to adapt to the economic and social development needs, can be well reflected in the quality education and vocational ability as the main line, knowledge ability quality reasonable structure, can meet the requirements of expected jobs, also need to adhere to the talent demand and industry or area every year graduates tracking survey, It can optimize the professional connotation according to the change of talent market demand and improve the employment competitiveness of graduates.

References

1. Z. Chao, J. Jingshan, Z. Yanxi, Thoughts on first-class application-oriented undergraduate talents training mode in the New Era. J. China Multimed. Netw. Teach. (The top 10) (03), 134–135 (2020)

2. F. Yong, T. Liwei, Network engineering professional qualification certification system and its improvement measures. Electron. Technol. Softw. Eng. (05), 20–21 (2017)
3. L. Shuai, H. Beautify, X. Kaixi, G. Guolin, Z. Zhongkui, Educ. Teach. Forum (11), 86–88 (2020) (in Chinese)
4. H. Debin, P. Xiaoning, et al, Construction of practical teaching system for network engineering specialty based on University-Enterprise Cooperation. Univ. Educ. 168–170 (2019)
5. Z. Yihua, Innovation and entrepreneurship theory research & practice **1**(09), 53–55 (2018)
6. L. Shuling, H. Minjuan. Research and practice of teaching quality assurance system based on OBE concept. Think Tank Times (11):163–164+224 (2020)
7. Xu. Yue, Ma. Xiaofeng, Research and implementation of student behavior comprehensive evaluation system based on block chain. Inf. Technol. Informatiz. **12**, 131–133 (2016)
8. Xu., Tao, Application and challenge of Block chain technology in education and teaching. Mod. Educ. Technol. **1**, 108–114 (2017)
9. H. Jiajia. College English MooC Evaluation System based on block chain. Educ. Obs. (2020)
10. H. Longwen. Research on the application of We-media Block chain technology in classroom teaching quality evaluation. China-Arab. States Sci. Technol. Forum (2020)
11. Ji., Bo, Study on the strategy of professional construction in higher vocational colleges characterized by the integration of production and education. Occupations **06**, 62–64 (2022)
12. Y. Mei, Z. Zhengzhu, Research on the promotion strategy of production-education integration policy in application-oriented undergraduate universities: a two-dimensional analysis based on policy tools and policy elements. China Univ. Sci. Technol. (3), 79–84 (2022). https://doi.org/10.16209/j.carol carroll nki cust. 2022.03.027
13. L. Lan, K. Yuan, Exploration and reflection on the construction of higher vocational teachers' team from the perspective of product-education integration. Reform Open.-Up (06), 44–49+55 (2022). https://doi.org/10.16653/j.cnki.32-1034/f.2022.006.007
14. S. Wei, Research on talent training mode of industry and education integration under the background of 5G+ industrial internet. South Agric. Mach. **06**, 162–164 (2022)
15. Z. Xintong, Research on teaching reform path of network engineering specialty education: the growth of man as man. In: Proceedings of the Conference on the Refinement and Practice of Principals' Educational Ideas (2022), pp. 658–661

Construction of Finance and Economics Discourse Parallel Corpus and Implementation of Learning Platform for Translation Teaching

Geyang Hu and Yiqin Zheng

Abstract This product aims at the current situation that the teaching contents of financial and economic translation major in Colleges and universities in China are old, the methods are traditional, and the students' knowledge and skills are single, which can't meet the needs of enterprises for financial translation talents. Based on the field of financial discourse, a special corpus and mobile learning platform are established to provide services for research and problem solving in the field of finance and economics. According to the three standards of corpus collection, this product uses Python to crawl the bilingual corpus of finance and economics from major authoritative newspapers or websites at home and abroad, clean the corpus, align and mark it with professional language software, manually proofread and beautify the typesetting, and then classify and arrange it, so as to complete the construction of dynamic knowledge atlas corpus based on MySQL and neo4j, and then design the function and basic construction route of mobile learning platform based on Wechat developer tools, in order to realize the multi-functional application platform of information retrieval, result visualization, user self-test, community communication and so on. This study is in line with the current professional hotspot of parallel corpus research, which can reduce the difficulty of financial translation and enhance students' practical skills. This study can contribute to the teaching, scientific research and social services of financial and economic translation teachers and students.

Keywords Translation teaching · Financial discourse · Parallel corpus · Mobile learning platform

G. Hu (✉) · Y. Zheng
School of Foreign Studies, Central University of Finance and Economics, Beijing 102206, China
e-mail: 2020312096@email.cufe.edu.cn

S. Patnaik and F. Paas (eds.), *Recent Trends in Educational Technology and Administration*, Learning and Analytics in Intelligent Systems 31,
https://doi.org/10.1007/978-3-031-29016-9_23

1 Introduction

In today's economic globalization and financial internationalization, financial translation has become a link to communicate and strengthen economic and trade exchanges among countries, and plays an important role in various economic activities. Financial translation is different from general translation. Financial vocabulary has the characteristics of "large professional vocabulary, many terms, fixed usage, standardized format, and common use of abbreviations, long sentences" [1]. In the rapidly changing and risky transnational financial transactions, the "difference of meaning" caused by the loss and deformation of financial information caused by careless translation will lead to unimaginable consequences [2]. Financial translation is difficult and risky. However, the students trained under China's traditional foreign language education mode have narrow knowledge, less interdisciplinary learning, and single knowledge and skills, which can't meet the needs of enterprises for financial translation talents [3]. Facing the increasing demand for translation, how to reduce the difficulty of financial translation and enhance students' practical skills is a key research topic of experts and scholars, in which the construction of parallel corpus has become a research hotspot.

With the rapid development of Internet technology, corpus-based teaching and research have been widely used, and the processing degree of corpus is becoming deeper and more detailed. The special corpus established based on professional fields can provide services for field research and problem solving. Practice has proved that the memory function and repeated-use function of corpus can reduce the difficulty of professional English translation in a certain extent, and plays an irreplaceable role in translation teaching and translator training [4]. Compared with the development and maintenance of large corpora that require a lot of manpower, material and financial resources, self-built small corpora have clear collection objectives and high efficiency. They can change flexibly according to the needs of teaching and scientific research. The development and maintenance cost are relatively low, and they can be opened to permission or used free of charge within a certain range, let students better control the translation quality with the help of corpus resources in professional English translation [5].

Therefore, more and more scholars point out that corpus is bound to enter the classroom of foreign language teaching in the future, and will also develop in the direction of specialization, sharing and intelligence [6]. Corpus-driven teaching reform can provide guidance and practical significance for the cultivation of students' financial translation ability [7]. However, there have less bilingual corpora built at present, and the corpora of applied styles such as economy and trade are even rare [8]. In this study, the corpus with finance or finance as the subject words was searched in CNKI. After manual screening and excluding irrelevant documents such as meetings, a total of 50 papers on the financial discourse corpus were obtained. Export the selected 50 papers, make statistics on the key words of the papers by using SATI software, and draw the chart by SPSS (Fig. 1).

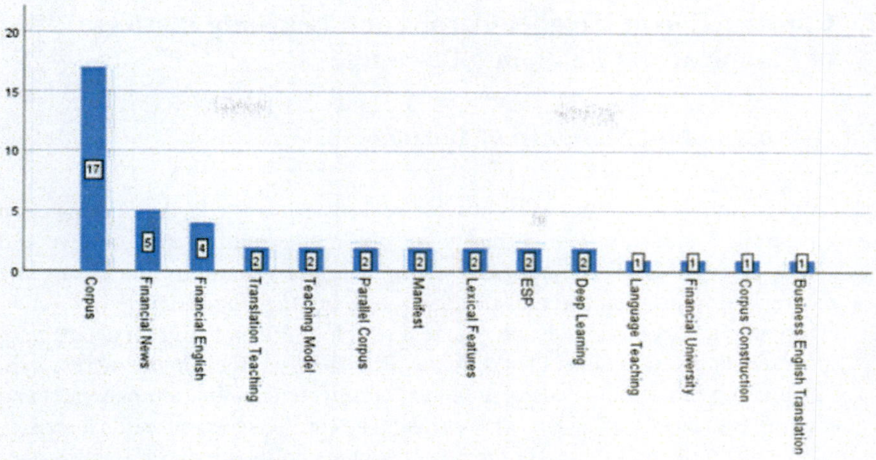

Fig. 1 Keyword frequency statistics of domestic financial and economic discourse corpus

As can be seen from Fig. 1, the research characteristics of financial discourse corpus in recent years are: corpus, financial English, financial news, financial English, lexical features, translation teaching, deep learning, visualization, teaching mode, parallel corpus, ESP, financial colleges, language teaching, corpus construction and business English construction. After merging the subject words, it is found that the domestic research on financial discourse corpus has the following two significant characteristics: First, financial discourse corpora are mostly English Chinese bilingual parallel corpora, and most of them are temporary corpora established by quantitative analysis of the language characteristics of Financial English news and economic discourse. Second, the exploration of Financial English teaching methods is a new research direction in recent years. The establishment of teaching oriented financial discourse parallel corpus is also a hot spot for domestic scholars. For example, Sun Yan proposed to establish a multimodal Financial English corpus, which is conducive to teaching because of its authentic corpus and flexible teaching form [9]. Wu Feina also pointed out that there have many shortcomings in the traditional financial English teaching design and corpus-based teaching model reform can improve teaching effect [10].

2 Construction of Parallel Corpus and Learning Platform of Financial and Economic Discourse

2.1 Construction Standard of Corpus

1. Authoritative. This corpus is to provide real materials for financial translation teachers and students. The corpus should be representative, authoritative and cutting-edge, focus on language expression in real situations, and ensure the accuracy of translation and the standardization of language use.
2. Practical. This corpus is rich and practical which should not only serve teaching, scientific annotation, but also provide data analysis for linguistic research. Considering the actual needs of students, certificate examination materials and enterprise materials are added to the corpus. The certificate examination materials mainly cover the real subjects of the professional qualification examination over the years, combined with the self-test module of the platform to help prepare for the examination. Enterprise materials mainly covers the annual reports of major commercial banks and international well-known enterprises, so that students can understand enterprise planning, business activities and other contents.
3. User friendly. The corpus is classified and arranged according to financial reports, certificate examination materials, enterprise materials and proper nouns. Financial reports are classified according to "theme", which is convenient to find with simple and convenient visual operation and excellent user experience.
4. Extensibility. The collection of corpus is a multi-party cooperation and cumulative process. With the change of international economic form, the emerging financial hot words and neologisms should be continuously incorporated into the corpus to ensure the cutting-edge and integrity of the corpus.

2.2 Construction of Corpus and Learning Platform

See Fig. 2.

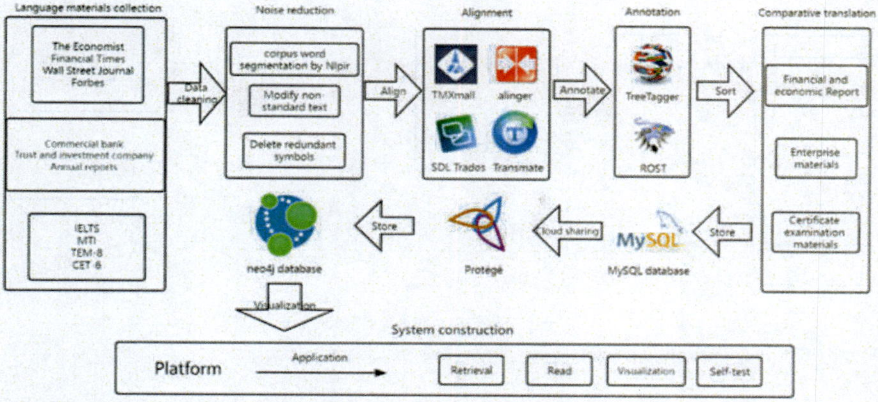

Fig. 2 Construction ideas of corpus and learning platform

3 Technical Implementation of Parallel Corpus and Learning Platform

3.1 Functional Design of Corpus

3.1.1 Conceptual Design of Corpus

This corpus is a special corpus of financial discourse for translation teaching and scientific research. In addition, according to the actual needs of students, this corpus also adds textual research materials and enterprise activity materials, which are placed in four different sub-databases according to the data types, namely "financial and economic reports", "enterprise materials", "certificate examination materials" and "proper nouns".

Building a conceptual model generally uses the E–R diagram (Entity–Relation model), which is most widely used in database design. It has the advantages that the data model is more general, more abstract and closer to the real world. It can more simply describe the concepts and relationships between various entities. It is mainly composed of entity set, attribute and relation. The E–R diagram can be used to clarify the contents of each database and the relationship between databases. The conceptual model of this corpus is shown in the Fig. 3.

3.1.2 Structure Design of Corpus

In order to realize the wide coverage, specialization and high quality of the corpus, MySQL database can be selected as the construction software. MySQL database has the characteristics of data structure, low redundancy, high sharing and easy expansion,

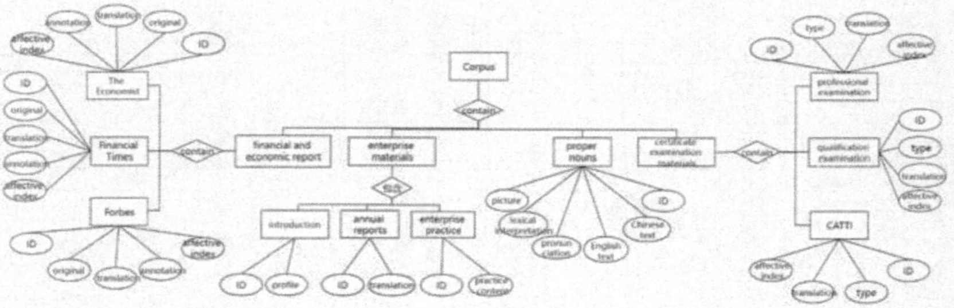

Fig. 3 Corpus conceptual model

Table 1 Data structure of corpus elements

Order	Name	Data type
1	Order number	SMALLINT
2	Chinese text	TEXT
3	English text	TEXT
4	English audio	MEDIUM BLOB
5	Physical picture	BLOB
6	Annotation	TEXT

and is uniformly managed and controlled by the database management system. It can scientifically organize and store data, and efficiently obtain and maintain the data environment.

Data table is the basis of database building. It is not only an important part of the database, but also the basic unit of storing data which is usually composed of table structure and table content. It is classified according to financial and economic reports, enterprise materials, certificate examination materials and proper nouns in the three parts. Finally, it is necessary to define the data type of each field in the table. Since the data type will restrict the data storage mode, in order to meet the storage requirements of the content, select the appropriate field type to build the data table according to the conceptual design. The data table construction of this library is shown in the following Table 1.

3.2 *Function Realization of Corpus*

3.2.1 Corpus Collection and Processing

(1) Language material collection. According to the three standards of "the correspondence between the original text and the translation; the readability of the

original text; and the translation should meet the expression habits of the target language", bilingual language material related to finance and economics are collected from major authoritative newspapers or websites at home and abroad, including the Economist, the Financial Times, the Wall Street Journal, MTI, IELTS, the real postgraduate entrance examination questions of colleges and universities over the years, as well as the enterprise annual reports of Bank of Shanghai, Agricultural Bank of China, Alibaba and other companies. Then use Python to crawl the data within the selected range, convert the results obtained by crawling into TXT text format, delete the non-standard and inaccurate text and symbols through manual screening, and store them in MySQL to obtain clean language material.

(2) Language material alignment. Tmxmall and Aligner software are used to align the sentences of the language material, delete the spaces and enters in the language material. Then SDL Trados and Transmate are used to extract, export and integrate the proper nouns in the language material. Each proper noun is recorded by the professor of phonetics, and the link is stored in the file and uploaded to the corpus to build an audio corpus.
(3) Language material annotation. Treetagger is used to annotate the language material, including the annotation of part of speech and punctuation. NlPIR is used to segment the corpus. Finally, Rost is used to analyze and annotate the emotion index of the language material in terms of sentences.
(4) Manual proofreading. Manually proofread all language material, check the wrong annotation, modify them with deviation expression and beautify the typesetting.
(5) Classified storage. The language material is divided into four parts: financial report, enterprise materials, certificate examination materials and proper nouns, which are stored in Excel and uploaded to MySQL database.

After completing the collection and processing of language material, use MySQL to establish a database on the network side, then use Navicat to operate the database, share the external interface of the database, and upload the collected language material to the terminal database. For example, RDS MySQL under Alibaba cloud is used to connect with Navicat to realize database sharing on the network side.

3.2.2 Visualization of Proper Nouns in Financial and Economic Discourse in Corpus

(1) Construction of dynamic knowledge graph based on ontology modeling

Knowledge graph originated from the concept of semantic networks proposed by Quillian in 1968 which is a structured representation of knowledge for semantic information and was first used as a data organization method of natural language processing [11]. On May 16, 2012, Google released the product of knowledge graph, which uses more complex structured data sources than ever before to serve natural language queries [12].

Knowledge graph is mainly used to describe the objective relationship between entities, concepts and events in the real world which usually adopts the expression method based on SPO triad to realize the structural characteristics. Knowledge graph is mainly divided into general knowledge graph and domain knowledge graph. It has the characteristics of easy to understand and simple structure and is suitable for the construction of professional terms such as economy, industry and company and construction of a relatively mixed framework in the field of finance and economics.

Due to the wide variety of finance and economics proper nouns, constructing the proper noun knowledge graph framework and adopting the top-down construction structure can more concisely show the meaning of words and the relationship with other proper nouns and enhance the readability and comprehensibility of database content. The visual construction of this database uses the ontology editing tool protégé developed by Stanford University which has the characteristics of diverse functions, convenient operation and easy to use. At the same time, it supports the relationship categories between user-defined entities, which can meet the construction requirements of maps with different complexity. The nodes and relationship types for building the knowledge map are shown in the Fig. 4 below (Table 2).

Taking the professional terms of finance and economics as an example, the visual graph is constructed as shown in the figure below, in which the blue line represents the Subclass relationship and the yellow line represents the Translate relationship. The usage of visual graph can better divide the structure of database proper nouns, and meet the needs of users to deeply understand the connotation of nouns and the relationship between proper nouns.

(2) Construction of knowledge graph in neo4j.

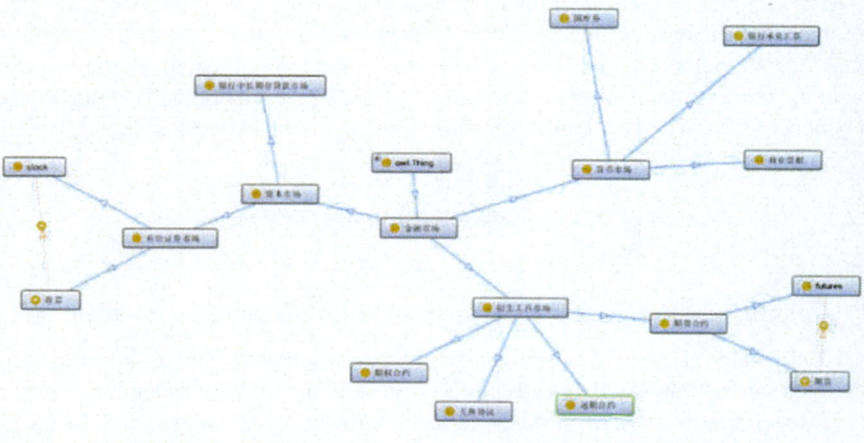

Fig. 4 Visualization of some terminology knowledge map ontology in the financial field

Table 2 Node and relationship types

Model data	Data structure	Function	Specific type
Entity	Label	Marked noun category	Chinese and English Vocabulary
	Attribute	Storage of proper nouns	Substantive noun text
Relationship	Type	Indicates relationship of entities	Subclass relationship, Translate relationship
	Attribute	Indicate relationship properties	Name of relationship

Neo4j is a high-performance graphic database, which can establish a relational network sub graph according to a certain topological relationship and store structured data, providing both good solution for database storage and data processing for graph model problems. Compared with the traditional relational RDBMS database, the graphic database neo4j has the characteristics of good portability and flexible graphic model, and is good at dealing with a large number of complex and low structured data [13]. Therefore, neo4j is more suitable for the visual construction of the database of financial and economic proper nouns. Based on neo4j, the platform constructs a proper noun visualization section to display the relationship and translation between nouns, making the database more readable and easier for users to understand.

Firstly, establish the relationship between proper nouns and build a clear SPO triad relationship. Some proper nouns and translated texts of financial market in financial and economic reports are shown in the Table 3 below:

Secondly, use neo4j to create a database, create a Chinese node to store Chinese text, an English node to store English text, and a parts node to create financial and economic reports of the categories to which all nouns belong. The programming statements are as follows:

```
create(n:Chinese) return n
create(n:English) return n
create(n:Parts) return n
```

Then create Chinese text and corresponding English text in Chinese node and English node respectively, and create financial and economic report category in Parts node. The programming statement is as follows:

```
create(n:Chinese{name:‘金融市场’})
create(n:Chinese{name: ‘资本市场’})
```

Table 3 SPO triad relationship

Head	Tail	Relationship
金融市场	资本市场	包括
金融市场	Financial market	翻译
资本市场	Capital market	翻译

create(n:English{name: 'financial market'})
create(n:English{name: 'capital market'})
create(n:Parts{name: '财经报道'})

Create the relationship between different terms in each node and the translation relationship between terms in different nodes. The programming statements are as follows:

MATCH (a:Chinese {name: '金融市场'}),(b:Chinese{name: '资本市场'})MERGE (a)-[:包括]->(b).

3.3 Function Realization of Learning Platform

3.3.1 Function Design of Platform

(1) Outline and functional design. With reference to the relevant literature of the corpus platform in CNKI, plan the construction ideas and policies of the platform, design the construction basis based on Python and Wechat Mini Program development software, and refer to the BFSU ParaConc language materials retrieval software developed by Beijing Foreign Studies University and various corpora at home and abroad to clarify the basic functions of the platform, And summarized as professional software outline design documents.
(2) User interface design and page interaction design. Use the Process On network flow chart and mind map to design the appearance of the user interface, clarify the appearance and various details of the platform, design various interactive pages, clarify the jump between pages, the jump between buttons and pages, and improve the response results of operation error response.

3.3.2 Function Realization of the Platform

With Wechat Mini Program as the front end and web development framework based on flask as the back end, the basic framework of the platform is built. Neo4j and RDS MySQL are used as data storage to realize the application system of front-end display, back-end business logic and data storage of the platform. So that the platform can run completely.

(1) The realization of retrieval visualization function. In order to meet the retrieval needs of users in different fields, the platform has set up retrieval functions in four sections respectively, and the search results only contain words in specific fields. For professional retrieval, the platform refers to BFSU ParaConc language materials retrieval software developed by Beijing Foreign Studies University to realize the function of excluding special displacement retrieval results when retrieving words. The specific operations are as follows:

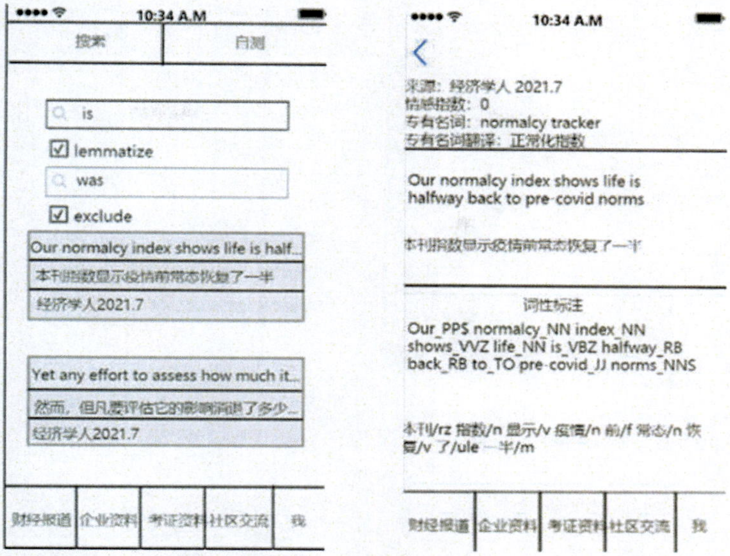

Fig. 5 Display of retrieval function interface

Enter the word that the user needs to find. For example, if enter ‘is’ in the search box in the first column, all aligned language materials containing ‘is’ will be displayed. After checking ‘lemmatiza’ will find all deformations of relevant words. Enter the deformation to be excluded in the search box in the second column. For example, after entering ‘was’ and checking ‘exclude’, only the results containing ‘is’ will be displayed. After clicking each corpus search result, all attribute information of relevant corpus in the database will be displayed (Fig. 5).

① Structure visualization. When users retrieve a proper noun, the platform will provide a knowledge graph of the noun in its field. Through the framework structure, they can understand the structural level of the term in the field. Taking the vocabulary of financial market as an example, when searching ‘money market’, the following knowledge graph will be shown, including the superior and subordinate structure of the field to which “money market” belongs and the translated version (Fig. 6):

② Visualization of general results. Users can find all proper nouns and their translations through the whole category, and can also view the relationship structure of proper nouns through the relationship graph. Taking some words in ‘financial market’ in financial and economic reports as an example, the visualization results are as follows (Figs. 7 and 8):

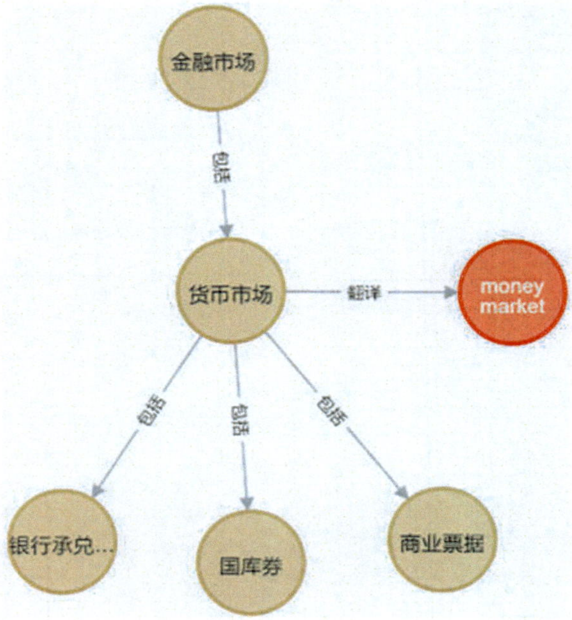

Fig. 6 Structural visualization of the term "money market"

③ Physical visualization. As financial terms include a certain number of unique physical nouns, which are relatively unfamiliar to none financial practitioners. In order to meet the cognitive needs of users, the platform creates a physical graph for proper nouns in the retrieval function, including term translation and pronunciation, and displays the physical appearance in graphic form. The results in the Fig. 9 below take 'check' as an example.

(2) The translation self-test function. Randomly extract the corresponding Chinese or English translations from the corpus in the relevant modules. After clicking the 'view translation' button, the platform will display various labels or attributes of the relevant language materials. After independent translation, users can view the authoritative translations in the database and compare them by themselves, so as to understand the laws of financial and economic translation and enhance the ability of financial and economic translation (Fig. 10).
(3) The reading function. The platform classifies topics and provides various articles in different topics for users to read. At the same time, it provides auxiliary reading function in combination with corpus annotation information. After users click the marked proper nouns, they will jump to the vocabulary visualization page, so that users can better understand the articles and enhance the utilization rate of corpus (Fig. 11).
(4) Community communication function. In order to meet the social needs of relevant user groups and increase communication channels between specific fields,

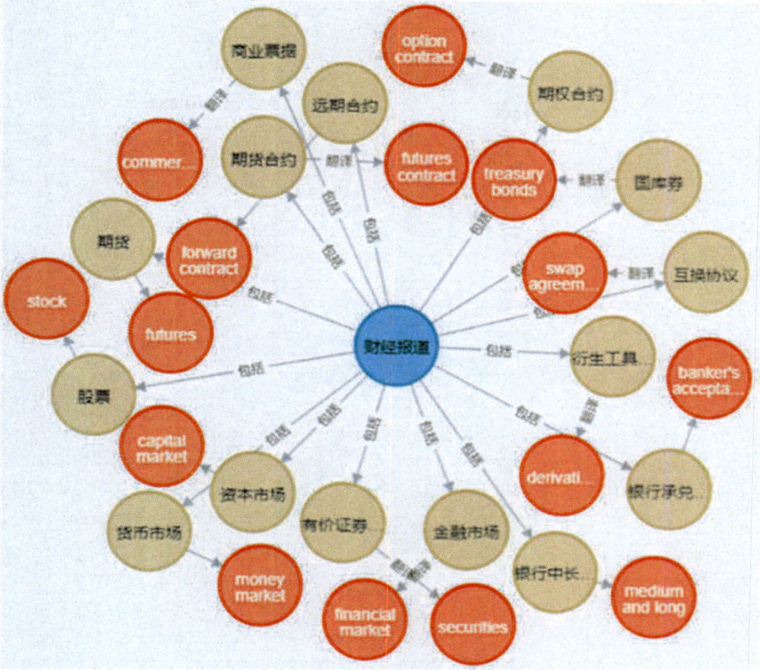

Fig. 7 Visualization of knowledge graph of "financial report"

the platform adds functions such as Wechat moments and posting. Users can independently publish text content and picture content. The platform has a like function and a reward mechanism to encourage users to share and exchange expertise in relevant fields within the platform. At the same time, users can also manage the content in the "my post" module (Fig. 12).

(5) Other auxiliary functions. In order to enhance the diversity of language materials sources, the platform adds the function of users' independent uploading materials, which can be uploaded to the platform for users after being reviewed and confirmed by relevant professional managers of financial translation. At the same time, set up the opinion feedback function to improve the platform according to the real-time feedback of users (Fig. 13).

4 Summary

Corpus is the basis of all natural language processing [14]. It plays a great role in improving translation efficiency, ensuring the accuracy of translation and the standardization of language use, and improving the translator's translation awareness.

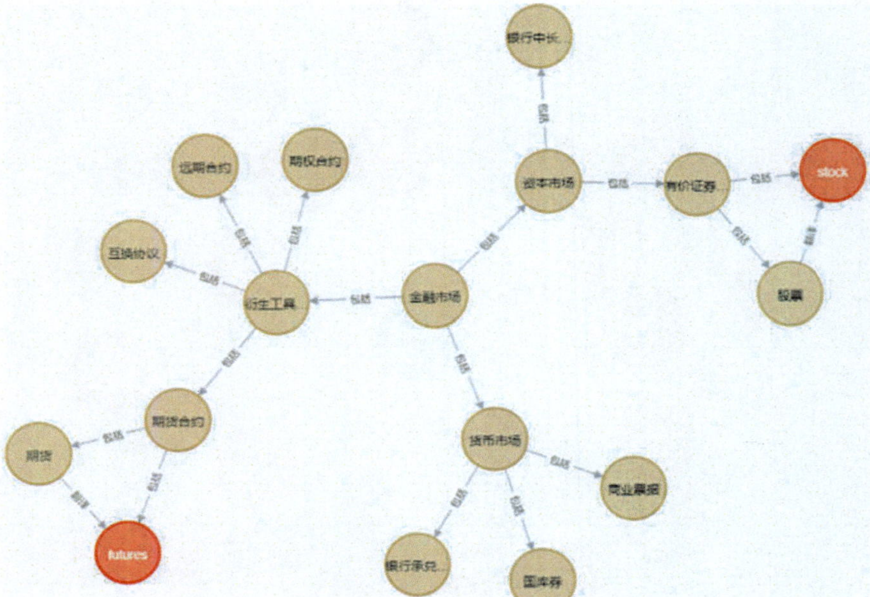

Fig. 8 Visualization of knowledge graph of all term relationships

Fig. 9 Knowledge card of the word "check"

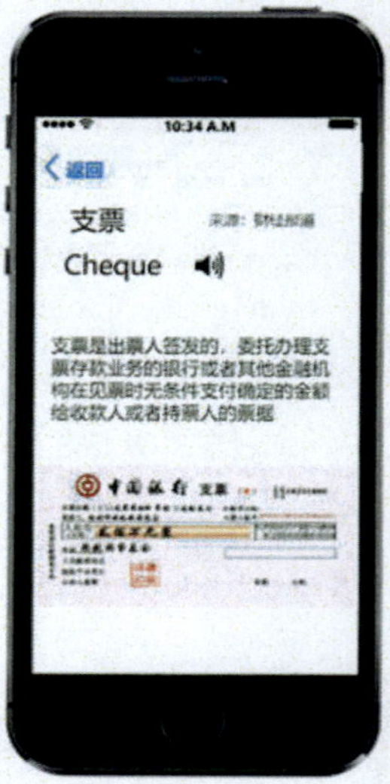

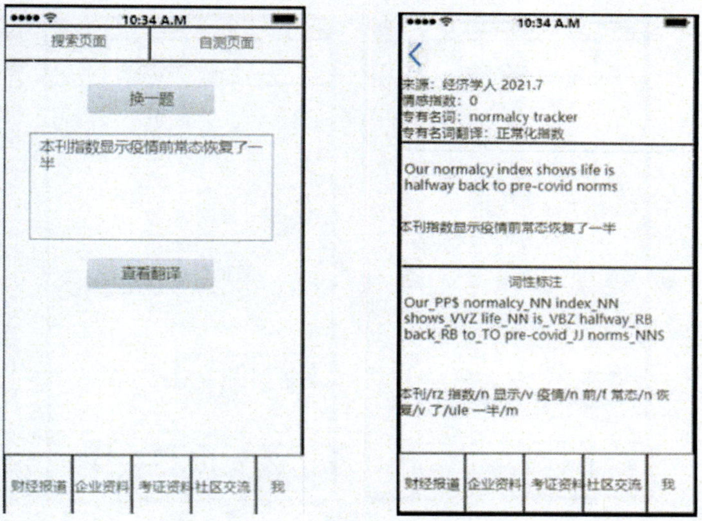

Fig. 10 Page display of self-test function

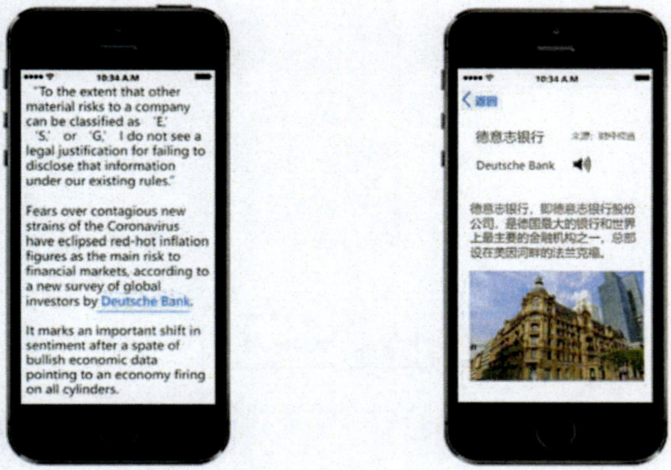

Fig. 11 Page display of reading function

Therefore, foreign language majors in domestic colleges and universities also include the use of corpus technology in one of the necessary skills of students.

This corpus learning platform has novel design, distinctive characteristics, strong pertinence and convenient use. It breaks the disconnection between the learning of financial translation major and the actual situation of the industry, innovatively introduces the advanced teaching means of financial and economic discourse corpus

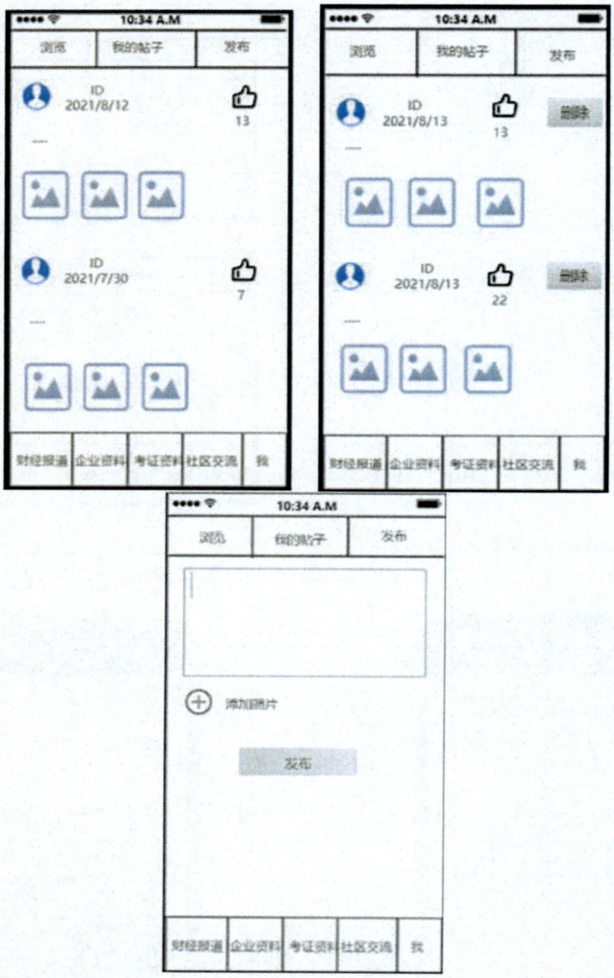

Fig. 12 Page display of community communication function

into the translation teaching mode and establishes an open classroom. Translation teachers can choose texts according to the syllabus, students' interests, students' majors and the degree of difficulty, which can be used as auxiliary reading materials for students; In the process of translation learning and practice, students can use the corpus to carry out rapid search and query the usage, collocation and subtle differences of the corresponding words in translation, so as to improve the efficiency of translation and the accuracy of expression. At the same time, this corpus can also provide students with professional qualification examination materials and various materials in enterprise activities for students' learning reference.

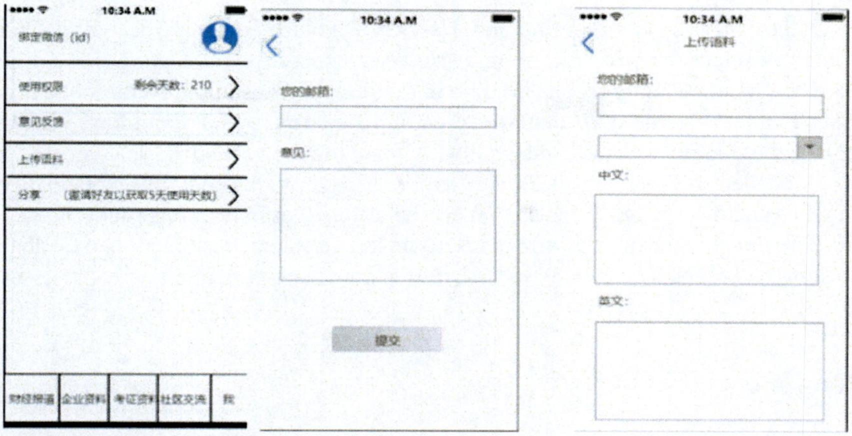

Fig. 13 Auxiliary function page display

In addition, students can also objectively analyze the phonetics, context and other elements of the text from the real text corpus of the corpus through special corpus translation software, so as to obtain rational research results. At the same time, with the extensive usage of corpus in teaching reform, empirical research based on econometrics and statistics can provide data support for the effectiveness of teaching reform. This corpus learning platform has accurate positioning, fits in with the hot spots of the industry and is easy to use, which can improve the translation awareness of social translators.

Therefore, the construction of financial and economic discourse parallel corpus is of great practical significance to the teaching, research and social service of linguistics. The mobile learning platform based on self-built corpus is easy to operate and run smoothly which also meets the requirements of users to use anytime and anywhere.

5 Prospect

For users to make more flexible use of corpus resources, provide them with diversified self-training methods for financial translation, and exercise their financial translation ability in an all-round way. The platform is different from traditional translation software, and it is planned to add diversified self-test function.

Considering the users' pronunciation of financial proper nouns and the differences in the meaning expression of different pronunciation in a specific context, the platform plans to add a listening self-test function to broadcast the voice through the platform. Users can recognize the audio and choose the correct translation form for training, so as to better remember vocabulary pronunciation. Secondly, in order to make more effective use of vocabulary resources, the platform plans to add a vocabulary memory training function. Users can create a personal vocabulary, independently select the words to be remembered, display the translation through the platform, independently recall and train in the form of translation. At the same time, using Python to write a vocabulary forgetting function, which combined with the feedback of memory degree, independently planning the vocabulary and vocabulary which needed to be reviewed. Finally, in order to provide users with personalized statistics and display functions and reflect the differences in learning effects between different individuals, the platform plans to add learning statistics function to feedback learning records and statistical data when it is used for a certain amount.

References

1. F. Changpu, Translation skills of financial English words and sentences. Era Lit. (Academic Edition) **05**, 31–32 (2007)
2. C. Can, W. Lifei, Facing the "one belt, one road" language service, promoting the reform of foreign language education. Lang. Educ. **6**(01), 2–6 (2018)
3. Z. Wei, Construction and application of parallel corpora for business English translation[J]. J. Hubei Open Vocat. Coll. **34**(01), 186–187 (2021)
4. H. Shuiqing, W. Dongbo, A review of corpus research in China. J. Inf. Resour. Manag. **11**(03), 4–17+87 (2021). DOI:https://doi.org/10.13365/j.jirm.2021.03.004
5. H. Enpei, Y. Lili, Development and prospect of language service industry in the 40 years of reform and opening up. Chin. Transl. **40**(01), 130–135 (2019)
6. L. Lili, On the cultivation of students' financial translation ability driven by corpus[J]. J. Lul. Univ. **4**(03), 88–90 (2014)
7. Y. Chenyue, Construction and application of Japanese Chinese bilingual corpus for Japanese finance. Mark. Community **51**, 294–295 (2019)
8. L. Rong, Construction and application of parallel corpus of Chinese tourism texts under the "One belt, one road" background[J]. Comp. Study Cult. Innov. **3**(25), 108–109 (2019)
9. S. Yan, Application of multimodal financial English corpus in college English teaching. Foreign Lang. Transl. **27**(02), 70–76 (2020)
10. W. Feina, Research on corpus-based Financial English (ESP) Teaching Mode[J]. Sci. Consult. (Sci. Technol. Manag.) **05**, 25–26 (2020)
11. M.R. Quillian, Semantic networks. Approaches Knowl. Represent. Res. Stud. **23**(92), 1–50 (1968)
12. A. Singhal, Official Google Blog: Introducing the Knowledge Graph: things, not strings. Official Google Blog 1–8 (2012)
13. L. Peng, Research and application of big data organization retrieval based on neo4j[D]. Southeast University (2015)
14. F. Delian, Y. Lingyun, Q. Chaochen, Construction and analysis of ethnic minority corpus for information processing. Wirel. Internet Technol. **16**(19), 77–79 (2019)

Evolution Model of Online Public Opinion in University and the Countermeasures Based on the Dynamic Field Theory in the BIG DATA Era

Pinghao Ye and Liqiong Liu

Abstract In the big data era, the frequent occurrence of online public opinion events in university harms the regular teaching activities of colleges. It is necessary to figure out the evolution and development trend of online public opinion in university and reduce the negative impact of online public opinion. Moreover, it is also imperative to establish a set of scientific and practical monitoring, early warning, and response mechanisms. Based on the dynamic field theory, this study establishes a model of the evolution of online public opinion in university. It uses NetLogo software to simulate and analyse the model. Results show that volunteers related to online public opinion events in university will play a guiding role earlier than public opinion leaders and participate in the governance of online public opinion. Public opinion leaders are vital forces for colleges to respond to online public opinion, but their role lags behind volunteers'. The development of online public opinion in university is a gradual process of calming down. Public opinion leaders and disseminators will interact for a long time until the public opinion ends.

Keywords Dynamic field theory · Online public opinion · Evolution model · Big data era · University

1 Introduction

In the big data era, the all-around digital transformation of the economy and society continues to change the internal environment of national governance and internet public opinion governance. At the same time, the guiding concept, work focus, and operational practice of internet public opinion governance have also undergone significant changes. As a barometer reflecting college teachers' and students' learning, work, and living conditions, online public opinion in university has attracted

P. Ye (✉) · L. Liu
School of Information, Wuhan Business University, Wuhan, China
e-mail: yezigege1977@163.com

S. Patnaik and F. Paas (eds.), *Recent Trends in Educational Technology and Administration*, Learning and Analytics in Intelligent Systems 31,
https://doi.org/10.1007/978-3-031-29016-9_24

increasing attention from all walks of life [1]. In recent years, frequent online public opinion incidents in university have continuously pushed the schools to the forefront of online public opinion. Considering that these incidents are closely related to college students, the outlook on life and the world of college students is not yet fully developed. Emergencies in university can easily cause resonance or irrational behaviour. Moreover, life events' continuous occurrence and fermentation can be catalysed and ignited easily, seriously impacting colleges' teaching order, image, and reputation [2].

The participation of netizens in the university is an essential basis for dividing the life cycle of such online public opinion. This participation is the main body to promote the development of online public opinion in university [3]. The behaviour of netizens (posting, following, and commenting) allows the development of the online public opinion of university from incubation to outbreak, peak, repetition, and digestion periods [4, 5]. Therefore, it helps to explore the evolution rules of university online public opinion by simulating the development of each participant in university. Moreover, the impact of emergency management measures on the development of public opinion can be analysed. This research provides a reference for university to reduce the influence of online public opinion.

This study takes the evolution law of online public opinion in university as the research object. In addition, the study uses the NetLogo simulation platform to conduct multi-agent modelling and simulation experiments on the evolution of in university. According to the actual situation, the experiment extracted the participants of the online public opinion, clarified the interaction between the subjects, and designed the interaction rules. The experiment also simulated the evolution process of online public opinion in university and verified the correctness of the simulation results [6]. Based on this, suggestions on managing in university are proposed. These suggestions help the university monitor and warn events scientifically and effectively. Furthermore, the suggestions are significant for the schools to reduce and respond effectively to the negative impact of online public opinion.

This paper consists of six parts. In Sect. 1, the author introduces the research background, research significance, and research content. In Sect. 2, the research and theoretical background of university online public opinion is summarised. Section 3 constructs the evolution model of online public opinion in university and determines the attributes and meanings of each subject. Section 4 analyses the evolution result of the model. Then, based on the model evolution results, Sect. 5 suggests how university respond to online public opinion. Finally, Sect. 6 includes the research limitations of this study and future research prospects.

2 Literature Review

2.1 *Individual Impact Model of Online Public Opinion*

In the field of public opinion research, an individual's attitude towards public opinion, that is, whether to participate in the propagation process and the emotional tendency towards public opinion events, is generally believed to be mainly affected by the attitudes of neighbouring individuals [7]. Moreover, other nearby individuals will also be affected once an individual joins the propagation process. This mechanism is called an impact model [8]. The construction of the impact model involves two complementary processes, namely, the evolution and propagation of public opinion. The former refers to the change process of public opinion, which occurs at the micro-level of individual-specific attitudes and changes. On the contrary, the latter refers to interactive behaviour that appears at the micro-level without considering the individual's specific attitudes or views [9]. This study mainly takes college students, teachers, and social netizens as research individuals to explore their behavioural changes in the propagation process in university. The objective is to examine the law of propagation in university.

2.2 *Impact of Online Public Opinion in University*

At present, several scholars have conducted research on the online public opinion of university and achieved certain results Kelling et al. [10]. The campus survey method is used to study the university. The results show that although most students use new media, they still express their views mainly in traditional ways, such as petitions. Hew [11] studied the willingness of college students and teachers to use Facebook in spreading public opinion on the Internet. The results showed that the primary purpose of college students and teachers in using Facebook is interpersonal communication and disclosing more personal information, which increases the risk of privacy disclosure in the process of propagation. Research in university mainly focused on the response and governance of negative online public opinion. Xuan and Lu [12] believed that, by improving the leadership and management ability, teachers' exemplary ability and students' pressure resistance ability of university could effectively deal with the negative public opinion of university. Fuji and Lijuan [13] believed that increasing the degree of information disclosure, reducing the attention of netizens, and reducing the role of the media can effectively improve the efficiency of university in responding. Cao and Song [14] found that the type of hot practice in university has an important influence on the popularity of public opinion and the spread of university reputation. In addition, the network search index of university is related to the popularity of public opinion events. Most of the existing literature studied the factors affecting the propagation in university focused on analysing public opinion.

Thus, judgment on the evolution trend of public opinion is limited, with insufficient guiding significance for university to deal with the change.

2.3 *Dynamic Field Theory*

The psychologist Lewin is known as the father of social psychology. He used the concept of "field" in the physics theory to describe a psychologically non-materialized "field," what Lewin called Life Space. His research showed that, although the relative importance of the individual and environment is different in varying cases, each behaviour depends on an individual's state and the environment [15]. The main content of Lewin's dynamic field theory is to regard human behaviour as the result of the combined effect of individual subjective and objective factors. Individual needs and motivations are the basis of human behaviour, and objective environmental factors are the conditions that simulate human behaviour. However, the objective environment has positive and negative effects. Individuals in a specific environmental system will generate the direction and motivation of individual behaviour based on various influential factors of the individual's psychological environment [16]. According to Lewin's dynamic field theory, combined with research models, this study describes the behaviour of individual netizens in the real society from three fundamental aspects: individual subjective factors, objective social factors, and behaviour coefficients.

3 Model Formation

3.1 *Impact Model of Online Public Opinion in University Based on Dynamic Field Theory*

Based on the multi-agent model, this study explores the issue of trust identification of the subject of information. Firstly, based on the traditional personalised trust model, emergency response subjects are divided into different status roles [17]. Secondly, considering the issue of individual heterogeneity, dynamic weighting factors are introduced through historical interaction information between Agents. Moreover, direct and indirect trust are combined to present a more effective trust evaluation, and a trust recognition model considering heterogeneity is constructed. The objective is to provide theoretical and methodological support for the effective trust identification of online public opinion individuals and the high-quality dissemination of rumour refutation information [18].

The agent-based modelling and simulation (ABMS) method is a modelling method from bottom to top that describes the macro behaviour of complex systems by characterising the behaviour of agents and their interactions (including the environment) [19]. This method defines the modelling object according to different individuals in the system and abstracts the key attributes to establish the individual agent model. The ABMS method also sets the agent attributes, behaviour, interaction rules, and related constraints. Furthermore, this approach investigates how individual behaviour affects the overall characteristics of the system through the system's operation. The simulation system based on the ABMS method is called a multi-agent system [20]. The ABMS method can better simulate individual heterogeneity. However, its disadvantage is that the available resources and data limit the detailed level and complexity of the model, and the simulation operation needs high computing resources, so the model verification is challenging [21]. This study designs the evolution model of online public opinion in university [22] on the basis of NetLogo's model library.

3.1.1 Individual Subjective Factors

As the management organisation of teachers and students, university have the responsibility and obligation to guide and regulate online public opinion. They are also the main body of publishing authoritative information on campus. Therefore, management departments in university should guide and control online public opinion and resolve crises, which is also a key factor in managing online public opinion in university. university must always pay attention to campus dynamics, control emergencies, and respond positively to the questions of netizens [23].

An individual's knowledge structure directly affects an individual's attitude and behaviour towards online public opinion in university. This structure is not only a simple educational background or education level but is also related to the correlation amongst online public opinion events, personal knowledge, and personal experience [24]. Individuals will make a fundamental judgment on the event based on their knowledge when they contact online public opinion events. When the events are highly related to themselves, their judgment will be significantly affected.

In public opinion propagation, netizens can independently express their own opinions. When events in university differ from netizens' values or involve their interests, netizens will represent their interests. Moreover, they will trigger a heated network discussion in a short time, which can easily detonate a crisis of online public opinion. Therefore, netizens are important in forming and disseminating in university [25]. At the same time, discussions between netizens will affect their attitudes and behaviour changes. This study uses authority values and immunity indicators to reflect the impact of the interaction between disseminators and recipients in university on the choice of acceptable behaviour. Authority value indicates the size of individual influence ability, that is, the more evidence an individual holds, the greater the authority value. Immunity refers to the individual's ability to refuse to participate in the discussion of events in university. Figure 1 depicts the information judgment process of the participants in the propagation of university.

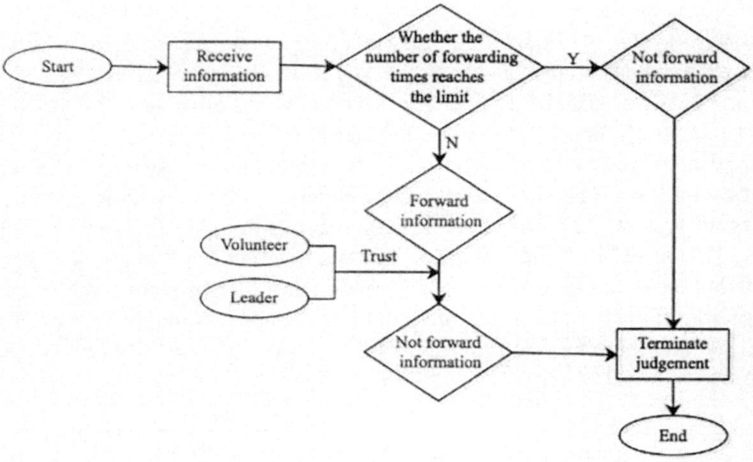

Fig. 1 Process of judging the information of the participants in the online public opinion of university

3.1.2 Behaviour Coefficient

Netizens have their behaviour coefficients in the process of disseminating in university, including behaviour, the scope of influence, and the number of contacts. Behaviour is the feedback presented by an individual based on comprehensive factors. The range of influence is the individual's radius of influence. Moreover, the number of contacts refers to the times (the number of communications) that an individual contacts in university. When the number of contacts reaches a certain value, immunity will decrease.

3.1.3 Public Opinion Leaders

Public opinion leaders are responsible for guiding and managing students' online public opinion dissemination behaviour during the evolution. Counsellors will play a crucial role as opinion leaders in disseminating online public opinion in university. The greater the students' trust in counsellors, the greater their influence, which is conducive to university to give full play to the role of counsellors and better manage.

The objective society in which netizens live is affected by the media, divided into traditional and online media according to different forms of communication. Traditional media mainly include television, radio, and newspapers, which are authentic, reliable, and authoritative. Online media include news websites, Weibo, forums, and WeChat, including other social software and platforms. Online media has features of fast dissemination, strong interaction, low authority, and extensive information. These features can easily cause widespread discussion and public curiosity once an

Table 1 Attributes and meanings of individual netizens

Agent	Variable name	Value range	Notes
Individual subjective factors	Number of transpond	[1, 3000]	Number of transpond
	Probability posting	[0.01, 1]	The initial propagation intention decreases linearly with time
	Volunteer	[10, 5000]	Number of netizens
	Relevance	[0, 1]	The degree of association between individuals and university online public opinion events
	Authority	[0, 1]	The ability of the individual to affect the transmitted individual during the communication process
	Immunity	[0, 1]	The ability of individuals to refuse to participate in the communication process
Behaviour coefficient	Behaviour	[−1, 1]	Netizen's final behaviour; −1 means not paying attention, and 1 means paying attention
	Probability transpond	[0, 1]	Probability transpond of netizens
	Duration	[1, 100]	Public opinion cycle
	Chance-recover	[0, 1]	The probability that netizens generate trust
Leader	Opinion leader	On or off	The probability of opinion leader existence

event with a strong value impact breaks out and is prone to "herd effect," which leads to the outbreak of an crisis. Table 1 shows the meaning of the three attributes, namely, individual netizens, behaviour coefficients, and leadership in this study.

4 Simulation Research on the Evolution and Communication Model of Network Public Opinion in University Based on the Netlogo Platform

4.1 Introduction to the Netlogo Platform

NetLogo is a programmable modelling environment used to simulate natural and social phenomena. In 1999, Uri Wilensky first launched this tool, subsequently developed at the Center for Connected Learning and Computer-Based Modelling at Northwestern University in the United States [26]. NetLogo's software, models, and related documents are available for free download on its official website. NetLogo is particularly suitable for modelling complex systems that evolve over time. This software comes with multiple agents, such as turtle, patch, and link, allowing modellers to issue instructions to hundreds of independent agents. This case makes it possible to explore the relationship between individual behaviour at the micro-level and macro models, which emerge because of the interaction amongst many individuals [27]. The bottom layer of NetLogo is implemented in the Java programming language, which can run on all mainstream platforms and run as Java Applets in the browser.

NetLogo has detailed documentation and teaching materials. NetLogo also comes with a model library, which contains many simulation models that have been written. These simulation models cover many fields of natural and social sciences, including biology and medicine, physics and chemistry, mathematics and computer science, economics and social psychology and others.

4.2 *Various Groups and Their Attributes in the Online Public Opinion of University*

This study categorises the people involved in the university into three categories: disseminators, volunteers, and opinion leaders. Disseminators are people who pay great importance to online public opinion in university. Volunteers are groups that are associated with online public opinion events in university. Opinion leaders mainly refer to college administrators, including student cadres and class teachers. Table 2 shows the attribute characteristics of each group.

Table 2 Attribute characteristics of the main groups of online public opinion in university

Attributes	Transpond	Volunteer	Leader
Event correlation	Low	Medium	Highest
Authority	Low	High	Highest
Immunity	Uncertain	Medium	No
Scope of transmission	Small	Medium	Widest
Proportion	High	Less	Less

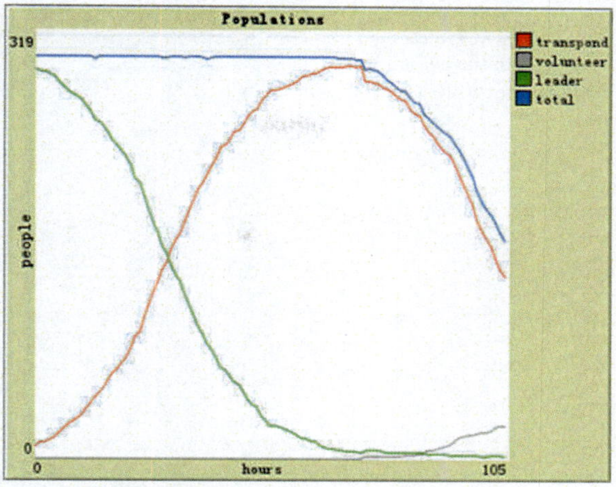

Fig. 2 The active number and trend of the evolution of public opinion after two hours

4.3 Model Simulation

4.3.1 Beginning and Dissemination of Online Public Opinion in University

The model's simulation results show the simulation results at different time nodes, as shown in Figs. 2, 3, 4, 5, 6, 7 and 8. In the figures, the red small people pattern represents public opinion disseminators, green represents public opinion leaders, and grey represents volunteers. The population graph shows changes in people participating in disseminating online public opinion in university over time. The red, green, grey and blue lines represent the change in public opinion disseminators, leaders, volunteers, and participants.

4.3.2 Impact of University Management on Results

For analysing the relationship between university management and the propagation, this experiment has introduced public opinion leaders to observe the evolution of public opinion events for recording and analysis. The simulation results show that at the initial stage of online public opinion in university, with the outbreak of public opinion, the leaders' response in university is lagging. Moreover, the red people pattern appears in large numbers, and the role of leaders is not evident, leading to an upward trend in public opinion. With the intervention of public opinion leaders, public opinion propagation shows a clear downward trend (Fig. 2). This trend is most evident in the first two hours of public opinion. Therefore, when public opinion on the school network appears, we should seize the critical period and play the role of leaders as soon as possible, achieving a better effect.

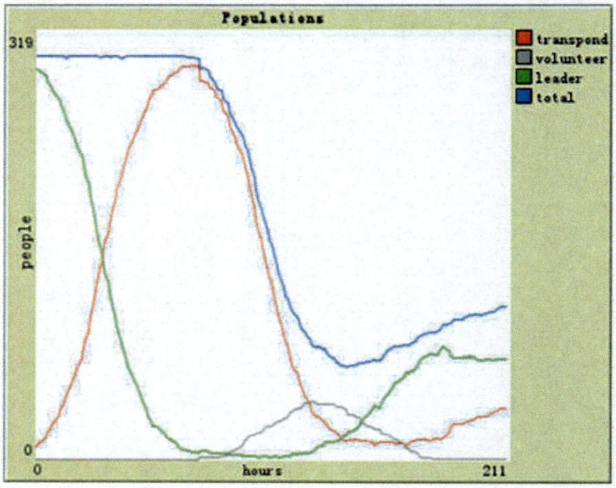

Fig. 3 The active number and trend of the evolution of public opinion after four hours

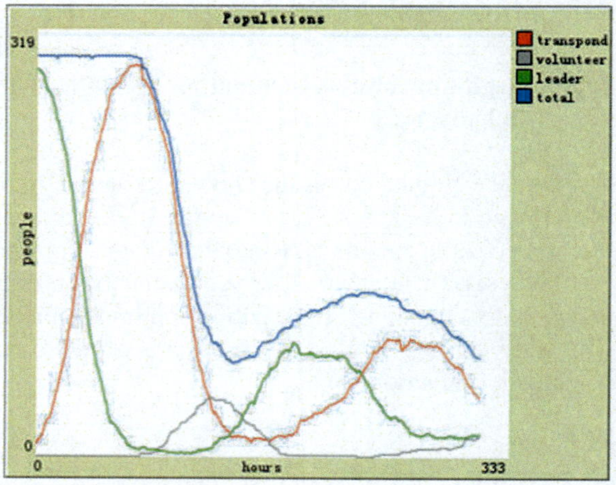

Fig. 4 The active number and trend of the evolution of public opinion after six hours

4.3.3 Impact of Volunteers on Results

The model studies the relationship between volunteers and the dissemination of online public opinion in university. Moreover, most volunteers are students and teachers related to the online public opinion in university. The simulation results show that volunteers play a leading role earlier than public opinion leaders because of the large number of volunteers and their wide distribution (Figs. 3 and 4). Moreover, the grey people pattern appears earlier, which serves as an early warning for the early spread of public opinion. When public opinion leaders begin to play a guiding role, volunteers and public opinion leaders jointly guide the propagation in university. However, as the role of public opinion leaders becomes stronger, the guiding role of volunteers tends to slow down. Therefore, university should pay full attention to the

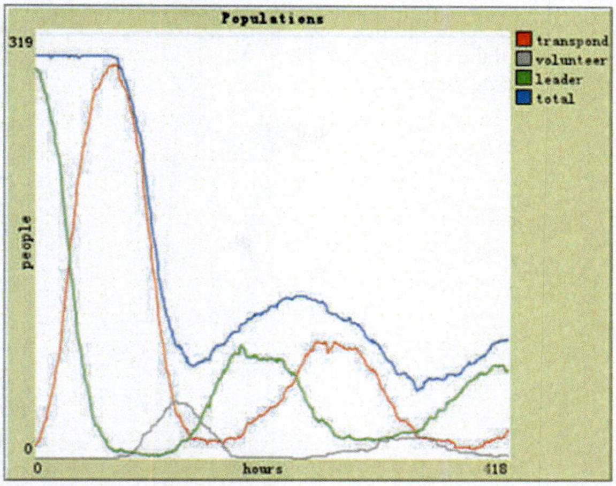

Fig. 5 The active number and trend of the evolution of public opinion after eight hours

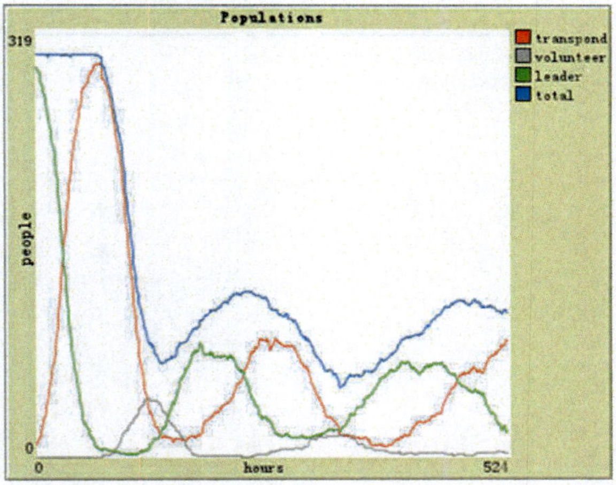

Fig. 6 The active number and trend of the evolution of public opinion after ten hours

behaviour of volunteers, play their early warning role and take timely measures to improve the efficiency of public opinion management and reduce losses.

4.3.4 Attenuation and End of Online Public Opinion in University

The development of online public opinion in university gradually declines with time, and the number of disseminators and leaders increases one after another (Figs. 5, 6, 7). Several red and green people patterns appear alternately, and the online public opinion in university gradually subsides in the mutual game of the patterns. Finally, only the green pattern remained (Fig. 8), which is the end of public

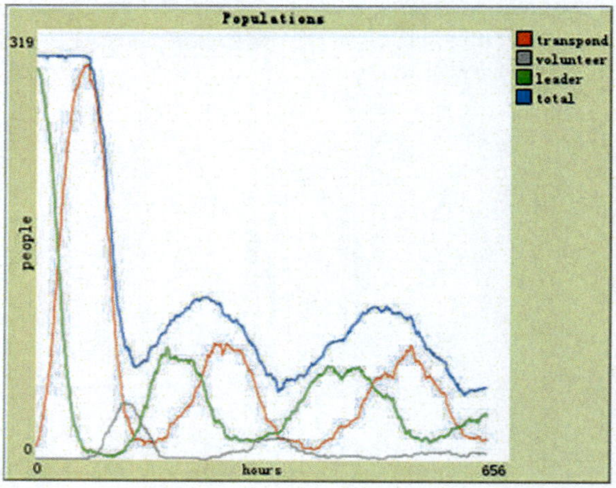

Fig. 7 The active number and trend of the evolution of public opinion after 12 h

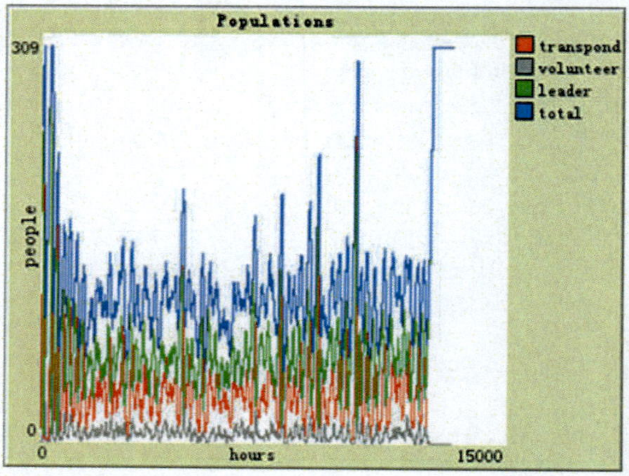

Fig. 8 The active number and trend of the evolution of public opinion after 24 h

opinion. Therefore, university need to carry out guidance work unremittingly, play a leading role and jointly solve the problem of propagation in university.

5 Conclusions

This study takes university as the main research object, builds a dynamic model of the evolution of online public opinion in university, and uses NetLogo software to simulate and analyse the model. The objective is to provide decision-making ideas for university to timely deal with online public opinion.

Information disclosure is the basis for responding to effectively, and the degree of information disclosure will significantly impact the credibility of university. In the network environment, the anonymity of information publication makes all kinds of news, even false news, published and quickly spread on the Internet. In this case, university should refute rumours to prevent a crisis of online public opinion. In this process, university have always played the leading role in controlling and resolving crises. Therefore, university need to increase the degree of information disclosure to improve credibility [28]. Therefore, university must first report the relevant event to society, including the cause, treatment plan, and punishment on the relevant responsible persons. Through this, the people may understand the actual situation of the event and prevent rumours from spreading on the Internet.

In the primary role of netizens, the influence of leaders will have an important impact on netizens' attention. The emergence, development, climax, and decay of online public opinion are all accompanied by the influence of leaders. Therefore, university should strengthen the management of social networks and formulate relevant policies to guide people to use social networks correctly. They should timely respond to rumours on the Internet, guide the netizens accurately from their emotional needs, empathise with netizens, interact with them on the Internet, answer questions for the public and calm down their negative emotions.

Amongst the main actors in the social environment, the force of volunteers is mainly affected by the impact of the event and the degree of harm. If the influence and the degree of harm are great, volunteers will increase their efforts to expose these events, increasing the probability of a public opinion crisis. At this time, university need to reduce the influence and harmful degree of events. University must deal with these events that may cause major crises and strong value shocks cautiously, keeping an eye on the trends of the events and eliminating them as soon as possible. For events that have occurred, university should respond positively, keep up with the spread of public opinion and strive for the initiative to reduce the crisis [23].

In the era of the big data, big data can completely record public opinion on social platforms. It contains rich connotations and a lot of regular information. It has risen to a new level of ideological and cultural connotations. Specifically, it can include big data thinking, digital literacy, data mining, data reorganization, etc., not limited to the technical attributes themselves. Online public opinion management, characterised by deep data mining and fusion applications, can make full use of data retrieval technology to integrate data information through specific strategies, transform complex data into usable decision-making information, and quickly find the cause of events. Understanding the source of information can help relevant departments reorganize the entire incident and put forward targeted preventive measures; a large database covering the whole information flow in China could be established to enhance the integration of data resources while reducing dependence on foreign data technology. "Big Data" optimizes and improves domestic supervision and monitoring mechanisms in all aspects and improves the accuracy of hot spot detection of various types of information.

Shortcomings and Expectations

This research has conducted a simulation study on the evolution of online public opinion in university. However, the integrity of the simulation model, the authenticity of the data, and the verification of the network platform still have many shortcomings and deficiencies, which mainly include the following:

The study of online public opinion in university is a complex social topic, and many factors affect in university. College students' group and individual characteristics will also affect the evolution in university. The model in this study simplifies reality to a large extent. However, some complex factors have not been suitably added to the model and need to be improved in future research.

This research conducts a simulation study on the evolution of online public opinion in university and is a partial qualitative study based on multiple assumptions. However, the study lacks real data support and does not effectively integrate the online public opinion data in university in actual society. In future research, we should introduce more real data of cases to verify the model.

This study conducts simulation experiments on the evolution of online public opinion in university. Limited by various constraints, the authenticity and effectiveness are not verified on a specific network platform. Future research requires testing on platforms, such as WeChat and forums.

Acknowledgements This work was supported by the Hubei Social Science Research Fund [19ZD074].

References

1. Z.O. Ebniya, The impact of religious values in the Jordanian political discourses on public opinion (field study on university students). Rev. Econ. Polit. Sci. ahead-of-print (2020)
2. Y. Li, Y. Zhang, Research on simulation and coping strategies of network public opinion resonance in university. J. Intelligence **38**, 107–113 (2019)
3. D. Zaptcioglu Celikdemir, G. Gunay, A. Katrinli, and S. Penbek Alpbaz, Defining sustainable university following public opinion formation process. Int. J. Sustain. Higher Educ. **18**, 294–306 (2017)
4. C. Ling, J. Feng, P. Wu, S. Zhang, A study on crisis responce of campus network public opinion based on SOAR model. Inf. Sci. **37**, 145–152 (2019)
5. M. S. Mohamad Saleh, N. Md Kassim, and N. Alhaji Tukur, The influence of sustainable branding and opinion leaders on international students' intention to study: a case of University Sains Malaysia. Int. J. Sustain. High. Educ. ahead-of-print (2021)
6. A. Jędrzejewski, K. Sznajd-Weron, Impact of memory on opinion dynamics. Phys. A: Stat. Mech. Its Appl. **505**, 306–315 (2018)
7. S. Fang, N. Zhao, N. Chen, F. Xiong, Y. Yi, Analyzing and predicting network public opinion evolution based on group persuasion force of populism. Phys. A: Stat. Mech. Its Appl. **525**, 809–824, (2019)
8. G. Jiang, S. Li, M. Li, Dynamic rumor spreading of public opinion reversal on Weibo based on a two-stage SPNR model. Phys. A: Stat. Mech. Its Appl. **558**, 125005 (2020)
9. H. Zhu, B. Hu, Impact of information on public opinion reversal—An agent based model. Phys. A: Stat. Mech. Its Appl. **512**, 578–587 (2018)

10. N.J. Kelling, A.S. Kelling, J.F. Lennon, The tweets that killed a university: A case study investigating the use of traditional and social media in the closure of a state university. Comput. Hum. Behav. **29**, 2656–2664 (2013)
11. K.F. Hew, Students and teachers use of Facebook. Comput. Hum. Behav. **27**, 662–676 (2011)
12. Z. Xuan, W. Lu, A study on the ability of university to deal with negative public opinion in the all-media era. Inf. Sci. **37**, 146–150 (2019)
13. C. Fuji, W. Lijuan, Research on the strategies of university network public opinion under the mobile environment based on System dynamics. Mod. Inf. **38**, 118–123 (2018)
14. X. Cao, Y. Song, A study on the influence of the hot issue of university education on university—a case study of the event of education public sentiment in 2014. Book Inf. Work. 100–104 (2015)
15. K. Lewin, Principles of topological psychology: Principles of topological psychology (1936)
16. M. Wallerberger, A. Hausoel, P. Gunacker, A. Kowalski, N. Parragh, F. Goth, et al., w2dynamics: Local one- and two-particle quantities from dynamical mean field theory. Comput. Phys. Commun. **235**, 388–399 (2019)
17. T. Kano, M. Iwamoto, and D. Ueyama, Decentralised control of multiple mobile agents for quick, smooth, and safe movement. Phys. A: Stat. Mech. Its Appl. **572**, 125898 (2021)
18. Y. Bao, F. Huo, An agent-based model for staircase evacuation considering agent's rotational behavior. Phys. A: Stat. Mech. Its Appl. **572**, 125923 (2021)
19. M. M. G. Correia, J. V. M. Barboza, and A. L. Espíndola, Sleeping sickness: An agent-based model approach. Phys. A: Stat. Mech. Its Appl. **582**, 126282 (2021)
20. S. H. Sabzpoushan and F. Pourhasanzade, A new method for shrinking tumor based on microenvironmental factors: Introducing a stochastic agent-based model of avascular tumor growth. Phys. A: Stat. Mech. Appl. **508**, 771–787 (2018)
21. T. Weron, A. Kowalska-Pyzalska, R. Weron, The role of educational trainings in the diffusion of smart metering platforms: An agent-based modeling approach. Phys. A: Stat. Mech. Appl., **505**, 591–600 (2018)
22. M. L. Netlogo. http://ccl.northwestern.edu/netlogo/models/
23. J. Chen, T. Zhang, X. Zhang, University network public opinion management innovation machine under mobile environment System and coping strategy. Inf. Sci. **35**, 57–62 (2017)
24. W. Han, Y. Lu, The evolution and propagation simulation model of network anti corruption public opinion based on dynamic field theory. J. Mod. Inf. **36**, 3–11 (2016)
25. J. Neñer, M. F. Laguna, Optimal risk in wealth exchange models: Agent dynamics from a microscopic perspective. Phys. A: Stat. Mech. Appl., **566**, 125625, (2021)
26. A. Fichera, A. Pluchino, R. Volpe, A multi-layer agent-based model for the analysis of energy distribution networks in urban areas. Phys. A: Stat. Mech. Appl. **508**, 710–725 (2018)
27. T. Kochanski, Toward teaching markets as complex systems: a web based simulation assignment implemented in Netlogo. Int. Rev. Econ. Educ. **11**, 102–114 (2012)
28. F. Chen, L. Weng, Research on the coping strategies of network public opinion in university under mobile environment based on system dynamics. J. Mod. Inf. **38**, 118–123,176 (2018)

Design and Implementation of Campus Information Aggregation Platform

Yuhang Huang and Jianxin Zhang

Abstract With the rapid development of informatization and digitization, the tide of digital services such as digital campus and digital government affairs is rising. However, many scattered, disordered and even redundant data exist on campus due to the non-interoperability of information used by various systems. In view of the current situation analysis, the B/S architecture was adopt in this paper, and the Vue.js, PHP, MySQL, Python and WeChat applet were used to develop a campus information aggregation platform. With the convenience of obtaining information on the Internet, information about life and learning on campus was collected and processed, which was convenient for teachers and students to quickly obtain the desired information.

Keywords Information aggregating · Information aggregation platform · Digital campus

1 Introduction

With the rapid development of informatization and digitization, the tide of digital services such as digital campus and digital government affairs is rising. Teachers and students mainly obtain information on campus by using the school website, there is a portal website developed based on desktop browsers in a department or a college. They need to open many portal websites when teachers and students want to find specific information involving multiple departments. However, nowadays, with the more popularity of mobile devices, mobile devices was used by more people for web browsing. It is obvious that the websites developed based on desktop browsers cannot meet the access needs of teachers and students. Therefore, a campus information aggregation platform was designed and implemented based on this situation, which includes a web platform and Wechat applet platform. Teachers and students can get

Y. Huang · J. Zhang (✉)
Nanfang College, Guangzhou, Guangdong 510970, China
e-mail: zhangjx@nfu.edu.cn

S. Patnaik and F. Paas (eds.), *Recent Trends in Educational Technology and Administration*, Learning and Analytics in Intelligent Systems 31,
https://doi.org/10.1007/978-3-031-29016-9_25

the best browsing experience on cross platform and cross device by using the front-end multi platform solution, and at the same time, the information you need can be easily and quickly obtained.

After investigation, the campus information aggregation platform should have the following functions:

(1) Campus news information. Users can search, sort and view news, which includes the news information released by all portal websites on campus.
(2) Campus address book. Users through the campus address book can quickly search for the contact information of public services on campus, as well as the public contact information of educational administration and counselors, which includes the public telephone and contact information of teachers in the school.
(3) Campus fraud prevention information. It includes all fraud case data and anti-fraud articles in the school. At the same time, the platform analyzes the fraud case data in real time and generates summary data and visual charts to show vividly the current telecommunications fraud situation in the school. Users can perceive the current situation of telecommunications fraud in the school and improve their awareness of fraud prevention by displaying and learning anti fraud articles.
(4) Information of Educational administration system. It includes semester schedule, today's schedule, credit information, score information, dormitory electricity information, etc. Through this information, the data of students' life and learning were covered in an all-round, so that students can understand clearly academic information and dormitory electricity consumption information.

2 Developing Browser-Side Single-Page Applications Based on Vue.js and T Design

The web platform of the campus information aggregation platform was developed by using the combination of Vue.js and TDesgin. Compared with traditional web development, Vue.js was used for componentization, the development of this platform and the separation of data and structure, the development efficiency is higher, Vue is a single page application, which speeds up the visit speed of the users and enables users to have a better experience when visiting the website. However, the website pages with a unified style can be quickly created by introducing the TDesign component library.

Vue. JS is a progressive framework for building user interfaces [1], and is also a set of front-end frameworks mainly used for building user interfaces [2]. HTML, CSS, and JavaScript were combined in a Vue file by Vue [3], dynamic website pages can be built quickly by using Vue.

TDesign is a set of enterprise-level design system precipitated by the business teams of the Tencent in the process of service business, which has unified design values, consistent design language and visual style to help users form a continuous and unified experience cognition [4].

The browser client page of the campus information aggregation platform with unified style and powerful performance was created by the comprehensive use of Vue.js and TDesign.

3 Developing Data Interface Based on CodeIgniter4

CodeIgniter is a set of application development framework and toolkit for PHP website developers [5]. Its version 4.0 and higher were used by the campus information aggregation platform as the data interface required by the back-end to build the front-end of the platform. The model using the CodeIgniter can quickly add, delete, modify and query the data in the database through its defined method, and also support the creation time and update time of records. Meanwhile, the data returned by CodeIgniter will be converted automatically into JSON format, which facilitates the data exchange between the front and rear end to the maximum.

4 System Design

This platform adopts the front and rear end separation development mode [6]. Compared with the traditional development mode, the web sites developed by the front and rear end separation development mode has faster response speed, better user experience, and high maintainability. The interface effect of the front-end only needs to be paid attention to when developing the front-end, and the rear-end only returns the data required by the front-end. The system deployment diagram of the campus information aggregation platform based on Vue.js, WeChat applet and CodeIgniter4 is shown in Fig. 1.

As can be seen from the above figure, the system consists of three parts—front end, rear-end and data capture.

The front end consists of dual platforms (Web platform, WeChat applet), and the web platform consists of TDesgin component library + Vue.js, which sends an asynchronous requests to the rear-end through axios to dynamically refresh website data; The WeChat applet sends an asynchronous request to the rear-end through wx.request to dynamically refresh the page information of the applet.

The rear-end uses the CodeIgniter4 framework + MySQL database to build a data interface, which is responsible for responding to the asynchronous requests initiated by the front-end. After the controller in the framework completes the corresponding operations, the data is returned to the front-end, the rear-end is also responsible for connecting the educational administration system and obtaining the data of the educational administration system.

Data capture is to use the request base in Python to regularly capture data from external data sources (educational administration system, power system and campus portal website), and use pymysql to connect to MySQL database to store the processed data into the database.

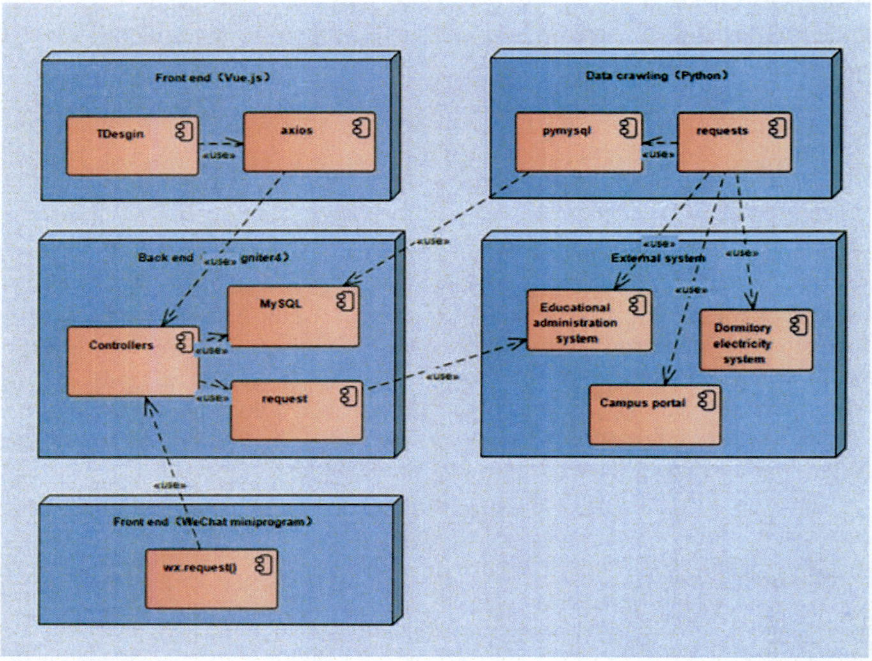

Fig. 1 System deployment diagram

4.1 Data Capture and Storage

The data capture was realized by using Python, capturing the news of campus portal website was taken as an example, and its processing process is shown in Fig. 2:

(1) All campus portal websites were captured every 1 h, the captured HTML text was matched by the bs4 base to get all the news list data.
(2) The obtained news list data will use pymysql to connect to the MySQL database and match the data in the database. If there is no matching result in the database, it will go to the next step, otherwise, the capture process will end.
(3) If there is no matching result in the database, and the news details page without matching results will be further crawled, and the HTML text of the news detail will be identified by using the bs4 base again, and only the text of the news will be filtered out. The screened news text was stored in the database by using the pymysql to complete the capture process.

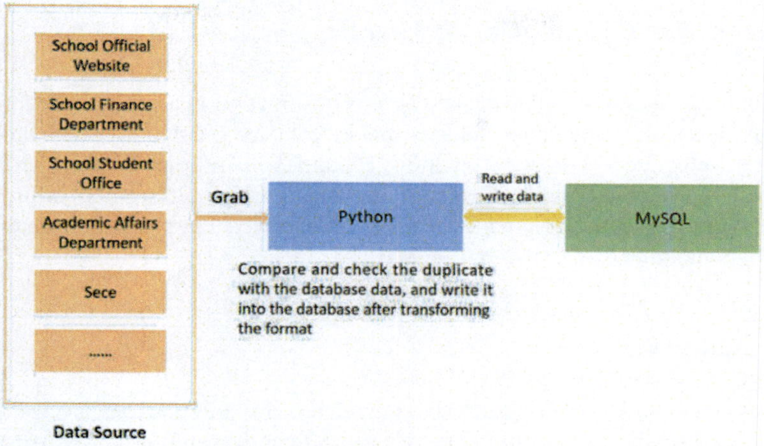

Fig. 2 Schematic diagram of data capture operation

4.2 Data Interaction with Educational Administration System

The data of the educational administration system was transmitted in JSON format and can only be obtained after identity authentication. The identity authentication was realized based on JSON web token (JWT). The following will explain in detail how the campus information aggregation platform interacts with the educational administration system:

(1) Obtaining the user credential JWT. After obtaining the user provides consent, the account and password entered by the user were obtained to the identity authentication interface of the educational administration system for simulated login, and the user-specific JWT was obtained. If the JWT was not obtained, it indicates that the account and password of the user were wrong or the educational administration system was under maintenance. At this time, the user will be prompted to try again later.
(2) Storing the account and password of the user and user credentials JWT. After obtaining the JWT, the account and password of the user and JWT will be stored in the localStorage of the user's browser.
(3) Requesting the specific data. When the user visits the page that needs educational administration data, the front-end Axios base will carry the JWT stored in the localStorage of the user's browser to request a rear-end specific interface, and the data reques was made by the rear-end controller to the educational administration system[7]. After obtaining the educational administration data, it is processed and returned to the front-end to be displayed on the client's browser. Since then, a complete process of data interaction with the educational administration system has been completed.

4.3 Detailed Design of Each Module

The main functional modules of the campus information aggregation platform are campus news, life (integrated address book, campus fraud prevention module), learning (integrated course information, examination information, teacher information module), aggregation Center (i.e. educational administration information module, integrated score, credit, dormitory electricity, class schedule and other functional modules), as shown in Fig. 3.

4.3.1 Campus News

Campus news consists of navigation bar, search box, list area and details page, which mainly displays the news information of most official websites of our university, so that users can visit all news by visiting one website. The following are the detailed description of each functional module:

(a) The navigation bar was classified by Department (College), and clicking it will refresh the news content in the list.
(b) The search box is to search for the news information under the current category, which is convenient for users to quickly find the content you need.
(c) In the list area, the title, release date and classification of the news were displayed in a card format, and clicking it will jump directly to the details page.
(d) The details page displays the news article content, article information, article source website, etc.

4.3.2 Address Book

The address book consists of a search box and an address book list, which mainly includes some contact information commonly used in the school. Users can find the contact information you want by searching or sliding browsing. Contact information includes name, remarks, short number, long number and email.

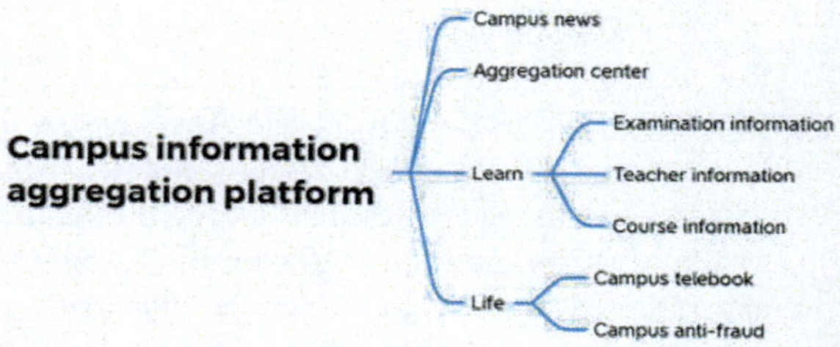

Fig. 3 Each functional module

4.4 Examination Information

Examination information consists of examination list, examination details and examination branches, which mainly helps users to quickly understand the examination and some conditions of the examination.

(1) National popular examination information was listed in the examination list, such as the National Civil Service Examination, the Putonghua Proficiency Test, the National Postgraduate Unified Examination, the National Computer Ranking Examination, etc. Each item roughly shows the name and introduction of the examination.
(2) The examination details include the examination name, examination introduction, examination official website and the latest examination information. The user clicks the latest examination information to jump to the examination branch details.
(3) The details of the examination branch include the name of the examination branch, official website, introduction, examination fee, time node and remarks. The time node includes node name, start time, end time, text time (during text time setting, start time and end time will not be displayed), and remarks.

4.5 Course Information

Course information consists of search box, course search result list and course details. Through the search box, users can use keywords to search for the course name, and then click the course you want to query in the course search result list to jump to the course details page, which displays the course information and teachers team information. The course information includes the course name and course name of the Department. The teachers team information includes the total number of teachers in the course, teachers name and teachers name of the College, Users can click on the teacher's name to jump to the relevant teacher information [8, 9].

4.6 Teacher Information

Teacher information is closely associated with course information. Teacher information consists of search box, teacher search results list and teacher details page. Through the search box, users can use keywords to search for the teacher name, and then click the teacher you want to query in the course search result list to jump to the teacher details page, displays the teacher information, data overview and the teaching courses list. The following is a detailed description:

(1) The teacher information includes the name and Department of the teachers.

(2) The data overview includes the total number of courses taught by teachers, the number of students, the average final score, the average overall appraisal score, the highest comprehensive score, and the comprehensive failure rate, which is convenient for users to quickly understand the teacher.
(3) The teaching course list includes the courses taught by teachers, and the information includes the course name, the data statistics year, the average final score, the average overall appraisal score, the failure rate and the number of students, which helps users understand the scoring orientation of teachers in the course and the final examination of students in the course.

4.7 Campus Fraud Prevention

With the rapid development of China's telecom network and finance, mobile phone, Internet and other telecom network platforms have greatly facilitated People's Daily life, but they have also spawned many telecom fraud criminal activities implemented by telecom network platform, and spread rapidly, and even become one of several major new crimes [10].

Telecom fraud also occurs on campus. Campus fraud prevention mainly displays case statistics (total loss and number of cases) and case details by semester. The fraud case data of each semester will be flagged and classified by this function, and the typical fraud situation in the semester will be displayed vividly by using visual charts. The users can quickly know the current types of cases that are more frequent, so as to improve their awareness of fraud prevention by browsing and learning fraud cases and viewing visual icon information.

4.8 Aggregation Center

The aggregation center mainly includes three functions, such as the functions of external data, credits, scores and dormitory electricity, which all require users to log in with the account and password of the educational administration system before the users can be used.

(1) Scores: Through the scores system, users can visit their own annual score data (academic year score point, academic year score), academic year score list and class score analysis. Through these functions, users can clearly understand their own learning situation and how to arrange courses in the future.
(2) Credits: Through the credit system, the users can view the credit structure, the list of courses to be recognized, the list of courses in progress, the list of courses completed and the list of failed courses. Through these functions, the users can be helped to plan scientifically the credits they take, and avoid the situation that they cannot successfully complete their studies due to wrong or missing credits.

(3) Dormitory electricity: Through the dormitory electricity system, the users can view the remaining electricity, payment records and electricity consumption analysis of the current dormitory; Among them, the payment record generates a visual recharge chart on a monthly basis. The electricity consumption analysis was combined with the electricity price for charging. At the same time, the histogram of the peak electricity consumption in recent 24 h will also be given to help users plan their electricity consumption scientifically.

5 System Implementation

The main modules of the campus information aggregation platform are campus news, address book, course information, teacher information, campus fraud prevention, aggregation center (i.e. educational administration information module, integrated score, credit, dormitory electricity, class schedule and other functional modules). The functional modules of the front-end multi-platforms (Web platform, WeChat applet) are basically the same, and the data used are all from the same database.

5.1 Web Side Effect

The effect of partial page on the Web-end (the implementation effect of the home page is shown in Fig. 4, and the implementation effect of the aggregation center is shown in Fig. 5) is shown in Fig. 4 and Fig. 5.

5.2 Small Program Side Effect

The effect of partial front-end page on the Wechat applet (home page, campus fraud prevention page and score details page respectively from left to right) is shown in Fig. 6.

5.3 System Development Environment

The system was written by using Visual Studio Code and Wechat developer tools, and Vue cli service was used to compile Vue.js code, PHP Study Pro was used as the backend (CodeIgniter4 framework) to run the server, MySQL5.7 database was used to store website data, and Python3 was used to complete the data scraping task.

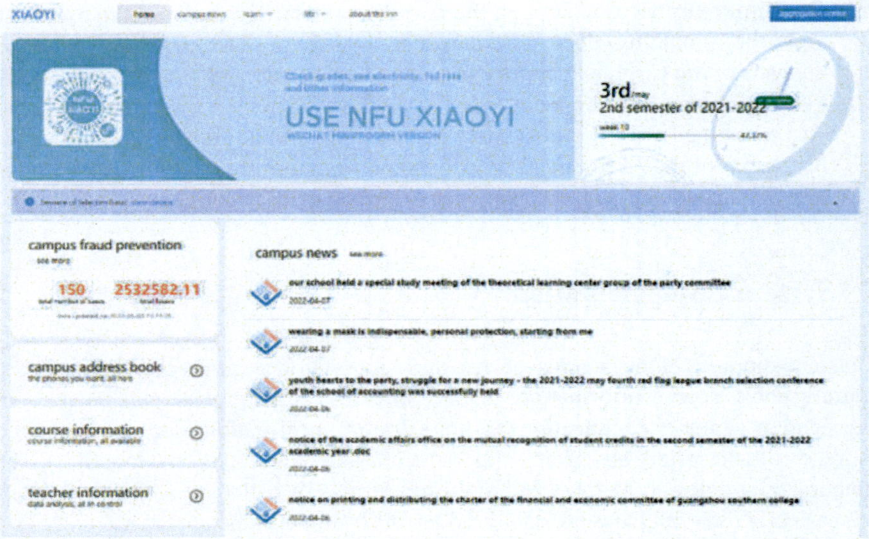

Fig. 4 The renderings of the home page implementation

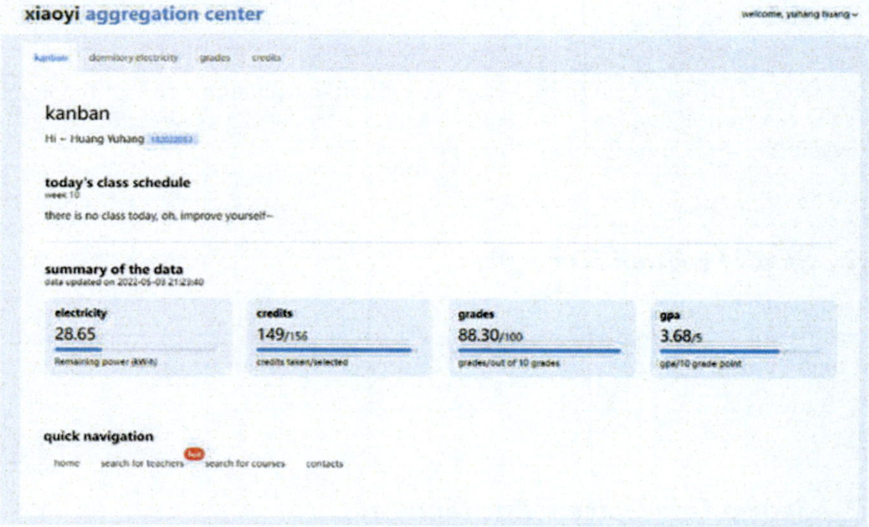

Fig. 5 The implementation effect of the aggregation center

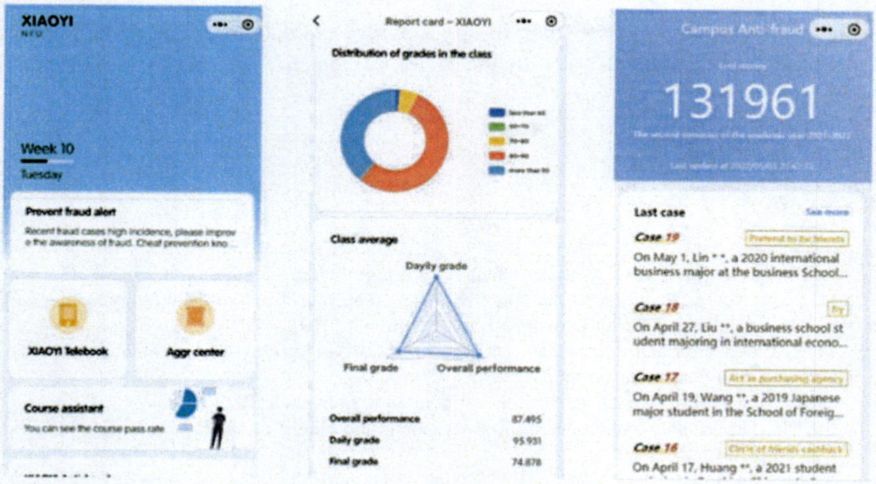

Fig. 6 The implementation effect of partial front-end page of applet-end

5.4 *System Deployment Environment*

Two Tencent Cloud lightweight application servers were used to deploy front-end resources and rear-end resources respectively. Among them, the compiled static resources were placed in the front-end server for deployment, and the python files were placed in the rear-end server. The Python files were placed in the rear-end server to execute the scraping task regularly by using a timer. A MySQL database server was used to ensure the stable storage of data, and the database server was connected to the rear-end server by using a dedicated intranet link to ensure the rapid transmission of data.

6 Conclusion

The emergence of campus information aggregation platform will open up many information gateway websites on campus and aggregate all scattered information together, which allows teachers and students to visit a website to understand the information of the whole school, the mobile device no longer needs to be zoomed in and out by using the front-end multi-platform solution, which reduces gradually the information gap on campus to help students quickly understand, find and discover the information scattered in every corner of the campus.

In the future, after maturing and has a certain user group, the platform operation will have a greater say to plan to communicate with the educational administration system of the school, and develop jointly a part of the data interface, which allows more teachers and students to enjoy the digital campus life brought by scientific and technological progress.

References

1. Vue. Introduction [EB/OL]. https://v3.cn.vuejs.org/guide/introduction.html,2022-03-27
2. R.L. Liu, Research and implementation of virtual scene construction technology for network attack and defense. Beijing Univ. Posts Telecommun. (2019)
3. M. Yue, Research and implementation of NB-iot In-road Parking Management System [D]. Beijing Univ. Posts Telecommun. (2020)
4. TDesign. About us. https://tdesign.tencent.com/about/introduce,2022-03-27
5. S.J. He. Design and implementation of a smart photo album for Android platform. Beijing Jiaotong University (2017)
6. X. Xiaohui, L. Jiangtao, G. Han, D. Yanyan, Development of gas recovery scheme system based on SpringBoot+Vue framework. Comput. Simul. **38**(06), 248–250+382 (2016)
7. Z. Jian, Design of rolling bearing online monitoring and diagnosis system based on cloud service. Dalian Jiaotong University (2020)
8. G. Xiaoyu. Chinese teaching management system based on Wechat public account. Beijing University of Posts and Telecommunications (2019)
9. L. Yuanming. Design and implementation of cloud classroom platform for primary and secondary schools based on Python [D]. Beijing Jiaotong University (2018)
10. Linkou County Court. Prevent telecom fraud knowledge propaganda [EB/OL]. https://m.thepaper.cn/baijiahao_12945596,2021-12-16

College English Blended Teaching Model Based on Mobile Learning Platform

Yali Qiang

Abstract The paper explores the construction of blended teaching model of college English based on the Super Star mobile learning platform. This study comprises six parts, introduction, related literature and studies, Super Star learning platform, methodology, experimental results and conclusion. Form the experiment, the conclusion has been reached that college English blended teaching model based on mobile learning platform not only can improve students' English ability more effectively but can arouse students' English learning interests.

Keywords Blended teaching model · Mobile learning platform · College English · Super star · English learning interests

1 Introduction

The deep integration of information technology and education in the era of "Internet+" has caused changes and accelerated the process of education informatization. The traditional classroom teaching model cannot meet the needs of education in the era of Internet+, and the development of the times requires reform and innovation in teaching. The Modernization of Education in China 2035 proposes to build an integrated, informative and intelligent teaching management and service platform to enhance teaching and learning, and to cultivate students' awareness and ability to carry out independent management, service and learning, and other related requirements [1].

With the rapid development of mobile network technology, mobile applications have penetrated into our lives. Cell phones, as the most frequently used intelligent mobile terminals, play an important role in daily life. Traditional media can no longer meet the needs of users to obtain information, and more and more people

Y. Qiang (✉)
Guangdong University of Science and Technology, Dongguan, China
e-mail: 254894038@qq.com

S. Patnaik and F. Paas (eds.), *Recent Trends in Educational Technology and Administration*, Learning and Analytics in Intelligent Systems 31,
https://doi.org/10.1007/978-3-031-29016-9_26

use cell phones or PADs and other mobile terminals to obtain information. In recent years, smart teaching tools and applications based on mobile terminals have been developed and used more and more widely, changing the teaching ways [2]. These mobile learning applications reflect the achievements of information technology in mobile learning. Although the functions are slightly different, if they are reasonably applied in different parts of the teaching process, they can play a positive role in improving learners' independent learning ability and also provide corresponding support for personalized learning.

Blended teaching is a mixture of online and face-to-face teaching, and a mixture of different teaching theories, teaching modes, different teaching media, classroom lectures and virtual environments [3]. Currently, research on blended learning in China is distributed in various fields such as school education and on-the-job training. The development of blended learning has been tested for more than ten years, and the current blended learning mainly combines traditional teaching and online learning to form a hybrid model with teachers as the leading and prior organizers and students as the main body, which can retain the systematic and leading advantages of traditional teaching, but also combines rich online learning resources with the goal of developing competence, providing a new teaching reform model [4]. It has become one of the research hotspots in the field of education.

With the advent of the "Internet + education" era, there have been some changes in terms of teaching resources, teaching ways and teaching methods, but at present, the traditional teaching model dominated by teachers still accounts for the majority. Its single teaching method cannot well stimulate the students' interest and motivation of independent learning under the current background, and it does not pay attention to individual differences in the design of teaching objectives and teaching content, which has many deficiencies.

2 Related Literature and Studies

By reviewing the relevant literature and studies, it can be found that there are not many domestic articles on blended teaching assisted by mobile learning platform, but the number of articles is gradually increasing, mainly focusing on 2018 to 2021. The content of the studies mainly discusses the teaching function of using the mobile learning platform, mainly focusing on the teaching effect, classroom teaching process, independent learning, the blended teaching and instructional design [5]. The research levels are mainly focused on higher education and basic research. Liu Xin applied the mobile learning platform to study blended learning in public courses in colleges, put forward the problems in the process of information-based teaching and designed the "3-3-2" blended teaching model. Shen Yanxia used the learning platform to design the learning mode of public courses in universities, taking the mobile Internet environment as a breakthrough point, and emphasizing the cognitive training, active learning and teaching feedback of students [6]. He Xiaowei adopted the Super Star mobile learning platform to emphasize the cultivation of students' abilities, construct student-themed, and make students the master of the classroom and the construction of knowledge. Feng Rongzhen built a hybrid teaching model based on "flipped classroom" and proposed that the construction of a smart classroom should comprise the construction of resource data, the analysis and mining of student data, and the actual measurement of teaching effects to promote students' personalized learning. Ye Bilan explored the interaction of the flipped classroom of college English speaking based on the Super Star mobile platform [7]. Zhou Changzhou proposed the "A + A" teaching model for the classroom teaching of art majors in universities, taking the Super Star learning platform as an example.

Through combing the literature, it is found that many studies exploring blended teaching based on mobile learning platform have some problems such as using a single function, no timely and effective learning evaluation, deficiencies in resource integration and optimization, unclear online and offline teaching, and insufficient use of technology. This study will apply the relevant data collected, design and develop relevant learning resources on its own, build a blended teaching mode of college English based on Super Star mobile learning platform, revise it through teaching practice to strive for better teaching effects, and make certain additions to the domestic research on relevant blended teaching modes.

3 Super Star Learning Platform

Super Star learning platform is designed and developed by Chaoxing Group and is a mobile teaching tool based on the concept of mobile terminal + classroom under the background of "Internet + education". The platform has four major features: massive resources, making it easier to obtain and share resources, timely feedback, which makes it easier for teachers to understand the learning situation, more communication, making it easier for teachers and students to interact and big data, making assessment and evaluation easier [8].

Using the Super Star learning APP, we can build a bridge between classroom teaching and online learning, and realize the interaction covering the whole teaching process. The main functions of Super Star are divided into two major modules, which are course management module and teaching activity module. By using the functions of each module wisely, it can provide a new interactive experience for the whole teaching process [9]. Before class, teachers use the App to push relevant learning resources and learning requirements to students' cell phones, and after students finish learning as required, they can get feedback on their learning and teachers can get feedback on students' learning. In classroom teaching, teachers use Super Star to conduct real-time teaching activities such as question answering, selection, and quizzes, which can add a more effective form of interaction to traditional classroom teaching. After class, teachers can deepen students' understanding of knowledge by providing some extra-curricular materials and organizing some discussion activities to broaden students' horizons [10]. At the same time, the statistical function of the learning platform will record the whole process of learning and provide complete data support for the whole teaching activities. In addition, more new functions are developed according to teaching needs, such as cloud reading, cloud disk, etc. Teachers can manage and expand the functions of the platform according to their teaching needs.

4 Methodology

This section comprises research hypotheses, participants, research process and experimental procedures.

4.1 Research Hypotheses

There are two research hypotheses designed in this research as follows:

Hypothesis 1: College English blended teaching model can improve students' English ability more effectively;

Hypothesis 2: College English blended teaching model can arouse students' English learning interests.

4.2 Participants

The participants of this research are 76 students of Guangdong University of Science & Technology. They are from two classes, students of class one are majors of digital management with 36 and students of class two with 40 are majors of accounting. The ability of their English is quite similar. Henceforth, class one is randomly selected as experimental class (EC) while class two as the controlled class (CC). What makes the two classes different is their learning model. The experimental class has their college English in blended teaching model based on Super Star mobile platform while the controlled students have English classes in traditional way.

4.3 Research Process

This study summarizes the connotation of blended teaching through the study of existing blended teaching theories, follows the process of linking theory with practical experience, verifying through teaching practice and designs a blended teaching model of college English based on mobile learning platform. The research process is shown in Table 1 below.

Table 1 Research process of college English blended teaching model

Time	Contents	Subjects	Methods
16 weeks	Blended teaching model based on mobile learning platform	EC	Tests (pre-test & post-test) observation questionnaire interview
16 weeks	Traditional teaching model	CC	Tests (pre-test & post-test) questionnaire interview

4.4 Experimental Procedures

Blended teaching model is an organic combination of traditional classroom teaching and the teaching model based on mobile learning platform. The purpose of designing the teaching activities is to arrange the use of the functions provided by the mobile learning platform to make the teaching activities more reasonable and efficient. The blended teaching model based on the mobile learning platform designed and built by the author encompasses three stages: before class, in class and after class and each stage contains three aspects, teacher's activities, mobile learning platform application and student's activities, which are combined with each other, in line with the characteristics of mobile learning. The concrete teaching procedure is presented as follows:

(1) **Pre-class Activity Design**
In the pre-class preparation, teachers should clarify the teaching objectives and key and difficult points of teaching from the results of front-end analysis, and make relevant learning resources according to the teaching content, including courseware, media resources, mind map, learning objectives, key and difficult

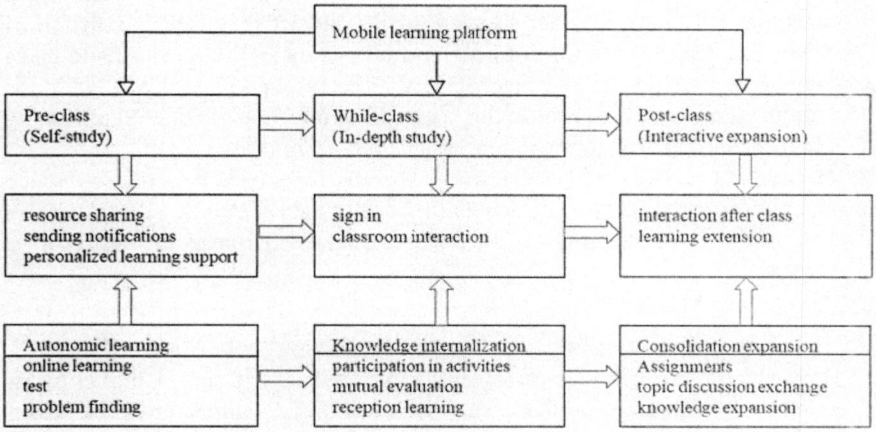

Fig. 1 College English blended teaching model based on mobile learning platform

points of learning, etc. The learning content should be visualized to help students establish a knowledge framework system and be posted to the students through the learning platform. After students complete the learning task, teachers will check the relevant data feedback on the platform, understand the learning effect of students based on the actual situation, and make reasonable adjustments to the next stage of classroom teaching.

Before class, students learn independently according to the learning content published in the platform under the guidance of the learning objectives, use the mind map to strengthen the grasp of the learning content, and establish a preliminary knowledge framework system. The problems and difficulties encountered in the learning process should be timely fed back and recorded, and teachers should monitor and guide the learning at this stage. At this stage, students complete the task of independent learning through the guidance of teachers, solve the possible problems and realize the preliminary construction of knowledge. This way of learning reflects the subjectivity of students and the dominant role of teachers. In this process, the Super Star learning platform plays a role of sharing resources, sending notifications and personalized learning support.

(2) **While-class Activity Design**

Traditional classroom teaching is mainly in the form of teacher explanation, and questions and answers between teachers and students. Its advantage is that through face-to-face teaching activities, teachers can make intuitive judgment of students' learning situation. However, the interaction between teachers and students in the traditional classroom is easily affected and the test and evaluation results in class cannot be timely fed back to students, since it takes a certain period of time for students to know the teacher's evaluation on them. In the construction of classroom teaching activities, existing problems should be fully considered, and teaching activities should be composed of both teachers and students [11].

In classroom teaching, teachers should answer the questions that students independently encounter before class, and then carry out efficient classroom teaching and discussion activities with the interactive teaching function of Super Star learning platform. Finally, both teachers and students should make comments on students and teachers assign corresponding tasks to guide the learning direction after class. In this stage, the application of Super Star learning platform enriches the forms of teaching activities, improves the classroom efficiency, and leaves detailed classroom records for students to review, which also keeps a record of each student's performance in class, saving time in non-teaching aspects such as attendance, answering questions, and scoring records.

(3) **After-class Activity Design**

The study after class mainly focuses on the interaction and extension of students. According to the Learning Pyramid theory, students will have a higher retention rate by teaching and discussing such learning activities to others. Abundant learning resources are needed in after-class learning, and reasonable arrangement and design of discussion and exchange activities are required [12].

Teachers' tasks include posting learning materials, diagnosing learning situations, initiating topic discussions, and testing after class. At this stage, the resource posted should focus on extended knowledge resources and extracurricular malleable knowledge resources. In topic discussion, both teachers and students can initiate relevant topics on the learning platform, and students and teachers can discuss together. In this process, teachers should act as the organizer of learning activities and arouse the enthusiasm of students. After class, the learning effect of students will be fed back to teachers and students will consolidate what they have learned. The test content will be posted through the learning platform and the deadline for task completion will be informed to students. After that, the platform will automatically collect scores according to the time, and sort out data reports for teachers to diagnose the learning situation, and whether the knowledge points should be strengthened. At this stage, through online discussion, expansion of learning content and testing, students can judge their own learning degree and timely check and fill in the gaps. Topic discussion can promote learners to improve each other and deepen their memory of knowledge. In addition, they can make use of the abundant digital resources of the learning platform to expand learning and achieve personalized learning.

(4) **Learning Evaluation**

In the traditional teaching evaluation system, the subject of evaluation is mostly teachers who have a certain authority, and students are in a passive state of acceptance. This single evaluation system is unfavorable to students' interest in learning, which is easy to cause students' resistance. Moreover, such evaluation also has a certain degree of subjectivity and one-sidedness, and the result cannot accurately reflect students' learning process. In the blended teaching evaluation system of mobile learning platform, it integrates a variety of evaluation subjects, including students' self-evaluation, teachers' subjective evaluation, platform automatic evaluation, student mutual evaluation and so on. Diversified evaluation can feedback students' learning effect and situation from many aspects and realize the change of evaluation focus from "result" to "process". This model evaluates the online learning process data, performance and homework evaluation from multiple perspectives. The evaluation content runs through the whole learning process, including diagnostic evaluation, formative evaluation and summative evaluation, and finally achieves the purpose of diversified evaluation. Formative evaluation mainly reflects the learning process.

Diversified evaluation can reflect students' learning state, stimulate and motivate students' learning behavior, and play an important role in improving learning effect and regulating learning state.

Diagnostic evaluation is an evaluation carried out at a certain stage to find possible problems, in order to revise the teaching plan and then enter a new round of learning. This mode mainly includes the information of self-learning before and after class and online testing. These data will be automatically generated by the system in the mobile learning platform to reduce the burden of teachers, and teachers can optimize the teaching in the next stage according to the diagnosis results.

Formative evaluation is a phased evaluation that evaluates students in the teaching process. The model mainly includes data such as class participation, test scores, homework completion and class check-in. The learning process and activities of each stage will be recorded accordingly to generate individual detailed three-dimensional data as one of the evaluation basis.

Summative evaluation is based on the value judgment made by the course designer after the completion of the course implementation, and the result of evaluation is taken as the achievement. This model mainly evaluates from the perspectives of self-evaluation, mutual evaluation (teacher evaluation, group evaluation), mid-term and final grades. Among them, self-evaluation is not only the observation of self-evaluation, but also one of the ways to improve the learning process initiative and enthusiasm. Through self-evaluation, students can reasonably set learning objectives and promote learning behaviors, so as to make learning more effective. Self-evaluation can be carried out under the guidance of teachers or arranged on mobile learning platforms. Mutual evaluation is a process in which learners put forward ideas and opinions through mutual communication and other means. Through the thought collision between peers, thoughts can be triggered to improve innovative thinking and learning effect. Mutual supervision can also arouse the enthusiasm of participation, cultivate the ability of communication, and better complete the study. Teacher evaluation means that teachers summarize and evaluate students' learning from multiple perspectives based on teaching objectives. As the guide of teaching, teachers with rich experience can provide more convincing basis for students to judge themselves.

5 Experimental Results

The data from the research has collected and statistical analysis has been made after one semester's blended teaching experiment.

5.1 Test Results

In order to compare the results of the two models at each stage of the test, the test before the experiment and the final test are chosen as a reference. Although the data of the two classes has been obtained from the two different teaching modes, other influencing factors are basically the same and there is no big difference in students' overall English ability, which could avoid the interference of irrelevant factors to a certain extent. The experimental data has been analyzed using the stage assessment scores of the two groups, and the data analysis software Microsoft Excel was used to analyze them. The experimental data has adopted the scores of two classes' tests and been analyzed by Microsoft Excel, as shown in Table 2.

Table 2 Results of the two classes' scores in the pre-test and post-test

	Mean		SD		MD	t	Sig
	EC	CC	EC	CC			
Pre-test	67.03	65.10	11.78	11.88	1.93	0.633	0.529
Post-test	73.73	67.23	10.10	10.97	6.50	2.982*	0.004

* $p < 0.05$

From Table 2, it could be seen that when the mean scores are compared in the pre-test, the mean score of EC is a little higher than CC and their mean difference is 1.93. The SD of the two classes is very similar which shows that their scores are equally distributed. From the value of T and P, there is no significant difference between the pre-test scores of EC and CC. In other words, the English proficiency of two classes is similar before experiment and the empirical study could continue.

After the 16-week experimental teaching practice, it could be presented that the mean of experimental class is 73.73 and the mean score of control class is 67.23, that the score of EC is higher than CC. The value of T and P verified there is a significant difference between the scores of the two classes. The mean score of CC has made no difference but the mean score of EC has improved a lot. Therefore, the experiment shows that college English blended teaching model can improve students' English ability more effectively.

5.2 *Questionnaire Results*

The questionnaire surveys the students' "learning effectiveness" and "recognition" on college English blended teaching model based on mobile learning platform. The t-test results of the questionnaire shows that students' "learning effectiveness" is significant at the 0.01 level ($t = 4.496, p = 0.000$), and the mean value of EC (3.97) is significantly higher than that of CC (3.35). The students' approval level is significant at the 0.01 level ($t = 4.651, p = 0.000$), and the mean of EC (4.04) is significantly higher than the mean of CC (3.43). From the above, it is clear that the different student samples have showed significant differences in "learning effectiveness" and "recognition". This indicates that students in EC have higher recognition of the blended teaching model based on mobile learning platform than CC, and students in EC have higher learning efficacy than CC. In addition, the experimental students' learning effectiveness and recognition are higher (mean value > 3.5), while CC students' learning effectiveness and recognition are at an intermediate level (mean value between 2.5 and 3.5).

5.3 *Interview Results*

Based on the questionnaire survey, five students are randomly selected from EC as interviewees, and face-to-face interviews are conducted to understand students' perceptions in four dimensions. The information collected from the interviews is as follows: the clever use of mobile learning APP in teaching has increased their interests in learning, improved their independent learning ability, and made them experience the changes brought by the development of information technology. During the learning process, they felt that students were the main body of learning and teachers played a guiding role, while using the mobile learning platform to obtain more high-quality learning resources. In the activities of quiz, topic discussion and group work, students actively participated, so they improved their abilities. However, there is a problem that the teacher cannot arrange the time of the activities well, which needs to be arranged reasonably. At the same time, the teacher should consider the learning characteristics of different students and use data analysis to give more opportunities to the less active students to motivate them to learn and enhance their self-confidence.

6 Conclusion

To sum up, the blended teaching model of college English based on the Super Star learning platform is feasible and can realize the complementary advantages of information-based teaching and traditional classroom teaching. The experimental results show that learners are more receptive to the blended teaching mode, who can effectively carry out learning, which makes learners feel the immersive learning experience, mobilize their interests in English learning, and cultivate their independent learning ability and creative thinking. They are more active in the learning process to complete the corresponding learning tasks, and they are improved in many aspects through active interaction with teachers and classmates to form effective information exchange.

References

1. The Modernization of Education in China 2035. People's Daily, 2019-02-24(001)
2. Y. Shuyuan, L. Fang. Practical exploration of blended teaching in universities based on mobile cloud platform. J. High. Educ. (16), 147 (2020)
3. W. Zhiying, Exploring the application of mobile learning in vocational education. J. Heilongjiang Teach. Dev. Inst. **39**(10), 12 (2020)
4. F. Xinye, W. Jian, Design of blended learning model based on "rain classroom". Softw. Guid. **18**(02), 194–197 (2019)
5. L. Xin, Research on blended learning of public courses in colleges based on super star APP. Inn. Mong. Norm. Univ. (2020)

6. Z. Jing, Exploring the design of mobile flipped teaching in college English. J. Hebei Radio Telev. Univ. **25**(1), 82 (2020)
7. Z. Dong, Practice and exploration of mobile learning in higher vocational colleges under the background of "Super Star Network Teaching Platform." J. Huaibei Vocat. Tech. Coll. **17**(4), 54–56 (2020)
8. M. Qing, Analysis of mobile teaching mode based on super star learning. Wirel. Internet Technol. **20**, 88–89 (2020)
9. Y. Bilan. A study of interaction in flipped classroom of college oral English assisted by super star mobile learning platform. Sichuan International Studies University (2020)
10. X. Jinge, Study on the influence of super star learning platform on the learning effect of students majoring in travel service in secondary vocational schools under the background of epidemic. Shenyang Normal University (2021)
11. L. Xin. Research on blended learning of university public courses based on super star learning platform. Inner Mongolia Normal University (2018)
12. W. Jing, Y. Zhuo, Blended teaching model design based on cloud classroom—a case study of cloud classroom. China Audio-Visual Educ. **4**, 85–89 (2017)

Research on the Implementation Path of Business School Enterprise Integration Teaching Mode Under the Background of Internet+

Boren Gao and Jingxian Wang

Abstract School enterprise cooperation is an important measure for higher vocational colleges to seek development and vigorously improve the quality of education, and the "order training" mode is an important part of school enterprise cooperation. Through order training, students' theory and practice can be fully combined, targeted learning and employment, and schools can also smoothly promote school enterprise cooperation towards deep integration. With the progress and development of society, the society requires that schools should not always cultivate talents in a unified mode, which puts forward a new topic for the reform of the educational system. At present, in order to get the right staff for their own enterprises, some powerful enterprises also hope to jointly cultivate customized talents with schools, so the "order training" mode has emerged., From the perspective of "Internet+ education to online courses", this paper demonstrates the necessity and feasibility of school enterprise cooperation in the construction of Higher Vocational online courses, explores the ideas and ways of co construction, and looks for a feasible entry point for the deep cooperation between schools and enterprises and the deep integration of production and education, with a view to changing the current situation of insufficient participation of enterprises in the vocational education industry. To innovate the integrated teaching mode under the school enterprise cooperation mechanism, schools and enterprises need to strengthen interaction, work together, find new problems, explore new ideas, build a new system, and ensure the smooth implementation of the integrated teaching mode. First, establish a "amphibious" teaching team. Both schools and enterprises can send professional teachers and technicians to jointly complete the professional teaching tasks, and the school establishes a stable team of professional instructors from enterprises; The second is to convert and decompose the production, management and other tasks of the enterprise into teaching projects or tasks, and prepare textbooks, teaching plans and curriculum systems according to the enterprise standards. The modern talents that our country needs are professional talents with comprehensive abilities, and vocational education is an important way to

B. Gao · J. Wang (✉)
Guizhou University of Commerce, Guizhou, China
e-mail: 201820769@gzcc.edu.cn

S. Patnaik and F. Paas (eds.), *Recent Trends in Educational Technology and Administration*, Learning and Analytics in Intelligent Systems 31,
https://doi.org/10.1007/978-3-031-29016-9_27

cultivate professional talents. The Ministry of Education has proposed that vocational education is market-oriented and aims to cultivate comprehensive talents. With the strong practicality of college majors and the continuous development of the market economy, the traditional teaching methods can no longer adapt to the development of modern economy. It is imperative to change the original teaching methods. At this time, the development of the Internet and its more and more extensive application in the economic field also brought new business opportunities for enterprises. Enterprises attach great importance to exploring and developing online marketing and online sales. Now enterprises gradually realize that developing e-commerce and network marketing have become the common choice of enterprises, thus generating the demand for relevant talents.

Keywords Internet · Business school enterprise integration

1 Introduction

Practical education is an important part of the training of high-quality applied talents. At present, there are still many deficiencies in the practice work of accounting schools in financial and economic colleges [1]. Therefore, how to understand the needs of students, connect with enterprises and reasonably arrange internship links has become a serious challenge and practical problem [2]. With the arrival of the mobile Internet era and the promotion of the informatization of colleges and universities, it is necessary for colleges and universities to build a mobile Internet platform [3]. 4G/5G campus and wireless campus are important foundations of mobile teaching in colleges and universities [4].

In today's society, science and technology have become an important factor in promoting the development of productive forces. Mobile Internet is different from the previous civil Internet. It is based on mobile devices, mobile networks and application software services, relying on intelligent terminal devices and intelligent operating systems, and has become an important entry channel for the Internet [5]. The rapid development of the information age has made the Internet more and more popular in all sectors of society. Its organic application in the field of education is one of the manifestations of network application in society [6]. As a compulsory basic course to help people develop and enrich their professional knowledge and enhance their computer practical application ability, computer course teaching is an important channel to promote the effective application and deepening development of computer network technology in society [7]. With the continuous development of China's education information technology, the function of the Internet has gradually penetrated into teaching [8]. The common trend of world education reform is to pay more and more attention to students' personality education and emphasize students' personality development. The deep cooperation between schools and enterprises is an important way to cultivate talents in modern vocational education. The joint construction of the curriculum system by schools and enterprises can effectively improve the quality of talent training. The school is required to actively provide the required courses, teachers and other resources for enterprises, so that enterprises and schools can form a community of common destiny [9].

Higher vocational colleges and enterprises should build a long-term cooperation mechanism, jointly establish a talent training program, jointly develop textbooks, build a curriculum system, and promote the integration of production and education [10]. The construction of the practice teaching system of the economic management specialty deeply integrated with enterprises needs to be laid out and developed from the aspects of defining the talent training objectives, reasonably optimizing the training process, creating the practice system of "school enterprise, enterprise secondary school", and strengthening the construction of the teaching quality monitoring feedback system. It is also necessary to innovate the school running system and mechanism, promote cooperative school running, cooperative education, cooperative employment and cooperative development, and enhance the vitality of school running. The order training mode refers to the school running mode in which vocational schools, according to the requirements of enterprises for talent specifications, jointly formulate talent training plans, sign employment contracts, cooperate in teachers, technology, school running conditions, etc., and jointly take charge of a series of education and teaching activities such as enrollment, training and employment. The "order training" mode is a new talent training plan formulated by the school in response to the requirements of the enterprise. Before starting the "order training" mode, the school needs to sign an agreement with the enterprise, in which the specific forms of school enterprise cooperation and ways are specified.

2 On the Development of School Enterprise Cooperation in the Context of "Internet+"

2.1 School Enterprise Cooperation Plan

Under the background of "Internet+ education", online courses have become a new learning mode of mankind. However, the biggest problem with enterprises' participation in teaching is that teaching plans and teaching methods are not flexible enough, and enterprise personnel's participation in teaching is a part-time behavior outside of their normal work. Therefore, the inflexible teaching plans and teaching methods bring about contradictions that can not be taken into account by enterprise personnel. Digitization is a process of "machine learning", which is based on a large amount of operational data (data recorded by the information system), mathematical modeling and optimization of the enterprise's operational logic (management experience), and in turn guides the daily operation of the enterprise.

Innovation is the soul of a nation's progress and the inexhaustible driving force for a country's prosperity. The United States, Japan, Europe and other developed countries attach importance to the cultivation of innovative talents. Gradually form a unique innovative talent training mode, mainly including dual system mode and ternary system mode. Among some surveyed students, most of them think that internship is of great significance for personal growth, and can enhance the ability to communicate with others (89.07%); Accumulate social experience and lay a foundation for the future (90.54%); Being able to exercise their professional ability, enhance their awareness of independent learning, and change their employment concept (83.02%); Understand the operation mode and system culture of the enterprise (73.1%); However, a small number of students thought it was formalism, which was not very useful, and they just obeyed the arrangement of the school (8.99%). The structure of school enterprise cooperation is shown in Fig. 1.

While cooperating with enterprises to build 5G campus, colleges and universities also began to promote the construction of their own brand wireless campus. The wireless network uses radio waves as the medium of data transmission, without any wires or transmission cables. The transmission speed is very fast. The wireless network basically covers all indoor places in colleges and universities, and the client network login software based on mobile devices is introduced. The phased achievements and practical experience obtained after in-depth research not only fully demonstrate the wisdom crystallization of all teachers and students in the college, but also provide reference for people to think about how to promote the construction and development of the information based teaching system. The threshold of information-based

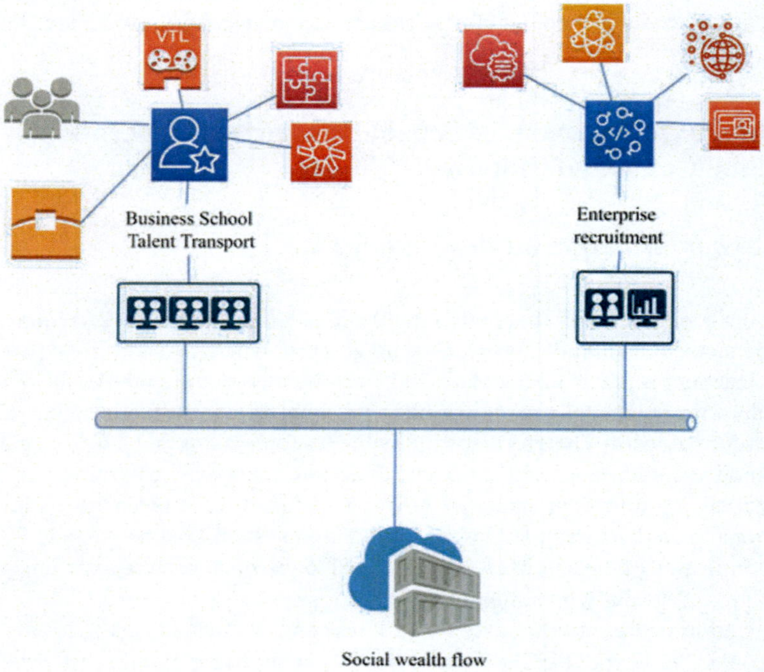

Fig. 1 Structure of school enterprise cooperation

teaching can be set as x, The specific variable of information-based teaching is set as k, and the relevant formula for information teaching is as follows:

$$P(\text{o}) = \frac{2\sum_{k=1}^{3} 7/x}{12} \tag{1}$$

P-Numerical value.

Students can learn in a virtual collective through the computer Internet. This collective is not limited to a class, a school or a region. It can be a national or even global collective. The curriculum system is the key to ensure the teaching quality. Schools and enterprises jointly build the curriculum system, establish curriculum resources, and jointly establish curriculum standards. The curriculum content is integrated into the professional standards, and the teaching process is connected with the production process. We should optimize the knowledge structure, enrich social practice, and strengthen ability training. Efforts should be made to improve students' learning ability, practical ability and innovation ability, to educate students to learn knowledge and skills, and to strengthen teaching infrastructure such as laboratories, practice bases inside and outside schools, and course textbooks.

2.2 *The Growth of School Enterprise Cooperation in the Context of the Internet*

Most of the economic management majors are applied majors, and they are also one of the fastest growing majors with the largest enrollment in recent years. Although many colleges and universities have carried out comprehensive reform pilot programs or implemented outstanding talent plans for economic and management majors, they have strengthened the construction of internal laboratories and external training bases. But in terms of practical application, the effect is not obvious. The main reason is that everyone is exploring and developing, lacking a set of practical teaching system construction methods and ideas for reference. We must take improving quality as the core, deepen the reform of education and teaching, optimize the professional structure, strengthen the construction of teachers, improve the quality assurance system, and improve the quality of talent training and school running level. Realize the mutual promotion between industry enterprises and schools, and the harmonious development of regional economic society and higher vocational education. According to this method of education and teaching reform or setting the function formula as, a formula can be designed, which sets ϖ as the deviation value and $\int$ as the threshold value. This formula can be established according to the ideas proposed in Formula 1 above. The Formula 2 can be obtained by combining the relevant design formula of education between schools and enterprises:

Table 1 Questionnaire on School Enterprise Cooperation

Cooperation compliance rate	Capital transportation	Talent transportation
School	34.78	76.39
Enterprise	66.78	56.75

$$12\mathrm{kl} = \frac{\varpi^{\mathrm{i}} + 89}{\int \sum_{\mathrm{j}}^{3} \mathrm{kl}} \quad (2)$$

kl-Booking scope.

The purpose of the "order training" mode is to enable the graduates in colleges and universities to have specific practical application ability, so the entire education system should be changed accordingly, and the curriculum should also be changed accordingly. On the basis of ensuring the traditional culture teaching, the curriculum should be more flexible and practical. Online courses under the background of "Internet+" can effectively solve the above contradictions. The enterprise in the survey said that the construction of online courses with professional teachers in higher vocational colleges not only solved the problem of limited on-site classes, but also integrated the resources of both schools and enterprises. At the same time, it can also put the latest developments of the industry into online courses to help students understand the development of the industry. Universities and enterprises also need to grasp the potential needs of the market and customers sensitively and accurately, use cross-border thinking and Internet thinking mode, constantly break through the bottleneck of traditional property model, create new business areas and quickly realize productization, which can organically combine traditional property services with digital technology, achieve technology empowerment, effectively promote business model research and development, improve operational efficiency, and achieve diversified expansion of service types. We should optimize the knowledge structure, enrich social practice, and strengthen ability training. The questionnaire on school enterprise cooperation is shown in Table 1.

All walks of life should focus on improving students' learning ability, practical ability and innovation ability, and explore ways to cultivate innovative talents through all kinds of education at all levels; We should encourage colleges and universities to jointly cultivate top-notch innovative talents and raise the cultivation of practical and innovative abilities to an unprecedented height. In the survey, the number of students who hope to engage in professional related internships exceeds the number of students who actually practice in their professional fields. The main reason why I can not engage in professional internship is that I have to find a job temporarily instead of a satisfactory job, which will become an obstacle and challenge for school enterprise cooperation.

3 Innovative Development of the School Enterprise Integration Teaching Model in Business Schools

3.1 *Future Development Plan Based on Teaching Mode*

The teaching method based on mobile Internet is an innovative teaching method based on mobile Internet and aimed at improving teaching efficiency. However, in today's explosive growth of information, timeliness gives users a sense of fatigue, and people are difficult to deal with timely big data. So on the basis of the "small world network" model based on the six dimensional space theory, social network services began to simulate user behavior and interaction between users, forming a virtual social network. The rapid improvement of the Internet and computer information technology has made the Internet more and more influential in public life, learning and work, which not only makes the society's demand for high-level computer talents increase sharply, but also puts forward higher requirements for the cultivation of professional computer talents. Internet teaching requires teachers to have computer application ability and Internet application ability. With the development of society and the updating of information, teachers should constantly revise textbooks and teaching plans, and supplement new content and materials at any time to meet the requirements of the times.

With reference to the national training standards for intelligent control technology talents, the professional curriculum system is constructed according to the professional field talent specifications. The professional construction team composed of industrial and enterprise engineering technology experts, teaching management experts and professional teachers has extensively investigated the relevant professional and technical fields of the profession. Through the job task analysis of post activities, the professional action fields related to the intelligent control technology profession have been formed, and the professional action abilities related to the profession have been defined. The training method of relevant talents can be improved according to the method mentioned above. For example, the improvement method of school enterprise combination project mentioned in Formula 1 and Formula 2 above can set ilk as the enterprise development progress rate and *xl* as the school enterprise combination practice achievement rate. The formula is as follows:

$$\mathrm{op} = \sqrt[2]{\mathrm{ilk} - 12^{\mathrm{x1}}} \tag{3}$$

op-Ordinal number.

The research on the status of practical teaching mainly focused on the year before 2010. The status of practical teaching is not clear enough. Scholars discussed the importance of practical teaching from different perspectives. Colleges and universities need to start from the four needs of social development, improving students' mastery of modern information technology, cultivating students' practical work

Table 2 Internet university teaching questionnaire

Rate of teaching quality reaching the standard	First-tier cities	Second-tier cities	Third-class cities
Freshman	45.67	67.94	65.39
Sophomore	78.81	57.91	87.31
Trainee	56.88	78.45	68.34

skills, and cultivating application-oriented talents. They believe that it is imperative to strengthen the construction of practical teaching. The enterprise personnel and professional teachers work together to develop the curriculum syllabus, prepare the theory and practical training materials, and realize the joint construction of the curriculum. Co build core courses. The application of the Internet in college teaching is shown in Table 2.

Relying on modern information technology and based on the school enterprise collaborative education e-commerce platform, it can become a professional software with more functions. It can run on the office terminal or mobile phone APP, and will be gradually improved and upgraded in operation. Its use combines the school and enterprise together to build a pattern of complementary advantages, resource sharing and close integration. In recent years, with the rapid development of China's economy, the employment environment has been greatly improved, and some enterprises have encountered difficulties in recruiting workers, which has brought great challenges to the stable production of enterprises. Through the integration of schools and enterprises, enterprises can establish employee reserves in vocational colleges, so that enterprises can obtain more stable human resources support. At the same time, enterprises can open internship channels to colleges and universities to enable students to obtain one-year internship. The enterprise provides students with comprehensive practical teaching projects such as special lectures, on-the-job training and job rotation practice. According to the needs of the cooperative enterprise, discuss with the enterprise to formulate new teaching materials needed for teaching, and timely introduce the part of teaching in the enterprise that is lacking in school teaching into the classroom. Of course, the content of basic courses should be included in the training system as a compulsory course for students of all majors, so that students who participate in the "order training" mode can be familiar with the management mode, equipment conditions and production methods of enterprises during their school years.

3.2 *Reflections on School Enterprise Cooperation Based on Internet*

The establishment of the system is the premise for the smooth implementation of the practice, and the signing of the contract is an important guarantee to ensure the rights and obligations of both parties. Both parties involved in the construction of online courses must reach an agreement on core issues such as cooperation content, cooperation methods and cooperation channels, and form a joint construction system and contract. Only in this way can we help the smooth implementation of the co construction of online courses. This integrated teaching mode under the school enterprise cooperation mechanism integrates theoretical learning, practical training, enterprise culture, etc. into school education, which is welcomed by teachers and students. The operation mechanism of school enterprise cooperation is not systematic and perfect. The common feature of the existing school enterprise cooperation model of higher vocational education in China is that schools are the main part. The school enterprise cooperation model, which is dominated by schools, seeks to pass all kinds of examinations and obtain all kinds of certificates. Students have passed the examinations and obtained certificates, but their actual working ability is poor. The learning of other knowledge takes less time, and the time for practical teaching cannot be guaranteed. Compared with theoretical teaching, the management of practical teaching involves a wide range of aspects, has multiple objectives, and the process is complex. It needs to coordinate the multi-level relationship between inside and outside the school, which is more difficult to manage. A strict and standardized management system must be established.For the actual situation of the combination of schools and enterprises, a data chart can be designed, which takes the quality assurance of school enterprise cooperation and long-term development of enterprises as the data distribution, sets the abscissa as the forward-looking economic development of enterprises, and the ordinate as the economic growth rate. The data chart obtained through the research and investigation of school enterprise cooperation data in different cities is shown in Figs. 2 and 3.

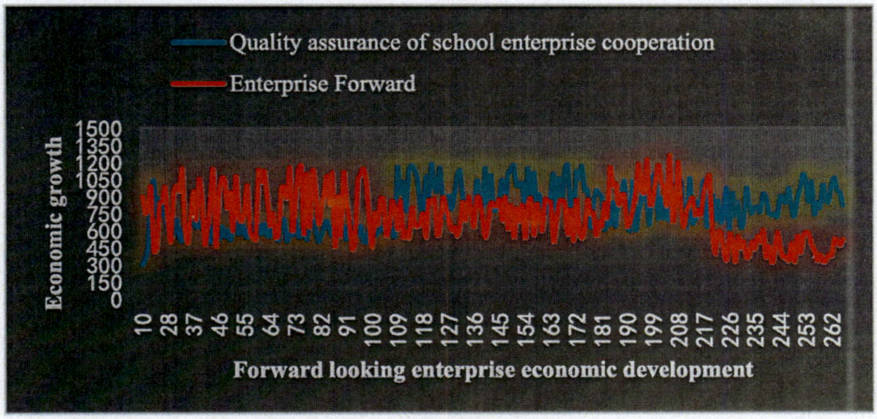

Fig. 2 Data of school enterprise cooperation in first tier cities

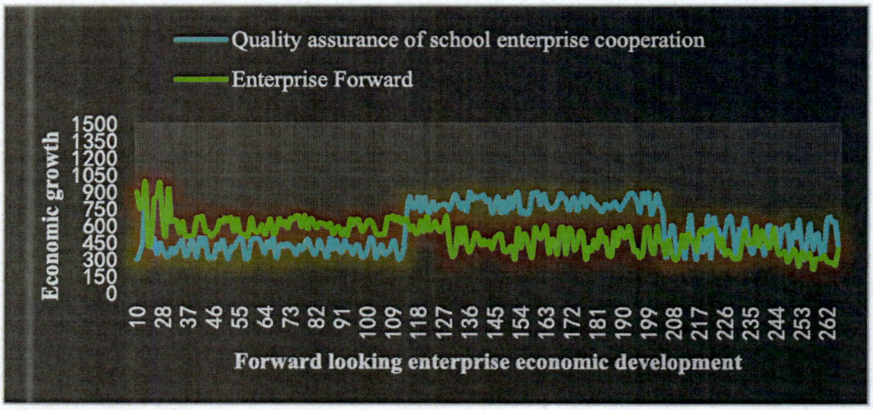

Fig. 3 Data of school enterprise cooperation in second and third tier cities

It can be seen from the comparison of Figs. 2 and 3 that the success rate of school enterprise combination in education practice is high in first tier cities, and the long-term development of some enterprises is also an important measure for the development of school enterprise combination projects, which will have a great impact on the future employment development of college students. For second tier and third tier cities, the application of school enterprise combination mentioned in Formula 1 and Formula 2 above can be added to this data chart, Fig. 3 shows that the development rate of enterprises is not very high due to the imbalance of urban economic development, which also demonstrates the correctness of the views discussed in this paper from the side. In order to achieve the training goal of applied innovative talents, we must explore and build a talent training system that serves the training of "strong practical ability, thick basic knowledge, strong learning ability, fast adaptability, high innovation quality, and good comprehensive quality". It is necessary to implement the integration of schools and enterprises and the cooperative education of production, teaching and research. It is necessary to establish and improve the internship system with goal orientation. The internship system includes the system of signing a training agreement between the school and the enterprise to define the training objectives, responsibilities and processes of the internship unit.

The school should take the initiative to communicate with the enterprise to jointly formulate the graduation practice teaching plan, from planning to organization, to process supervision, incentive and achievement assessment, so that both parties can reach a consensus. Teachers can also rely on social network services to establish a group of teachers and learners as members. Instructors can release Internet learning materials such as text, pictures and videos related to teaching at any time. Learners can use mobile devices to view, download and share relevant information at any time and anywhere, and can feed back problems to instructors at any time. After sorting out the relevant questions, the instructor will release them in the form of topics, so that all learners can enter the topics of interest and participate in the discussion.

Teaching informatization mainly refers to the educational form in which teachers reasonably use advanced information technology and diversified new media equipment to improve the efficiency of curriculum teaching when carrying out curriculum teaching activities. It has great practical significance to vigorously promote the optimization and innovation of curriculum teaching mode. What is more important is that the Internet collects a lot of resources. Therefore, Internet teaching can enable students to break through the book centered limitations and expand their learning content. In particular, the Internet has no geographical boundaries and time and space constraints, which can enable students to receive high-quality distance education and lay a good foundation for lifelong learning.

4 Conclusions

School enterprise cooperation is an inevitable choice for vocational education reform, and the depth of enterprise participation needs to be strengthened. Most school enterprise cooperation is aimed at students' employment and appointment, and further cooperation is needed in the determination of curriculum content, the establishment of curriculum system, and the formulation of enterprise participation in talent training programs. The main reason is that there is no long-term cooperation mechanism between schools and enterprises. The research on the construction of practical teaching system is the focus of scholars' research, but due to different angles, the proposed methods are also different. The training of application-oriented talents is not a training mode that simply adds practical application links to teaching. To improve the training of application-oriented talents, the first thing is to optimize the training process reasonably. To achieve two-way interaction between theoretical teaching and practical teaching, and combine the theoretical training of the school with the practical guidance of the enterprise. The mode of school enterprise cooperation requires both parties to deeply integrate, develop in an integrated way, actively adapt to the needs of regional industrial structure upgrading, and timely adjust the professional structure; Deepen the reform of diversified talent training modes such as order training and work study alternation, formulate training plans by referring to the requirements of professional posts, and introduce the technical standards of industrial enterprises to develop professional courses.

In recent years, the school has explored a lot of cooperation methods in the process of cooperation with enterprises. By regularly organizing teachers to go to the front line of enterprise production to learn specific information, the school has accumulated teaching materials in practical teaching, summarized the courses and skills required by the "order training" model, and effectively solved the problem of building the school's teaching staff. Schools and enterprises jointly build a path to improve the application of online courses. The successful application of online courses in the teaching process and the achievement of better learning results are important links to achieve the goal of building online courses jointly by schools and enterprises. There is still a gap between the talents trained in traditional e-commerce courses and the current market demand, and the graduates trained can not directly meet the needs of enterprises when entering the enterprise and employment market. Under this teaching mode, students' knowledge is biased towards theory and their ability to practice and operate is weak, which leads to our professional graduates finding high-tech positions. It is difficult for us to cultivate e-commerce talents needed for enterprise development. The number of employees who meet the professional talent specifications in the market is far from meeting the needs of the enterprise. The enterprise recommends this work platform to employees, which can make up for and improve their professional level, and is more conducive to their business expansion. The implementation of collaborative education in higher vocational schools can not only improve students' practical ability, but also cultivate the talents needed by enterprises. The improvement of students' practical ability and the cultivation of their innovative consciousness can not be separated from the accumulation of theoretical knowledge, but the updating of theoretical knowledge lags behind the development of practice. The practical experience is summarized, refined, discussed and demonstrated, and then entered the teaching plans and classes of university teachers. In the process of promoting the deep integration of schools and enterprises, and cultivating practical and innovative talents in economics and management, of course, we have accumulated a little experience and some own thinking, However, with the rapid development of social economy and the dynamic change of market demand for talents, there is still a large space for research on the cultivation of practical and innovative talents in economics and management.

References

1. C. Lin, W. Luo, Y. Na et al., Workplace well-being: challenge and crafting in internet context. Hum. Resour. Dev. China **12**(9), 7 (2018)
2. M. Guy, S. Aviv, G. Rakesh, Personnel crisis communications management system: WO, WO2014143602 A1 **12**(8), 11 (2017)
3. K. Nisula, Holistic Business Learning Environment: Bringing practice and integration to business education 77(17), 16 (2019)
4. B. Hu, The Cultivation Mechanism of Innovative Talents Based on School-enterprise Cooperation, **67**(7), 19 (2019)
5. X. Fang, Study on "Internet+" class teaching mode based on school enterprise integration—a case study of "E-Commerce." J. Zhejiang Int. Marit. Coll. **7**(89), 71 (2017)
6. S. Chen, Q. Wang, S.O. Business, Practice of "School-Enterprise Integration, Task Oriented" teaching mode in chain management specialty: taking chain store development and design course as an example. Value Eng. **17**(79), 31 (2017)
7. X.L. Liu, Y.L. Pei, J.Y. Ren et al., Research on evaluation system of postgraduates' innovation and entrepreneurship ability under the school-enterprise integration mode. Value Eng. **12**(72), 9 (2019)
8. J.F. Yin, F.Y. Chen, B. School et al., Research on the inner mechanism of the expansion of the enterprise merger and acquisition within the background of internet. Technoeconomics Manag. Res. **17**(87), 12 (2017)

9. T. Yang, Q. Li, The analysis on the new mode of school-enterprise cooperation and integration in sports industry. Fr. Acad. Press. **13**(71), 17 (2019)
10. W. Ding, H. Wang, Exploration on the talent training mode of the industry education integration and school enterprise cooperation of applied undergraduate majors 2021 2nd information communication technologies conference (ICTC). **54**(8), 17 (2021)

New Blueprint of Labor Education: Intelligent Labor Education Based on the Concept of STEAM Education

Bai Jing, Yang Xiao-Hong, Li Qi, Zhang Ming-Juan, and Su Qiang

Abstract Under the background of the change of labor form in the new era and the rapid development of new technology, we thoroughly analyze the value implication and existing predicament of labor education. The concept of smart labor education based on the concept of STEAM education is proposed for the first time, and the conceptual framework and curriculum system of smart labor education are constructed. In addition, the implementation path is proposed. The concept of intelligent labor education provides a new direction and draws a new blueprint for the future development of labor education.

Keywords Labor education · The concept of STEAM education · Intelligent education · Wisdom labor · Artificial intelligence

1 Introduction

In the era of remarkable informationization achievement and rapid development of science and technology, China has issued a series of policies and policies concerning labor education. In September 2018, General Secretary Xi pointed out at the China Education Conference that the next generation of socialist successors should pay more attention to the comprehensive cultivation of morality, intelligence, body, beauty and labor [1]. In March 2020, the "Opinions on Comprehensively Strengthening Labor Education in Colleges, Primary and Secondary Schools in the New Era" stated that labor education should use multiple approaches and integrate multiple perspectives. It is incorporated into the learning process of students in each learning period, and explores the labor education mode with Chinese characteristics [2]. Under the guidance of these policies, it is enough to highlight the important position of labor education in the march of education. All schools attach great importance to labor

B. Jing (✉) · Y. Xiao-Hong · L. Qi · Z. Ming-Juan · S. Qiang
Northwest Normal University, Lanzhou 730070, China
e-mail: 2021213055@nwnu.edu.cn

S. Patnaik and F. Paas (eds.), *Recent Trends in Educational Technology and Administration*, Learning and Analytics in Intelligent Systems 31,
https://doi.org/10.1007/978-3-031-29016-9_28

education, and actively explore effective teaching content, methods and strategies of labor education. However, it still appears in the process of labor education and teaching, which is boring, simple and lacking the new idea of keeping pace with The Times. Therefore, the article first deeply analyzes the value implication and existing predicament of labor education in the new era. On this basis, the intelligent labor education based on the concept of STEAM education is proposed, which draws a new blueprint for labor education. The innovation of this paper lies in the following work:

(1) It first gives the value implication of labor education in the new era.
(2) A detailed analysis of the existing difficulties in current labor education.
(3) The concept of intelligent labor education based on the concept of STEAM education is given for the first time, and the corresponding curriculum system is proposed.

2 The Value Implication of Labor Education in the New Era

In his report to the 19th CPC National Congress, General Secretary Xi pointed out that China has entered a new era. At the new starting point, labor education should adapt to the changes of The Times, grasp the pulse of the development of The Times, and actively explore and update the content of labor education. On this basis, labor education ability better play the mission.

2.1 The Source of Value of Labor Education

Labor provides necessary conditions for human development and is the source of value of labor education. Labor is the cornerstone of human development. In the Encyclopaedia of China (Philosophy Volume), labor is interpreted as a basic social practice unique to human beings. At the same time, Marx emphasized that labor is an activity process in which human beings use nature to transform themselves [3]. Labor is an effective activity carrier for the externalization, exercise and development of people's brain and physical strength, and it highlights individual value. If labor education is absent, the goal of cultivating high-quality talents with all-round development will be out of the question. To a large extent, it is difficult for human beings to realize the objectification of intelligence and physical strength. Therefore, labor is the fundamental foundation of the standard value of labor education, and it is a required course for students to grow and develop. Cultivating people's labor quality should be the important purpose and basic content in the educational itinerary.

2.2 *Key Tasks in Labor Education*

It is the key task of labor education to cultivate people's good labor value recognition. Suhomlinsky stressed that training qualified workers is not the ultimate mission of labor education, but labor can make people truly realize his value in society. It is this recognition of labor value that makes the implementation of labor education have a guiding direction. Through the integration of all aspects of social resources, the benign development of labor education can be formed. Only in this way can it be more beneficial to the comprehensive and lifelong development of students, and become an indispensable wealth for students to benefit from throughout their lives. We should pay attention to the spirit of labor, the spirit of model workers, the spirit of craftsmen, to cultivate students to have a good sense of value as the key task of labor education.

2.3 *The Important Objectives of Labor Education*

Labor can make people develop in an all-round way, which is an important goal of labor education. The simultaneous development of "five educations" is an important way to cultivate people's all-round development, among which labor education plays an indispensable role [4]. Labor education promotes talent training through value orientation and practical activities. In the spring breeze of the new era, only based on practice and ambition, can we become a new talent in line with the needs of social development. Only in this way can we realize the all-round development of talents, create value for society and enhance happiness for the people.

3 The Existing Predicament of Labor Education

3.1 *Families, Schools and Society Lack of Understanding of the Value of Labor Education*

Labor education has a unique role in educating people, but the concept of family, school and society for labor education has not been iterated with the development of the society. Firstly, it is widely believed that the training of labor skills is the ultimate goal of labor education. Second, people lack the dynamic understanding of labor education. The goal, content, method and evaluation of labor education are not invariable. With the needs of society and the development of science and technology, we should pay attention to the dynamic nature of labor education. Third, people lack the cultivation of labor literacy, such as labor spirit, labor values, labor thinking and other important roles in human growth.

3.2 *Emphasizing Physical Strength Over Intelligence in Labor Education*

Due to the traditional concept of labor, when people refer to labor education, the term "work" will always be defined, that is, a simple manual labor. In fact, with the development of The Times, labor is not so.

Now farmers have got rid of the traditional manual labor, more rely on science and technology operations, such as automatic harvesting, automatic irrigation, automatic sowing and so on. Therefore, we should attach importance to the combination of current real labor, not to let students simply sweep the floor, wipe the table, tidy up the library and so on. We should focus on the combination of physical strength and brain, the integration of technology and labor, and do intelligent labor education.

3.3 *Labor Education Lacks Teachers, Places and Systematic Courses*

At present, society and family do not play a real role in labor education, and most still rely on school education. In the exam-oriented environment, many parents unilaterally believe that improving students' academic performance is the primary task. Many parents think that other things will be done for their children, which is obviously not to mention the cultivation of labor literacy. Labor education also became the school "self-directed from the" activities. However, schools also marginalize labor education due to the lack of professional teachers, venues and systematic courses. First, the above phenomenon leads to labor education without a clear goal, the content of the old, unable to keep pace with The Times. Second, the labor course lacks the top-level design, not enough system. Each learning period can not be effectively connected, the implementation process is not interest oriented, students learn boring. Third, some schools do not meet the relevant provisions of the Ministry of Education, do not conform to the characteristics of the age of students. Lack of labor literacy parents involved in too much, resulting in students become a formality. Fourthly, some schools or families simply dissimilate labor education as a means to punish students [5].

4 Intelligent Labor Education Based on the Concept of STEAM Education

4.1 STEAM Education Concept

The concept of STEAM education refers to a form of education that takes problem solving as the core and integrates humanistic and artistic content [6]. As an interdisciplinary educational concept, STEAM education focuses on creating valuable problem scenarios. In the multi-disciplinary integration, series of sub-activities support, multimodal interaction to cultivate students' problem-solving ability, communication and cooperation ability, higher-order thinking ability [7].

4.2 Wisdom Education of the Basic STEAM Education Concept

Combined with the theory of smart education and the status of labor education, we propose smart labor education based on the concept of STEAM education. Guided by the STEAM concept of education, we intelligently permeate science and technology in the whole process of labor education, using new technologies and tools such as the Internet of Things, big data, artificial intelligence, learning and analytics. Finally, the labor education system of "combination of learning and use" and "combination of wisdom and labor" will be formed. The concept framework of smart labor education based on the STEAM education concept is learner-centered and based on the integration of disciplines. The main purpose is to solve problems and cultivate learners' comprehensive quality. The idea is to focus on problems for students' academic learning and classroom activities. The focus is to realize the four elements of collaborative learning, process inquiry, sharing and communication, and outcome output for practice, as shown in Fig. 1.

4.3 Construct the Curriculum System

The function of labor education should play a substantial role in the new era, which will inevitably make the practice of labor education move from a single form to an integrated and three-dimensional implementation path. Curriculum is the key factor to carry out labor education activities. The construction of intelligent labor education curriculum system under the concept of STEAM education is also the requirement of The Times and the need for students to acquire labor knowledge and skills.

The Intelligent Labor Education Curriculum System (ILECS) under the concept of STEAM education is shown in Fig. 2. ILECS focuses on the cultivation of scientific

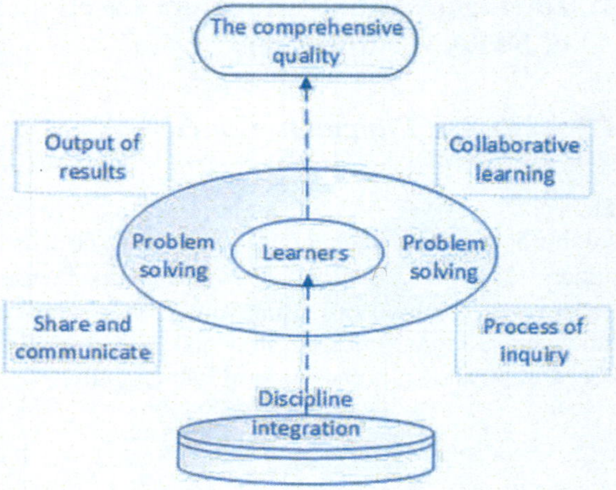

Fig. 1 Framework of intelligent labor education based on STEAM education concept

labor values, labor skills and labor spirit through the integration of disciplines. ILECS is also the integration of labor resources and the innovation of labor activities with the help of intelligent technology. Among them, the activity design module is an important content of the implementation of smart labor education, focusing on the comprehensive consideration of the integration of disciplines and intersections. In terms of the integration of disciplines, the knowledge should be modular, task-based and activity-based, centering on thematic activities. ILECS can flexibly combine the knowledge and way of thinking of the current curriculum to design sub-activities. ILECS focuses on the top-level design, so that the curriculum design of primary school, middle school and university is progressive.

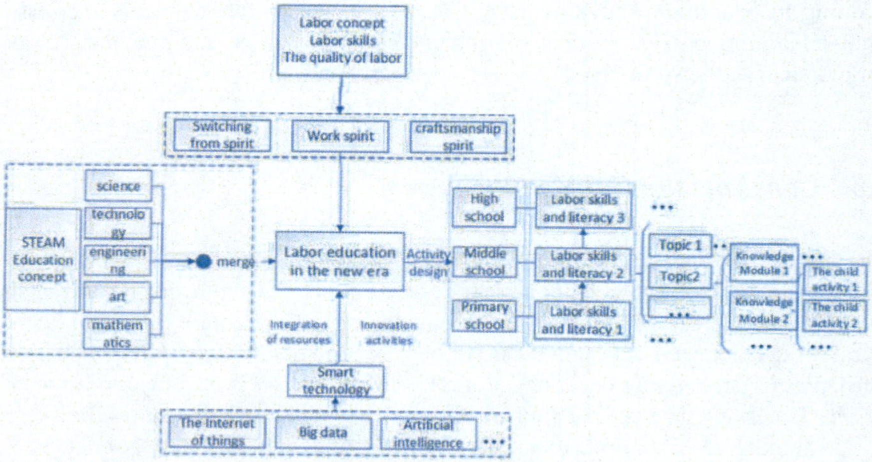

Fig. 2 Intelligent Labor Education curriculum system under the concept of STEAM education

5 Implementation Path of Intelligent Labor Education Under the Concept of STEAM Education

When constructing the STEAM education concept of smart labor education, we should fully understand the connotation of smart labor education, analyze the characteristics of students in each learning period and the actual situation of the current curriculum. From both horizontal and vertical aspects, interdisciplinary integration and learning segment cohesion should be integrated. Finally, we should fully organize and mobilize all kinds of resources to provide a strong guarantee for the development of labor education activities.

5.1 *Horizontal Perspective: Interdisciplinary Integration*

From the perspective of horizontal and interdisciplinary integration: First, we should integrate labor education into the "Five Education" and make labor education regular and diversified. Second, in the implementation of labor education, it is necessary to integrate the knowledge, skills and ways of thinking of other disciplines, so that students can apply what they have learned and make labor education more valuable. In order to realize the effective integration of disciplines, it is necessary to create theme-based intelligent labor education projects centering on labor skills and comprehensive practical literacy, so that students can experience the process of labor in real situations. Among them, labor skills focus on the results of labor projects completed by students with new technologies, new tools and new ideas. The goal of education is to acquire labor skills, acquire knowledge and share labor fruits. Labor literacy focuses on the whole learning process of labor theme project activities. The goal of education points to the formation of labor spirit and labor concept. The theme items of labor education can include "experience", "skill training", "inquiry learning", "innovation results" and so on. According to different educational objectives, each activity can be divided into several sub-activities, pointing to multiple knowledge points. At the same time, there are different needs for teaching places and teaching methods.

The completion of labor-themed projects depends on multiple sub-activities and points to the acquisition and cultivation of labor skills and qualities. Labor practice comes from life, but should be higher than life, in line with the students' learning and age characteristics of the objective law. In addition, we should make the best use of the knowledge gained by students in the current course and the way of thinking cultivated, so that students can find problems from many aspects, solve problems and think out of the box. In actual teaching, this kind of case is not rare. Humanities subjects such as Chinese and politics often include "hard work and saving", "cherish the fruits of labor", "labor is the most glorious" and so on. Science and engineering subjects such as mathematics, chemistry and physics often include "data calculation," "soil pH," and "leverage principle." These similar subject contents are compatible and interlinked with labor education. In addition, the prevailing maker education also provides ideas for the implementation of smart labor education courses under the concept of STEAM education. For example, 3D printing, small machine cutting and so on can be used as a new tool of labor.

In making use of knowledge and ways of thinking from other disciplines, the basic premise is that students have a good grasp of these knowledge. In this way, it can avoid the superimposition of new knowledge of multiple subjects, increase the learning burden of students, and deviate from the main channel of labor education. In addition, the integration of labor education courses into other disciplines can realize the integration of labor education into the system of students' overall development.

5.2 *Longitudinal Perspective: Learning Section Connection*

Learning period cohesion is a factor that must be considered in curriculum construction. According to the particularity of students in different classes, the corresponding training direction of labor skills and accomplishment should be set. For example, primary and secondary schools can use small machine tools, laser cutting and other machinery to learn traditional crafts, and make wooden small tables and chairs. In the form of labor keynote speech, labor project report, labor creation results display, we can increase students' sense of gain and joy of labor. Colleges and universities can choose modern agriculture, industry and service industries to build their capabilities and lay the foundation for future career development.

Each study section of labor education courses has its pertinence, but is not independent. From primary and middle schools to universities, the content is rising step by step, from simple to complex, and accompanied by creativity. The content of the same learning section should be themed, project-oriented and modular. The order of learning is not fixed. Learning can be adjusted according to the learning style, cognitive level and existing resources of educators and educatees.

5.3 *Integrating Smart Labor and Education Resources*

Labor education resources are the important guarantee and main carrier for the implementation of smart labor education, including teachers, courses and places. Enriching and integrating labor resources, on the one hand, is to integrate and develop digital resources, so as to enrich the curriculum and solve the problem of lack of teachers. On the other hand, it is necessary to flexibly develop practice sites to solve the shortage of labor education sites.

In today's informationization and digitalization of education, rich digital teaching resources such as MOOC, micro-course, flipped classroom and national cloud platform for intelligent education of primary and secondary schools have been spawned. Firstly, it can enrich the curriculum resources of labor education. Secondly, it can be used to assist teachers with lack of professional knowledge to carry out labor education to supplement teachers. Thirdly, students can acquire labor knowledge at any time to cultivate good labor values. In addition, the emergence of new technologies such as big data, artificial intelligence and cloud computing, as well as new concepts such as STEAM education and maker education, also gave birth to some new places for labor practice.

At present, through the front-line survey of teaching, many schools have specially set up maker classrooms, STEAM classrooms, etc. These include small machine tools, 3D printing, laser cutting and woodworking gadgets. This is enough to meet the practical activities of labor education. At the same time, it can also achieve a multi-purpose, but also the corresponding disciplines can be integrated. In addition, the appropriate use of Internet, virtual and reality technology in labor education is also a good way to expand the workplace.

Labor education resources are rich and colorful. It also fully combines multiple forces, such as campus and off-campus, online and offline, real and virtual, school and society, to promote the integration and sharing of resources. This can solve the shortage of labor education resources, labor education over-dependence on schools and other related problems.

6 Conclusion

Labor education must stand in the perspective of national development and the future of the nation, and cooperate with schools, families and social forces to cultivate talents needed by The Times. Labor education should focus on new technologies, new concepts and new forms of business. Only in this way can students master advanced labor skills and knowledge, and form good labor values. This study analyzes the problems and causes of labor education from many angles and accurately grasps the connotation of labor education in the new era. At the same time, through the construction of intelligent labor education curriculum system, it provides a reference case for the subsequent better development of labor education. In addition, the implementation path is proposed. The proposal of labor education will also make labor education pay more attention to the development of intelligent labor education curriculum and the development of practical activities.

Acknowledgements This work was supported by Graduate Research Funding Project(Intelligent Labor Education-a new mode of labor education), College of Education and Technology, Northwest Normal University.

References

1. Ministry of Education. Adhere to the education of socialism with Chinese characteristics development road culture art of morality, intelligence and physique full scale development of the socialist builders and successors [EB/OL]. http://www.moe.gov.cn/jyb_xwfb/s6052/moe_838/201809/t20180910_348145.html, 10 Sept 2018
2. CPC Central Committee, State Council. Opinions on comprehensively strengthening labor Education in Colleges and Middle schools in the New era. People's Daily (01) 27 March 2020
3. The Selected Works of Marx and Engels: Volume 2. Compilation and Translation Bureau of Marx, Engels, Lenin and Stalin Works of the Central Committee of the Communist Party of China, Compiled. Beijing: People's Publishing House, 2012:169
4. Labor education to promote the all-round development of human beings, http://ex.cssn.cn/gd/gd_rwhn/gd_ktsb/srxxxjpgyxsdmzgzdzyls/202012/t20201209_5230601.shtml, 21 May 2022
5. C. Tan, It is urgent to strengthen and improve labor education – the problems, reasons and countermeasures of current Chinese labor education. People's Education **20**, 30–31 (2018)
6. J. Michael, Mathematics in a STEM Context. Mathematics Teaching in the Middle School **18**(6), 324 (2013)
7. J.W. Ban, The connotation characteristics and practice path of "New" labor education. Educ. Res. **40**(1), 21–26 (2019)

Optimization and Innovation of College Collaborative Education Platform Based on "Internet+"

Liu Xiu

Abstract In the context of the "Internet+" era, higher vocational colleges need to constantly strengthen the coordination and linkage among various subjects, establish coordination goals, build new education models, determine scientific education contents, select coordination methods, explore effective education paths, and form a collaborative education pattern of innovation and entrepreneurship according to the coordination requirements. The collaborative education between colleges and families presents the situation of "main" being strong and "auxiliary" being weak, passive family participation, poor communication channels, and imperfect mechanism and system. Based on the analysis of the feasibility of "Internet+" collaborative education between colleges and families, combined with the work practice in the process of deep integration of "Internet+" and education, the concept of collaborative education mechanism between colleges and families under the background of "Internet+" is proposed.

Keywords Internet+ · College collaborative education · High quality teaching resources

1 Introduction

The collaborative education of course transfer in colleges and universities refers to that colleges and universities use the advantages of the online course transfer platform to unify their thoughts and understanding based on morality and education, and realize the education of all members, all processes and all directions [1]. As an important way to improve the quality of practical teaching, school enterprise collaborative education is an important way to enhance the practical ability and

L. Xiu (✉)
Big Data Management and Application Research Institute, Nanchang Institute of Technology, Jiangxi 330044, China
e-mail: 418928222@qq.com

S. Patnaik and F. Paas (eds.), *Recent Trends in Educational Technology and Administration*, Learning and Analytics in Intelligent Systems 31,
https://doi.org/10.1007/978-3-031-29016-9_29

innovation ability of college students. However, whether it is the form, content or effectiveness of cooperation, China's school enterprise collaborative education is at a shallow level compared with developed countries. Many enterprises, especially well-known enterprises, lack the motivation to actively participate, and the phenomenon of "hot schools, cold enterprises" is widespread. Enterprise collaborative education emphasizes the coordination between all members of the main body of the online classroom, as well as the overall process and all-round online ideological and political education (IPE) for college students through the online classroom [2]. However, in the context of "Internet plus", the ideology of collaborative education in colleges and universities is lacking, the education of all staff, the whole process and all directions is insufficient, and the guarantee mechanism of collaborative education is imperfect [3]. The needs of enterprises and industries guide the cultivation of engineering innovation ability of applied talents, while school enterprise collaboration helps to promote the improvement of innovation ability of applied talents. Although "Internet plus" is or will have a significant impact on the cultural industry, "school enterprise cooperation education" in university education has become an important measure for many local universities to achieve applied talent training [4].

The impact of "Internet plus" on college education mode is growing day by day. Under the background of "Internet plus", university collaborative education mode is becoming a potential education mode among universities. Exploring the school enterprise cooperation education model is of great significance for breaking the bottleneck of the mismatch between supply and demand of school enterprise talents and cultivating special compound talents [5]. With the development of society, the combination of production, teaching and research and collaborative education has become a new trend for colleges and universities to cultivate high-quality talents [6]. The integration of industry and education can realize the effective connection between schools and enterprises, work and learning, maximize the positive role of complementary advantages between schools and enterprises, and achieve satisfactory results that both parties can benefit from [7]. The growth of college students requires comprehensive and personalized development. The single development of the original college health education and IPE can no longer meet the development needs of college students, nor can it effectively cultivate new era talents with all-round development of morality, intelligence, physical, art and labor [8]. However, with the emergence of the concept of "big health" and the clear direction of the development of the "big politics" model, the task of cultivating talents for all-round and healthy development is particularly important. The second classroom activities in universities are usually independent, relatively independent of the first classroom, and lack of support for talent training. Strengthening the training coordination between the internal education administration department and the student engineering department, and strengthening the coordination between the supply and demand of talents in colleges and universities and the demand for talents in the industry have become an important topic in the reform of talent training in colleges and universities. The second classroom construction is deeply integrated with the talent training system in colleges and universities. IPE plays an important role in establishing correct ideas for students. As an important base for training professional talents, colleges and universities must strengthen students' ideas. Only in this way can students avoid detours in their future work and life [9]. The school enterprise cooperation and collaborative innovation applied talents training mode is a school running mechanism guided by the market and social needs. School enterprise cooperation is composed of four factors: School, students, enterprises and market. It is a teaching method in which schools and enterprises participate in the talent training process on the basis of making full use of school enterprise resources. It is one of the important ways to reflect the characteristics of colleges and universities. At this stage, with the continuous development of Internet technology, it has brought new opportunities for the coordinated development of party building and IPE in colleges and universities [10].

2 Optimization and Innovation of College Collaborative Education Platform Based on "Internet+"

2.1 How to Correctly Combine the Internet with Education

Construct the educational mechanism and platform of "Internet plus IPE"; Cultivate professional teams to achieve the goal of "Internet plus IPE". However, most local governments equate the construction of "Internet plus+ government service" with the construction of "online government hall", seriously ignoring the importance of improving the service capacity of offline entity government service hall, blindly pursuing the "speed" of "online", but forgetting the "temperature" of "offline". In order to cultivate innovative and applied talents, school-enterprise cooperative education is put forward. This teaching mode trains students' ability of communication and coordination and practical innovation to cultivate professionals with strong application ability.

The school enterprise cooperative education can be carried by the school enterprise cooperation council, which can jointly formulate talent training programs. Therefore, online and offline services are uncoordinated, processing standards are inconsistent, and other "two pieces of skin" are serious, reducing service efficiency and increasing the cost and burden of public enterprises. All colleges and universities need to adhere to the training mode of school enterprise cooperation, closely adhere to the basic requirements of resource sharing, cooperative school running, cooperative education and cooperative development, promote colleges and universities to jointly formulate training objectives, design curriculum systems, develop high-quality textbooks, organize teaching teams, and jointly build a practice platform with relevant departments and industry enterprises, and promote the combination of theory and practice. The questionnaire on school enterprise collaborative education and teaching is shown in Table 1.

The organizational education in universities mainly refers to that the party and league organizations in universities develop the party and league members and

Table 1 Internet teaching questionnaire

Quality of school enterprise combination education	First-tier cities	Second-tier cities
College teaching	76.35	75.43
Social teaching	88.36	81.42

educate them, so that the advanced elements among college students can play an exemplary role in learning, work, life and other aspects, and then achieve the goal of education. In order to enable more students to enjoy advanced teaching concepts and high-quality teaching resources, we choose excellent front-line teachers to "support", which seems to be a big measure of equal education, but we just ignore the role of students and teachers as ordinary "people", ignoring the adaptability and degree of adaptation as "people". These functions can be located:

$$\text{iod} + 6\phi(xy + 17) \tag{1}$$

iod-Value threshold

It emphasizes the development form of deep integration of the two, cultivates new products and business forms of education, creates new education service models, and realizes self-transformation, transformation and upgrading.

2.2 Thoughts and Plans on the Internet

At present, the curriculum system and teaching mode of adult education based on general higher education cannot meet the individual learning needs of students, and the limitations of classroom teaching mainly hinder the development of adult higher education with the times. The questionnaire on the implementation of technology teaching in universities is shown in Table 2.

In recent years, "Internet plus" has gradually penetrated, affected and changed all walks of life. Integration under "Internet plus" is the development trend. Higher vocational colleges are demonstration bases for economic and social innovation and entrepreneurship. Only when they work together to promote the growth of students and train qualified socialist builders and successors can the value of education be maximized. Actively explore the new mode of school enterprise collaborative education, and cultivate high-quality technical and skilled talents that meet the needs of

Table 2 Technical questionnaire

High tech usage	Universities	Society
Scientific usage	72.19	67.51
Scientific compliance rate	87.33	54.38

Table 3 Questionnaire on coordinated education in universities

Education qualification rate	Freshman	Sophomore	Junior and senior students
Primary account	56.77	77.19	88.31
Other accounts	62.34	76.19	76.93

industry development. School enterprise collaborative education can provide reference ideas for the current application oriented transformation of local undergraduate universities and the innovation of talent training. To cultivate high-quality applied talents who can serve the local social and economic development is the current goal of talent training, and is also the current trend of talent development. The development mode of school enterprise cooperation education can make talents output with higher quality and efficiency. Some formulas can be changed as follows:

$$g(k) = \Re - 117\frac{\sum_{k=1}^{3} 43}{\lambda} \tag{2}$$

g-Range

"Yiban" is a national college student network interaction demonstration community integrating ideological education, education and teaching, life services, and cultural entertainment. Collaboration means mutual cooperation and synchronization. Collaborative education means that all education subjects achieve the goal of talent training by establishing a unified education goal, sharing resources and interacting effectively. The questionnaire on collaborative education in universities is shown in Table 3.

In the education stage, we should establish the "great health concept" of the educates, and ideologically enable college students to turn their passivity in obtaining health into initiative, their passivity into positivity, and their compulsion into self-consciousness. Innovation and entrepreneurship education aims to lay a solid foundation for students' career choice, employment and entrepreneurship by cultivating young students' entrepreneurial awareness, entrepreneurial psychology, entrepreneurial knowledge and skills and comprehensive abilities, and cultivate creative talents to meet various needs for the country. We should shape independent and self-discipline health behaviors, improve health literacy, improve political status, and cultivate a sense of responsibility for the healthy development of individuals, families and the country. The construction of the second classroom collaborative education system requires a network platform that can be recognized by universities of different levels and sizes. It can connect different universities, different departments within universities, and universities and enterprises, so as to achieve educational coordination and symbiotic growth.

3 Reflections on Internet Teaching in Universities

3.1 The Continuous Development of Internet Teaching in Universities

The construction of the Party has had a profound impact on improving the level of IPE. They are interdependent and inseparable in the process of development. Some relevant policies and regulations will also have a far-reaching impact on college online education. The cooperative education in colleges and universities is a gradual process. This paper takes the reform rate of colleges and universities as the abscissa, the number of participants in the reform of colleges and universities as the ordinate, and the utilization rate of educational resources and the implementation rate of education as the data distribution. The data graph obtained is shown in Fig. 1 and 2.

As can be seen from Fig. 1, due to the small number of college reform participants, the low efficiency of college reform, and the lack of attention and reform awareness of relevant departments, the utilization rate of college education resources and the

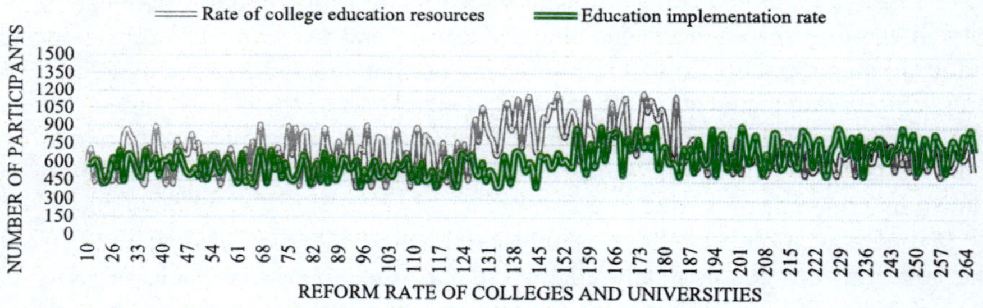

Fig. 1 Growth period

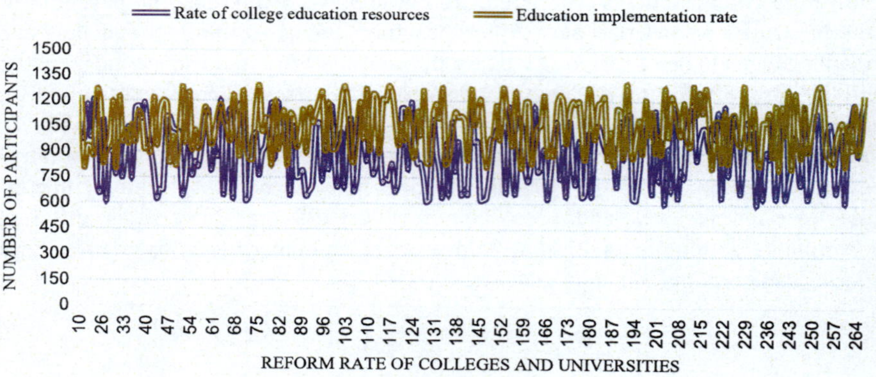

Fig. 2 Prosperity

implementation rate of education are at a disadvantage; It can be seen from Fig. 2 that the utilization rate of educational resources and the implementation rate of education in colleges and universities are in a thriving state, which is far higher than the data in Fig. 1. This is also due to the continuous innovation and expansion of collaborative education in colleges and universities. Therefore, the utilization rate of educational resources and the implementation rate of education in colleges and universities are also constantly improving. This result also proves the correctness of continuous innovation and optimization of collaborative education in colleges and universities. The Internet is open and anonymous, which affects the development of students to a certain extent. At present, there are still some problems in higher education in China, such as insufficient cultural value guidance ability, weakened network culture inheritance function, weak network culture penetration, and insufficient educational function. In the era of knowledge economy in the market economy, school enterprise collaborative education is not only an educational activity, but also an economic activity. It is a win–win and equal cooperation. Cooperation must be based on the equal status and balanced interests of both sides.

Therefore, in order to effectively carry out the cooperative education between schools and enterprises, it is necessary to cultivate a positive and healthy university network culture. More importantly, the production mode of logistics services has undergone tremendous changes, and the strategic thinking of "Internet plus" has further promoted and promoted the continuous upgrading of the logistics production mode. At the same time, higher requirements are put forward for the education background, professional qualification, cultural quality and ability and skills of logistics personnel. The service concept of logistics management personnel, the training of digital logistics information management, the use level of PC and other information terminals, and the control ability of logistics digital management platform have become the basic quality requirements of logistics management personnel. Let the classroom turn from closed to open, make the ideological and political lesson keep up with the pace of the development of the times and theoretical innovation, and make the IPE teaching more in line with the needs of student development. At the same time, the equality and interaction of network communication help to create an atmosphere of equal communication between teachers and students, and effectively play the main role of students. It is very important for enterprises to make reasonable choices for collaborative education. Colleges and universities should choose enterprises scientifically and rationally in the spirit of being responsible for students and sustainable development of the school. Because in the school enterprise collaborative education, the degree of heterogeneity and complementarity of resources and capabilities of both parties determines the strength of the cooperation motivation of both parties. If either party's strength is too weak, the other party's cooperation motivation will be weakened, and the cooperation will be difficult to sustain. Various online learning and communication platforms can not only guide college students to actively pay attention to current political news, the latest ideas of the Party and China's latest development achievements.

School enterprise collaboration often refers to the cooperation in scientific research. School enterprise collaboration in talent training is limited to the construction of off campus practice bases. And one of the most important responsibilities of universities is talent training, so school enterprise collaboration should be integrated into the whole process of talent training. That is, the university enterprise association can not only solve the core common problems facing industry development, but also in this process, let the university enterprise cooperation become a booster to improve students' engineering practice ability and innovation ability, and also can meet the autonomy of college students, the demand for interactive personalized learning in the context of "Internet+ education", and finally realize the positive guidance of college students' thoughts. Establish a high-quality resource sharing platform. The formula can be positioned as:

$$\mathrm{ki} - 17 = \sqrt[14]{i + x} \tag{3}$$

ki-Fixed value

Higher vocational colleges are the most active in the development of school enterprise cooperation teaching mode in China. With the continuous strengthening of the popularization of higher education and the diversification of talent demand, higher vocational education, as a component of higher education, has developed rapidly and occupied half of the country. For the sake of survival and development, higher vocational colleges always regard students' employment as the lifeline of the school. In the era of "Internet plus", "integration, reshaping, inclusiveness and openness" affect the innovation and entrepreneurship education of higher vocational colleges. Enterprises focus on "using" rather than "cultivating" talents. The purpose of this education mode is to rapidly export high-quality and practical talents with strong comprehensive ability to enterprises. To this end, it will quickly bring market competitiveness and higher benefits to enterprises.

The government should publish some relevant policies to support enterprises participating in school enterprise cooperation in terms of talents, funds, equipment and other resources, so as to promote the development of school enterprise cooperation. In order to meet the requirements of talent competition in the context of social and economic development and enhance the compatibility of talent training with the background of the "Internet plus" era, higher vocational colleges should explore innovation and entrepreneurship education reform, Integration of "Internet" + "The spirit of the times, through the full participation and integration of industry and education, strengthen the coordination and contact between the government, enterprises and schools. The school enterprise collaborative education model is different from the general school enterprise collaboration model, and collaboration is to optimize and integrate all elements in the process of education to form a positive impact on the educated, and achieve the optimization of the overall education effect. This is the mainstream model of modern engineering education, but also a breakthrough in traditional engineering education. The best way to model. Synergy refers to the process in which two or more individuals or elements work together to achieve the same goal. The goal of collaboration is to obtain certain synergy or value. In short, it is to achieve "$1 + 1 > 2$".

3.2 Future Development Methods of Internet Teaching

The combination of production and education is an innovative way of educating people in modern universities, which can promote the rapid and efficient development of regional economy. From other perspectives, the collaborative development of integration of industry and education belongs to connotation development, which is quality oriented. At this stage, universities are engaged in a battle to improve the quality and level of IPE, emphasizing the need to tackle problems from the aspects of ideas, teachers, textbooks, teaching methods and mechanisms, and paying attention to students' satisfaction and sense of gain, which puts forward higher requirements for the management and teaching of universities. All universities should issue education evaluation reports for relevant functional departments to serve the school's teaching and student work management; One is to issue quality appraisal report for students who are about to graduate, serving employment management; The second is to provide the employer with a recommendation report and carry out a survey on the matching degree of talent demand to serve social management; The third is to provide a stage for learning and exhibition among universities, and expand the effect of social collaboration education. The textbooks of these four courses are relatively slow in updating, and their content is mainly the explanation of theoretical knowledge. Therefore, it is usually difficult to guarantee students' enthusiasm for learning.

In addition, there is a lot of theoretical knowledge in the textbooks and the relevant content of politics in senior high school are repeated, and students will reduce their learning efficiency due to lack of freshness in the learning process, which is difficult to meet the requirements to ensure the quality of learning. At present, under the Internet environment, there are certain deficiencies in the diversified integration of network culture and correct value guidance of network education in universities. With the continuous development of the network and the development of the era of innovation, the diversity of the network has led to the continuous progress and innovation of the network culture of universities. While innovating, it has also enhanced the integration and connection of ideology and culture. Under the situation of diversified integration, it highlights the lack of cultural guidance ability and lack of executive power of universities.

4 Conclusions

This teaching mode has played an important role in the macro environment of economic transformation. One of the important tasks of the transformation of local undergraduate colleges and universities is to serve the regional economic development, which means that local colleges and universities urgently need to build a contingent of teachers with advanced technical strength. Diversified teaching methods. The Internet provides students with a picture of sound, image and text, which can innovate the teaching methods of IPE. Network IPE: From the perspective of space, it has realized the extension from classroom to extracurricular, from inside to outside the school; From the perspective of time, it realizes the extension from the classroom to the classroom, from the whole time to the dispersed time. The ideological and political work in colleges and universities can only accurately grasp the changing law of college students' ideological behavior, firmly grasp the leading force and initiative of network education, and realize the reconstruction of the party's cooperative education model, and the college league class under the background of "Internet plus education", if it adheres to the combination of truth and fiction and the actual life of students. Fang University should define its own position and development path, take this as an opportunity, take market demand as the guide, and take industry university research cooperation as the path, build a school enterprise cooperation platform, build an innovative school enterprise collaborative education mechanism, let enterprises deeply participate in the construction of the training system and talent training, and cultivate application-oriented high-quality talents that can serve the local social and economic development.

How to make every student enjoy high-quality education and teaching resources, how to better promote the equal development of education, how to make the "human" factor in education no longer be ignored, and become a new era. "Internet plus" education has also brought a new type of education maker education for university innovation and entrepreneurship education. Maker education is the direct result of the coupling development of information technology and innovation and entrepreneurship education. From the starting point of improving the quality of talent training, we should build a school enterprise collaborative education mechanism to meet the cooperation requirements of "joint education of talents, joint construction of bases, resource sharing, personnel sharing, and quality evaluation". It emphasizes the educational functions of innovation seminars, maker spaces, social laboratories, smart small business entrepreneurship bases and other new public entrepreneurship spaces. In the process of using digital tools to achieve creativity, students are encouraged to "learn in practice and start businesses in innovation". Colleges and universities should carefully and comprehensively study and analyze the characteristics of such enterprises' demand for talents, establish the awareness of cooperation and service, and actively cultivate talents for them, so as to promote the deep integration of schools and enterprises. Only in this way can colleges and universities provide services, make contributions and create value for the needs of enterprises in many aspects, ensure that enterprises can get the corresponding benefit guarantee in the collaborative education as far as possible, and mobilize and stimulate the motivation of enterprises to participate in this mode. The form of this mode should not be limited to classroom teaching, but should be diversified.

Acknowledgements This work was supported by "Jiangxi Educational Science 'Fourteenth Five Year Plan' Research Project" (Fund number:22YB270).

References

1. W. Zhang, The collaborative model research between college student entrepreneurship education and ideological political education. **17**(8), 71 (2018)
2. X. Wang, Design and implementation of college Internet+ English education system platform based on HTML5 CIPAE international conference on computers, Inf. Process. Adv. Educ. **9**(17), 73 (2020)
3. L. Xin, Application skills of education in aesthetic education of minority college students in the era of "Internet plus." Guizhou Ethn. Stud. **71**(40), 3 (2019)
4. W. Jiang, Exploration of college students' entrepreneurship and employment education under the background of "Internet plus." Basic Clin. Pharmacol. Toxicol. **7**(72), 8 (2020)
5. Chen J B, Zhong, Y X, Department F L, On the college English collaborative mode for students of brilliant engineer plans in the "Internet + " Era. J. Hainan Radio & TV Univ. **5**(14), 17 (2017)
6. X. Yan, 51.Research on the innovation and development of college English based on the "Internet Plus" Era. **62**(17) 91–102 (2017)
7. S. Jun, W. Yongzhi, L. Huiwang, Research on the architecture of intelligent education support platform under the "Internet+" environment. Comput. Appl. Softw. **12**(34), 11 (2017)
8. J. Wang, W.U. Shujuan, Y. University, Ways of collaborative education of college students' social responsibility. J. Yangzhou Univ. (High. Educ. Study Ed.) **17**(17), 76 (2018)
9. R. Wegerif, Introduction. Education, technology and democracy: can internet-mediated education prepare the ground for a future global democracy. **17**(98), 17 (2017)
10. Zheng H, Internet education solution for education equality. Francis Academic Press **17**(66), 15 (2020).

Research of Professional Diagnosis and Improvement Index System in Higher Vocational Colleges Based on CIPP

Yong Chao Xie, Jian Feng Huang, and Ji Cheng Duan

Abstract Professional education level is the key to the development of higher vocational colleges, and professional diagnosis and improvement can ensure the continuous improvement of professional education quality. Based on the CIPP evaluation model, this paper constructs the professional medical reform system of higher vocational colleges by analytic Hierarchy Process (AHP) and selects the applied electronic technology major of the author's school to carry out empirical research, which provides certain reference for the professional medical reform of higher vocational college.

Keywords CIPP · Professional diagnosis and improvement · Applied Electronic Technology major · Higher vocational colleges

1 Introduction

At present, China's higher vocational education pays attention to connotation construction, and higher vocational education has achieved high-quality development. Teaching, diagnosis and reform can effectively promote the level of higher vocational colleges, so as to promote the quality of talent training. With the continuous development of vocational education and teaching reform, higher vocational colleges have formed a regular and periodic teaching reform system and orderly operation, which promotes the perfection and development of the internal quality assurance system of higher vocational colleges. The notice of the General Office of the Ministry of Education on establishing the Diagnosis and Improvement System for Teaching Work in Vocational Colleges defines "Diagnosis and Improvement of Teaching Work in Vocational Colleges" as follows: Refers to the school according to their own school-running ideas, school-running orientation, personnel training

Y. C. Xie · J. F. Huang (✉) · J. C. Duan
Hunan Railway Professional Technology College, ZhuZhou, China
e-mail: 570129627@qq.com

S. Patnaik and F. Paas (eds.), *Recent Trends in Educational Technology and Administration*, Learning and Analytics in Intelligent Systems 31,
https://doi.org/10.1007/978-3-031-29016-9_30

target, focus on professional Settings and conditions, and construction, curriculum system and reform, teachers, school management and classroom teaching and practice system, university-enterprise cooperation and innovation, quality control and effective personnel training work elements, find the shortage and improvement to improve the working process.

The reform of professional diagnosis in higher vocational colleges is to take "professional in higher vocational colleges" as the entry point to study "diagnosis and improvement of teaching work in vocational colleges", in order to reverse find scientific methods of running a professional school, reduce the blindness of professional construction, enhance self-consciousness and autonomy. Promote the continuous spiraling development of the quality of professional education, so as to make the professional, curriculum, teachers and students develop together, and constantly improve the quality of professional talent training. The ultimate goal of teaching diagnosis and improvement in higher vocational colleges is to continuously improve the quality of professional personnel training. Specialty is the basic unit of personnel training in higher vocational colleges. Therefore, it is imperative to carry out professional diagnosis and improvement in higher vocational colleges.

2 Research Status of Professional Evaluation Based on CIPP Evaluation Model

American scholar Stufflebeam formed CIPP model in 1967 on the basis of reflecting on Taylor's behavioral goal model. The CIPP model consists of four assessment activities: Context evaluation(C), Input evaluation(I), Process evaluation(P), and Product evaluation(P), CIPP is the Product evaluation model. These four assessments provide information on different aspects of decision making, so the CIPP model is also known as the decision oriented assessment model.

Gao Yueqin, Chen Ling, Wu Sijian et al., from Guangdong Communications Vocational And Technical College, made an in-depth analysis of the compatibility between CIPP evaluation model and internal evaluation value orientation of higher vocational colleges in view of the current practical problems of lack of theoretical research on professional evaluation, lack of correlation between evaluation indicators and biased value orientation. Built on the basis of CIPP evaluation model of higher vocational colleges professional internal evaluation index model (as shown in Fig. 1), and with the target of the construction of the vocational colleges professional design as the criterion, the implementation of the resource configuration optimization in the process of specialty construction, running implementation testing, results, performance evaluation, based on the occurrence and development of logical things, Evaluate key factors throughout Context evaluation(C), Input evaluation(I),

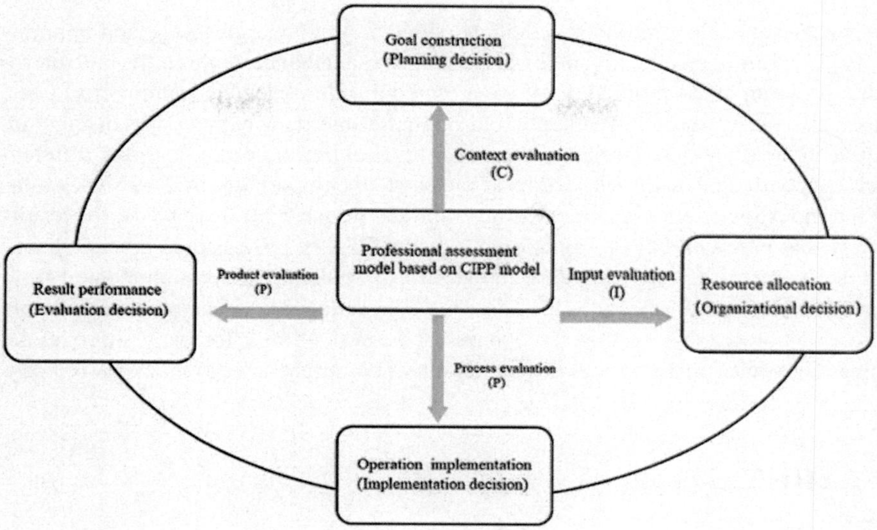

Fig. 1 Schematic diagram of theoretical model of professional internal evaluation in higher vocational colleges based on CIPP evaluation model

Process evaluation(P), and Product evaluation(P), Building construction by professional construction target, resource allocation in the specialty construction, professional construction operation and professional application results 4 first-level indicators and "professional setting and adjustment" and "talent training objective and specification" 12 secondary indicators and the corresponding observation points based on CIPP evaluation model of higher vocational professional internal evaluation index system.

Xu Huamei of Anhui Vocational College of Grain Engineering evaluated the quality of professional personnel training in mechatronics by CIPP evaluation model. Based on the "diversity" of teaching reform and the "flexibility" of teaching process, a professional evaluation index system of the CIPP evaluation model for mechatronics specialty was constructed, and the evaluation practice of the mechatronics specialty based on CIPP evaluation model was carried out. The results show that: Through the implementation of electromechanical integration major based on CIPP model, problems existing in the process of teaching and school-enterprise cooperation can be found in real time, and then dynamically adjust the teaching management regulations and teaching methods according to the evaluation results. College of mechanical and electrical integration of high quality of personnel training, conform to the development of higher vocational college concept, for the society to transport a number of high-quality mechanical and electrical integration professional personnel.

Song Yueqin, Pan Jian et al. from Huangshan University constructed a CIPP integrated practice evaluation system for forestry majors. The first-level evaluation indexes include background (C), input (I), process (P) and achievement (P), and the second-level evaluation indexes include demand, problem, advantage and opportunity (13). Tertiary evaluation index to cover the demand of ability training and internship program of the rationality of selecting the matching degree, training resources, such as security, terrain, weather, 27, through the enterprise expert panel discussion, students, teachers, symposia, etc., the four level of evaluation index gives different weight coefficient, at the same time, according to the importance of 27 tertiary evaluation index to weight assignment. And empirical research for three years, the results show that based on CIPP evaluation system of forestry professional practice based on self-expectation of cultivating the ability of students, fully implemented from practice plan/mode selection, the selection of training resources to practice whole process (holographic) on the effectiveness of the evaluation, effectively improves the objectivity and impartiality of the evaluation of comprehensive practice for forestry.

3 AHP Technology

The analytic hierarchy process (AHP) is to arrange the complex objects of professional diagnosis and improvement in higher vocational colleges into an orderly hierarchical structure, and then compare and judge each item in pairs to calculate the relative importance coefficient of each item, namely weight. In this study, the weight of professional diagnosis and improvement indicators in vocational colleges based on CIPP evaluation model was constructed by single criterion weight construction method.

3.1 Proportional Scale System

The core problem of analytic hierarchy process is to establish a reasonable and consistent judgment matrix, the rationality of judgment matrix is affected by the rationality of scale. The so-called scale refers to the quantification concept of the difference of importance level of each evaluation index by the evaluator. The common methods to determine the quantitative standard of the importance of indicators are: proportional scale method and exponential scale method. The scale method is based on the evaluation criteria for qualitative differences of things. Generally, 5 discriminant levels are used to represent qualitative differences of things. When higher accuracy is required for evaluation and analysis, 9 discriminant levels can be used to evaluate, as shown in Table 1.

Table 1 Proportional scale value system difference

The value meaning	1–9 scaling	5/5–9/1 scaling	9/9–9/1 scaling
Compare A and B, the two factors are equally important	1	1 (5/5=)	1 (9/9=)
Compare A and B, A is slightly more important than B	3	1.5 (6/4=)	1.286 (9/7=)
Compare A and B, A is obviously more important than B	5	2.33 (7/3=)	1.8 (9/5=)
Compare A and B, A is more important than B	7	4 (8/2=)	3 (9/3=)
Compare A and B, A is extremely important than B	9	9 (9/1=)	9 (9/1=)
The middle value of two adjacent scales	2, 4, 6, 8	1.222 (5.4/4.5=) 1.875 (6.5/3.5=) 3 (7.5/2.5=) 5.67 (8.5/1.5=)	1.25 (9/8=) y(9/6=) 2.25 (9/4=) 4.5 (9/2=)
Reciprocal	While comparing, two adjacent scale values are the reciprocal of each other		

3.2 The AHP Model

When AHP is used to solve practical problems, the AHP model is constructed and the components related to the problems are divided into several levels. The top layer is the target layer with only one element, the bottom layer is the scheme layer, and there can be multiple layers in the middle, known as the criterion layer or index layer, as shown in Fig. 2.

3.3 Calculate the Single Weight of Each Index Under a Single Subsystem

Suppose the total weight of the subsystem Ak ($k = 1, 2, \ldots, n$) is *ak*, and there are r indicators associated with Ak, denoted as $B_1^{(k)}$, $B_2^{(k)}$, ……, $B_r^{(k)}$; The weights are denoted as $b_1^{(k)}$, $b_2^{(k)}$, …, $b_r^{(k)}$.

Firstly, the ratio of the relative importance of Bi index to Bj index bij should be determined through expert evaluation. The specific determination method is as follows:

If Bi and Bj are considered to be equally important, bij = 1, bji = 1;
If Bi is considered to be slightly more important than Bj, bij = 3, bji = 1/3;
If Bi is considered to be significantly more important than Bj, bij = 5, bji = 1/5;
If Bi is considered more important than Bj, bij = 7, bji = 1/7;
If Bi is considered to be absolutely more important than Bj, bij = 9, bji = 1/9;

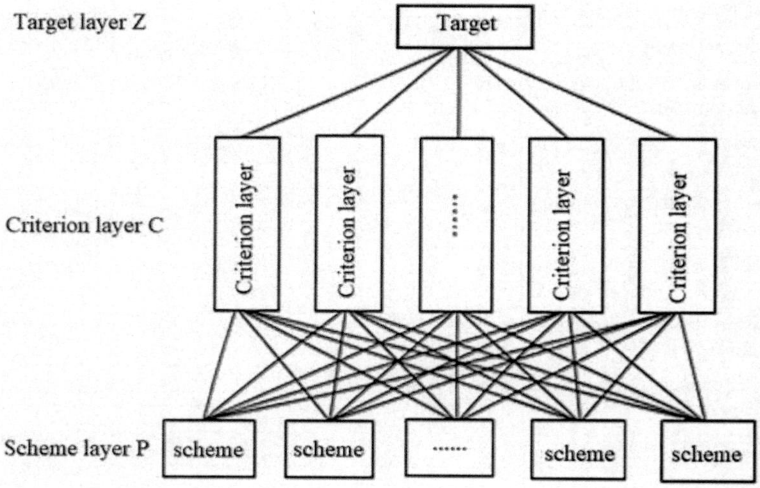

Fig. 2 Structure model of AHP

It is considered that the importance of Bi is between two adjacent odd numbers than Bj, so the value of bij is 2,4,6,8, and the value of bji is 1/2, 1/4, 1/6, and 1/8.

After determining the values of bij and bji, a pin-pin-comparison judgment matrix can be formed, as shown in Table 2.

Each expert carries out judgment analysis according to the above judgment rules and can write the comparative judgment matrix layer by layer. Since this matrix is written according to the individual judgment of each expert, we call it the individual judgment matrix. Due to the different experts based on the analysis of understanding may be some bias or differences, will often appear some extreme judgment (i.e. deviation from normal results or to judge the opinions of the majority of people), give reasonable weight an adverse impact, therefore, need to individual judgment matrix in the extreme judgment information effectively, and then integrated into a group judgment matrix.

With the comprehensive judgment matrix, the maximum eigenvalue λmax and the corresponding normalized eigenvector W(K) can be calculated by the sum method.

Among them, $W^{(K)} = (b_1^{(k)}, b_2^{(k)}, \ldots, b_r^{(k)})^T$; It needs to satisfy the following conditions: $(A_K - B) \times W^{(K)} = \lambda\max \times W^{(K)}$.

Table 2 Judgment matrix

Ak	B1	B2	…	Br
B1	b11	b12	…	b1r
B2	b21	b22	…	b2r
…	…	…	…	…
Br	br1	br2	…	brr

Table 3 Random consistency indicators of different orders

Order number	1	2	3	4	5	6	7	8	9	10	11	12	13	14
RI_K	0	0	0.58	0.90	1.12	1.24	1.32	1.41	1.45	1.49	1.52	1.54	1.56	1.58

As the comparison matrix constructed by experts may have certain errors, the maximum eigenvalue $\lambda max\ (k)$ of the r-order comparison judgment matrix constructed by experts may not be equal to R. In order to limit such errors, the relative error between $\lambda max\ (k)$ and R is taken as the consistency index of the comparison matrix, denoted as:

$$\mathrm{CI}_K = \frac{\lambda\mathrm{max}(K) - 1}{r - 1} \tag{1}$$

Considering the error caused by experts' different understanding of the problem, the consistency index CI_K was multiplied by the coefficient $1/RI_K$. Where, $1/RI_K$ is the random consistency index of comparison matrix of different order, and its value is shown in the Table 3 Shown below.

When the judgment matrix satisfies:

$$\mathrm{CR}_K = \frac{CI_K}{\mathrm{RI}_K} < 0.1 \tag{2}$$

It is considered that the comparison matrix has satisfactory consistency and the calculated eigenvectors (i.e. single weight) are acceptable. Otherwise, it indicates that the comparison matrix constructed by experts has a large error that exceeds the allowable range and needs to be adjusted.

3.4 *Calculate the Total Weight*

Since each element of layer A is directly corresponding to the total target G, its total weight, namely its single weight, is equal to the eigenvector of (G-A) judgment matrix in numerical value. Each indicator of other levels corresponds to its upper level, and its total weight is relative to each element of the upper level. Therefore, its total weight should be the sum of its single weight and the total weight of each element of the upper level, that is, the single weight and the total weight of each element of the upper level should be weighted. The formula for calculating the total weight of each index of layer B is:

$$b_i = \sum_{i=1}^{n} a_k \times \mathrm{b}_i{}^{(k)} i = 1, 2, \ldots, n \tag{3}$$

The calculation formula of indicators at other levels is the same.

Consistency test should also be carried out for the calculation results of total weight. First, calculate:

$$\mathrm{CI} = \sum_{i=1}^{n} a_k \times \mathrm{CI}_k \ \mathrm{RI} = \sum_{i=1}^{n} a_k \times \mathrm{RI}_k \tag{4}$$

When meet:

$$\mathrm{CR} = \frac{\mathrm{CI}}{\mathrm{RI}} < 0.1 \tag{5}$$

It is considered that satisfactory consistency is achieved and the total weight calculated can be confirmed. Otherwise, it indicates that some information elements in the judgment matrix still have large deviations and the judgment matrix still needs to be adjusted.

4 Construction of Professional Diagnosis and Improvement Index System

According to the CIPP evaluation theory, based on the current situation and future development of professional construction in higher vocational colleges, based on the "two-effect and four-core" professional diagnosis and improvement framework system, with efficiency and benefit as the value orientation, the evaluation of "four-dimension" is changed into background diagnosis, input diagnosis, process diagnosis and outcome diagnosis. With core objectives, core resources, core tasks and core development as the focus of the evaluation of the "four cores", based on the full investigation of professional diagnosis and improvement in higher vocational colleges, this study formed the professional diagnosis and improvement index system of higher vocational colleges based on the CIPP evaluation model with the support of experts and practice.

The design of professional diagnosis and improvement index system based on CIPP evaluation mode mainly uses literature analysis method and Delphi method. First of all, the relevant literature of major construction is comprehensively sorted out. For example, the evaluation index system of professional group construction of higher vocational education includes six first-level indicators: personnel training system construction, curriculum system construction, practical training system construction, information teaching resources construction, "double teachers" team construction, management system and operation mechanism construction. The evaluation index system of famous brand majors in colleges and universities is composed of five first-level indexes: faculty, teaching conditions, teaching management and reform, level and quality, benefit and characteristics. The evaluation index system of characteristic specialties of the Ministry of Education consists of eight first-level indexes, including construction objectives and support, teaching staff, teaching conditions, talent cultivation plan, teaching management, curriculum and textbook construction, practical teaching, talent cultivation quality and social reputation, and characteristics and innovation.

Secondly, carry out in-depth expert interviews. In this study, relevant experts were invited to discuss the design of the index system for higher vocational colleges. The brainstorming method and Delphi method were respectively used to discuss the design of the index system framework for professional diagnosis and improvement based on CIPP evaluation model. Combined with the results of literature analysis, a relatively reasonable professional diagnosis and improvement index system based on CIPP evaluation mode is formed.

Finally, the professional diagnosis and improvement index system based on CIPP evaluation model is revised and formed. On the basis of the professional diagnosis and improvement index system of higher vocational colleges based on CIPP evaluation model, the author revised and formed the professional diagnosis and improvement index system of higher vocational colleges based on CIPP evaluation model through comprehensive front-line investigation and in-depth analysis of survey results. See Tab.4.

5 Application of Professional Diagnosis and Improvement Index System in Higher Vocational Colleges

Using the professional diagnostic reform system of higher vocational colleges, the applied electronic technology major of Hunan Railway Vocational and Technical College was selected as the empirical research object to carry out the practical application research of the professional diagnostic reform system. In professional change clinical practice, invited seven constructions in the major of higher vocational education and evaluation experts, three industry experts, 10 experts to the application of electronic technology professional diagnosis, main reference data are provided by the professional documents, teaching management platform of real-time dynamic data, for each evaluation index scores, experts rated independently. After the experts scored, we collected the scores of 10 experts and relevant suggestions. At the same time, the evaluation tables of 10 experts and relevant suggestions were processed to obtain the average value of each indicator. According to the calculation, the final score of applied electronic technology major is 92.56, and the final evaluation conclusion is “excellent”.

Table 4 Index system of professional diagnosis and improvement in higher vocational colleges

First-level indicator	Second-level indicator	First-level indicator	Second-level indicator
B1 Specialty positioning and construction planning(8%)	C1 The construction target(10%)	**B5 Professional teaching team(14%)**	C51 Number of part-time teachers(5%)
	C2 Major Construction Planning(40%)		C52 Number of backbone teachers(5%)
	C3 Professional construction standards(30%)		C53 Proportion of part-time teachers(5%)
	C4 Professional construction mechanism(30%)		C54 Ratio of double-qualified teachers(10%)
B2 Personnel training program and curriculum system(18%)	C5 Professional Research Report(10%)		C55 Whether the course teachers meet the needs of talent cultivation(5%)
	C6 Talent Training Program(20%)		C56 The ratio of full-time teachers to teachers (5%)
	C7 Total academic hours(5%)		C57 Proportion of senior titles(5%)
	C8 Total Credits of major(5%)		C58 Proportion of teachers with master's degree or above(5%)
	C9 Professional core courses offered(10%)		C59 Age structure (proportion of teachers aged under 45)(5%)
	C10 Course name, code specification(2%)		C60 Number of full-time teachers exercising on site this academic year(10%)
	C11 Number of theoretical hours(3%)		C61 Teacher refresher training times this academic year(10%)
	C12 Number of practical hours(3%)		C62 Number of teachers awarded prizes (above city hall level)(5%)
	C13 Proportion of class hours in public courses(2%)	**B6 Social service ability(8%)**	C63 Technical Service Items(20%)
	C14 Proportion of professional course hours(5%)		C64 Amount of technical service(20%)

(continued)

Table 4 (continued)

First-level indicator	Second-level indicator	First-level indicator	Second-level indicator
	C15 Proportion of teaching hours of elective courses(5%)		C65 Social training items(20%)
	C16 Duration of on-post internship(5%)		C66 Social training Day(20%)
	C17 Total teaching hours per semester(5%)		C67 Social Training amount(20%)
	C18 Whether to offer courses on innovation and entrepreneurship(5%)	**B7 School-enterprise cooperation and international exchange(12%)**	C68 Number of cooperative enterprises(10%)
	C19 Whether to offer quality education courses(5%)		C 69 Number of in-depth cooperation enterprises(10%)
	C20 Whether all courses are offered according to the talent training plan(10%)		C70 This academic year, the professional construction Steering Committee will carry out activities(10%)
B3 Curriculum construction and teaching reform(12%)	C21 Number of course standards(5%)		C71 Number of modern apprenticeship pilot classes(10%)
	C22 Teaching material to choose(5%)		C72 Number of students in modern apprenticeship pilot classes(10%)
	C23 Development of college-level and above textbooks(10%)		C73 Number of jointly developed courses(10%)
	C24 National professional teaching resource library(5%)		C74 Number of jointly developed textbooks(10%)
	C25 Total teaching Resources(20%)		C75 Number of students accepted by cooperative enterprises for internship(5%)
	C26 Teaching reform courses at all levels(10%)		C76 Total value of donated equipment (5%)
	C27 School - level teaching results(5%)		C77 Number of international exchange projects(5%)

(continued)

Table 4 (continued)

First-level indicator	Second-level indicator	First-level indicator	Second-level indicator
	C28 Provincial teaching Achievements(3%)		C78 Number of international students(5%)
	C29 State-level teaching Achievements(2%)		C79 Foreign personnel training Day (5%)
	C30 Project teaching reform curriculum(10%)		C80 Overseas Exchange Day for Teachers(3%)
	C31 Reform course of assessment method(10%)		C81 Student Overseas Exchange Day(2%)
	C32 Blended teaching courses(10%)	**B8 Talent training effect and social reputation(8%)**	C82 Number of enrollment plans(10%)
	C33 Teaching reform awards(5%)		C83 Actual acceptance rate for the current year(5%)
B4 Teaching equipment and training conditions(12%)	C34 Total cost of the equipment(5%)		C84 First choice application rate of this year(5%)
	C35 Average instrument and equipment value per student(5%)		C85 Student enrollment rate this year(5%)
	C36 Number of campus training bases(15%)		C86 The graduation rate of this year's graduates(5%)
	C37 Number of training stations per batch(15%)		C87 Employment Rate of graduates(10%)
	C38 On-campus training room utilization rate(5%)		C88 The major matching rate of this year's graduates(10%)
	C39 Training program according to the rate of personnel training program(5%)		C89 Job satisfaction of this year's graduates after six months of employment(10%)
	C40 Build (expand) productive training base(5%)		C90 Average pass rate for all courses (5%)
	C41 New (expanded) virtual simulation training room(5%)		C91 Number of scholarship winners(10%)

(continued)

Table 4 (continued)

First-level indicator	Second-level indicator	First-level indicator	Second-level indicator
	C42 The training base (room) implements 6S management ratio(5%)		C92 Student Skills Competition (provincial and national)(10%)
	C43 Number of off-campus training bases jointly built by schools and enterprises(10%)		C93 Student Innovation and Entrepreneurship Awards (provincial and national level)(10%)
	C44 The proportion of professional practice hours in total professional hours(5%)		C94 Student Literature and Sports Awards (provincial and national)(5%)
	C45 Whether the professional training conditions meet the needs of personnel training(5%)		C95 Other awards (provincial and national)(10%)
	C46 Number of internship students(10%)	**B9 Characteristics and Innovation (8%)**	C96 Total number of teaching achievements above provincial level(20%)
	C47 Attendance rate of on-duty internship (5%)		C97 The total number of students at or above provincial level winning skills competitions(20%)
B5 Professional teaching team(14%)	C48 Number of professional leaders(15%)		C98 Total number of courses at provincial level or above(20%)
	C49 Number of professional teachers(5%)		C99 Total number of teachers awarded prizes at provincial level or above(20%)
	C50 Number of full-time teachers(5%)		C100 Total number of professional honors above provincial level(20%)

6 Conclusion

China's higher vocational education pays attention to connotation construction, and higher vocational education has achieved high-quality development. Teaching, diagnosis and reform can effectively promote the level of higher vocational colleges, so as to promote the quality of talent training. Based on CIPP evaluation model and AHP, this paper constructs a professional diagnosis and improvement index system in higher vocational colleges. The system consists of 9 first-level indicators, which are specialty positioning and construction planning (B1), personnel training program and curriculum system (B2),curriculum construction and teaching reform (B3),teaching equipment and training conditions (B4),professional teaching team (B5), Social service ability (B6),school-enterprise cooperation and international exchange (B7),talent training effect and social reputation (B8) and characteristics and Innovation (B9). Finally, the applied electronic technology major of the author's school is selected to carry out an empirical study. The results show that the professional diagnosis and improvement index system can evaluate the major construction and provide improvement suggestions, which is conducive to promoting the quality of the major construction in higher vocational colleges.

Acknowledgements Project of education and teaching reform in Hunan Vocational Colleges in 2021(No. ZJGB2021170).

References

1. Z. Jun, Diagnosis and improvement of teaching work in vocational colleges based on quality improvement. Vocat. Tech. Educ. China **26** (2015)
2. X. Yongchao, S. Jinyan, Construction of curriculum diagnosis and Improvement Model in higher vocational colleges from the perspective of participant heterogeneity. Lib. Educ. **13** (2020)
3. L. Hai, Teaching diagnosis and improvement: Endogenous motivation for Quality Improvement in vocational colleges. Vocat. Tech. Educ. **18** (2016)
4. S. Jinyan, X. Yongchao, Curriculum diagnosis and improvement in higher vocational colleges. Lib. Educ. **14** (2020)
5. L. Feng, L. Yanyun, Study on diagnosis and Improvement Mechanism of classroom teaching quality in Higher Vocational Colleges. J. Hubei Inst. Ind. Technol. **9** (2016)
6. Y. Da. A review of diagnostic studies on classroom instruction abroad. World Educ. News **12** (2015)
7. L. Luping, T. Guotong, Construction of higher vocational classroom teaching diagnosis and improvement system based on five perspectives. Vocat. Tech. Educ. **20** (2017)
8. S. Jinyan, Z. Kechang, X. Yongchao, F. Fanghong, Construction and application of Higher Vocational Curriculum Diagnosis and improvement index system based on AHP. Lib. Educ. **22** (2020)

A Study of Student Behavior Analysis Based on Campus Big Data

Chen Ge and Huang Chao Feng

Abstract With the construction of digital campus, colleges and universities will produce a large amount of data every day, such as study results, book information, dormitory access card information and campus card consumption, etc. How to properly organize and merge the stored data, use relevant technology to mine and analyze them, so that teaching managers can get rid of the dilemma of "abundant data but lack of information". The problem of improving the teaching management level of higher education institutions has become an issue to be solved by universities. In this paper, we use Apriori algorithm, K-means clustering algorithm and DBSCAN clustering algorithm to mine and cluster the data of a university, and use strong correlation to predict and warn students' behavior. The results of the analysis can provide an important basis for rewarding outstanding students and warning for unqualified students. This method can provide a scientific basis for school administrators to make decisions and provide good guidance for students' healthy development.

Keywords Data mining · Behavior analysis · K-means algorithm

1 Introduction

Colleges and universities have independent management systems for grades, books, computer rooms, student registration, finance, personnel, assets, laboratories, students, and various platform software such as network teaching, access control card, campus card, security monitoring, office, etc. The data information of these system platforms contains various behavioral data of students in school, and there is no intersection between these isolated data, resulting in information silos, which leads to a large amount of student data information being buried. The information is buried. This simple data management method cannot realize comprehensive

C. Ge · H. C. Feng (✉)
National University of Defense Technology, Changsha, China
e-mail: Earrychen@163.com

S. Patnaik and F. Paas (eds.), *Recent Trends in Educational Technology and Administration*, Learning and Analytics in Intelligent Systems 31,
https://doi.org/10.1007/978-3-031-29016-9_31

analysis of all-round data, nor can it find the implied laws among stored data, nor can it provide scientific basis for policy making and policy adjustment support for teaching management of universities. How to properly organize and merge the data stored in the existing system, use relevant technology to mine and analyze them, so that teaching managers can get rid of the dilemma of "abundant data but lack of information" and improve the teaching management level of colleges and universities has become a problem to be solved. The traditional technology can no longer realize the processing and analysis of massive data, and the multidimensional fusion and mining of various kinds of data from a wide range of sources and complexities has posed a great challenge to the efficiency of high-quality data pre-processing and analysis algorithms.

Using big data mining technology [1], we can analyze these massive data from multiple perspectives, so as to analyze the characteristics of students' school behaviors. Jingwen Ni and Liangzhong Shen analyzed the college students' college entrance examination results [2] by using decision trees, and several papers studied the abnormal behaviors of college students, centralized hot water use behaviors, poor students' counseling evaluation, and consumption behaviors respectively [3, 4], but they only analyzed the students' behaviors individually, and did not analyze the correlation between each behavior. Zhu QuanYin et al. proposed a multi-weighted adaptive student learning analysis method to discover the impact of student learning styles on academic performance [5]. Major universities abroad have jointly established educational datamining web pages for data sharing, an initiative that can use more data to mine more accurate characteristics of student behavior and make significant advances in the study of student behavior and faculty behavior in colleges and universities, notifying their counselors for help when students' behavior is found to be abnormal. For example, Purdue University in the United States has developed a course alert system, Wollongong University in Australia has developed a social network visualization tool that pays more attention to the behavior of college students, and they also analyze the behavior of primary and secondary school students; Tavares et al. [6] use association rule discovery to analyze students' motivation to learn programming, and Battaglia et al. [7], Bahr et al. [8] proposed an unsupervised clustering method to analyze students' behaviors, which can effectively improve the efficiency of the analysis of students' behaviors and has good results; for college students, club activities are an essential social activity, Bahr PR, Bielby R et al. analyzed the behavior of college students participating in clubs by clustering and found out what types of activities college students like to participate in the most and analyzed the behavior of students participating in activities. Charles and Angelpreethi [9] studied the effect of IQ on learning in the same learning environment and found that students with high IQ were more efficient in learning, but the proportion of the effect of IQ on learning performance was not 100%, and the effect of learning habits on learning performance was also greater. The impact of study habits on academic performance is also greater. The analysis of college students' behavior includes not only e-learning [10] but also daily life, for example, Bicikova [11] analyzed students' travel behavior and analyzed the overall behavior of students.

In this paper, we use the data pre-processing technique to process the raw data to meet the analysis standard, and then use the association rule Apriori algorithm to learn the intrinsic relationship between the data of grades, book circulation and campus card consumption (meal card consumption).

The K-means algorithm [12] is used to cluster the relationships among the three-dimensional data such as students' grade data, book borrowing data and campus one-card consumption data to discover the relationship between each data dimension of students. Using the relationship characteristics of the data, we can find students with excellent grades but with difficult families, and provide financial support for the truly poor students to help them complete their studies successfully, and also use the characteristics possessed by students with unqualified grades to provide convenient and scientific management solutions for college administrators.

The DBSCAN algorithm [15] was used to cluster the academic performance of students with high academic performance and students with low academic performance, the number of books borrowed, and the consumption data of campus card. The parameters of the algorithm were adjusted during the clustering process to achieve the best clustering results. The clustering analysis reveals the behavioral characteristics of the students with good academic performance and the behavioral characteristics of the students with failing academic performance, and uses these behavioral characteristics to predict the academic performance of students and to provide guidance and assistance to students who may fail.

2 Data Pre-processing

2.1 Data Mining

How to extract certain laws and features from the disorganized data is the core of data mining [16]. The whole process of data mining is shown in Fig. 1, in which the original data is first selected, pre-processed and converted, then the data is mined using the corresponding techniques [17], and finally the mining results are visualized and the analysis results are displayed visually in the form of graphs.

2.2 Data Integration

Data integration is the integration of data schemas for data in different data sources, in the process of integration, we should pay attention to the naming and format of each data source, detect and solve the problem of conflict of data values (attribute values may be different between source data, data storage rules are different) to correct one by one, and deal with redundant data (one attribute can be obtained from another attribute).

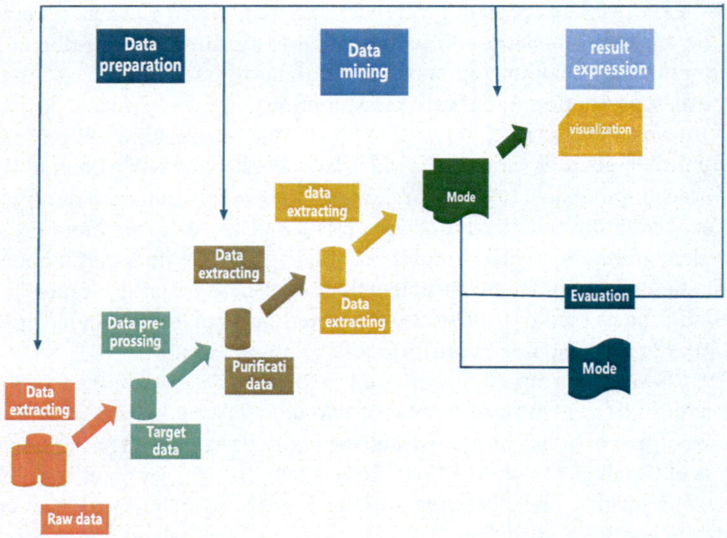

Fig. 1 Data mining flow chart

The correlation measure to determine whether an attribute is redundant is shown in Eq. (1):

$$\gamma_{A,B} = \frac{\sum(A - \overline{A})(B - \overline{B})}{(n-1)\sigma_A \sigma_B} \tag{1}$$

where n is the number of tuples, $\overline{A}$ and $\overline{B}$ are the averages of A and B, σ_A and σ_B Standard deviation of the properties A and B.

$$\sigma_A = \sqrt{\frac{\sum(A - \overline{A})}{(n-1)}} \tag{2}$$

$$\sigma_A = \sqrt{\frac{\sum(B - \overline{B})}{(n-1)}} \tag{3}$$

If there is a positive correlation between the attributes of $\gamma_{A,B} > 0$, A, B. This indicates that the value of A increases with the increase of B, the greater the value, The greater the probability that one attribute contains another attribute, one of the attributes can be removed as redundant if it is large enough.

If there is a negative correlation between the attributes $\gamma_{A,B} < 0$, A, B, This means that the value of A decreases as B increases, i.e. a genus Sex prevents the appearance of another attribute.

If $\gamma_{A,B} = 0$, A, B are independent of each other, they are not related.

If the attributes of the two data are independent of each other, then the two data attributes do not need to be deleted; If there is a positive correlation between two data attributes, then one of the data attributes needs to be deleted; If the data between two attributes is negatively correlated, the attributes of the two attributes are preserved.

2.3 Data Transformation

(a) Min—Maximum normalization

The most common data transformation method is normalization, that is, the attribute data is scaled proportionally so that it falls into a specific range.

For a given numeric property A, [$\min_A$, $\max_A$] is the value area [new_$\min_A$, new_$\max_A$] is the normalized value interval, min–max normalization normalizes the value v of A to v' according to the following equation, as shown in the formula (4) in:

$$v' = \frac{v - \min_A}{\max_A - \min_A} Value + new\_\min_A \tag{4}$$

$$Value = \left(new\_\max_A - new\_\min_A\right) \tag{5}$$

(b) Zero—Mean normalization

For a given numeric property A, $\overline{A}$, σ_A are the mean, standard deviation, zero—mean normalization of A according to the following equation to value v is normalized to v', as shown in Eq. (6):

$$v' = \frac{v - \overline{A}}{\sigma_A} \tag{6}$$

For a given numeric property A, max$|A|$ is the maximum absolute value of A, j is the smallest integer satisfying Eq. (7):

$$\frac{\max_{|A|}}{10^j} < 1 \tag{7}$$

and decimal scaling normalizes the value of A v to v' according to the following equation, as shown in Eq. (8):

$$v' = \frac{v}{10^j} \tag{8}$$

Table 1 Library circulation data attributes

Property name	Property meaning
XSKH	Student card number (unique)
XM	Name
QKJE	Amount owed
JSRQ	Borrowing date
HSRQ	Book return date
SM	Book name

Table 2 Consumer information data attributes

Property name	Property meaning
XYKZH	Student card number (unique)
XM	Name
CZSJ	Consumption time
CZJE	Spending amount
YE	Balance

Table 3 Achievement data attributes

Property name	Property meaning
YKTZH	Student card number (unique)
XM	Name
CJ	Achievements
XF	Credits
SJ	Year
XN	Semester

2.4 Data Profile

This study is based on the behavioral data generated by students at a university, and most of the data in this paper are recorded by students' campus cards. The original data of this study includes 30,848 book borrowing records, 24 million card spending records and 4 million grade records of the whole university. The following table shows the extracted attributes of each type of data (see Tables 1, 2 and 3).

2.5 Data Cleansing

Different data cleansing methods are used for these raw data from different databases [18]. The cleaned data can be used to directly calculate the number of students borrowed, their grades and the amount of money spent in a month. From the data

Table 4 Results of 10 students, number of loans consumption amount

ID	Grades (points)	Number of borrowed books (books)	Consumption amount (RMB)
1	79	22	468
2	80	18	300
3	74	16	584
4	70	14	403
5	72	13	301
6	74	13	294
7	74	12	473
8	84	11	200
9	91	11	170
10	80	11	423

of book borrowing records, it is found that the student's student ID number or employee's work number will appear once for each book borrowed by both students and staff, so this paper calculates the number of books borrowed based on the number of appearances of student ID numbers. The consumption data of undergraduate students are extracted according to the characteristics of their student ID numbers using the kettle data preprocessing tool. A weighted average of credits and grades is used to calculate student grades. Table 4 lists the 10-digit student data set at the end of the integration of grades, book borrowing, and consumption data.

3 Behavior Correlation Study

After data pre-processing, the student data has the ability of data analysis, firstly, using the discretized data and combining the association rule algorithm to find the association relationship between student behavior data, and secondly, using the association relationship to predict the student behavior.

3.1 Apriori Algorithm

The Apriori algorithm is the most influential algorithm for mining frequent itemsets of Boolean association rules. Its core is based on the recursion of the idea of two-stage frequent itemsets. The basic idea of the algorithm is:First find all frequent sets that occur at least as frequently as the predefined minimum support. Then strong association rules are generated from the frequent itemsets, which must satisfy the minimum support and minimum confidence. The frequent itemsets are then found to generate the desired rules, generating all rules that contain only the items of the set, where each rule has only one item in the right-hand part, and the definition of the middle rule is used here. In this study Apriori algorithm is used, which is the first algorithm on association rules for data mining. It is the first to use support-based pruning techniques and to systematically control the exponential growth of the candidate set. the Apriori algorithm generates the frequent item set pseudocode as shown below.

```
K=1
Fk={i|iI ({i} N min sup)} //find the set of frequent items
Repeat
        K=k+1
        Ck=apriori-gen(Fk-1) // Generate the set
        For every routine tT do
Ct=subset(Ck,t) //find the set of all terms containing t
For every candidate set cCt do
end for
        end for(c)=(c)+1//Support plus 1
        Fk={c|cCk ({c} N min sup)}
        //extract the set of frequent items k-
Until Fk=Ø
        Result= Fk
```

3.2 *Correlation Analysis of Behavioral Data*

In order to be able to have the dataset satisfy the application of the algorithm of association rules, the raw data of students after pre-processing are discretized (see Table 5).

The discretized data were analyzed by Apriori algorithm for correlation of the data set. The experimental results are shown in Tables 6, 7 and 8.

Table 5 Student data discretization

ID	Achievements	Borrowing number	Spending amount
1	B	Very high	d
2	B	More	d
3	B	More	d
4	C	More	e
5	C	More	d
6	B	More	d
7	B	More	e
8	C	More	e
9	B	More	c
10	B	More	e

Table 6 Reliability 0

Affiliation rules	Confidence level (0.5)
B => few	0.992845
C => few	0.995729
d => few	0.993557
e => few	0.994379
d, few => B	0.505935
d => B	0.506640
d => few, B	0.502675
B, d => few	0.992175
c, d => few	0.995028

Table 7 Reliability 0.7

Affiliation rules	Confidence level (0.7)
B => few	0.992845
C => few	0.995729
d => few	0.993557
e => few	0.994379
B, d => few	0.992175

Table 8 Reliability 0.9

Affiliation rules	Confidence level (0.9)
B => few	0.992845
C => few	0.995729
d => few	0.993557
e => few	0.994379
B, d => few	0.992175

The table shows that the probability of borrowing less is 99.57% when the student has a passing grade, and there is a strong correlation rule between passing grade and borrowing less books, and the probability of borrowing less is 99.22% when the student has a good grade and low consumption amount, and there is also a strong correlation rule between a good grade and low consumption amount and borrowing less books. When the student's borrowing quantity is low and the consumption amount is very low, the student's grade can be predicted to be good, and when the student's borrowing quantity is low, the student's academic performance can be predicted to be qualified, and the correlation relationship between the student's behaviors can be used to predict the student's academic performance, and when the grade is unqualified, the tutor can provide guidance to urge the student to study carefully and improve the passing rate of the student's exam.

4 Clustering Analysis of Student Behavior Based on K-Means Algorithm

The K-means algorithm [13] is a division-based clustering algorithm that uses iteration to gradually improve the clustering effect and has the advantage of being simple to understand and easy to implement. The algorithm was originally proposed independently by Steinhaus, and its ability to be applied to many fields of scientific research has led to the idea that each cluster centroid is represented by the average of all objects in the cluster. k-means first selects k objects at random, each representing the center of mass of a cluster [14]. For each of the remaining objects, the object is assigned to the cluster that is most similar to it, based on its distance from the center of mass of each cluster. Then, a new center of mass is calculated for each cluster. The above process is repeated until the criterion functions converge. The commonly used criterion function is the squared error criterion function.

The objective of a clustering algorithm is generally represented by an objective function. the similarity of the K-means algorithm is determined by the Euclidean distance, and its measure of clustering quality is determined by the sum of squares of errors. A dataset containing n data objects $D = \{x_1, x_2, \ldots, x_n\}$. Define the set of categories generated after the cluster analysis performed by this algorithm as $C = \{C_1, C_2, \ldots, C_k\}$. The equation for the objective function SSE of this algorithm is as follows.

$$SSE(C) = \sum_{k=1}^{K} \sum_{X_i \in C_k} \|X_i - C_k\|^2 \tag{9}$$

Equation (1) is the centroid of the ck cluster Ck, which is calculated as follows.

$$c_k = \frac{\sum_{X_i \in C_k} X_i}{|C_k|} \tag{10}$$

The final result of the K-means clustering algorithm is to find the clustering result that minimizes the SSE, and this optimization problem is carried out to find the optimal solution.

Student behavior data includes student achievement data, book borrowing data and campus card consumption data in three dimensions.

4.1 Clustering Analysis of Student Achievement and Circulation

As shown in Fig. 2, the horizontal axis represents students' grade information and the vertical axis represents students' borrowing quantity information in a semester. The K-means clustering algorithm was used for the analysis, and the number of data clusters k was set to 4. The Euclidean distance was used to calculate the distance from the data object to the central cluster, and the clustering results obtained are shown in Fig. 1. In the figure, four different colors are used to represent four different data clusters, where the green part represents the number of books borrowed by students with scores of 0–65 in a semester is between 0 and 5, and the number of borrowed data of most students is 0; the red part represents the number of books borrowed by students with scores of 65–75 in a semester is between 0 and 13; the black part represents the number of books borrowed by students with scores of 75–80 in a semester The black part represents the number of books borrowed by students with scores of 75–80 in a semester, and the blue part represents the number of books borrowed by students with scores of 80–100 in a semester, which is between 0 and 22. The results of the comprehensive cluster analysis show that the number of books borrowed by undergraduate students is small, and most students do not borrow books. The visualization of the clustering results showed that the number of books borrowed by students with good grades was about two times more than that of students with failing grades. Therefore, the number of books borrowed can also tell how good a student's grades are.

The clustering results shown in the above figure are a self cognitive process for students. From the figure, we can see the category of a certain class of students' scores and the number of books borrowed, and find out the gap between ourselves

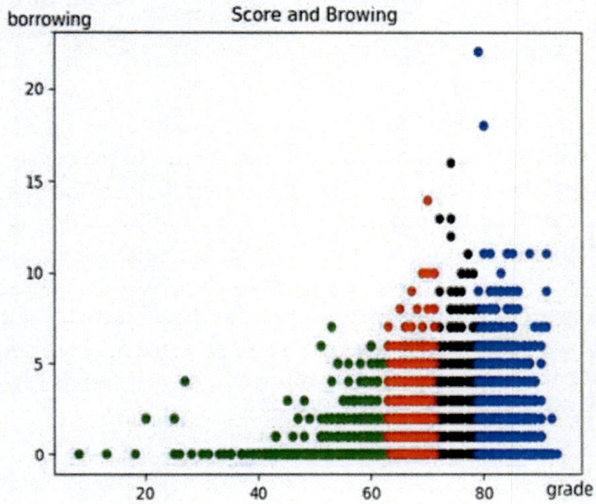

Fig. 2 Cluster analysis of student performance and circulation

and other students. Let students understand that they have little reading or poor academic performance, gradually narrow the gap with other excellent students, and make themselves better. Teachers can recommend books related to the major or course to help students learn, improve their knowledge and make students have a deeper understanding of the course, so as to help students improve their course performance. It is an indicator light for student managers to monitor and predict students' behavior, predict students' academic performance according to the number of students who borrow books, and focus on students who do not borrow books. Once they find that there is a risk of failing to pass the exam, they will help them, and let students with excellent scores coach them, so that they can pass the exam with excellent results.

4.2 Clustering Analysis of Student Achievement and Spending Amount

As shown in Fig. 3, the horizontal axis represents the students' academic performance and the vertical axis represents the amount of money spent by the students in a month at school. The K-means clustering algorithm was also used to analyze the data, and the number of clusters was set to 4. The Euclidean distance was used to calculate the distance between data objects to obtain the final clustering results as shown in Fig. 3. The clustering of the data in the figure is divided by the amount of student consumption, and the blue part of the figure represents the amount of student consumption around 0–250 RMB; the black part represents the amount of student consumption around 250–400 RMB; the green part represents the amount of student consumption around 400–500 RMB; the red part represents the amount of student consumption around 500–2000 RMB. From the visualization results of the cluster analysis, it can be found that there is no obvious intrinsic relationship between student achievement and campus one-card consumption data, yet it can be understood the activity trends and consumption characteristics of most undergraduate students in school, which are classified into the four categories as shown in the figure.

As shown in the above figure, the clustering result is a self cognition process for students. From the figure, it can be seen that the consumption quota of a certain class of students belongs to higher, lower or medium categories; At the same time, we also understand the difference between their consumption level and other students. For example, help students with high consumption quota to understand specific consumption details and let them pay attention to their own consumption behavior. For student logistics managers, it can be used as the indicator of the increase in the price of meals in the canteen. For example, most students have a high monthly consumption quota. It can be seen that the food price in the canteen is high, which requires the canteen manager to supervise and improve.

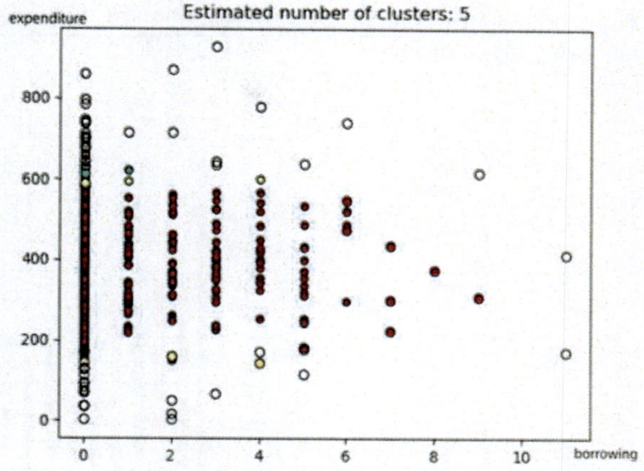

Fig. 3 Clustering of student achievement and consumption data

4.3 Clustering Analysis of Book Borrowing and Consumption

As shown in Fig. 4, the horizontal axis represents the number of students' borrowings and the vertical axis represents the amount of students' spending in a month at school. The K-means clustering algorithm was also used to analyze the data, and the number of data clusters k was set to 4. The clustering results obtained by using Euclidean distance to calculate the distance between data objects are shown in Fig. 4. The clustering of the data in the figure is divided by the amount of student consumption, and the blue part of the figure represents student consumption of about 0–250 RMB; the black part represents student consumption of about 250–400 RMB; the green part represents student consumption of about 400–500 RMB; the red part represents student consumption of about 500–2000 RMB. From the visualization results of the cluster analysis based on the division, it can be found that students with consumption amounts around 400–500 RMB borrowed significantly more books than those in other consumption amount categories; most students with monthly consumption amounts below 250 RMB and above 1000 RMB did not borrow books. The clustering results, shown in Fig. 3, revealed a relationship between the amount of student spending and book borrowing. Students who spent around 400–500 RMB borrowed more books, and those who spent less or more borrowed fewer books. According to this result students can judge whether they read more or less according to their consumption amount, and if less they can increase their reading amount to ensure their knowledge storage.

As shown in the clustering results above, the relationship between student consumption quota and book borrowing is found. Students with a quota of about 400–500 RMB borrow more books, while students with a lower or higher consumption quota borrow fewer books. According to the result, students can judge whether they read more or less according to their consumption amount. If they read less, they can increase their reading amount to ensure their knowledge storage.

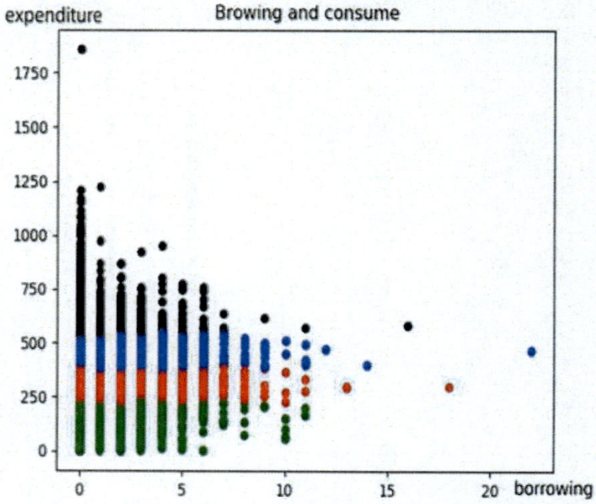

Fig. 4 Clustering of student borrowing and consumption data

5 Clustering Analysis Based on DSCAN Algorithm

In the previous section, certain behavioral characteristics of current undergraduate students, such as higher grades and higher number of books borrowed than other students, were found through the analysis of a division- based K-means clustering algorithm, which led to an intrinsic relationship between consumption amount, academic performance, and number of books borrowed. In this section, the density clustering based DBSCAN algorithm is used to cluster and analyze the specific behavioral characteristics of students with high and low grades.

5.1 Clustering Analysis of High-Achieving Students Based on DSCAN Algorithm

The student behavior data contains three-dimensional data, i.e., students' grades, the number of books borrowed and the amount of money spent on the Campus One Card. The clustering results are shown in Figs. 5, 6 and 7, which show the clustering results between the number of books borrowed and the number of students with good grades (weighted average above 85); the clustering results between the number of books borrowed and the amount of students' monthly spending; and the clustering results between the number of students' grades and the amount of students' spending. The clustering results are shown in the figure below.

As shown in Fig. 4, the horizontal axis represents the students' grades and the vertical axis represents the number of books borrowed by students in a semester. The clustering results are shown in Fig. 5. The distance radius and point threshold of the clusters are set to 0.2 and 5, respectively. The clustering results are shown in

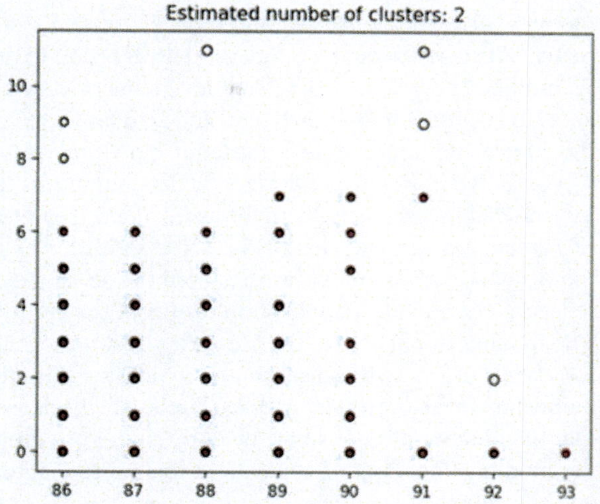

Fig. 5 Clustering of grades and number of books borrowed for high-achieving students

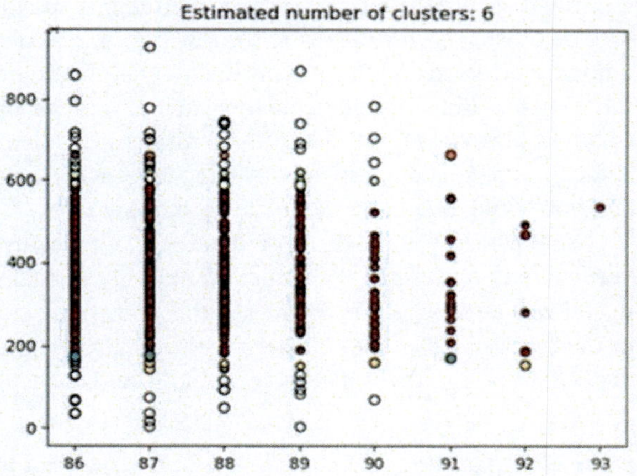

Fig. 6 Clustering of borrowing and consumption data of high-achieving students

Fig. 5. The clusters are divided into 2 data clusters, and the data clusters in red, The clusters in white indicate that the DBSCAN algorithm treats the number of books borrowed by high-achieving students as noise points during their school years. The number of books borrowed by students is low, and the number of books borrowed by high-achieving students is significantly lower, which is probably due to the fact that students prefer electronic devices and other readers to paper books.

The clustering results as shown in Fig. 5 are a process of continuous awareness and improvement for high achieving students, who can understand their own shortcomings from the clustering results, find out the gap between themselves and other high achieving students, recognize the reasons for reading books, and if they read less or no books, increase their borrowing amount appropriately, gradually narrow the gap between themselves and other outstanding students, and make themselves better. The teacher can recommend several books to the students. The instructor can recommend several books related to the major or the course to help students read, so as to improve their knowledge, and also let them have a deeper understanding of the course and improve their academic performance.

As shown in Fig. 6, the horizontal axis represents the number of books borrowed by students in a semester, and the vertical axis represents the monthly consumption of students in school. The DBSCAN algorithm was used for cluster analysis, and the distance radius and point threshold of the data clusters were set to 0.28 and 6, respectively, and the visualization results of the cluster analysis are shown in Fig. 6. The number of clusters in the clustering result is 5, where the blank points indicate the data clusters with noise points, which are directly excluded by the DBSCAN algorithm; the data clusters with red circles are the denser data clusters, which indicate that the number of books borrowed by students with good grades is about 1–9, and the monthly spending of students in this borrowing data range is about 400 RMB, and the monthly spending of students with good grades is about 400 RMB. The monthly spending of heterogeneous students is around $200–600, which is more densely distributed and the majority of students spend around 400 RMB.

The clustering results, as shown in Fig. 6, revealed a relationship between the amount borrowed and the amount spent by high-achieving students, finding that The monthly spending of high-achieving students is around $400, and the average amount of books borrowed by students in a semester is 2–4 books, from this clustering result students can judge their own borrowing level according to their consumption amount and borrowing quantity. and make corrections.

As shown in Fig. 7, the horizontal axis represents the students' grades and the vertical axis represents the students' monthly consumption in school. The clusters in the red circles in the figure are the noise clusters, which are directly ignored in the analysis. The data clusters with red circles in the figure are the most dense data clusters, and these data objects indicate that the monthly consumption amount

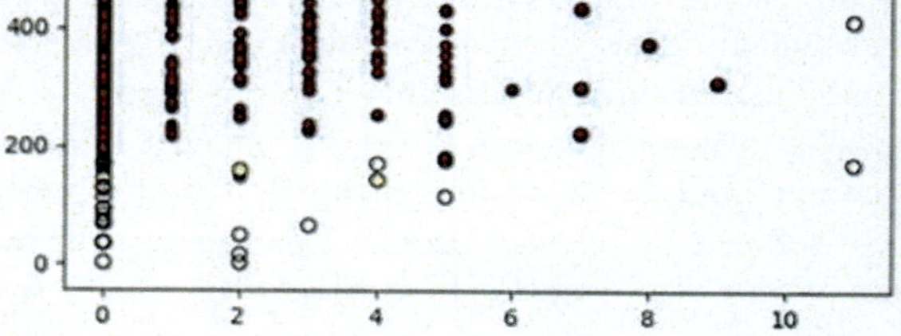

Fig. 7 Clustering of high-achieving students' performance and spending data

of high-achieving students in school is relatively stable and most of them have a monthly consumption amount of about 250–600 RMB, among which those with a monthly consumption amount of about 400 RMB are the most dense data clusters. The distribution of students is dense, and during the study it was also found that students with high academic achievement spent less than $1,000, and few and almost no students did not spend money at school.

The clustering results shown in Fig. 7 show that students with good grades spend more steadily each month during school, and there are few students who do not spend money at school, i.e., students with good grades eat more regularly during school and go to the student cafeteria at meal times. There are also students who spend more than $600 in the graph, but this is also a small percentage, and it is possible that they spend more in supermarkets. A very small number of high-achieving students did not spend any money during the school year. Therefore, it can be assumed that this group of students eat off-campus or use the online ordering APP to eat. Using the visualization results of this cluster analysis, it can be seen that students with good grades are more rational in their spending. The clustering data in the yellow part of the graph shows that these students have good grades, but their monthly spending is less than $200. The school can understand the actual living situation of these students and give financial support to help these students with good academic performance but difficult family conditions, so that their studies will not be affected by financial problems, and the national grants can be fully used. For student managers, it is an indicator to monitor and predict students' behavior, to determine whether they are in the habit of eating on time, based on their monthly spending in school.

5.2 *Clustering Analysis of Students with Unsatisfactory Grades Based on DSCAN Algorithm*

The student behavior data contains three-dimensional data, i.e., students' grades, the number of books borrowed and the amount spent on the Campus One Card. Cluster analysis is performed separately. The behavioral analysis of the students who failed to pass the grade can get a good grasp of certain behavioral characteristics, which can be used by the student managers to understand whether the students need help or not, and provide convenience for the college managers. The distance radius and point threshold of the data clusters are set according to the results of several experiments, and the specific values and clustering results are shown in Figs. 7, 8 and 9. The figures show the clustering results between the grades of students with unqualified academic performance and the monthly consumption amount in school; the clustering results between the number of borrowed books and the consumption amount of campus card; and the clustering results between the academic performance and the number of borrowed books. The specific clustering results are shown in the figure below.

As shown in Fig. 8, the horizontal axis represents the students' academic performance and the vertical axis represents the students' monthly consumption at school.

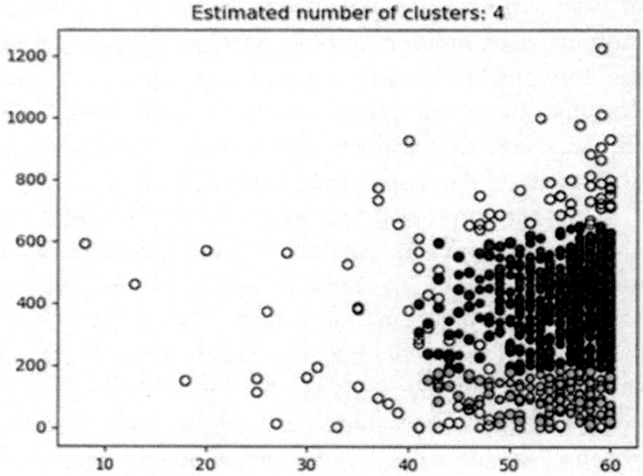

Fig. 8 .Clustering of failing students' grades and spending data

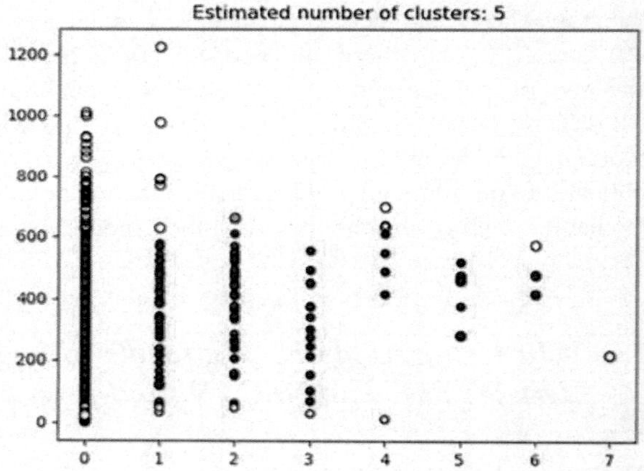

Fig. 9 Clustering of borrowing and spending for students with failing grades

The clustering analysis is performed by DBSCAN algorithm, and the distance radius and point threshold of the data clusters are set to 0.25 and 5 respectively. From the figure, the number of clusters of the algorithm is 4, and the blank circles are the clusters with noise points. The cluster in red is the cluster with the largest density distribution, which means that most of the students failing in the 40–50 range and the number of students scoring between 50 and 60 is biased. more; most of the students with failing grades spend around 200–600 RMB per month; the data clusters in yellow are density The second cluster of the distribution, this data cluster indicates that students with grades around 40–50 spend around 0–200 RMB per month in school. The clusters in green represent students with scores between 50 and 60 who spend 0RMB per month at school, meaning that they do not spend any money at school.

The clustering results, as shown in Fig. 8, are a cognitive process for students with failing grades to recognize their deficiencies, understand where they fall in the category of students with failing grades, and understand where their monthly spending falls in the overall spending amount. The fact that the spending amount of students with failing grades is more than 1000 RMB and rarely exists in the overall student distribution indicates that the spending amount of students with failing grades is high; there are also students with failing grades whose spending amount is around 1–200 RMB, indicating that most of the students with failing grades have the habit of not spending at school, and their spending amount is low at school and high outside of school.

As shown in Fig. 9, the horizontal coordinate represents the number of books borrowed by students in a semester, and the vertical axis represents the monthly consumption amount of students in school. The distance radius and point threshold of the clusters were set to 0.15 and 4.8 respectively, and the final clustering results are shown in Fig. 9. The number of clusters in the figure is 5, where the blank circles indicate the clusters with noise points, and the clusters in red are the most densely distributed, representing the monthly consumption of students who borrowed 0–6 books in a semester, which is around 200-600RMB, among which the large The majority of students spend around 400-600RMB per month. The graph shows that the number of students with failing grades and not borrowing books is high.

As the clustering results in Fig. 9 show, the number of students with failing grades and not borrowing books is overrepresented. Students should pay attention to this phenomenon, as failing grades and not borrowing books indicate that students spend their time in other places, and their thoughts and actions are not on studying. These students need the supervision of teachers and classmates to help them develop a normal diet and borrow books, so that they can pass exams successfully and not fail. Compared with the results of the clustering of monthly consumption and grades of high-achieving students and the results of the clustering of grades and number of books borrowed by low-achieving students, the behavior of high-achieving students and low- achieving students differed significantly.

6 Conclusion

In this paper, we focus on the grade data in the undergraduate grade system, the borrowed book data in the library data system, and the swipe records of the campus one-card during the school year. In this paper, we perform pre-processing on the raw data collected from the behavioral activities of undergraduate students, i.e. data cleaning, data integration, data conversion and data statute on the students' grade data, book borrowing data and campus one-card consumption data, calculate the weighted average of students' grades, classify the number of books borrowed from the library and the amount of money spent by students in one month during a semester, and The data is integrated according to the student's student number, so that the pre-processed data can meet the requirements of data analysis. The pre-processed student data are discretized, and the Apriori algorithm is used to analyze the association rules for student grades, book circulation data and student campus card swipe data, and strong association rules are derived. The clustering analysis was performed on the undergraduate student data using the division-based K-means algorithm to analyze the relationship between student grades and student borrowed books data, between student grades and the monthly consumption amount in school and between borrowed books data and the monthly consumption amount. The method of this paper can find the correlation between student achievement and book borrowing data and monthly consumption amount, and discover the behavioral characteristics of students with good grades and students with bad grades, and use these characteristic values to make scientific guidance for teaching management.

References

1. X. Wu, X. Zhu, G.Q. Wu et al., Data mining with big data. IEEE Trans. Knowl. Data Eng. **26**(1), 97–107 (2013)
2. X. Jian. *Analysis of the Correlation Between Consumption Behavior and Achievement based on Campus Card Data* (Nanchang University, 2010)
3. Z.L. Kozina, S.S. Iermakov, Analysis of students' nervous system's typological properties, in aspect of response to extreme situation, with the help of multi-dimensional analysis. Phys. Educ. Stud. **19**(3), 10–19 (2015)
4. A.L. Murray, I. Obsuth, M. Eisner, et al., Disaggregating between-and within-classroom variation in student behavior: a multilevel factor analysis of teacher ratings of student prosociality and aggression. J. Early Adolesc., 0272431618797005 (2018)
5. Q.Y. Zhu, E.Q. Shen, Y.P. Qian, et al., K-means cluster-based multi-weight self-adaptive student learning behavior analysis method. CN 105913353A (2016)
6. P.C. Tavares, E.F. Gomes, P.R. Henriques, Studying programming studentsmotivation using association rules, in *10th International Conference on Computer Supported Education*, pp. 514–520 (2018)
7. O.R. Battaglia, B.D. Paola, C. Fazio, A new approach to investigate students' behavior by using cluster analysis as an unsupervised methodology in the field of education. Appl. Math. **07**(15), 1649–1673 (2016)
8. P.R. Bahr, R. Bielby, E. House, The use of cluster analysis in typological research on community college students. New Dir. Inst. Res. **2011**(S1), 67–81 (2011)
9. D.S. Charles, A. Angelpreethi, An impact of intelligent quotient and learning behavior of students in learning environment. Artif. Intell. Syst. Mach. Learn. **7**(1), 1–6 (2015)
10. K. Umbleja, M. Ichino, *Predicting Students' Behavior During an E-Learning Course Using Data Mining*. Interactive Collaborative Learning (Springer International Publishing, 2016), pp. 12–100
11. K. Bicikova, Understanding student travel behavior: a segmentation analysis of British university students. J. Travel Tour. Mark. **31**(7), 854–867 (2014)
12. X. Zhong, D. Enke, A comprehensive cluster and classification mining procedure for daily stock market return forecasting. Neurocomputing **267**(12), 152–168 (2017)
13. C. Cortes, V. Vapnik, Support-vector networks. Mach. Learn. **20**(3), 273–297 (1995)
14. C.F. Zheng, L. Jiang, L.Q. Jiang et al., Application and research of Bayesian network in data mining. Adv. Mater. Res. **532–533**, 738–742 (2012)
15. B. Rmbrmb, G. Zhongren. Research based on K-means clustering algorithm. J. Southwest Univ. Natl. **01**, 1–3 (2009)

16. S.W. Draper, M.I. Brown, Increasing interactivity in lectures using an electronic voting system. J. Comput. Assist. Learn. **20**(2), 81–94 (2004)
17. M.A.A. Dewan, F. Lin, D. Wen, M.Murshed, Z. Uddin, A deep learning approach to detecting engagement of online learners, in *IEEE International Conference on Internet of People, Guangzhou* (2018)
18. Z.-H. Zhou, A brief introduction to weakly supervised learning. Natl. Sci. Rev. (2017)

Visualization Analysis of Blockchain Technology in Education Arena Based on Citespace

Linwei She and Liqi Lai

Abstract Blockchain has brought tremendous value to all fields of today's society, and it has also brought enormous challenges, which have attracted great attention from all walks of life. Blockchain also has emerged as an important concept at the interface of ICT and education. The article uses the information visualization software Citespace to study Blockchain in education in the Web of Science and CNKI database from 2016 to 2022, from macro to micro to the representative countries of the literature, keywords, and co-cited documents. Through visualization analysis, the article clarifies the key research directions, key documents, and hot spot frontiers in the field of blockchain in education research, forecasts the future development trends in this field, and compares the research situation at home and abroad, to provide readers and other researchers with certain reference and help.

Keywords Blockchain · Education · Visualization analysis

1 Introduction

The first blockchain concept appeared in 2008, when an academic with the pseudonym "Satoshi Nakamoto" proposed a blockchain-based Bitcoin system, which was the world's first blockchain application [1]. In a narrow sense, a blockchain is a decentralized shared ledger that combines chronologically ordered blocks of data into a specific data structure that is cryptographically guaranteed to be tamper-proof and

L. She
International Business School, Jinan University, Zhuhai 519070, China

L. Lai (✉)
Modern Education Technology Center, Jinan University, Zhuhai 519070, China
e-mail: LaiLQ@jnu.edu.cn

L. She · L. Lai
GBA and B&R International Joint Research Center for Smart Logistics, Jinan University, Zhuhai 519070, China

S. Patnaik and F. Paas (eds.), *Recent Trends in Educational Technology and Administration*, Learning and Analytics in Intelligent Systems 31,
https://doi.org/10.1007/978-3-031-29016-9_32

capable of securely storing simple, sequentially related data that can be verified within the system [2]. Blockchain technology, broadly defined, is a new paradigm for decentralized infrastructure and shared computation that uses crypto-chain block structures to verify and store data, distributed node consensus algorithms to generate and update data, and automated scripting code (smart contracts) to program and manipulate data [3, 4]. Blockchain, an emerging decentralized architecture and distributed computing paradigm that underpins Bitcoin and other cryptocurrencies, has recently attracted intense attention from governments, financial institutions, high-tech companies and capital markets [5, 6]. In the education field, blockchain technology has been hailed as the “new cornerstone of education informatization” due to its strong innovation potential and application prospects [7]. On this basis, the focus of this study is on research hotspots and research frontier issues, research frontiers, and keyword clustering analysis of blockchain education application fields in China draws a knowledge map of blockchain education application by visual citation analysis software CiteSpace 5.7.R2 and analyzes the retrieved literature by content analysis, to make a detailed sorting and visual analysis of blockchain application in education.

This research takes the documentation of blockchain in education from 2016 to 2022 in Web of Science and CNKI databases, and analyzes representative nations and keywords from macro level to micro-level by using information visualization software Citespace. And this paper draws the hot frontiers of blockchain in education research and the possible future development trends by comparing domestic and international research. These studies not only provide scholars in the field of knowledge mapping and visualization analysis with references to relevant tools and research methods but also provide great guidance for scholars engaged in blockchain in education research to understand this field and alleviate some of the workloads of relevant researchers.

2 Research Method and Data Source

2.1 Research Method

This study focuses on the use of the Citespace software developed by Chaomei Chen of Drexel University’s School of Information Science and Technology and the WISE Lab at the Dalian University of Technology. A Java application for analyzing and visualizing common citation networks is Citespace. It is citation visualization and analysis software developed in the context of scientometrics and data visualization. It is software for the visualization and analysis of citations developed in the context of scientometrics and data visualization, focusing on the analysis of the fundamental knowledge contained in scientific analysis. Two complementary perspectives are realized based on the concept of temporal duality between the “knowledge base” and “research frontiers” of information science: the focus perspective and the time zone perspective. Since the structural, model and distribution of scientific knowledge are presented in a visual way, the visual map analyzed in this way is also known as the "knowledge domain map" [8].

2.2 *Page Numbers Data Source*

All data in this article are mainly from the database of CNKI and Web of Science. CNKI was established in 1999 and is currently the largest literature database in China. First, we used "blockchain" as the keyword and set the keywords of "blockchain + education" related topics as "blockchain * (education + teaching + learning + teachers + students + schools + teaching resources + curriculum + teaching evaluation)" for further search, and the time range was set as 2016–2022, and the researchers needed to manually remove irrelevant data when selecting records, and finally got 176 search results. After selection, click "Export/References", select "Refworks" format, then click "Export", save the file as "download_CNKI.txt", and then use the data converter in Citespace to perform the conversion. In Web of Science, first we also used "Blockchain" as the keyword, then set the subject keyword related to "Blockchain + Education" as "Blockchain * (Education + Teaching + Learning + Teachers + Students + Schools + Teaching Resources + Curriculum + Teaching Evaluation)" for further search. The time range was set to 2016–2022, the language was selected as English, the literature type was selected as an article, and irrelevant data were removed when selecting records, and 222 search results were finally obtained. The format of the file was "other file format", and the name of the file was "download_wos.txt". The data was downloaded in April 2022.

3 Comparative Analysis of the Literature

Based on the 398 domestic and foreign-related papers obtained from the search, we set time slice, node type, threshold selection, etc. in Citespace to find out the research hotspots and frontier issues of blockchain in education by analyzing the publication time, keywords, important literature, clustering, and research frontier.

3.1 *Time Distributional Characteristics*

Trends in the number of published papers can indicate the trend of the field from a macro perspective. Thus, in this study, the annual paper counts on CNKI and Web of Science in this area of blockchain in education were conducted. Results are displayed in Fig. 1.

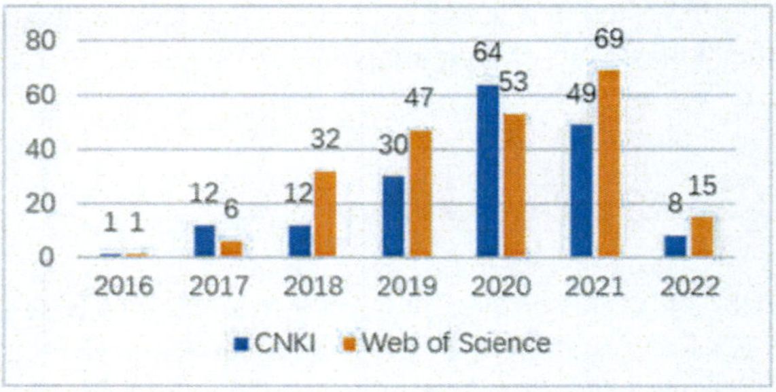

Fig. 1 Bar chart of the number of published papers

In 2022, the current data are 8 and 15 articles, which cannot be used as a reference for quantitative trends. The histogram of annual publication volume shows that the literature published with "blockchain" and "education" as keywords initially appeared in 2016, and macroscopically, from 2017 to 2021, the number of papers shows an increasing trend year by year. The trend of annual publication volume shows that the research on "blockchain + education" has also received more and more attention.

3.2 *Country Distribution Characteristics*

The publication of country analysis can disclose the cooperative relations among countries, offer a novel viewpoint for assessing the country's academic influence, contribute to the discovery of noteworthy national institutions, and get a macroscopic view of the spatial distribution of countries in the research area. The node type is set to country, the time slice threshold is set to top50, the network pruning algorithm is chosen as the minimum spanning tree algorithm, and the visualization method is chosen as the static view. Citespace was run to visualize and analyze the literature related to blockchain in education to obtain a visual map of the network in the study countries, as illustrated in Fig. 2.

Fig. 2 Distribution map of the study area

Table 1 Research field of high frequency country table

No	Country	Frequency	No	Country	Frequency
1	China	47	6	Saudi Arabia	11
2	Spain	23	7	Portugal	10
3	India	21	8	Indonesia	7
4	USA	20	9	Pakistan	7
5	England	12	10	Romania	7

Within Fig. 2, every circle is used as a node and every node stands for a nation. A node's size indicates the amount of information; the bigger the size of the node, the more files are issued by that country. The connections among nodes represent the cooperation between countries. The more connections, the closer the cooperation among each country. The graph shows that the threshold for displaying the labels is set to 7. This figure displays the names of countries with a number of published papers greater than or equal to 7. It is clear from the graph that the circle of nodes is significantly larger in China than in other nations. The number of posts is far ahead with a rate of 47. This is followed by Spain with a frequency of 23, the frequency in India is 21, the United States with a frequency of 20, and the United Kingdom with a frequency of 12. The number of papers published by other countries is shown in Table 1. it is evident that China, Spain, India, and the United States are leading in the area of educational study on the blockchain.

3.3 Literature Keyword Frequency Analysis

Keywords are important indicators to characterize the research content of related literature, and the analysis of keywords can sort out the research hotspots of blockchain applications in education. The keywords that appear more frequently in Web of Science and CNKI are display in Table 2. In addition to the three thematic keywords of blockchain, blockchain technology, and education, the keyword comparison shows that the similarity is that "smart contract" is a common focus of international and domestic research, appearing 24 times in foreign countries and 10 times in China, ranking third and fifth, respectively. The differences are. (1) From the education level, domestic vocational institutions focus more on the application of blockchain technology, while foreign countries focus more on higher education of blockchain technology. (2) Compared with foreign countries, domestic focus on the combination of keywords such as "big data" and "artificial intelligence" indicates that the close combination of blockchain technology with artificial intelligence, big data and other technologies can provide accurate education and smart education, thus promoting the change of education and teaching.

Table 2 High-frequency keywords

Keywords (WOS)	Frequency	Centrality	Keywords (CNKI)	Frequency	Centrality
Blockchain	140	0.37	Blockchain	115	0.76
Education	45	0.32	Blockchain technology	34	0.34
Smart contract	24	0.15	Credit bank	13	0.18
Higher education	19	0.12	Artificial intelligence	12	0.17
Technology	17	0.17	Smart contract	10	0.08
Blockchain technology	15	0.08	Vocational education	7	0.04
Security	9	0.06	Federated learning	9	0.04
Bitcoin	8	0.04	Big data	6	0.06
Ethereum	8	0.05	Common governance	6	0
Internet	8	0.09	Learning outcomes certification	5	0.02

3.4 *Literature Keyword Importance Analysis*

The keyword analysis can yield the evolution pattern, research hotspots, and clustering of existing research. The emergent words in the keywords have emergent nature, reflecting the surge of keyword attention, indicating that the keywords become the research hotspots at that stage; the central words in the keywords have central nature, reflecting the importance of the status of the keywords in the network, and the high-centered keywords can distinguish the clusters in the network; the clustering analysis of keywords can exhibit the clustering relationship of the research, the update of the literature and the interplay. In this paper, we will analyze four aspects of keyword burstness, centrality, clustering, and knowledge evolution of blockchain in education research.

3.4.1 Burstness Analysis

The burstness of an emerging field may foreshadow the direction of the topic, and the results obtained by the emergent algorithm can identify trends in the topic over time. There are six keywords with emergent nature for international studies and four for domestic studies. As display in Table 3, as the table shows that in 2016, domestic research on regional education governance and shared governance showed a proliferation trend, in 2017 foreign research on the Internet of Things began its initial exploration, and in 2018, domestic research focused more on smart contracts, while international research focused more on both authentication and the Internet. in 2019, domestic research focused on the issue of lifelong learning, and the Blockchain

Table 3 Burstness words at home and abroad

Burstness words (WOS)	Burstness strength	Begin	Burstness words (CNKI)	Burstness strength	Begin
Internet of thing	0.95	2017	Regional education governance	0.97	2016
Learning analytics	1.88	2018	Shared governance	0.74	2016
Certification	0.93	2018	Smart contract	1.39	2018
Internet	0.6	2019	Lifelong learning	0.89	2019
Blockchain system	0.75	2019			
Blockcert	0.75	2019			

System and Blockcert become prominent concerns. The keyword burstness nature can reflect the difference in the frontier issues that domestic and foreign scholars focus on.

3.4.2 Centrality Analysis

Intermediary centrality depicts the bridge between 2 unrelated nodes established by a node, and having a high intermediary centrality indicates the importance of the node in the structure. In this paper, keywords with centrality greater than 0.1 (including 0.1) are selected for analysis, as shown in Table 4, international studies focus on blockchain, education, smart contracts, higher education, and technology, while domestic studies focus on blockchain, blockchain technology, credit banking, and artificial intelligence.

Table 4 Central words

Central words (WOS)	Centrality	Begin	Central words (CNKI)	Centrality	Begin
Blockchain	0.37	2017	Blockchain	0.76	2017
Education	0.32	2017	Blockchain technology	0.34	2016
Smart contract	0.15	2018	Credit bank	0.18	2017
Higher education	0.12	2018	Artificial intelligence	0.17	2017
Technology	0.17	2019			

3.4.3 Cluster Analysis

The clustering analysis of the domestic study formed five clusters through the keyword clustering timeline graph, as shown in Fig. 3, with a reasonable clustering of Q value 0.7144 and S value 0.9143 in the clustering timeline graph. The literature co-occurrence keywords are expanded horizontally with the chronology in the figure, and the clusters are arranged along the vertical direction. The darker the color of the timeline and the cluster labels indicate the earlier the start of a cluster, and the starting and ending position and length of the time line indicate the period of the literature in a cluster. The clustering results show that the research on blockchain in education in China is divided into the following aspects: category 0 is federal learning, and from 2017-to 2021, scholars continued to focus on the federal learning problem of blockchain, which includes support vector machines, intrusion detection systems, autoencoders, smart campus, machine learning, classification prediction, patent quality, and patent quality analysis. Category 1 is blockchain technology, which has been a hot issue for research since the emergence of blockchain in 2016; Category 2 is big data, and from 2016-to 2021, based on big data, scholars have conducted research at different levels from the perspectives of education big data, education technology, intelligent education, smart technology, and blockchain platform; Category 3 is credit certification, which originated in 2016, as long as it includes issues such as credit banking, trust system, open education, learning records, qualification framework, results in transformation, mutual recognition of credits and learning outcomes; Category 5 is integration, with a time span of 2016–2020, and mainly includes issues such as online learning, digital badges, learning outcomes certification, learning process assessment, smart contracts, Internet infrastructure and consensus mechanisms.

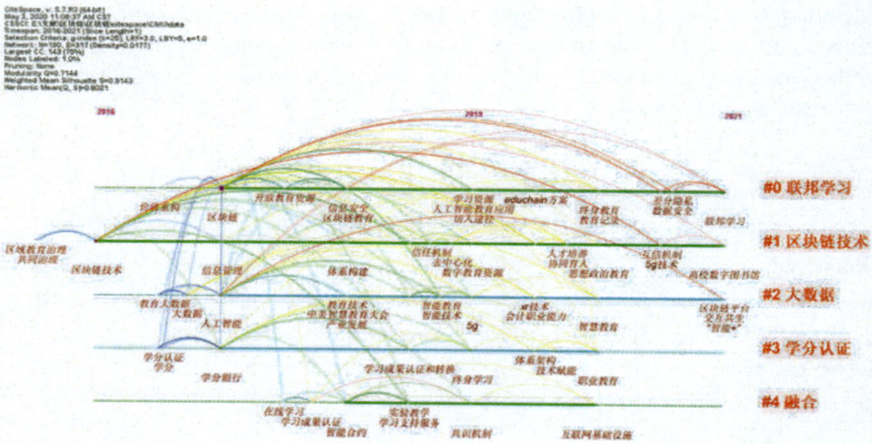

Fig. 3 Domestic research keyword clustering time mapping

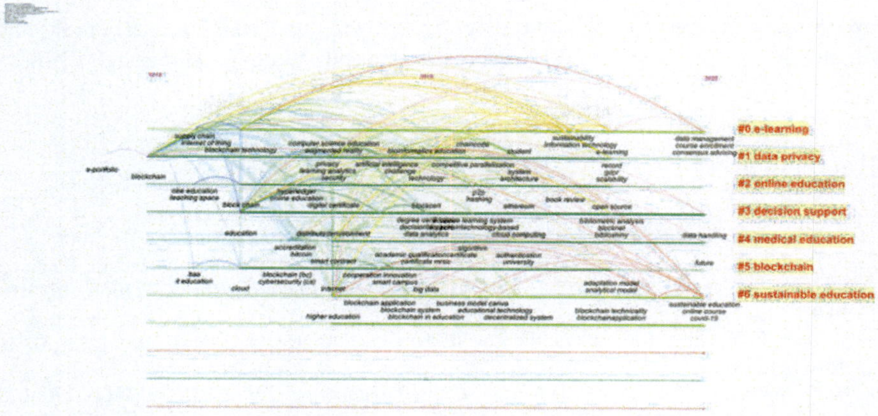

Fig. 4 International research keyword clustering time mapping

The clustering analysis of the International study formed five clusters through the keyword clustering timeline graph, as shown in Fig. 4, with a reasonable clustering of Q value 0.5577 and S value 0.8132 in the clustering timeline graph. The clustering results show that the research on blockchain in education in China is divided into the following aspects: category 0 is e-learning, which mainly includes blockchain technology, course enrollment, hybrid system, scheduler, smart contract, study planner, consensus, system, data management, and so on. Category 1 is data privacy, which includes scalability, informal learning, formal learning, lifelong learning, fog computing, data management, records, and so on; Category 2 is an online education, as we can see, it includes digital certificates, credentials, blockchain technology, digital innovation; content capsule, learning economy, online education, and so on; Category 3 is decision support, it includes decision support, degree verification, data analytics, labor market, curriculum optimization, curriculum optimization, accreditation; learning object, learning process and so on;

4 Conclusion

The aim of this study is to offer a systematic review of the literature, knowledge evolution, opportunities, and trends in domestic and international research on blockchain in education. Using both quantitative bibliometric methods and qualitative literature review, 398 core journal articles from 2016 to 2022 in the field of the enterprise innovation ecosystem, both domestic and international, were reviewed using Citespace software. The main contribution is to outline the emerging themes of research through keyword importance analysis, for example, literature with high emergence can be used to Secondly, quantitative studies of specific periods allow tracking the evolution of themes, i.e., which themes are growing and which are declining; and finally, the analysis captures different research trends and general research themes.

References

1. S. Nakamoto, Bitcoin: a peer-to-peer electronic cash system. Decent. Bus. Rev., 9 (2008)
2. F. Loukil, M. Abed, K. Boukadi, Blockchain adoption in education: a systematic literature review. Educ. Inf. Technol. **26**(5), 5779–5797 (2021). https://doi.org/10.1007/s10639-021-10481-8
3. Y. Yuan, F.-Y. Wang, Blockchain: the state of the art and future trends. Acta Autom. Sin. **42**(4), 481–494 (2016)
4. R. Raimundo, A. Rosário, Blockchain system in the higher education. EJIHPE **11**(1), 276–293 (2021). https://doi.org/10.3390/ejihpe11010021
5. M. Turkanović, M. Hölbl, K. Košič, M. Heričko, A. Kamišalić, EduCTX: a blockchain-based higher education credit platform. IEEE Access **6**, 5112–5127 (2018). https://doi.org/10.1109/ACCESS.2018.2789929
6. D. Amo, M. Alier, F. García-Peñalvo, D. Fonseca, M. J. Casañ, Privacidad, seguridad y legalidad en soluciones educativas basadas en Blockchain: Una Revisión Sistemática de la Literatura. RIED **23**(2), 213 (2020)https://doi.org/10.5944/ried.23.2.26388
7. L. Lan, F. Wu, R. Shi, Visualization analysis of 'Blockchain + Education' research in China–taking 160 core journal papers related to 'Blockchain + Education' as sample literature. Mod. Educ. Technol. **31**(10), 23–31 (2021). https://doi.org/10.3969/j.issn.1009-8097.2021.10.003
8. C. Chen, CiteSpace II: detecting and visualizing emerging trends and transient patterns in scientific literature. J. Am. Soc. Inf. Sci. **57**(3), 359–377 (2006). https://doi.org/10.1002/asi.20317

The Application of Bayesian Network in Higher Vocational English Application Ability Examination

Yiyi Chen

Abstract At present, the status of English application ability test in vocational College is getting higher and higher in practice teaching, and gradually becomes the main means of testing the English level of non-English majors. It can not only accurately judge the students' learning situation, but also provide effective information for English teaching in vocational college. Therefore, in the context of the era of big data, scholars propose to mine valuable content from massive data, so as to provide support services for the decision-making management of higher Vocational English application ability test and meet the needs of higher education for information system. On the basis of understanding the concept of data mining, this paper studies the experimental analysis of classification model by using Bayesian network classification tool and ID3 algorithm according to the theoretical basis of Bayes. The final results show that the model based on Bayesian network can play a positive role in the vocational College English application ability test.

Keywords Bayesian network · Classification model · Vocational English · Ability to test

1 Introduction

Under the trend of economic globalization, Chinese education attaches more importance to cultivating students' English skills, which aims to help students adapt to the market environment of all-round development and contribute excellent talents to the construction and innovation of various fields [1–3]. Especially for higher vocational education, it is necessary not only to get rid of the limitation of traditional teaching ideas, but also to propose professional examination items based on practical teaching content. Higher vocational English application ability test as all students to attend

Y. Chen (✉)
Guangxi Vocational Institute of Technology, Guangxi, China
e-mail: chenyiyignu@outlook.com

S. Patnaik and F. Paas (eds.), *Recent Trends in Educational Technology and Administration*, Learning and Analytics in Intelligent Systems 31,
https://doi.org/10.1007/978-3-031-29016-9_33

and pass an exam, is mainly used to test the students' English learning and application level, it can help teachers adjust teaching plan, and can be positive effects for students to enter society, so is the probe into the main content of the higher vocational English teaching. PRETCO Essentially, The Practical English Test for Colleges and Universities, also called PRETCO, is a proficiency test approved by the Education Department and the Examination board for practical English, where there are two aspects And it can provide effective basis for practicing higher vocational education according to the basic content of the current college English ability test syllabus. This examination needs to be carried out by the local teaching departments, and in line with the teaching examination requirements of adult higher vocational colleges, colleges and universities, and higher vocational colleges. It is usually divided into two types: A and B, and was formally and widely used in 2000. Because the traditional form of vocational English application ability test is too single, need to spend too much time and resources, so under the background in the era of big data, based on data mining technology advantage, reasonable choice of appropriate algorithm model, such not only can quickly classification processing of various data, and can enhance the level of higher vocational English application ability test. In view of the application of data mining technology, it is regarded as the knowledge discovery process in the database. Because it involves multiple disciplines, researchers from various countries have proposed a number of research topics based on visualization, artificial intelligence, database and so on. Theoretically, data mining can operate on different types of data, such as text data, data warehouse, relational database, streaming media data, etc. When dealing with various data sources, different mining methods and tools should be selected to improve the efficiency of data mining. This paper studies the Bayesian network proposed in the Vocational College English Application Ability Test, which belongs to a probability network. Figure 1 below is the simplest Bayesian network structure diagram, because A leads to B, and A and B lead to C, so there are:

$$p(a, b, c) = p(c|a, b)p(b|a)p(a)$$

Probability reasoning is simple, the use of a variety of variable information to other probability information process, based on probabilistic inference of Bayesian network is mainly used to solve the problem of incomplete or uncertain, by processing the complex equipment relevance plays a positive role or the uncertainty caused by the failure, therefore has wide application prospects in the current scientific research to explore.

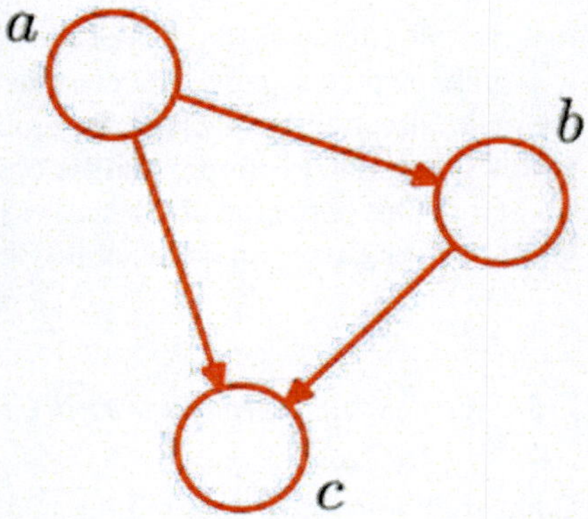

Fig. 1 Bayesian network structure diagram

2 Method

2.1 Data Mining

Data mining is mainly divided into three steps: first, data preparation. In the application ability test of Higher Vocational English, storing large quantities of data information is the basic condition of data mining, which can be obtained from the existing transaction processing system or data warehouse. Second, data collation. Since the content acquired in the data collection stage is not standard and the data itself has defects, it is necessary to generalize the data and obtain richer information on the basis of the original data. Third, data mining. Is to use a variety of data mining methods for analysis; Fourth, evaluation of results. Because part of the data mining results are not practical significance, and there are differences with the actual demand, so it is necessary to verify the correctness of the model according to the test experience or actual data, so as to get more valuable content; Fifth, analyze the decision. The purpose of this operation is to provide effective basis for decision-making of higher vocational English examination [4–6].

2.2 Classification Problems

Classification, as the most common research topic in data mining, aims to find a conceptual description of a category, which represents the overall information of this kind of data. This pattern maps tuples in a database to a given collection of categories. Generally speaking, the process of data classification can be divided into two kinds: on the one hand, it refers to the construction of classification model, which is based on the database tuple described by attributes. This learning process is also regarded as supervised learning; On the other hand, it refers to model application. The accuracy of model prediction should be evaluated first. If the model is considered to meet the requirements, the model can be used to classify and process the data targets with unknown class tags. The classification method based on Bayes technology is the main direction of machine learning research, which not only has strong model representation and learning ability, but also takes Bayes theory as the main basis. Therefore, this paper mainly discusses the application of Bayes network classifier in vocational English application ability test.

2.3 Bayesian Networks and Classification

Compared with other data mining application methods, Bayesian network has the following advantages: First, when dealing with the classification or regression problems with correlation, the correlation between variables is not the main factor of the standard supervised learning algorithm. In the case of missing values of relevant variables, the actual predicted values will have large deviations. However, Bayesian network can obtain more intuitive probability correlation and effectively avoid the occurrence of errors. Second, the integration with other technologies helps to deeply explore the causal relationship and facilitate accurate prediction; Thirdly, data information and prior knowledge can be fully integrated to ensure the effectiveness of network modeling. Fourthly, it can effectively avoid data overfitting.

The Bayesian network based classification model is a probabilistic model, which has the following characteristics in application: Firstly, Bayesian classification cannot provide a target to a certain category absolutely, but obtains the probability of a certain category by means of calculation, and the class with the maximum probability is the class to which the target belongs. Secondly, in Bayesian classification, all attributes have potential effects. In other words, it is not one or a few attributes that determine the classification, but all attributes participate in the classification. Finally, the target attribute of Bayesian classification has the characteristics of mixture, continuity and dispersion.

2.4 Prediction Model

The research content of this paper involves examination subjects and scores in different regions, in which the gender and source of students can be directly obtained, but the scores should be effectively handled according to the grade division of the region, and the students' majors should also be classified according to various types of nature. The selected data in this study contains 2994 instances, 5 conditional attributes and 1 category attribute, all of which belong to discrete values. The category attribute is the classification of English GRADE B, including excellent, pass and fail.

In order to ensure the accuracy and effectiveness of the model, two thirds of the data information should be randomly selected as the training set, and the rest belongs to the test set. Assuming that the category variable is $B = \{b_1, b_2, b_3\}$, the feature variable is $Y = \{A, S, T, E, R\}$, the range of all features is $Val(Y_i), i = 1, ..., m$, and the value of features is $y_i (i = 1, ..., m)$, then for a certain strength, the purpose of classification is to learn a certain training sample set D and define its corresponding category label B. Thus, bayesian classifiers use the following formula Max {P (bk | x)} to determine categories:

$$P(b_i|x) = \frac{P(b_i)\prod_{j=1}^{m} P\left(x_j\middle|b_i; \pi\left(x_j\right)\right)}{P(x)}$$

In the above formula, π (xj) represents all parent nodes of node Yt except class node B, and xj represents the value of the JTH feature of instance X. In this process, the task of learning Bayesian network classification is to learn the probability distribution function in the training sample set, as shown below:

$$P(b_i), P\left(y_j\middle|b_i; \pi\left(y_j\right)\right), i = 1, ..., I; j = 1, ..., m$$

In this model, if all feature variable nodes are independent with respect to category nodes, it can be obtained:

$$P(b_i|x) = \frac{P(b_i)\prod_{j=1}^{m} P\left(x_j|b_i\right)}{P(x)}$$

According to the accumulated experience of vocational College English Application Ability Test, the model obtained from the above research can only be applied to practical work after modification. Therefore, the prediction model of Bayesian network is obtained as shown in Fig. 2.

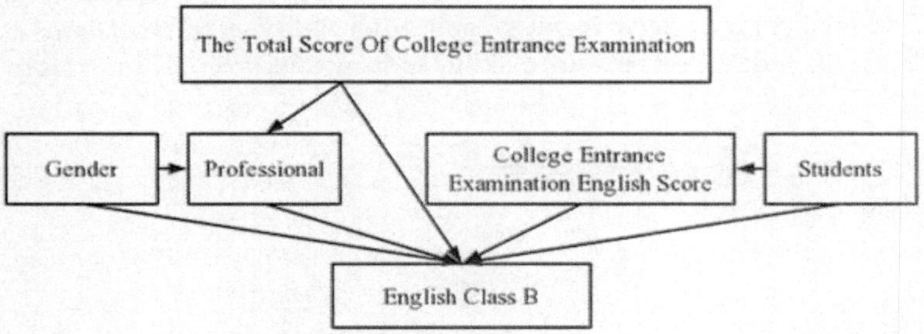

Fig. 2 Prediction model of Vocational College English application ability Test based on Bayesian network

3 Result Analysis

According to the above experimental model analysis, more than 3,100 original data were collected and sorted out. After pretreatment, the evaluation factors and specific indicators of the corresponding model were obtained, as shown in the following Table 1.

According to the analysis of network structure model graph, it is found that all variables studied in this paper are visible, so network training has directness. In order to predict and analyze the categories of unknown samples, P (X | Bi) P (Bi) can be calculated for all categories Bt. Sample X belongs to category Bt and only meets the condition that P (X | Bi) P (Bi).

For B = 1, 2, 3, we can assume:

$$Px1 = P(X|B_1) \times P(B = 1)$$
$$Px2 = P(X|B_2) \times P(B = 2)$$
$$Px3 = P(X|B_3) \times P(B = 3)$$

Combined with the knowledge of classifier, if the calculated result of Px2 is the largest, the predicted result of taking English LEVEL B examination on behalf of sample X is qualified.

In order to better verify the experimental results, this paper uses the Bayesian network toolkit and ID3 algorithm for comparative analysis. The specific flow chart is shown in Figs. 3 and 4.

Comparing the two prediction results, it is found that under the same sample data experiment, the classification accuracy of the algorithm can reach 82.30%, while the accuracy of the Bayesian network classifier can reach 89.03%, which proves that the application efficiency of the latter is higher than that of the former, so it can be used reasonably in the vocational College English application ability test [7–9]. The prediction model of English application ability test of vocational college students based on Bayesian network classification model can guide students to make

Table 1 evaluation factors and index analysis

The variable name	Indicator and value status
Major (S)	Science and Technology (1), Foreign Affairs (2), Literature and History (3)
Gender (A)	Male (1), Female (2)
Student source (R)	City (1), village (2)
College entrance examination English (E)	Good >= 90 (1), Medium >= 60 (2), Poor < 60 (3)
Total score (T)	Good >= 420 (1), Medium >= 350 (2), Poor < 350 (3)
English grade B (B)	Good (1), Qualified (2), Unqualified (3)

Fig. 3 Flow chart of Bayesian network

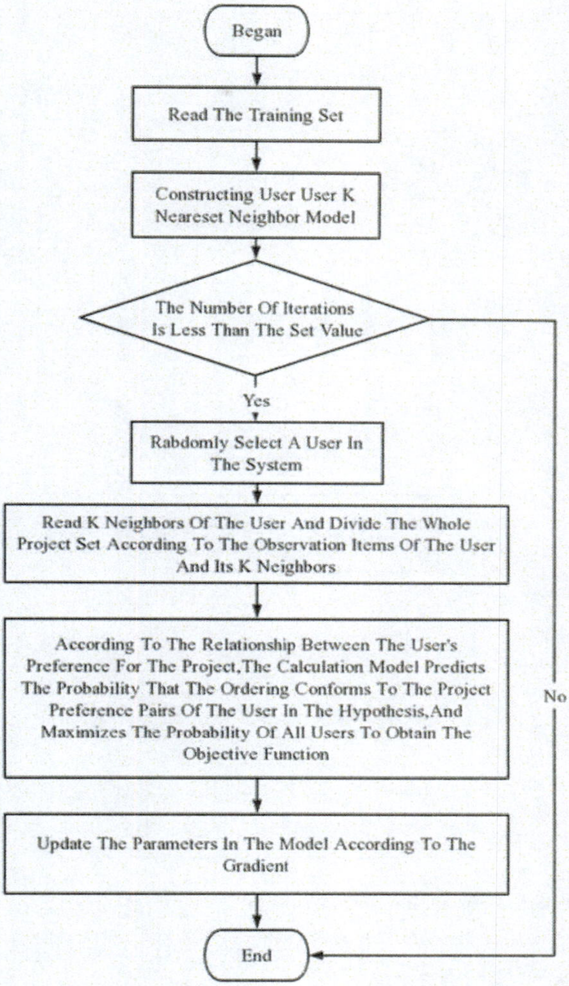

reasonable arrangement according to their own situation according to the knowledge of English application ability test and college entrance examination, thus providing effective basis for educational innovation [10].

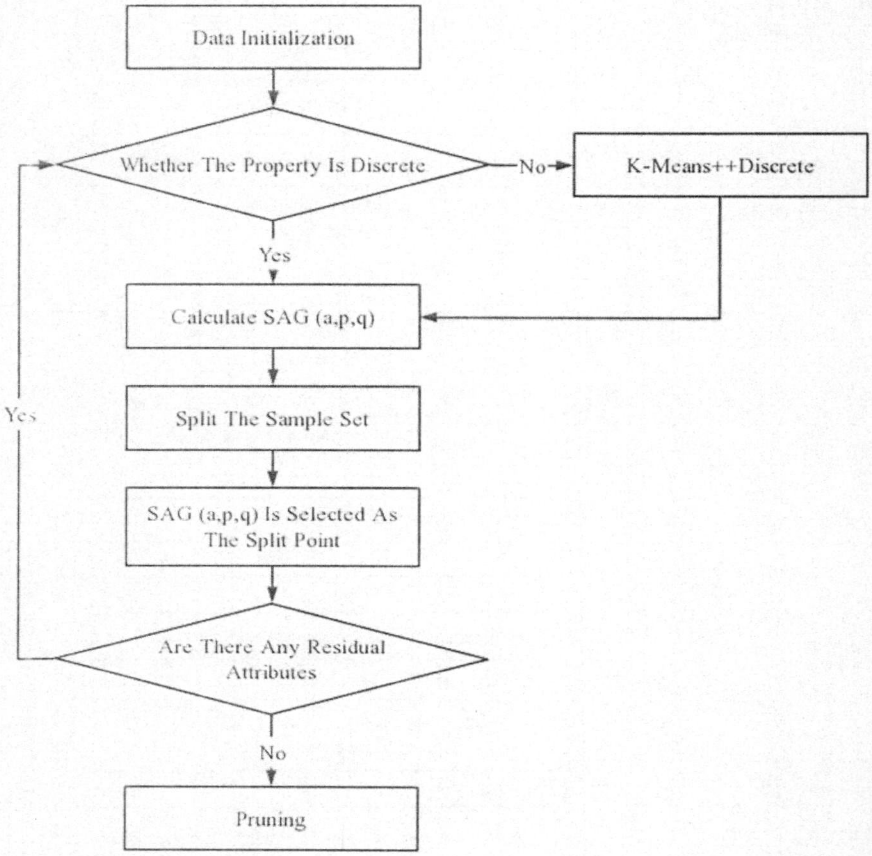

Fig. 4 Flow chart of ID3 algorithm

4 Conclusion

To sum up, on the basis of defining the prediction model of vocational College English applied ability test, this study selected sample data for experimental analysis. After comparative discussion, it was found that the prediction model based on Bayesian network classifier had more application value, and the actual classification accuracy could reach 89.03%. Therefore, in the future technology discussion, researchers should further analyze the application advantages of Bayesian network in higher vocational education, compare and study the classification models in other data mining, and constantly improve the current examination and management models of higher vocational education, so as to better meet the needs of practical education. At the same time, to strengthen the training of technical personnel, and actively introduce advanced technology concept, and constantly adjust the leadership and the ability to English teachers quality, to ensure that they can recognize the data mining technology, mastering the skills used in the technical concept, in order to give full play to the positive role of higher vocational English application ability test.

References

1. J. Li, G. Li, Z. Ding, Application of CS-PSO algorithm in Bayesian network structure learning. J. Meas. Sci. Instrum. **11**(41(01)), 98–106 (2020)
2. F. Chen, P. Xu, Application of ordered Bayesian network in obesity data. J. Changchun Univ. Technol. Nat. Sci. **42**(5), 7 (2021)
3. S. Wang, C. Shi, B. Teng, et al., Application of Bayesian network model in physical examination results analysis. Chin. J. Health Stat. **37**(6), 4 (2020)
4. X. Zhuang, Y. Zhong, S. Zhai, et al., Application of dynamic Bayesian network model in prediction of other infectious diarrhea. Mod. Prev. Med. **48**(9), 6 (2021)
5. Q. Zeng, H. Zheng, S. Wei, Gas turbine health status assessment method based on fuzzy theory and Bayesian network. Sci. Technol. Eng. **20**(11), 7
6. M. Xu, H. Wang, Y. Liang, et al., Application research of computers **39**(2), 5 (2022)
7. Y. Chen, R. Jin, Y.C. Zha, Research on schedule delay risk of large-scale public projects based on Bayesian network. J. Zhengzhou Univ. Eng. Sci. **43**(2), 7 (2022)
8. S. Wei, Q. Zeng, Y. Chen, Bayesian network parameter learning method based on AHP/D-S evidence theory. J. Naval Univ. Eng. **33**(6), 6 (2021)
9. L. Li, H. Qin, J. Can, et al., Study on epidemiological characteristics of adverse drug reactions induced by Traditional Chinese medicine injections based on Bayesian Network and decision tree. Chin. Hospital Drug Eval. Anal. **21**(1), 5 (2021)
10. Z. Tong, T. Lu, Z. Qin, Rolling bearing fault diagnosis based on PSO-VMD and Bayesian network. J. Henan Polytech. Univ. Nat. Sci. **40**(1), 10 (2021)

Study on Learning Method of Logistic Regression Classification for Class Imbalance Problem

Yucai Zhou, Si Chen, Yamei Zhong, and Xiaowen Deng

Abstract In the innovation and development of information technology, the growth rate of data information in various fields is getting faster and faster. How to realize intelligent data processing and extract important information resources from large quantities of data information are the main problems discussed in the field of data mining at present. Data classification is the focus of technical research in the field of data mining. Although the traditional classification method has achieved excellent results in the processing of balanced data sets, in practice, the traditional algorithm does not have application advantages because most of the data combination is unbalanced. Therefore, this article studies in the class imbalance problem and based on the concept of logistic regression, based on the class imbalance problem put forward the corresponding logistic regression learning algorithm, and using the way of data preprocessing, comparison and analysis the RBLR, GBLR, FBLR with traditional logistic regression, owe sample logistic regression, logistic regression index of these three methods. The final results show that the new logistic regression algorithm effectively improves the performance of each index while maintaining the high accuracy of logistic regression.

Keywords Class imbalance problem · Logistic regression · Classification learning · Data mining · Data classification

1 Introduction

In the era of big data, data analysis technology faces new opportunities and challenges due to the increasing demand for quantitative analysis of the objective world. In order to obtain more valuable contents from big data information, researchers

Y. Zhou (✉) · S. Chen · Y. Zhong · X. Deng
Guangzhou Nanyang Polytechnic College, Guangzhou, China
e-mail: zhou278642297@163.com

S. Patnaik and F. Paas (eds.), *Recent Trends in Educational Technology and Administration*, Learning and Analytics in Intelligent Systems 31,
https://doi.org/10.1007/978-3-031-29016-9_34

conducted empirical studies based on quasi-imbalance problems during data analysis, such as credit fraud, troubleshooting, and cancer detection. In essence, this kind of problem is also regarded as a rare class problem, which is characterized by the fact that the number of instances of one class is lower than the number of instances of another class, the former is positive class, and the latter is negative class. Traditional classification learning algorithms mainly use maximum scale accuracy to learn the optimal classifier model, such as cost sensitive, two-stage rule learning, etc. [1–3]. This kind of learning strategy usually meets two conditions: on the one hand, the number components of the training data set are consistent in all classes; On the other hand, the cost of misclassifying each species was roughly the same. It should be noted that in practice, most of the above two assumptions are not tenable. For example, in medical cancer diagnosis and detection, how to correctly identify a few cancer patients in a large number of data is the focus of practical technical exploration; In network attack detection, the probability of network intrusion is only 0.1%, other data collection belong to the normal content, so the technology to achieve 99.9% accuracy, but this is the actual problem and no practical significance, only correctly identify which belongs to the normal communication packets, which belongs to the abnormal communication packets, Only in this way can the security of network communication be further improved. Therefore, the distribution of most data in practical application is not balanced, and the error classification costs of all categories are not consistent. In this context, if we continue to use the traditional classification algorithm, it is difficult to get high accuracy prediction results. Especially in the innovation and development of modern science and technology, the application scope of classification algorithm is more and more wide. Therefore, in order to correctly deal with the class imbalance problem, it is necessary to deeply explore the logistic regression classification learning algorithm and pay attention to the accumulation of experience according to practice to improve the generalization performance of the algorithm. At present, researchers in various countries have proposed a variety of methods to deal with the class imbalance problem, which can be divided into the following types: first, index data pretreatment, second, specific algorithm, third, prediction revision method, and finally, mixed method. The specific model is shown in Fig. 1 [4–6].

In the process of dealing with unbalanced data sets, how to accurately predict a few class instances is very critical, but the traditional statistical classification method only uses the maximum likelihood estimation method to build a learning model, so as to obtain the accuracy of maximum generalization. However, this form is difficult to meet the class imbalance problem, so this paper mainly uses the logistic regression classification learning algorithm for empirical research.

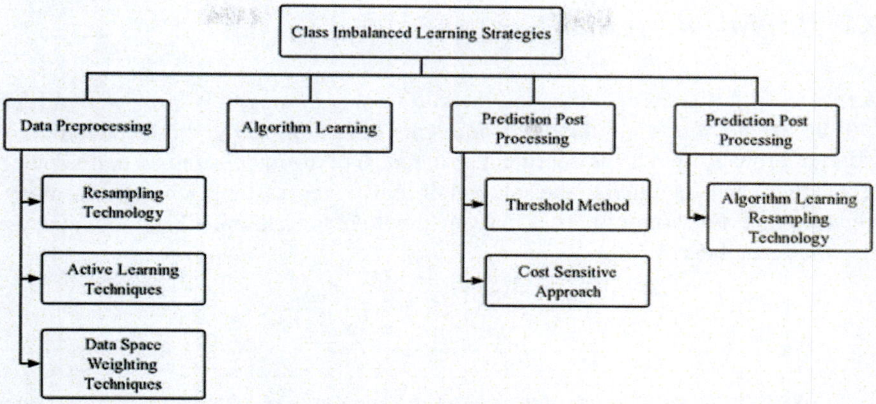

Fig. 1 Model structure diagram of the application method

2 Method

2.1 *Knowledge Analysis*

First, unbalanced data sets. Unbalanced class distribution means that in some data sets, there are too many instances of a class and too few instances of other classes. In other words, the number of instances of different classes differs too much. This kind of problem is also regarded as unbalanced classification. The reason why this kind of problem is difficult to solve is that the class distribution has an unbalanced degree, which is regarded as the ratio of the number of minority class instances to the number of majority class instances. For the balanced data set, the majority and minority classes can provide balanced information for the classifier, and the classifier treats each instance of the class fairly. However, when dealing with the data set with imbalanced class distribution, a few class instances can provide too little information to the classifier, while most classes can provide more information to the classifier, so it is difficult to guarantee the balance of the final processing [7–9].

Second, data classification. Traditional classification algorithms are directly used in unbalanced classification problems, so they can't get excellent results. At present, researchers have proposed two methods to solve the problem of unbalanced classification, one is the application method based on data, the other is the application method based on algorithm. The former mainly uses sampling technology to make the data combination become relatively balanced, and then uses the traditional classification method to learn and process; The most common algorithms of the latter include cost sensitive learning, local clustering and combinatorial methods.

2.2 Logistic Regression

As the most common classification method in data mining, this kind of algorithm will be extended to many kinds of problems according to the classification algorithm of probability. The most critical is that most of the unconstrained optimization techniques can be applied to the solution of logistic regression. Generally speaking, logistic regression algorithm uses line function to fit logarithmic likelihood ratio, the specific formula is as follows:

$$\ln \frac{p(y=1|x)}{p(y=2|x)} = w^T x$$

In the above formula, x represents the instance and W represents the unknown parameter. When selecting the unknown parameter W, it is necessary to ensure that the sum of the probability that the instance belongs to all types is equal to 1. The specific formula is as follows:

$$\sum_{k=1}^{2} p(y=j|x) = 1$$

According to the above formula, it can be obtained:

$$p(y=1|x) = \frac{\exp(w^T x)}{1+\exp(w^T x)}$$

$$p(y=2|x) = \frac{1}{1+\exp(w^T x)}$$

Parameter W can be calculated using maximum likelihood estimation, and the corresponding logarithmic likelihood function is shown as follows:

$$L(w) = \sum_{j=1}^{2} \sum_{i=1}^{N_j} \ln p\left(x_i^{(j)} | y=j; w\right)$$

In the above formula, Nj represents the number of class j instances, x (j) I represents the class label of instance xi is J, which can be used as:

$$p\left(x_i^{(j)} | j=m; w\right) = \frac{p\left(x_i^{(j)}\right) p\left(y=j \middle| x_i^{(m)}; w\right)}{p(y=j)}$$

Therefore, the corresponding objective function is:

$$L(w) = \sum_{j=1}^{2}\sum_{i=1}^{N_j} \ln p\left(y = j \middle| x_i^{(j)}; w\right) + c$$

In the above formula, c stands for constant.

2.3 Objective Function

Generally speaking, after the objective function is clear, the following optimization methods should be used to solve the analysis: first, the fastest descent method; Second, Newton's method; Third, quasi Newtonian method.

Since the traditional logical discrimination directly uses maximum likelihood method to estimate model parameters, it does not really understand the application value of different classes, so this paper will redefine the objective function of logistic regression, and focus on analyzing the application characteristics of different classes. Assuming that the set of class 1 instances is $C_1 = \{x_i | y_i = 1\}$ and the set of class 2 instances is $C_2 = \{x_i | y_i = 2\}$, the following formula can be obtained:

$$P_1 = \sum_{x_i \in C_1} p_{i1},\ P_2 = \sum_{x_i \in C_2} p_{i2}$$

In the above formula, where the condition $p_{ij} = p(y = j | x_i)$ is met, P1 represents the number of instances correctly classified into class 1 and P2 represents the number of instances correctly classified into class 2.

According to this definition, three objective functions studied in this paper are analyzed:

First, recall based objective function (LRM). This kind of function utilizes logistic regression and recall rate to maximize the maximum likelihood estimation while focusing on analyzing the recall rate of a few classes. At this time, the recall rate calculation formula is as follows:

$$R_1 = \frac{P_1}{N_1} = \frac{\sum_{x_i \in C_1} p_{i1}}{N_1},\ R_2 = \frac{P_2}{N_2} = \frac{\sum_{x_i \in C_2} p_{i2}}{N_2}$$

In the above formula, N1 represents the number of instances of class 1 and N22 represents the number of instances. Therefore, the objective function with maximum likelihood estimation and recall rate as the core is shown as follows:

$$LRM = L(w) + r * (R_1 + R_2)$$

Second, the objective function (GBM) based on G-mean. This kind of function needs to be analyzed by using the index G-mean, which is the main content of evaluating the performance of unbalanced classification algorithm, and its specific definition is as follows:

$$g - mean = \sqrt{\frac{n_1}{N_1} \times \frac{n_2}{N_2}}$$

In the above formula, Nj represents the total number of instances of class J, and Nj represents the number of correctly classified instances of class J. Because of monotone consistency, the following definition formula can be obtained:

$$GBM = \ln(J(w)) = \ln p_1 + \ln p_2$$

Third, objective function (FBM) based on F-measure. This kind of function is also analyzed by using indicators, which is a common content to evaluate the performance of unbalanced classification algorithms. Assuming that class 1 represents a minority class, the specific definition formula is as follows:

$$\begin{aligned} f - measure &= \frac{2 \times the.recall.rate \times precision}{the.recall.rate + precision} \\ &= \frac{\frac{2n_1}{N_1} \frac{n_1}{n_1+N_2-n_2}}{\frac{n_1}{N_1} + \frac{n_1}{n_1+N_2-n_2}} \\ &= \frac{2n_1}{n_1 + N_2 - n_2 + N_1} \end{aligned}$$

Meanwhile, the corresponding objective function is:

$$FBM = \frac{2P_1}{P_1 + N_2 - P_2 + N_1} = \frac{2\sum_{x_i \in C_1} p_{i1}}{\sum_{x_i \in D} p_{i1} + N_1}$$

2.4 *Logistic Regression Objective Function Analysis Based on Class Imbalance Problem*

For the three objective functions of the above research, the corresponding algorithms are regarded as RBLR, GBLR and FBLR in this paper. The specific descriptions are shown in Tables 1, 2 and 3.

3 Result Analysis

In this paper, 16 data sets were randomly selected from the machine learning library, and the algorithm performance of data combination was studied by cross-validation of predators. After repeated experiments, 50 models were finally constructed. In order to correctly classify the unbalanced data set, after preprocessing the data set, the information content of the numbers described in Table 4 is obtained.

Before the experimental analysis, two situations were set up in this paper: one was to verify the influence of hyperparameter C in the objective function on the algorithm performance; the other was to compare and analyze the performance comparison results of the algorithm proposed in this paper with traditional logistic regression, oversampling logistic regression and undersampling logistic regression, where the hyperparameter was set at 0.55.

The influence of hyperparameter C on algorithm performance is shown in Fig. 2.

The performance comparison results of various algorithms focus on the analysis of accuracy, as shown in Table 5.

Table 1 Description process of RBLR algorithm

1—RBLR algorithm
The stage of training
Input: D- Training data set
Output: fitting parameter W
Methods:
1. Initialize $w^{(1)}$, given allowable error $\varepsilon > 0$
2. Set $H_1 = In$ (identity matrix) and k = 1 to assign the hyperparameter r
3. Do {
To calculate
$\nabla LRM = \sum_{x_i \in C_1} p_{i2}x_i - \sum_{x_i \in C_2} p_{i1}x_i + r\left(\frac{\sum_{x_i \in C_1} p_{i1}p_{i2}x_i}{N_1} - \frac{\sum_{x_i \in C_2} p_{i1}p_{i2}x_i}{N_2}\right);$
To calculate $p^{(k)} and\ q^{(k)}$;
To calculate H_{k+1};
Make $d^{(k)} = H_k \nabla LRM$
Find the step size λ_k, such that
$LRM\left(w^{(k)} + \lambda_k d^{(k)}\right) = \min_{\lambda \geq 0} LRM\left(w^{(k)} + \lambda d^{(k)}\right)$
$w^{(k+1)} = w^{(k)} + \lambda_k d^{(k)}$
$\}\ while\left(\left\| \nabla LRM\left(w^{(k+1)}\right)\right\| > \varepsilon\right)$
4. Return w

Table 2 Description process of GBLR algorithm

1—GBLR algorithm
The stage of training
Input: D- Training data set
Output: fitting parameter W
Methods:
1. Initialize $w^{(1)}$, given allowable error $\varepsilon > 0$
2. Set $H_1 = In$ (identity matrix) and k = 2
3. Do {
To calculate
$\nabla GBM = \sum_{j=1}^{2} (-1)^{j+1} \frac{\sum_{i \in C_j} p_{i1} p_{i2} x_i}{\sum_{i \in C_j} p_{ij}}$;
To calculate $p^{(k)} and\ q^{(k)}$;
To calculate H_{k+1};
Make $d^{(k)} = H_k \nabla GBM$
Find the step size λ_k, such that
$GBM\left(w^{(k)} + \lambda_k d^{(k)}\right) = \min_{\lambda \geq 0} GBM\left(w^{(k)} + \lambda d^{(k)}\right)$
$w^{(k+1)} = w^{(k)} + \lambda_k d^{(k)}$
$\}\ while\left(\left\|\nabla GBM\left(w^{(k+1)}\right)\right\| > \varepsilon\right)$
4. Return w

Combined with the above experimental results, it is found that the three algorithms studied in this paper can effectively improve the classification performance of a few classes while ensuring the overall accuracy [10]. However, logistic regression in the traditional sense cannot be directly used in the classification of unbalanced problems. Although oversampling and undersampling logistic regression can improve the corresponding values, but reduce the accuracy of processing. Therefore, when dealing with the class imbalance problem, the objective function of logistic regression should be optimized to get a more effective application algorithm [11].

4 Conclusion

To sum up, in the modern social economy develop steadily, with class imbalance problem, promote scientific research scholars studied in application of traditional classification algorithms at the same time, according to the particularity of unbalanced data set, more high-quality processing algorithm is put forward, so that can not only guarantee the accuracy of the identification, also can have more valuable content. As the most common classification method in data mining, logistic regression algorithm can spread to multiple types of problems according to probability

Table 3 Description process of FBLR algorithm

1—FBLR algorithm
The stage of training
Input: D- Training data set
Output: fitting parameter W
Methods:
1. Initialize $w^{(1)}$, given allowable error $\varepsilon > 0$
2. Set $H_1 = In$ (identity matrix) and k = 1
3. Do {
To calculate
$\nabla FBM = \frac{\sum_{x_i \in C_1} p_{i1} p_{i2} x_i}{\sum_{x_i \in D} p_{i1} + N_1} - \frac{\sum_{x_i \in C_1} p_{i1} \sum_{x_i \in D} p_{i1} p_{i2} x_i}{\left(\sum_{x_i \in D} p_{i1} + N_1\right)^2}$;
To calculate $p^{(k)} and\ q^{(k)}$;
To calculate H_{k+1};
Make $d^{(k)} = H_k \nabla FBM$
Find the step size λ_k, such that
$FBM\left(w^{(k)} + \lambda_k d^{(k)}\right) = \min_{\lambda \geq 0} FBM\left(w^{(k)} + \lambda d^{(k)}\right)$
$w^{(k+1)} = w^{(k)} + \lambda_k d^{(k)}$
} $while\left(\left\| \nabla FBM\left(w^{(k+1)}\right) \right\| > \varepsilon\right)$
4. Return w

Table 4 Information analysis of experimental data set

Datasets	Rarity	Features	Instances
Balance-Scale	0.15	5	337
Blocks-Classification	0.01	11	4941
Breast-Cancer	0.17	10	242
Credit-Approval	0.10	16	426
German-Creidit	0.19	21	859
Heart-Statlog	0.16	14	178
Hepatitis	0.14	20	143
Horse-Colic	0.22	23	299
Horse-Votes	0.16	17	316
Ionosphere	0.14	35	263
Kr-Vs-Kp	0.14	37	1936
Letter-Image	0.14	17	942
Pima-Diabetes	0.09	9	551
Promoters-Gene	0.16	58	63
Vehicle	0.21	19	276
Yeast	0.04	9	483

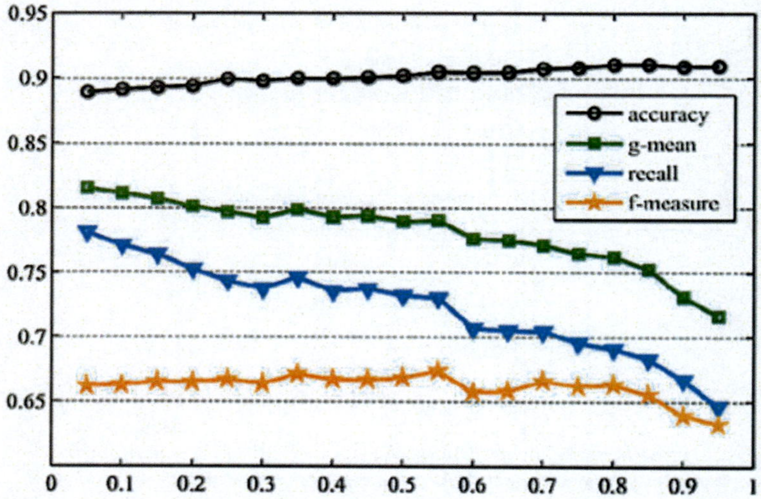

Fig. 2 Influence of hyperparameter C on algorithm performance

when dealing with unbalanced problems. Therefore, most unconstrained optimization techniques will use this kind of algorithm to solve and analyze. This paper studies the characteristics of unbalanced data set based on logistic regression and proposes three objective functions LRM, GBM and FBM, and obtains algorithms RBLR, GBLR and FBLR for unbalanced data classification based on various indexes and traditional logistic regression algorithms. According to the final experimental results, the three algorithms proposed in this paper can comprehensively improve the classification performance of logistic regression on the basis of guaranteeing the classification accuracy. Therefore, when exploring imbalance problems, researchers in various countries should propose more effective processing methods according to the application advantages of logistic regression classification algorithm, so as to not only get more valuable algorithm models, but also explore more data information in deep exploration, so as to provide effective basis for practical application.

Table 5 Comparison results of accuracy of various algorithms

Datasets	RBLR	GBLR	FBLR	LR	OSLR	USLR
Balance-Scale	92.27 (5.60)	92.27 (5.60)	92.27 (5.60)	91.39 (6.43)	92.27 (5.60)	91.10 (5.00)
Blocks-Classification	99.64 (0.21)	99.64 (0.21)	99.64 (0.23)	99.66 (0.21)	92.43 (2.62)	95.56 (2.13)
Breast-Cancer	77.17 (7.82)	72.25 (10.16)	70.18 (9.90)	80.10 (7.13)	58.60 (7.42)	54.07 (8.04)
Credit-Approval	91.57 (4.26)	89.93 (5.15)	90.85 (3.38)	91.78 (3.70)	85.94 (6.04)	75.39 (6.37)
German-Creidit	78.70 (5.25)	70.44 (6.43)	72.77 (5.09)	80.56 (3.94)	65.67 (6.09)	68.92 (7.43)
Heart-Statlog	89.87 (4.43)	79.71 (10.14)	83.69 (9.67)	88.17 (5.63)	79.80 (8.29)	77.48 (8.43)
Hepatitis	85.90 (10.18)	81.00 (10.22)	83.81 (10.23)	83.81 (8.97)	79.80 (8.29)	65.14 (11.40)
Horse-Colic	82.26 (3.92)	83.97 (4.61)	84.62 (5.91)	80.94 (5.20)	79.05 (9.83)	70.62 (8.03)
Horse-Votes	94.92 (2.71)	94.29 (3.89)	95.55 (4.83)	94.27 (3.95)	94.61 (2.18)	89.86 (5.39)
Ionosphere	94.32 (4.08)	93.55 (3.10)	94.33 (3.60)	90.88 (2.63)	91.68 (4.17)	87.07 (6.26)
Kr-Vs-Kp	96.75 (1.24)	94.99 (1.87)	96.38 (0.88)	97.31 (0.72)	96.38 (1.72)	94.32 (1.38)
Letter-Image	96.60 (1.86)	93.41 (2.61)	95.86 (1.46)	97.24 (1.34)	95.01 (2.11)	93.31 (2.55)
Pima-Diabetes	88.01 (3.87)	69.50 (5.37)	85.10 (4.93)	89.84 (2.11)	71.49 (4.11)	73.67 (5.43)
Promoters-Gene	84.29 (19.61)	84.29 (17.97)	74.52 (18.58)	91.19 (18.17)	87.38 (6.74)	72.38 (19.41)
Vehicle	97.13 (2.26)	95.66 (3.30)	97.12 (2.27)	96.03 (8.19)	96.76 (4.60)	96.76 (3.55)
Yeast	97.52 (1.31)	96.48 (2.95)	97.52 (1.63)	97.73 (2.05)	86.12 (6.41)	76.58 (8.16)
Average	90.43	86.96	88.39	90.68	84.5	80.2

Acknowledgements Project name and Number (Yes): Guangdong General University Youth Innovation Talent Project, Project Name: Research and development of Unbalanced classification system based on logical regression No.: 2018GkQNCX118; "Innovative Strong School" youth Innovation Project of Guangzhou Nanyang Polytechnic Vocational College, Project Name: Unbalanced data classification study based on Monte Carlo algorithm, No.: NY-2020CQ1QNPY-11.

References

1. G. Zheng, *Research on Logistic Regression Classification Learning Algorithm for Class Imbalance Problem* (Xinyang Normal University, 2017)
2. J. Guo, *Research on Network Loan Default Warning based on Improved SMOTE Method* (Jilin University, 2021)
3. D. Yang, *Research and Application of Unbalanced Data Processing Algorithm* (Jiangsu University of Science and Technology, 2021)
4. A. Gong, *Mobile Application Risk Assessment Technology based on Multidimensional Android Features* (Nanjing University, 2021)
5. L. Zhang, *Research on Multi-class Unbalanced Classification Algorithm based on Decomposition Strategy* (Hunan University, 2020)
6. S. Du, *Research and Application of Boosting Self-iterative Weighted Ensemble Classification Method* (Northern University for Nationalities, 2021)
7. J.J. Li, *A Kernel Method for Constructing a New Classifier based on Density Ratio Model* (University of Science and Technology of China, 2021)
8. W. Fangfang, *Research on Software Defect Prediction Method based on Active Learning* (Nanjing University, 2019)
9. M.Z. Tang, *Classification Methods and Applications of Data Sets with Unbalanced Classification and Misclassification Costs* (Central South University, 2012)
10. F. Sun, *Research on Classification Method of Urban Unbalanced Data Set based on Machine Learning* (China University of Geosciences, 2020)
11. W. Haomin, *Online Credit Evaluation Method and Application based on Cost Sensitive and Integrated Learning* (University of Electronic Technology of China, 2020)

Verification Analysis and Value Mining of Statistical Report Data of Higher Education

Meng Chen and Xuebo Li

Abstract Under the new situation, in order to ensure the basic guarantee status of educational statistics in educational development, better serve the "double first-class" construction and improve the quality of educational statistics in grass-roots units. High quality education statistics play an important role in understanding and exploring the development law of higher education and promoting the formation and development of new ideas and theories of higher education. As the grass-roots main unit of educational statistics, colleges and universities scientifically carry out educational statistics, verify the logical relationship of statistical data, improve the quality of statistical data, mine the value of statistical data, and transform static statistical data analysis into dynamic detection of school running standards, which is helpful to provide data support for their scientific management and scientific decision-making, Better make due contributions to the connotative and high-quality development of higher education. Through its vertical and horizontal comparison, colleges and universities can better manage. In view of the current problems of higher education statistics, it is of practical significance to find out and improve the verification analysis of education statistics quality and excavate the value of higher education management.

Keywords Higher education · Statistical report data · Check analysis and value mining

1 Introduction

Higher education statistics is faced with problems in the whole statistical process, statistical management system and data value. How to do well in higher education statistics plays an important role in the country, local governments, local education

---

M. Chen (✉) · X. Li
Shandong University of Traditional Chinese Medicine, Jinan, China
e-mail: 56028325@qq.com

S. Patnaik and F. Paas (eds.), *Recent Trends in Educational Technology and Administration*, Learning and Analytics in Intelligent Systems 31,
https://doi.org/10.1007/978-3-031-29016-9_35

competent units and colleges and universities themselves [1]. It is an important part of the management of colleges and universities, a basic work that provides data support for leaders at all levels to make plans, make decisions and guide work, and an important basis for colleges and universities to carry out audit and evaluation, prepare and report enrollment plans, change the key investment direction of funds, and evaluate the management level and school running efficiency of colleges and universities [2]. Statistics is an important tool for understanding phenomena. Educational statistics is an important channel for people to understand education, understand education and manage education. Under the new situation, in order to ensure the basic guarantee status of Education Statistics in education development, better serve the "double first-class" construction and improve the quality of Education Statistics in grass-roots units [3]. High quality education statistics play an important role in understanding and exploring the development law of higher education and promoting the formation and development of new ideas and theories of higher education. Since then, every year, colleges and universities have to make a comprehensive statistics on the majors, enrollment, teachers, student management, assets, infrastructure, equipment, books and materials, which is called the statistics of grass-roots undertakings of Higher Education [4].

Colleges and universities, as the grass-roots main units of educational statistics, scientifically carry out educational statistics, check the logical relationship of statistical data, improve the quality of statistical data, tap the value of statistical data, and transform static statistical data analysis into dynamic testing standards for running schools, which will help to provide data support for scientific management and scientific decision-making, and better make due contributions to the connotative and high-quality development of higher education [5, 6]. The Ministry of Education has uniformly recommended four campus network editions to the whole country, and some provinces and cities have also developed their own network collection systems in due course, and the data collection methods have been continuously optimized. In the past 30 years, education statistics not only objectively evaluated the running conditions of colleges and universities with scientific statistical index system, but also became an important basis for colleges and universities to understand themselves, find out the gaps and make horizontal and vertical comparisons [7, 8]. Therefore, it is an important and necessary work to improve the problems in statistical work and improve the quality and application value of statistical data. Massive data in the data center are interrelated, from the top to the bottom of the data, that is, from national data to provincial data to school data, the Ministry of Education has direct access to every school [9].

2 Current Situation of Educational Statistics in Colleges and Universities

2.1 The Caliber Uniformity of Data Before Filling is not Enough

The operation of big data platform can avoid repeated statistics, reduce workload, manage, mine and analyze data scientifically and effectively, and promote the quality of data statistics [10]. In the statistical work of colleges and universities, in addition to higher education statistics, there are many special statistics organized by different levels of education administrative departments and different functional offices of the same education administrative department. As shown in Table 1.

These special statistical reports are independently designed by the competent authorities according to their needs. There are certain repetitions in content, but there are differences in caliber. This phenomenon increases the workload of grass-roots data collection stage, which is easy to cause unclear ideas of data providers and directly affect the quality of statistical data. Although the application of big data technology may require more funds, in the long run, the introduction and construction of professional big data platform is beneficial and harmless for the development of

Table 1 Comprehensive statistics and special statistics completed by public higher vocational colleges in 2020

Comprehensive statistics	Statistics of educational undertakings in the academic year, annual data collection of personnel training status in higher vocational colleges, and annual monitoring of educational modernization
Party-mass department	Annual statistics of inner-party statistics, investigation and handling of cases by education discipline inspection and supervision institutions, and statistics of complaints and reports
Personnel department	Annual statistics of public institution staff, annual salary statistics of public institution staff and national teacher information management system
Financial department	Annual charge report and charge statistics
Industry-University-Research department	Annual statistics of technology market transactions in colleges and universities, statistics of economic and social development of science and technology and social science services in colleges and universities
Students' League Committee Department	Annual statistics within the group and statistics of school aid data
Educational administration department	Statistics of annual education international exchange data, statistics of university laboratory information in academic year

higher education. Educational statistical indicators refer to the comprehensive quantitative characteristics and specific values of the observation points in the basic table of the higher education grass-roots statistical questionnaire. At present, the statistical data collection of educational undertakings regularly carried out by colleges and universities every year is basically based on the interpretation of various statistical indicators in the statistical questionnaire of colleges and universities (Institutions) prepared by the development planning department of the Ministry of Education [11]. In order to give full play to the powerful function of big data, it is necessary to collect and sort out data in time in the process of statistical data, and maintain the continuous update and improvement of big data platform system, so as to improve the reliability of data statistics and analysis.

2.2 The Authenticity of the Data in the Report Needs to be Strengthened

Accuracy is the embodiment of the quality of statistical data in the objective authenticity of statistical information, and it is a requirement for those who participate in statistical procedures. Data analysis plays a very important role in the development direction and long-term development of higher education. However, many schools have not set up special management departments, and data mining, analysis and collation can not be timely and effective, which leads to the decrease of availability and accuracy of data analysis, which seriously affects the efficiency of data, and is not conducive to data better serving higher education and promoting the development of higher education [10]. The authenticity of data lies in the following aspects: data analysis plays a very important role in the development direction and long-term development of higher education, and the role of this relationship can be divided into direct function and indirect function, as shown in Fig. 1.

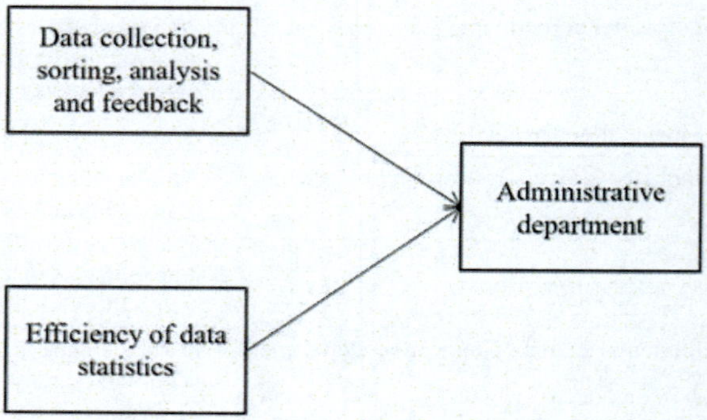

Fig. 1 Value Mining of data authenticity

Firstly, the concept of mutual information is introduced, as shown in formula (1)

$$I[x; y] = \int dx\, dy\, p(x, y) \log_2 \frac{p(x, y)}{p(x)p(y)} \tag{1}$$

where, X and Y represent two associated random variables, that is, the attention of two service value characteristics; P (x, y) is the joint probability density distribution function, which is difficult to calculate, and then it is solved by Monte Carlo idea, as shown in formula (2)

$$I[x, y] \approx I[x, y] = \sum_{X,Y} p(X, Y) \log_2 \frac{p(X, Y)}{p(X)p(Y)} \tag{2}$$

Since Monte Carlo method needs a large amount of data to solve the problem of calculating probability density distribution, and the experimental data just meet the conditions, MIC value can be calculated, as shown in formula (3)

$$MIC[x, y] = \max(|X||Y| < B) \frac{I[X, Y]}{\log_2(\min(|X|, |Y|))} \tag{3}$$

In Eq. (3), the threshold value B is 0.6 or 0.55 power of the total data, and the mic value range is [–1,1]. According to the correlation classification method of the national standard, the following rules can be known: (1) mic = 0, X and y are irrelevant; (2) Mic = (0, ±0.2), very weak correlation between X and Y; (3) Mic = (−0.2, −0.4) or mic = (0.2, 0.4), weak correlation between X and Y; (4) Mic = (−0.4, −0.6) or mic = (0.4, 0.6), moderate correlation between X and Y; (5) Mic = (−0.6, −0.8) or mic = (0.6, 0.8), there was a strong correlation between X and Y; (6) Mic = [±0.8,1], there is a strong correlation between X and y.

In order to ensure the quality of higher education, the Ministry of education has formulated a series of school running indicators and monitoring indicators, such as "teaching administrative room per student", "instruments and equipment per student", "books per student", etc. whether the school running conditions meet the standards can be calculated through the statistical data of High-based statements. Therefore, it is very important to build a special management department. Within the Department, there should also be a clear division of labor, such as data collection, sorting, analysis and feedback, so as to improve the efficiency of big data statistics and better serve higher education. In order to meet their own interests and ensure that the statistical table of school running conditions is qualified, the probability of false reporting, concealment and tampering with statistical data increases, which directly affects the authenticity of data. To sum up, it is mainly divided into basic information indicators, personnel indicators, school running conditions indicators and some other indicators. All kinds of statistical indicators are independent and interrelated. Colleges and universities do not have enough understanding of the importance of statistical work, regard statistical work as a job to deal with, and the filling personnel are arbitrary and changeable. As a result, the authenticity of data is affected due to the lack of professionalism, multiple counting, lack of logic and other reasons.

3 Application Value of Verification Analysis and Value Mining of Educational Statistical Report Data in Colleges and Universities

3.1 Reasonable Design of Statistical Reports and Expansion of Statistical Data Information

In the development of colleges and universities, educational statistics and educational management are interrelated. Relying on the annual statistical report system, improve the report indicators, centrally arrange and collect the same time points and indicators, avoid multiple arrangement and repeated statistics, reduce the workload of statistical reporting units, and ensure the consistency and accuracy of statistical data. Through the statistics of students in different majors, we can see the demand points of market economy construction, which also provides an objective basis for further resource allocation of the school; Through the statistics of teaching staff, we can intuitively see the composition of talents, disclose the deficiencies of talents at all levels, and provide help for further optimizing the talent structure; Through the statistics of assets and fund use, we can clearly and concretely see the basic conditions of the school and the annual investment in all aspects. It is a means to monitor the school running conditions. Expanding the information content of statistical data is a suggestion to the designer of educational statistical statements, that is, the external level of data. The content of Education Statistics stipulated by the Ministry of education mainly includes four categories: comprehensive data, students, teaching staff and basic school running, as shown in Fig. 2.

On the basis of investigating and summarizing the perplexity of grassroots statisticians faced with some statistical indicators in the process of data collection, after

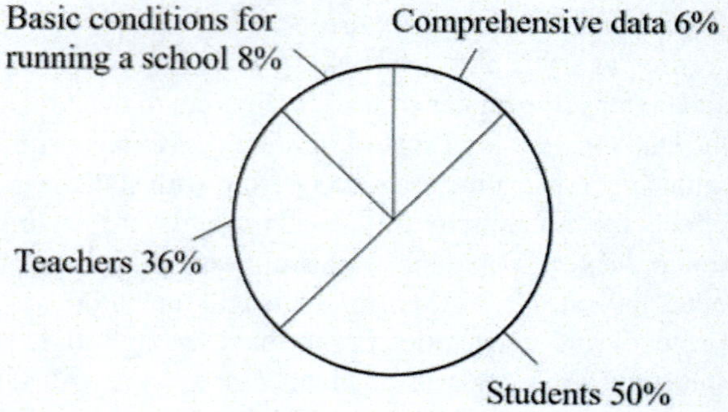

Fig. 2. Proportion of statistical content of higher education grass-roots level

carefully studying the explanation of statistical indicators in the Statistical Questionnaire of Higher Education Schools (Institutions), in order to help grassroots statisticians accurately grasp the connotation of statistical data, collect statistical data with high quality and carry out statistical work smoothly, some flexible data indicators with deviations in understanding are specially interpreted, as shown in Table 2.

With the construction of digital government, especially digital campus, students' information collection system, personnel management system, educational administration system, all-in-one card system, asset management system, financial management system, distance education information system, library management system and so on have been established, which makes the internal data of higher education institutions constantly expand and improve. Expanding the information content of statistical data is a suggestion to the designer of educational statistical report, that is, the external level of data. Promote the construction of comprehensive information processing platforms for education authorities and universities, so that independent resources of universities and internal departments can be transformed into public resources. At present, the mobility and sharing of higher education statistics are far from enough. At present, the social service function of colleges and universities is constantly strengthened. Increasing the statistics of market information and scientific research transformation can comprehensively examine the present situation of colleges and universities, which is conducive to their better development.

3.2 Analyze Data Flexibly and Strengthen the Connection Between Static Data

College education statistics is a report filled in at the beginning of the school year, which is to collect and summarize the statistical data of the total amount of indicators in the operation of colleges and universities in that year. This process does not involve the historical indicators of previous years, so that the data in the statistical information does not have historical continuity. In order to give full play to the powerful function of big data, it is necessary to collect and sort out data in time in the process of statistical data, and maintain the continuous update and improvement of big data platform system, so as to improve the reliability of data statistics and analysis. When carrying out statistical data reporting, if the grass-roots statisticians understand the relationship between relevant base table data, they can predict whether the collected reporting data is reasonable in advance to avoid basic data errors. This article only lists the inter table relationships between some base tables, as shown in Table 3.

In the stage of statistical analysis, we should not only look at the static data of the current year, but also deal with the development data of recent years. We should not only look at the development of a certain piece of special data, but also comprehensively analyze the data situation in combination with all aspects of school data. In the near term, statistical data can improve the quality of decision-making; In the long run, statistical thinking can help managers become leaders of enterprises.

Table 2 Interpretation of statistical data indicators

Serial number	Data indicators	Connotation interpretation
1	A first-year student in school	It refers to the students who have student status and have registered at the beginning of the academic year. This data is greatly affected by the registration time of student status, and the statistical data will be larger than the actual student value. It should be checked with reference to the enrollment number
2	Residential students among students	It refers to the students who live in the student dormitory (apartment) under the unified management of the school. In general, the data should be $\leq$ the number of students in school
3	Number of graduates	Including students who graduated late in the past and graduated early under the credit system. It shall be checked with reference to the expected number of graduates in the previous year's statement
4	Estimated number of graduates	Generally, it should be consistent with the number of students in the graduation grade. It should be combined with the year system and checked with reference to the number of students in the highest grade
5	Enrollment	Generally, it should be $\leq$ the number of first grade students in the school. It shall be cross checked with the number of students admitted in the current year
6	Admission number	Generally, the number of enrollment in the current year should be ≥ 1. It should be cross checked with the enrollment number
7	General preparatory students	Generally, the data of ordinary provincial colleges and universities is 0. It shall be checked in combination with the enrollment policy of the current year
8	Hire off campus teachers	Including retired teachers of the school. Under normal circumstances, the total number of off campus teachers employed should not exceed the number of full-time teachers in the school 1/4。

(continued)

Table 2 (continued)

Serial number	Data indicators	Connotation interpretation
9	Full time teachers are not allowed to teach in this academic year	Normally, this item cannot be 0. It shall be checked in combination with the actual course arrangement of the academic affairs office and other departments

Table 3 Relationship between statistical data tables

Base table no	Statistical indicators	Base table no	Statistical indicators	Relationship
G112	Residential students of this and junior college students	G321	Total of students of this and junior college by age	≤
	Number of participants in physical health test in last academic year	G331	Number of students in the undergraduate and junior college statements at the beginning of the academic year	≤
G311	Total enrollment	G322/G943	Enrollment of general junior college students	≤
G312	Total enrollment	G322/G944	Enrollment of ordinary undergraduates	=
	Total number of students in school	G321	Total of ordinary undergraduate students by age	=
	Number of graduates	G331	Reduce the number of graduates in the number of students	=
G331	Drop out + drop out in reducing the number of students	G332	Total of main reasons for students' withdrawal	=
G411	Number of full-time teachers in teaching staff	G421	Full time teaching teachers + full-time non teaching teachers in this academic year	=
		G423	Total of full-time teachers by age	=
		G424	Total number of full-time teachers by discipline	=
G941/G942	Number of ordinary college and undergraduate students admitted	G943/G944	Enrollment of ordinary college and undergraduate students	≥

The cycle of planning, implementation and inspection can be an effective method to cultivate statistical thinking.

4 Conclusions

Combining higher education with statistical report data, verifying, analyzing and updating the value mining of management mode, so as to make higher education management more scientific and convenient. It reflects the symbiotic relationship between people and science and technology, which not only promotes the development of university management and higher education, but also provides power for the development of big data. Educational statistics objectively evaluates the school running conditions of colleges and universities with a scientific statistical index system, which is an important basis for horizontal and vertical comparison of colleges and universities. This paper first excavates the value of higher education and statistical report data, then verifies and analyzes its value, so that more educators can better understand the help of reading materials for mathematics teaching, and finally analyzes and summarizes its value mining, which has some enlightenment and help for more educators. The quality of statistical data can be improved through two external "establishment" and three internal improvement of data; Rational design of reports, flexible analysis of data and training of statistical thinking can improve the application value of data. These measures can effectively improve the problems in statistical work and have practical significance for higher education management.

References

1. D. Xichen, Technical analysis of educational statistical data quality analysis and verification tool development. Heilongjiang Sci. Technol. Inf. **000**(026), 64–65 (2019)
2. D. Xiaohua, The value, characteristics and application strategies of the material analysis questions in the self-study examination questions of higher education. Mod. Educ. Manag. **000**(002), 53–56 (2019)
3. L. Jianhui, R. Shuili, L. Yongxin, Evaluation and prediction of regional higher education development based on data analysis. Sci. Technol. Promot. Dev. **014**(006), 533–540 (2018)
4. D. Xichen, L. Xiaobing, Analysis of problems in college education statistics and data utilization under the background of information technology. Vocat. Technol. **000**(10), 4–5 (2019)
5. H. Lin, Looking at the construction of the teaching staff of colleges and universities from the statistical data of education **000**(012), 80–84 (2021)
6. C. Jing, J. Cheng, et al., Problems and countermeasures of effective utilization of educational statistics in the background of big data——based on the thinking of basic statistics work of higher education. Value Eng. **36**(v.36, 476), 190–191 (2017)
7. H. Jianjia, Deeply digging system functions to improve data reporting accuracy and high efficiency. Think Tank Times **000**(35), 2–6 (2019)
8. Z. Mi, L. Kebin, Research on the quality management of university education statistics data. Univ. Educ. **000**(10), 3–9 (2020)

9. D. Xiaohua, The value, characteristics and application strategies of material analysis questions in the self-study examination questions of higher education. Mod. Educ. Manag. **000**(2), 4–5 (2019)
10. D. Zhigang, Analysis, mining and effective utilization of big data in university education. China Inf. Technol. **295**(11), 67–68 (2018)
11. R. Liang, Research on the statistics of higher education undertakings. Sci. Consult. **000**(17), 2–3 (2020)

Evaluation Method of Children's Quality and Ability Development Based on Cloud Platform

Gege Liang

Abstract In view of the children's quality and ability development evaluation analysis from multiple perspectives, because individual score will be influenced by subjective factors, so the combination of cloud computing platform is critical to build a comprehensive evaluation system, help each evaluation index is objective, will score vector as input, and effective processing algorithm through the network. Therefore, based on the systematic understanding traditional preschool education during the development of ability quality assessment measure, according to the existing technical framework and project reflect theory to build the perfect evaluation system, and carried out in accordance with the function module partition, children's ability quality is estimated based on the Bayesian posterior expect quality assessment report, build rich information surrounding the core development to design more auxiliary functions. The final system test proved that the system interface screenshots and core codes not only standardized the development of children's ability and quality, but also improved the overall work efficiency of the kindergarten while realizing the system functions.

Keywords Cloud platform · Young children · Ability quality · Evaluation · Posterior expectation estimation

1 Introduction

With the steady development of China's social economy, the number of urban born population is increasing year by year. With the comprehensive popularization of the first Five-year education, more and more parents begin to invest in their children's preschool education. According to the Social Survey, most of the parents surveyed rejected free-range parenting, with 56.5% saying early education was critical. So

G. Liang (✉)
Montelisso Kindergarten, Ezhou City, Hubei Province, China
e-mail: gegeliang@hotmail.com

S. Patnaik and F. Paas (eds.), *Recent Trends in Educational Technology and Administration*, Learning and Analytics in Intelligent Systems 31,
https://doi.org/10.1007/978-3-031-29016-9_36

under the background of information age, in the traditional sense of the early childhood education mode and the development of ability quality evaluation methods have been unable to meet the demand of modern education, which requires combining advanced technology concept to build a new management mode, convenient to real time control of the young children's ability quality management personnel changes, to strengthen the management of communication between parents, teachers and students. At present, the evaluation results of children's ability and quality level development is an important reference standard for early childhood education. It can not only master the comprehensive development ability of children, but also clarify their advantages and disadvantages, so it has received the attention of children's parents and education staff. Under the background of rising social competition pressure, the education industry is also innovating accordingly. At this time, the scientific and perfect evaluation method of children's ability and quality development can build a broader space for children's development, but it also stimulates the research and development of children's ability evaluation system. Starting in the 1990s, the computer information technology has been used in the field of early childhood education, as the further integration of information technology and the education idea and explore, work more and more on the evaluation of early childhood education software construction, national policy also speed up the development of the industry, prompting early childhood education sustained rise in the number of all kinds of software systems. The current market for the development of children's ability and quality assessment system contains two kinds of content, one is to refer to the educational software, the other is to refer to the special assessment system. From the practical point of view, educational software mainly helps children to learn independently at home or at school, providing them with targeted learning directions and programs, while special evaluation systems pay more attention to understanding and improving children's abilities, such as language skills, cognitive level, emotional and social skills, etc. There are also some assessment systems that will analyze the ability and quality recovery of children with disabilities. In this paper, on the basis of systematically mastering the existing assessment system and methods of children's ability and quality development, the cloud platform technology software is used for in-depth discussion, which is helpful to realize automatic management of children's assessment work, and comprehensive assessment of children's growth and ability and quality changes at different stages [1, 2].

2 Method

2.1 Requirement Analysis

In this paper, we study the system all the operations will be made online, so choice B S structure design, the user can through the browser processing part of the page rendering logic, present the related things in front, and the server main things data

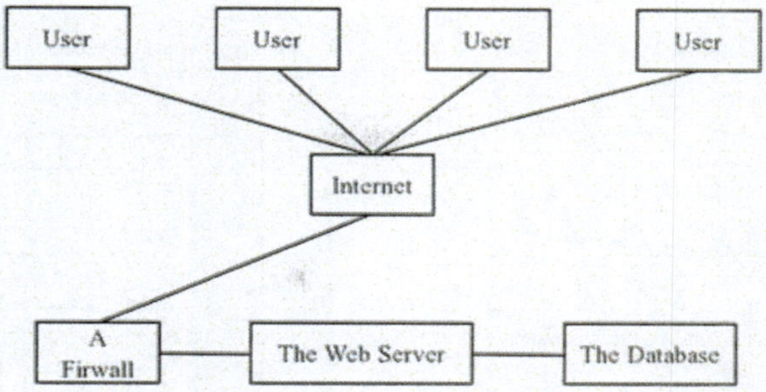

Fig. 1 System topological structure diagram

processing logic, the user can according to children's ability quality measurement system to the system server data operation command, The server will process the request in a timely manner, and then transmit the result to the browser and present it to the user, which can not only reduce the user's operating pressure, but also reduce the user's hardware configuration requirements. The topology diagram of the specific system is as follows (see Fig. 1).

Analysis of the data flow is refers to the use of graphic drawing, to recognize the data from the cloud platform in the development of children's ability quality measurement system directly, the core of the whole operation is to clear all the process of input and output information, help to users and developers to discuss the functional requirements of the system operation, accurate grasp the interaction pattern between the various functional modules, Effectively simplify the application difficulty of the overall design. The data flow structure of the top layer of the actual system is shown as follows [3–5] (see Fig. 2).

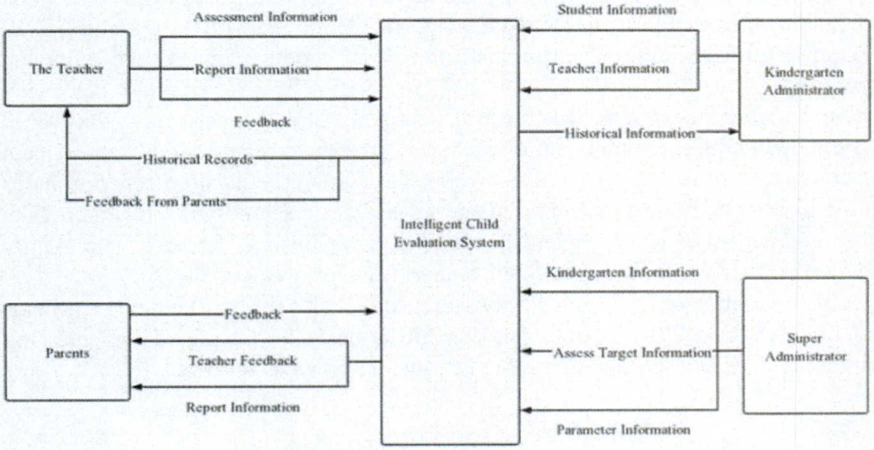

Fig. 2 Data flow structure diagram of the top layer of the system

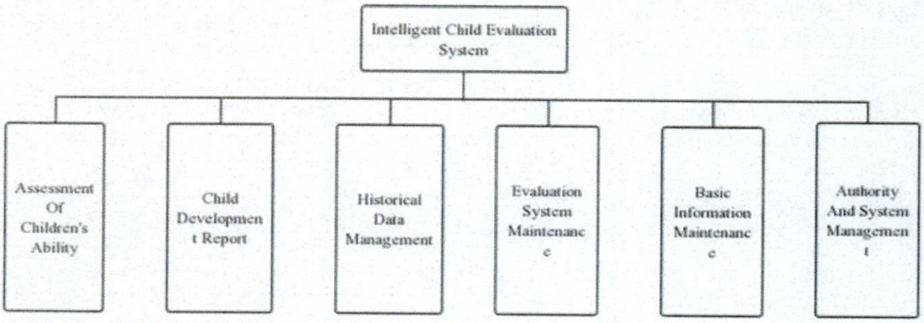

Fig. 3 Overall frame diagram

Combined with the system users and data flow analysis results of the above research, it is shown that the system module can be designed according to the following structure to construct the evaluation method of children's ability and quality development based on the cloud platform, as shown in Fig. 3.

2.2 *System Design*

On the one hand, database model design. Combined with the above system demand analysis results show that the corresponding structure diagram should be constructed according to the entity, attribute and association relationship contained in the system, and the relational database SQL-Server should be used to manage the system data. The structure diagram of the database model is as follows (see Fig. 4).

On the other hand, functional module design. Combined with the structural analysis of the model built above, the main functional modules involve the following contents:

First, ability assessment. This functional module belongs to the core of the overall system operation, and mainly helps teachers or parents to carry out daily assessment of children at different stages. Meanwhile, the home co-education function helps guide parents and teachers to participate in relevant research work together. Take class assessment as an example. The users of this work are teachers, who mainly operate in daily work. The specific flow chart is as follows (see Fig. 5).

The calculation of children's ability score value is also the core content of systematic research. Usually, two-parameter Logistic model is used to regard item guessing parameter as 0, and the corresponding formula is shown as follows:

$$p_i(\theta) = \frac{1}{1 + e^{-Da_i(\theta - b_i)}}(i = 1, 2, 3......n)$$

In this paper, we use bayesian posterior expectation archaic method to study the related values of competence. Assuming that there are n items and the value of tested

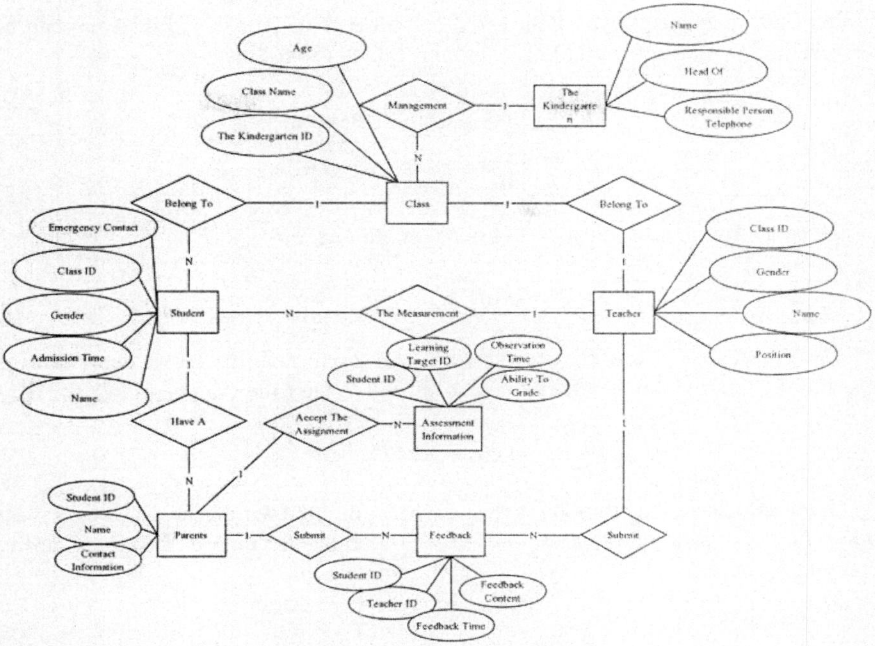

Fig. 4 Structure diagram of system database

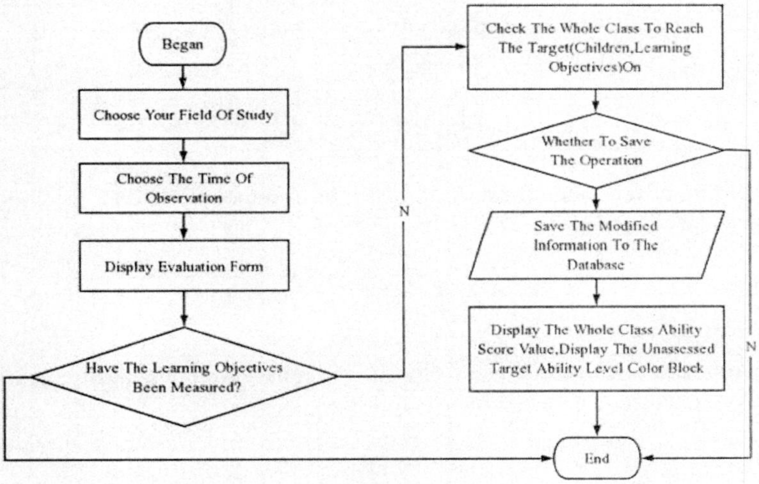

Fig. 5 Operation flow chart with class evaluation as the core

ability is θ, then for a given item I, the response of the subject to the item can be represented by random variable UI, and the value is as follows:

$$u_i = \begin{cases} 1\ correct\ answer \\ 0\ wrong\ answer \end{cases}$$

Thus, the project response matrix can be obtained:

$$U(u_1, u_2, ..., u_i)$$

The probability of correct answer is Pi, and the probability of wrong answer is Qi, and the relationship between them conforms to the following formula:

$$Q_i = 1 - P_i$$

Given that the item parameter is $\xi = (a, b)$, combined with the analysis of Bayes' theorem, it can be found that the posterior probability formula of the tested ability and quality is:

$$P(\theta|U, \xi) = \frac{P(U|\theta, \xi)P(\theta)}{P(U)}$$

Based on the hypothesis analysis of local independence proposed by IRT theory, the following results can be obtained:

$$P(U|\theta, \xi) = \prod_{j=1}^{m} P_j(\theta)^{u_j} Q_j(\theta)^{1-u_j}$$

According to the random selection of subjects subject to ability, the following results can be obtained:

$$P(U) = \int_{-\infty}^{+\infty} P(U|\theta)P(\theta)d\theta$$

In combination with the above formulas, the expectation of capacity parameters is calculated, which can be obtained as follows:

$$E(\theta|U, \xi) = \frac{\int_{-\infty}^{+\infty} \theta P(\theta) \prod_{j=1}^{m} P_j(\theta)^{u_j} Q_j(\theta)^{1-u_j} d\theta}{\int_{-\infty}^{+\infty} P(\theta) \prod_{j=1}^{m} P_j(\theta)^{u_j} Q_j(\theta)^{1-u_j} d\theta}$$

The integrals in the above formula choose the myopia estimation of gauss-Hermite numerical integration algorithm. Bock, Mislevy et al. proposed the specific calculation formula in their study:

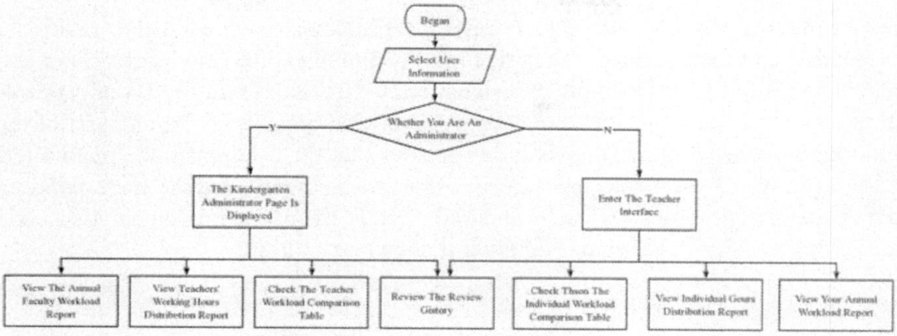

Fig. 6 Flow chart of data management

$$E(\theta|U,\xi) = \frac{\sum_{k=1}^{q} X_k L(X_k) A(X_k)}{\sum_{k=1}^{q} L(X_k) A(X_k)}$$

In the above formula, θ represents the estimated value of the capability parameter being tested, and Q represents the number of points.

Second, growth report. This module helps teachers and parents to jointly understand the growth situation and ability quality of children at various stages. The practical application is mainly divided into two types of reports, one is evaluation report, the other is monthly report of children's growth. It should be noted that teachers can check all children's reports, while children's parents can only check their own children's reports.

Third, data management. This module is mainly used to record the test results of traditional children's ability and quality, and form the historical data analysis report to provide high-quality services for education personnel or management personnel. The specific operation flow chart is as follows (see Fig. 6).

Fourth, the maintenance function. This module mainly helps system managers to build management evaluation system. The evaluation system of the research system in this paper contains multiple learning areas, and each area involves a large number of knowledge scope and learning objectives. And basic information protection function and authority system management function should further maintain and manage the system.

3 Result Analysis

3.1 System Implementation

This paper Outlines that the system should be combined with project reflection theory to study, on the basis of constructing scientific evaluation methods and convenient management operation, effectively solve the problems existing in the traditional evaluation and measurement, strengthen the communication between teachers and parents of children, and promote the evaluation of children's ability and quality development to become more accurate and effective. Combined with the analysis of the function modules outlined above, it can be seen that the user enters the main interface of the system after entering the user name and password, and the corresponding data codes are designed for each function. Among them, the implementation code of children's ability assessment function is shown as follows:

```
Record.Set Assignmentmodifytime(Now);
Record.Set Assihnmentobservetime(Observetime);
Record.Set Classid(Assign.Get Classid());
Record.SetFieldid(Assign.GetFieldid());
Record.SetStudentid(Stuid);
Record.Set Teacherid(Assign.Get Teacherid());
Save Attence(Record);
List<Assignment By Child View>ParamMap.Get(Stuid);
BigDecimal DlScore = BigDecimal.Value Of(Algo Util.Get Ability Score(Param
Algo Util,Level_TYPE_DEFAULT));
Score Tbl Sco=New Score Tbl();
Sco.SetFieldid(Assign.GetFieldid();
Sco.Set Observetime(Observedate);
Sco.Set Studentid(Stuid);
Sco.Set Dlscore(Dlscore);
Sco.Set Capability(Assignmentmapper.Get_EarnedNumByChild(Sco.GetFieldid()
Sco.GetStudentid().Observetime));
If (Score TblMapper.Isexis(Sco.Get Fieldid().Stuid,Observedate)>0);
Score TblMapper.UpdateByPrimaryKey(Sco);
}Else{
Score TblMapper.Insert(Sco);
}
}
```

3.2 System Test

In order to ensure the safe and stable operation of the system, after the completion of the design code, the software test of the overall system should be carried out, so as to grasp the problems existing during the operation of the system. There are two methods of software testing. One is white box testing, which mainly studies code logic, and the other is black box testing, which mainly analyzes system functions. This paper mainly explores the black box testing technology, in other words, the child assessment system will be regarded as a black box, on the basis of not studying how to achieve the internal system, focus on the analysis of whether the system can normally receive output data, whether there are problems. The following table shows the specific results of the class-by-class assessment and system login test (see Tables 1 and 2).

Combined with the information obtained from the above table, it is found that the test task of this system involves every stage of system development and design,

Table 1 Assessment results by class

steps	Operation description	The expected results	The actual results
1	Click to select study area and observation time	The page shows the current observation time of the whole class assessment table, the children's scores are normal	The results were as expected
2	Click the full screen button	The page is displayed in full-screen mode	The results were as expected
3	Click on the learning objective in the evaluation table	The learning object details page is displayed	The results were as expected
4	Check the mark (children, learning objectives) Yeah, click Save	The page is refreshed, and the system displays that the file is saved successfully The scores of evaluated children are updated, and the ability levels of unevaluated children are updated	The results were as expected

Table 2 System login test results

Steps	Operation description	The expected results	The actual results
1	Enter the correct user name, password, and verification code	Normal login to the main screen	The results were as expected
2	Enter incorrect verification code, correct user name and password	Do not enter the system; The system prompts that the verification code is incorrect	The results were as expected
3	Enter the correct verification code: Wrong user name and password	Do not enter the system: The system prompts: The user name does not exist	The results were as expected
4	Enter the correct verification code: user name and incorrect password	Unable to access the system: The system prompts that the password is incorrect	The results were as expected

which requires designers to conduct in-depth research strictly in accordance with the test cases proposed in advance. Although some problems were found during the system test, such as charts and formats, they could be effectively solved in practice and exploration, and the functional modules of the system also reached the expected requirements. Now, all functions of this system platform have been basically realized, and it has been widely used in some kindergarten education management. After a period of trial operation, relatively high-quality feedback information has been obtained. Thus, combined with cloud and big data information technology on the basis of mastering the development characteristics of children's ability quality, using the theory of project reflect build perfect children ability quality evaluation system, not only can reduce the working pressure of the practice education, can guarantee the test work and process become more standardized, chooses the data information is more reasonable [8].

4 Conclusion

To sum up, for the children's quality and ability development of a platform evaluation methods are discussed, in clear to system operation requirements and the overall framework technology, on the basis of the practice and the function module user management is analyzed, and using the Bayesian posterior expectation estimation method for children's ability quality estimation calculation, not only can get accurate measurement results, It can also provide effective information basis for practical education management. In technological innovation, developing new era, the evaluation method of children's quality and ability development and research work will develop towards mobile end, such as tablet, mobile phone and other mobile devices are the basis of the test platform as a carrier, such not only can further promote the measurement system and related work, can also according to different cases of children's abilities to conduct a comprehensive analysis. Therefore, relevant researchers should continue to study the evaluation method and application system of children's ability and quality development based on cloud platform on the basis of actively learning advanced scientific and technological concepts.

References

1. L. Shi, F. Yang, H. Wang, Research hotspot and prospect of artificial intelligence application in preschool education: based on CiteSpace knowledge graph analysis. Early Educ. **43**, 7–10 (2021)
2. B. Li, Dance skills training strategies for preschool teachers under the Background of Internet +. Chin. Writ. Artis. (08), 109–110 (2021)
3. X. Yaolin, Three "clouds" to promote regional education quality–practice and exploration of pre-school education informatization in Haicang District, Xiamen City. Fujian Educ. **29**, 30–31 (2021)

4. L. Chen, Z. Zhang, Educ. Commun. Technol. (03), 15–18 (2021)
5. Q. Li, Research on the path of "Internet +" application in kindergarten care service. Educ. Obs. **10**(08), 55–57 (2021)
6. M. Kuang, Micro-curriculum construction of children's art activities in intelligent environment. Early Educ. **07**, 12–13 (2021)
7. Y. Gao, Using smart Education cloud platform to effectively improve the level of kindergarten teaching management. Fujian Commer. Trade Assoc. **209**, 4. (in Chinese)
8. X. Hou, The company of special time–thoughts on family education in holidays. Parents **01**, 183–184 (2021)

Research on the Application of Student Information Management System in the Management of Higher Vocational Colleges

Jiasi Huang

Abstract As an effective countermeasure to ensure the quality and safety of the educational environment, the management of higher vocational colleges, combining information technology to construct a student information management system, is an important subject of practical innovation research, which directly determines the internal management level of higher vocational colleges and the integration of students Quality. Generally speaking, the management of college students involves education, basic management, rewards and subsidies, employment and accommodation, etc. As the overall work is more complex, it is difficult to implement informatization construction. After understanding the current problems faced by higher vocational colleges during the education management period, this article discusses in depth how to build a diversified system module that meets the development needs of the times based on the construction of the student information management system. The final use results prove that the use of student information management system during the management of higher vocational colleges can solve the problems of traditional education management on the basis of improving overall work efficiency.

Keywords Higher vocational colleges · Education management · Student information management system · System requirements

1 Introduction

The student information management system of higher vocational colleges is also called SIS system. It is an indispensable content during the period of practical education management. Especially under the conditions of continuous development and expansion of higher vocational colleges, the number of students on campus continues to rise. On the basis of systematically perfecting and mastering student information,

J. Huang (✉)
Guangdong Mechanical & Electrical Polytechnic, Guangdong, China
e-mail: gas-wong@163.com

S. Patnaik and F. Paas (eds.), *Recent Trends in Educational Technology and Administration*, Learning and Analytics in Intelligent Systems 31,
https://doi.org/10.1007/978-3-031-29016-9_37

it is important for the development of student management to quickly retrieve the content of the forms needed for the work during the practical education management. The construction of a student information management system can help relevant staff to query and store basic information. In the traditional sense, student information management is mainly processed by manual or basic software such as Excel tables. There are many problems in practical operation, such as poor integrity, low application efficiency, and cumbersome practical operations. These problems will affect student information management. Quality constitutes a direct impact. It can be seen that in the context of the development of the new era, scientific research scholars must use network technology to build a corresponding management system, so as to obtain required data anytime and anywhere in fast retrieval and convenient screening. This will not only reduce costs, but also improve practice. Work efficiency and quality. At the same time, the use of unified data interfaces can enhance the linkage of data and information. For example, the educational administration management system and the student information management system can directly transmit basic information such as student performance and daily performance in an effective connection, thereby comprehensively strengthening. The management level of higher vocational colleges. Since its development in the mid-1950s, computers have been widely used in management work. Because computers play an active role in information processing, scientific research scholars begin to construct corresponding data models according to the management development needs of higher vocational colleges in practical exploration, and start from the perspective of overall development to help managers make effective decisions. From the perspective of practical development, foreign information management systems have gone through three stages: First, it refers to the single data processing stage, also known as the electronic data processing stage. At this time, the development conditions of computer software and hardware are limited, although data processing Computers can be used to replace manual labour, but only simple data processing, such as calculation and statistics registration, etc.; secondly refers to the data processing stage, also known as the transaction processing system stage, when the computer software and hardware are all obtained Comprehensive improvement, the computer began to be used in the control subsystem, and has the feedback function, the corresponding technical methods began to face the terminal for online real-time processing; finally refers to the management information system stage, at this time the need to use computer management of various subsystems Construct a comprehensive computer information system, and realize the sharing of data and information based on operations research and data methods [1]. Based on the current management situation of higher vocational colleges, this article deeply discusses the student information management system, which helps to build a real-time and unified data management system on the basis of getting rid of the stand-alone operation mode. This can not only improve the efficiency of practical work, but also form a systematic student information management work [2, 3].

2 Method

2.1 Demand Analysis

The current student information management system constructed by the management of higher vocational colleges must meet four requirements: first, it refers to the management personnel of the school and the Youth League committee, which mainly handles basic information management, remote log-in office, and student psychological communication; secondly, it refers to students. Mainly inquire about personal information, understand school announcement documents, download relevant learning materials, etc.; again, refer to staff of each department, mainly understand the latest announcement content, upload relevant reports and data, download information materials, and inquire about the learning status of students in the department; Finally, it refers to system administrators, who mainly manage work information, backup and restore data, and maintain and manage system security [4, 5].

2.2 System Design

The system structure of the student information management system in higher vocational colleges should be clarified according to the object, development status, internal and external process relationships of the application software. According to the analysis of the system development flowchart shown in Fig. 1, the main content is divided into the following three modes: First, it refers to the stand-alone mode. The software design of this mode is mainly used to process personal affairs and help the application system prepare various basic information, such as multimedia materials, data materials, text forms, etc.; secondly, it refers to the client (server) Mode, the client runs the user service request program and transfers relevant content to the server, while the server manages data resources, responds and processes the request made by the client, and finally transmits the calculation result to the client. In this model structure, the system divides the computer and the application system into two parts based on the local area network. On the one hand, it refers to the server, which mainly provides the required functions and data for each application module. On the other hand, it refers to the customer part., Which mainly provides users with interface, logic processing and various resources; finally refers to the browsing server mode, which is suitable for wide-area and larger information sharing, combined with the analysis of the software configuration structure diagram shown in Fig. 2. The client only needs to install a browser to access multiple server software distributed in the network. The browser interface is uniform, which helps keep training time and costs to a minimum. At the same time, the hardware and operating system of the client only need to support the browser software, and the actual application time is longer [6, 7].

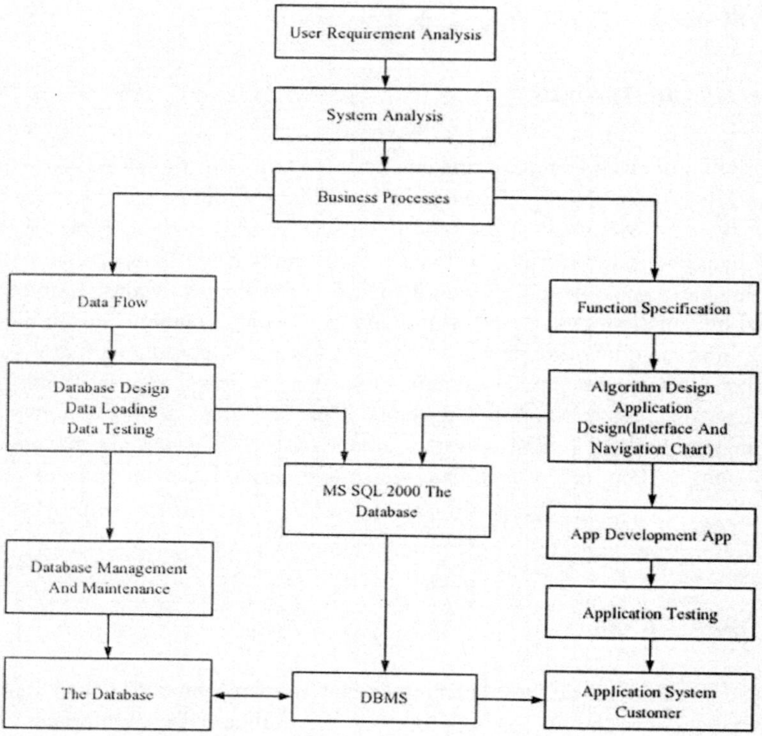

Fig. 1 Flow chart of system development

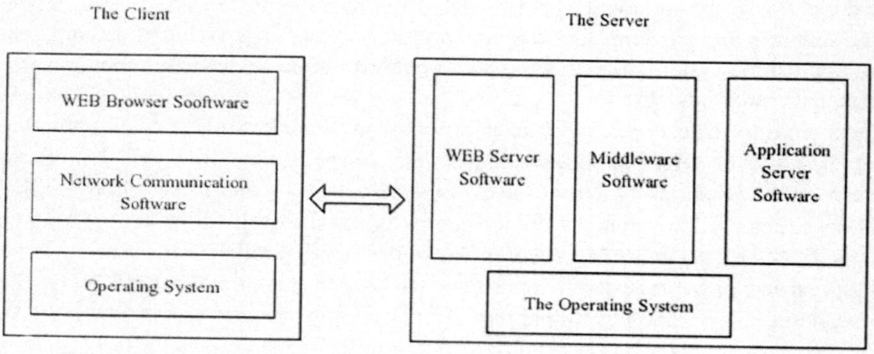

Fig. 2 Mode structure diagram of browsing server

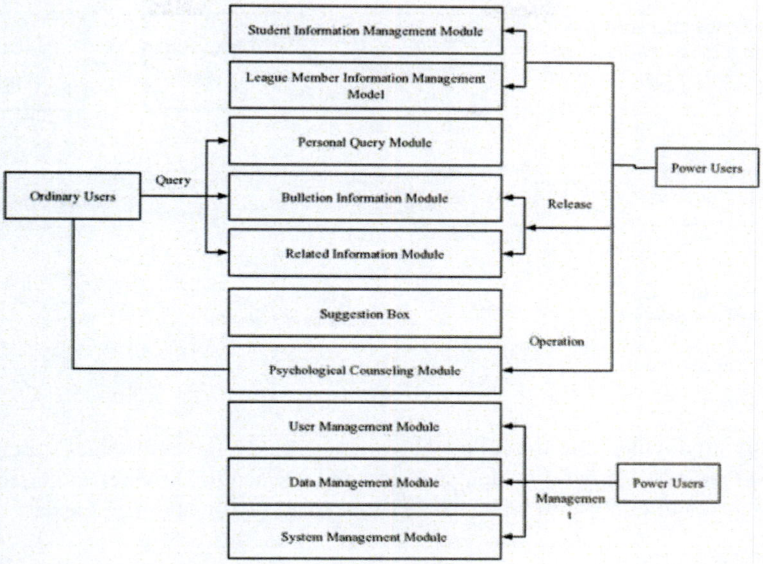

Fig. 3 Student information system module in higher vocational colleges

2.3 Functional Design

According to the above-mentioned demand analysis, it is proved that the basic requirements and data flow of the practical system are clear, and users at various levels have different permissions and tasks. According to the analysis of the higher vocational student information system module shown in Fig. 3, it can be seen that the front-end page contains two types of content. Any verification; while advanced users enter the internal management module page after login authentication, which involves students, announcement information, psychological consultation, data management, etc. Super users, that is, system administrators must pass identity verification to enter the user management page, data management page, and system maintenance management page [8, 9].

2.4 Database Design Security

As the core content of the student information management system, the database will organize the large quantities of data within the system according to a certain mode, and design various functions such as retrieval, maintenance and storage, so that the information system can be quickly and conveniently obtained during the operation of the time. Combined with the analysis of the database conceptual structure design diagram shown in Fig. 4, it can be seen that the design steps of the practical logical

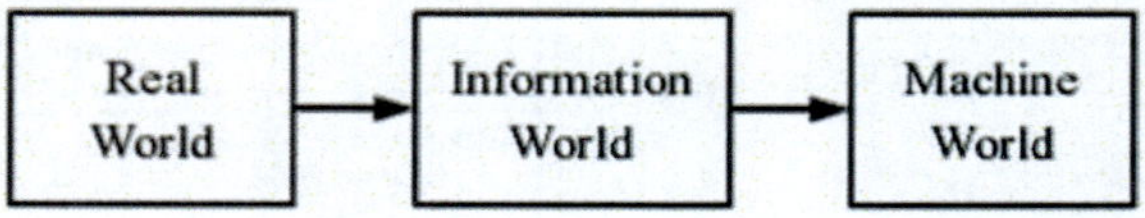

Fig. 4 Conceptual structure diagram of the database

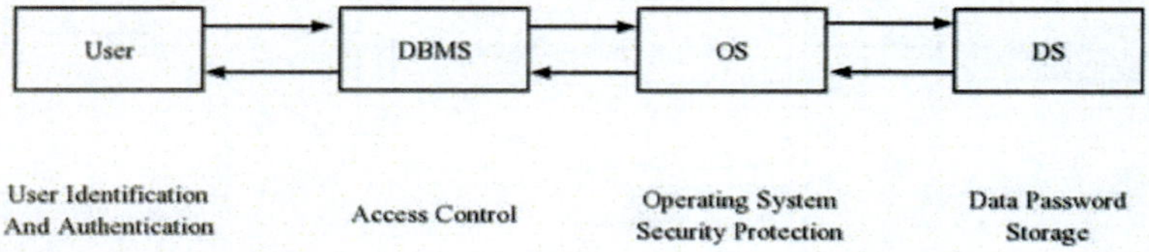

Fig. 5 Security model diagram of computer system

structure are divided into the following contents: firstly, the conceptual structure is transformed into a general model, and secondly, the acquired model is transformed into data supported by a specific DBMS Model, and finally optimize the data model [10].

Considering the security of the database helps to ensure that there will be no unauthorized data access at different levels of the information system. Combined with the analysis of the computer system security model diagram shown in Fig. 5 below, it can be seen that the computer system security measures are designed level by level. Before entering the computer system, users must complete identification according to the input user ID. Only users who meet the requirements can Users who have entered the computer system and have entered the system must use the DBMS for storage control and can only perform system operations that meet the requirements. The operating system and its own protection countermeasures, the stored data will be stored in the database in the form of a password, the most common logical security mechanism involves data encryption, tracking review, user authentication and other content.

3 Result Analysis

3.1 Environment Configuration

The system development outlined in this article builds the Zydatabase database as a back-end storage container, and puts all data information tables in the Zydatabase database. The correlation between the data information tables is not strong, so the data information tables will become independent content during development and design. In order to ensure the safe operation of the system, you can use the ODBC data source connection form to connect, define the name of the data source as Zydata,

Table 1 ASP program

Conn=Server.Create Object("ADODB.Connection")
Conn.Open("DSN=Zydata;UID=Sa;PWD=Sa")

Table 2 The actual program

SqlStr="Select*From Sihnaltab Where Code Like'%X%'"
Rs=Conn.Execute(SqlStr)
Use the RecordSet properties and methods to display the results

and connect the database as Zydata. The ASP program to open the database is as Table 1 follows.

After executing SQL commands to connect to the database, you can perform basic operations on the database. For example, in the data table signaltab, the query code contains X records, the actual program is as Table 2 follows.

3.2 *Function Realization*

First, the verification of user login. The realization of this function requires the user login verification interface to receive the user login information submitted by the page form, and directly compare and verify the content with the advanced user information table in the database.

Second, student information management. This system mainly authenticates the users who meet and submit. If you are already logged in, you need to browse this interface, otherwise it will be redirected to the system login interface. The specific code is as Table 3 follows.

Third, the public information system. This module is divided into two parts. On the one hand, it refers to the information column that ordinary users can directly view, and on the other hand, it refers to the information management module that advanced users enter after passing identity verification.

Fourth, psychological counselling module. The realization of this module function should be realized by means of CGI, Applet, etc.

Fifth, data management. The data upload of this functional module is completed by the file upload component active file, and the data title and attribute content are

Table 3 Code

If Isempty (Session("ID"))Then
Response.Redirect"../Login.Asp"
End If
Establish a connection with ODBC data source
Set Conn=Server.Creatteobject("Adodb.Connection")
Conn.Open"Zydata","Sa","Yourpassword

stored in the data table of the database. The specific content of the data is uploaded to the data folder of the system, and all contents will be in different forms. store.

Sixth, system management. In the main page, assuming that the SendTo value is System Manager, then it means that the user has to enter the System_manage.asp in the system management page. At this time, the system security should be ensured on the basis of judging the user's identity; if the field is True, then the user belongs to the system Managers can directly enter the management interface, and will provide a parameter System Order to enter the system management interface [11].

According to the above functional design and implementation, the verification analysis found that when the system has security problems, the data information can be regularly backed up and restored to ensure its integrity, and the data management of advanced users can only be processed within the authorized scope, which is important Information resources should be backed up regularly. This shows that the use of student information management system in the management of higher vocational colleges has a positive effect.

4 Conclusion

In summary, according to the current accumulated experience analysis of the management work of higher vocational colleges, we will continue to optimize the design content of the student information management system, focus on innovating the realization of functions in combination with the existing technology concepts, and build a fully functional and highly secure student information Management system is the main topic discussed by education management and scientific research scholars at this stage.

References

1. C. Mao, Design of student information management system in higher vocational colleges based on RFID technology. Inf. Rec. Mater. **22**(09), 167–168 (2021). https://doi.org/10.16009/j.cnki.cn13-1295/tq.2021.09.078
2. Y. Wang, Problems and countermeasures of student information management system in higher vocational colleges. Sci. Technol. Innov. Her. **16**(35), 165–166 (2019). https://doi.org/10.16660/j.cnki.1674-098X.2019.35.165
3. M. Li, Research and development of higher vocational student information management system. Digit. Technol. Appl. 37(08), 81+83 (2019). https://doi.org/10.19695/j.cnki.cn12-1369.2019.08.40
4. B. Zhang, Problems and countermeasures of student information management system in higher vocational colleges. Comput. Prod. Circ. (06), 214 (2019)
5. Y. Cui, The design and implementation of the information management system for student funding in higher vocational colleges based on the WEB platform. Firew. Technol. Mark. (02), 203 (2019)
6. J. zhao, Y. Sun, X. Ma, Problems and countermeasures of student information management system in higher vocational colleges. J. Qingdao Vocat. Tech. Coll. **30**(04), 33–37+82 (2017)

7. W. Wang, *Design and Implementation of Student Information Management System in Higher Vocational Colleges Based on Web* (Beijing University of Technology, 2017)
8. L. Wang, *Research and Implementation of Student Information Management Platform based on LAMP Architecture* (North China University of Technology, 2017)
9. Z. Liu, *The Design and Implementation of the All-in-One Card Student Information Service Management System of Sichuan Normal University Vocational College* (Beijing University of Technology, 2016)
10. X. Zhu, *Design and Implementation of Student Information Management System in Jiujiang Vocational University* (University of Electronic Science and Technology of China, 2016)
11. H. Chen, The design and application of comprehensive information management system for students in higher vocational colleges. Sci. Technol. Outlook **26**(24), 150+152 (2016)

Construction of Industry-Education Integrated Ecosystem of Vocational Education Based on Computer Artificial Intelligence

Luyan Dong

Abstract In order to improve the ecological evaluation level in practice education under the background of advanced modern man–machine intelligence, and enhance the ecological construction ability of the combination of industry and teaching practice. in practice and engineering education under the background of advanced modern man–machine intelligence, a design method of the combination of industry advantages of educational resources engineering education based on association rules is proposed. Based on JBPM (JBoss Business Process Management, JBPM) workflow engine framework, this paper designs an integrated ecosystem of production, education and practice and engineering education under the background of advanced modern man–machine intelligence, the information characteristics of advanced modern man–machine intelligence is analyzed by using association rule mining algorithm, and realizes the system integration structure of advanced modern man–machine intelligence by adopting the method of comprehensive practice system construction of data exchange and collaborative office and ecological construction integration is taken. A variety of web data formats are adopted to realize the data loading and program compiling control of the integration of production, and it is connected to the practice and engineering education and employment management system of the school to realize network management. The test results show that under the background of advanced modern man–machine intelligence, the integrated ecosystem of production, education and practice and engineering education has high stability and reliability.

Keywords Computer technology · Artificial intelligence technology · Education · Integration of production and education · Ecosystem · Comprehensive practice

---

L. Dong (✉)
Shandong Vocational and Technical University of International Studies, RizhaoShandong Province 276826, China
e-mail: dongluyan@163.com

S. Patnaik and F. Paas (eds.), *Recent Trends in Educational Technology and Administration*, Learning and Analytics in Intelligent Systems 31,
https://doi.org/10.1007/978-3-031-29016-9_38

1 Introduction

In terms of specialty setting, talent training scheme, technological innovation and achievement transformation. According to the understanding of the concept of practice base, it is known that there are three main modes of practice base at present, namely, on-campus practice base, public shared practice base and off-campus practice base. First of all, the practice base on campus is the basic environment for college students to enter the social post practice training in the future, so that students can get theory combined with practice teaching, a certain skill training, basic ability training, etc. [1]. Secondly, the practice base of public sharing is based on the sharing of resources in the region, and more professional vocational training is established on the basis of basic theoretical knowledge and practical operation study on campus. Only when schools and enterprises cooperate, can they complement each other to achieve win–win cooperation. Finally, the off-campus practice base is the real enterprise work environment for students, and the projects that students handle are the real work projects to be carried out in professional posts, so they can learn and practice from strict vocational training. Off-campus practice base is an extension of on-campus practice and public shared practice, which constitute a complete teaching system [2].

In order to enable students to be immersed in their professional posts and realize the zero distance between the training process and their jobs, which not only cultivates students' learning and practice ability, but also enables students to quickly solve and flexibly respond to problems in the actual production process, so as to achieve the deeper teaching goal of strengthening their professional quality and ability. At present, there are some problems in China's human capital investment that restrict the development level of social productive forces, such as insufficient total investment, perfect structure and large regional differences. Therefore, we must increase the investment of human capital. As a major way of investment in education, human capital is defined as: the expenditure of enterprises and the expenditure on educating or training students between schools, and this expenditure cost is human capital. Higher vocational colleges are an important base for training technical and skilled talents and an important way to realize the value-added of human capital. Many researchers also study this part of the content. Tian [3] used literature research, logical reasoning and other methods to explore the internal and external factors of the system, analyze the constituent elements of the system, and build a college innovation and entrepreneurship education ecosystem model based on vocational education at four levels, including the construction of innovation and entrepreneurship education teaching objectives, to provide theoretical guidance for training high-quality innovation and entrepreneurship talents and technical skills. Zhang [4] established an information platform integrating industry and education to provide data reference services for teaching, so as to build a smart education ecosystem for vocational education.

2 Overall Design and Module Structure Analysis

At present, there are some problems in China's human capital investment that restrict the development level of social productive forces, such as insufficient total investment based on behavior interaction, combining with the visual parameter analysis of advanced modern man–machine intelligence [5], through the collection of environmental parameters and physical information parameters, the association rule mining analysis model of the ecological construction of industry-education integration in practice and engineering education under the background of advanced modern man–machine intelligence is constructed. Combining with the information interaction and multi-parameter combination control of the ecological construction of the integration of production, and the MVC design and top-level structure design of the ecological construction of the integration of production, education and practice and engineering education under the background of advanced modern man–machine intelligence are carried out by combining the methods of REST, AJAX, JSON and other web data structure characteristics analysis, and a three-tier architecture design is adopted [6]. Construct the domain model object structure of the integration ecosystem of production and education in practice and engineering education under the background of advanced modern man–machine intelligence, and the overall structure model is shown in Fig. 1.

In the three-layer decoupling unit of the integrated ecosystem of practice and engineering education, industry and education under the background of advanced modern man–machine intelligence shown in Fig. 1, the bottom module realizes the process and adaptive control of the information of the integrated ecosystem of practice

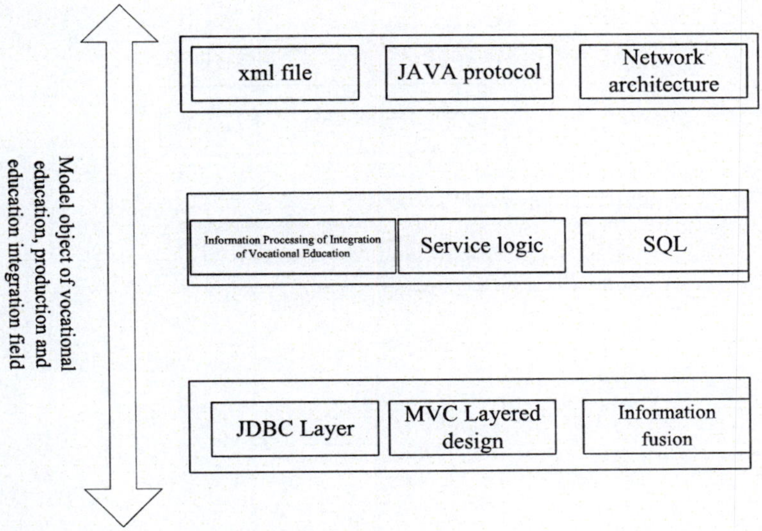

Fig. 1 Overall structure model

and engineering education, establishes the information processing terminal module [6] of the integrated ecosystem of practice and engineering education under the background of advanced modern man–machine intelligence, and uses business logic control and SQL statements as control instructions to realize the cross-compilation of the information of the integrated ecosystem of practice and engineering education, thus obtaining the functional system of the integrated ecosystem of practice and engineering education, industry and education under the background of advanced modern man–machine intelligence.

According to the functional module structure design in Fig. 2, the system integration structure under the background of advanced modern man–machine intelligence is realized by adopting the method of integration of comprehensive practice system structure of data exchange and collaborative office and ecological construction, combined with the method of system identification of high-level business users and experts in specific fields [7, 8].

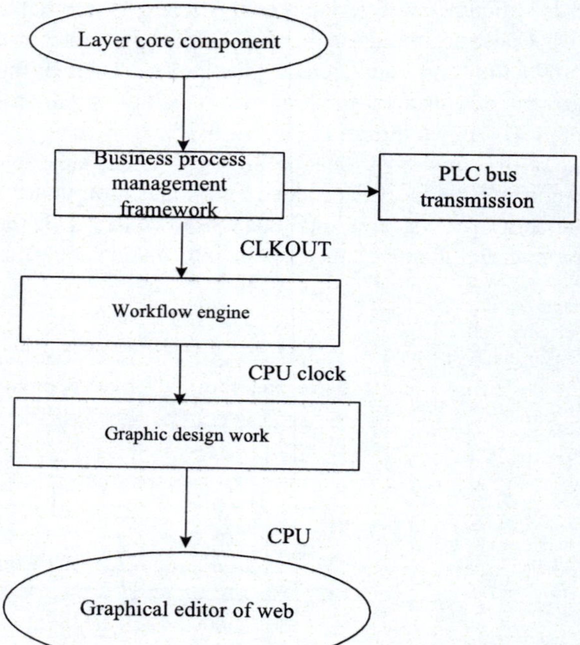

Fig. 2 Functional architecture of the integration ecosystem under the background of advanced modern man–machine intelligence

3 Mining Algorithm of Information Association Rules in Ecological Construction of Industry-Education Integration

3.1 Detection of Information Association Rules of Integration of Production and Education with Ecological Construction

On the basis of the overall design of the system, it is necessary to carry out the information fusion and association rules mining of the ecological construction of industrial education integration in practice and engineering education under the background of advanced modern man–machine intelligence. The sparse characteristic analysis method is used to randomly measure and decompose the association rules data of the information management of the ecological construction of industrial education integration, and the distribution of explanatory variables and parameters of the ecological construction of industrial education integration in practice and engineering education under the background of advanced modern man–machine intelligence is shown in Table 1.

Table 1 Distribution of explanatory variables of ecological construction of integration of production and education in practice and engineering education under the background of advanced modern man–machine intelligence

Explanatory variable	Practice teaching	Strengthen theoretical study	Development component
Data sample1	37.153	4.037	6.209
Data sample2	90.166	4.612	0.278
Data sample3	54.529	4.759	6.579
Data sample4	124.475	4.526	13.318
Data sample5	103.320	4.176	7.850
Data sample6	122.776	4.964	13.152
Data sample7	134.631	4.789	1.919
Data sample8	111.278	4.690	5.448
Data sample9	79.033	4.425	10.540
Data sample10	58.261	4.338	7.338
Data sample11	154.129	4.321	3.642
Data sample12	130.465	4.842	6.161
Data sample13	120.788	4.420	13.484
Data sample14	101.302	4.092	8.454
Data sample15	5.999	4.933	11.200
Data sample16	63.642	4.936	1.567
Data sample17	15.738	4.456	5.370
Data sample18	8.770	4.420	5.757

According to the distribution of sample parameters in Table 1, a regression analysis model of information parameters of the combination of industry and teaching practice is constructed [9], and the calculation formula of regression analysis is obtained:

$$\begin{cases} \dot{m}_i(t) = -m_i - \alpha^{rep} \frac{p_j^n(t-\sigma)}{1+p_j^n(t-\sigma)} + \alpha^{rep} \\ \dot{p}_i = -\beta^{rep}(p_i - m_i(t-\tau)) \end{cases} \tag{1}$$

where, m_i represents regression coefficient, α^{rep} represents attenuation coefficient, p_j^n represents education ecosystem, t represents iteration times, σ represents regression error, β^{rep} represents characteristic attenuation coefficient, p_i represents education goal, τ represents fitting times.

According to the characteristic attenuation of the information of the ecological construction of the integration of production, education and practice and engineering education under the background of advanced modern man–machine intelligence, the association rule set of the ecological construction of the integration of production, education and practice and engineering education under the background of advanced modern man–machine intelligence is obtained by using the normalized root mean square error analysis method:

$$E = \exp\left(-\frac{\sigma^{\alpha^{rep}} p_j^n}{\cos(\pi\alpha^{rep}/2)}\right) \tag{2}$$

According to the information feature distribution detection of the combination of industry and teaching practice, the bottom module realizes the process and adaptiveness is:

$$E_n = \sum_{m=-\infty}^{\infty} [x(m)w(n-m)]^2 \tag{3}$$

where, $x(m)$ represents the information feature distribution function, $w(\cdot)$ represents the distribution weight, n represents the fitness, and m represents the information feature. Combined with statistical analysis of panel data, the help model provided by enterprises for schools in the process of integration of production and education is established, and the distribution model is shown in Table 2.

The fuzzy feature estimation model of the ecological construction of the combination of industry and teaching practice. in practice and engineering education under the background of advanced modern man–machine intelligence is obtained as follows:

$$f_M(z) = (f(z), f_x(z), f_y(z)) \tag{4}$$

Using fuzzy segmentation clustering method, the characteristic distribution set is expressed as follows:

Table 2 The help provided by enterprises to schools in the process of integration of production education and education

Option	Frequency	Percentage
1. Talent training program and curriculum system construction	0.182	95.096
2. Professional and technical personnel to participate in practical teaching	0.173	95.005
3. Participate in the construction of training base	0.171	95.700
4. Improve the professional skills of school teachers	0.129	95.969
5. Help provide and solve students' employment	0.181	95.876
6. Others	0.152	95.991

$$F = \{f_1, f_2, \ldots, f_n\} \tag{5}$$

Therefore, the information association rule detection of industrial-educational integration ecological construction is realized.

3.2 *Integration of Production and Education of Ecological Construction Information Association Rules*

Construct the fuzzy relation distribution model of information management of the ecological construction of industry-education integration [10], and use the method of big data modeling to obtain the correlation fusion distribution function of the ecological construction of industry-education integration in practice and engineering education under the background of advanced modern man–machine intelligence:

$$\min(f) = \sum_{i=1}^{m}\sum_{j=1}^{n} C_{ij}X_{ij} \tag{6}$$

$$\text{s.t}\begin{cases}\sum_{j=1}^{m} X_{ij} = a_i, i = 1, 2\ldots m\\ \sum_{i=1}^{m} X_{ij} = b_i, j = 1, 2\ldots n\\ X_{ij} \geq 0, i = 1, 2\ldots m, j = 1, 2\ldots n\end{cases} \tag{7}$$

where, C_{ij} represents distribution coefficient and X_{ij} represents fusion coefficient. Using the method of multi-dimensional sample variance detection, the fuzzy correlation coefficient of information parameter estimation of ecological construction of practice and engineering education integration of production and education under the background of advanced modern man–machine intelligence is obtained:

$$
\begin{aligned}
u_i^1 &\leftarrow \lambda\, u_i^1 + (1-\lambda)\,\mu^1 \\
\sigma_i^1 &\leftarrow \sqrt{\lambda\,(\sigma_i^1)^2 + \lambda(1-\lambda)\,(\sigma^1)^2 (u_i^1 - \mu^1)}
\end{aligned}
\tag{8}
$$

Combining with the gray-scale feature reorganization, the multidimensional probability density function of sample data of the ecological construction of practice and engineering education integration of production and education under the background of advanced modern man–machine intelligence is obtained as follows:

$$
r^i = \sqrt{s} \times \begin{cases} 1 & \frac{1}{s} \\ 0\; 1 - \frac{1}{s} \end{cases} \tag{9}
$$

Using linear correlation analysis method, the quantitative evaluation result of production and education is obtained as follows:

$$
F(x) = \sum_{j=1}^{n} \sum_{i=1}^{m} C_{ij} X_{ij} \tag{10}
$$

To sum up, the association rule mining model of ecological construction of integration of production and education in practice and engineering education under the background of advanced modern man–machine intelligence is constructed, and the algorithm design of integration of production and education in practice and engineering education under the background of advanced modern man–machine intelligence is realized according to the information mining results.

4 Database Construction and Realization of Functional Components

Through SSH architecture protocol design of practice and engineering education-industry-education integration ecosystem under advanced modern man–machine intelligence, the integration of practice and engineering education-industry-education integration ecosystem under advanced modern man–machine intelligence is realized. The designed system meets the advantages of easy maintenance and fast running efficiency, and the XML format configuration of practice and engineering education-industry-education integration ecosystem under advanced modern man–machine intelligence is realized under Spring and Hibernate frameworks. The overall technical architecture diagram of practice and engineering education-industry-education integration ecosystem under advanced modern man–machine intelligence is shown in Fig. 3.

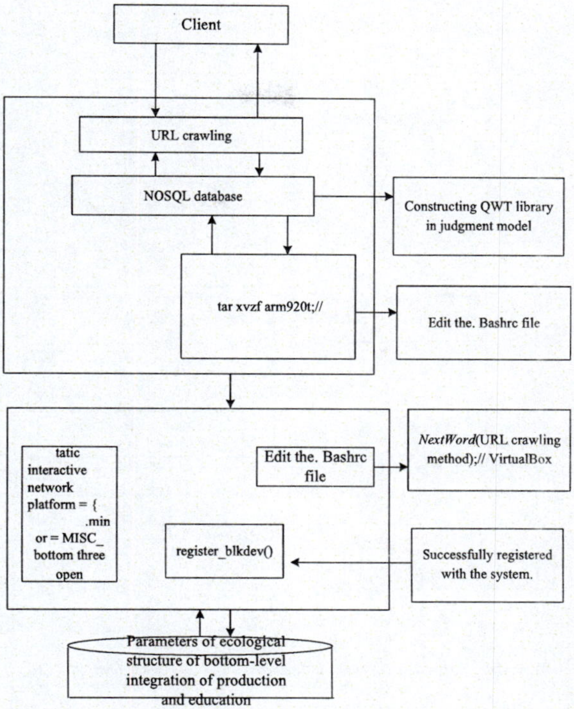

Fig. 3 The overall technical architecture of the integrated ecosystem system of practice and engineering education, production and education under the background of advanced modern man–machine intelligence

According to the overall technical architecture diagram of practice and engineering education-industry-education integration ecosystem system in the background of advanced modern man–machine intelligence in Fig. 3, five functional submodules are designed for users to choose, namely, information registration module, reimbursement module, information management module, attendance management module and access registration module, etc. Combined with the management of industry-education integration ecosystem construction and user approval circulation management, the functional module structure design of the system is shown in Fig. 4.

According to the above system component design, the MVC hierarchical design idea is adopted to realize the automatic hierarchical control of the ecological construction of the combination of industry and teaching practice. in practice and engineering education under the background of advanced modern man–machine intelligence, and a variety of web data formats are adopted to realize the data loading and program compiling control of the combination of industry and teaching practice. in practice and engineering education under the background of advanced modern man–machine intelligence, so as to realize the software development and design of the combination of industry and teaching practice. in practice and engineering education under the background of advanced modern man–machine intelligence.

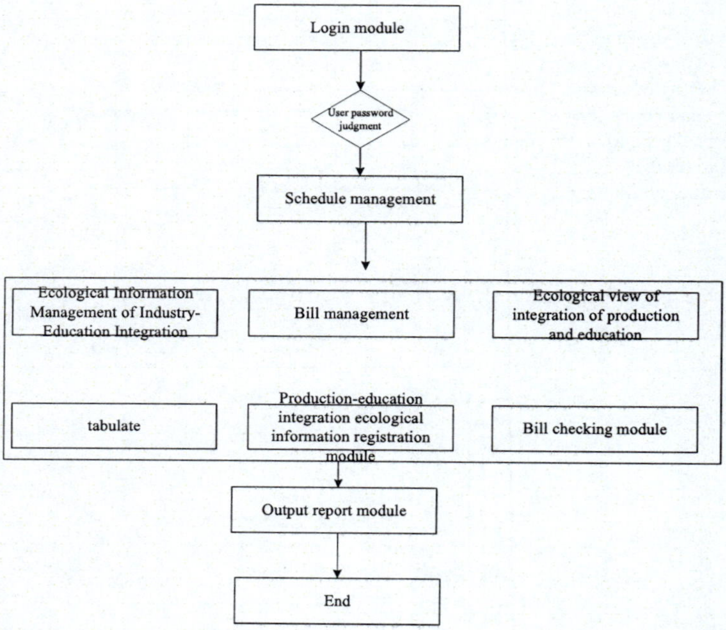

Fig. 4 Structure design of functional modules of the system

5 Test and Simulation

Using the method of data analysis and simulation, the experimental test design of the ecological system of integration of production and education in practice and engineering education under the background of advanced modern man–machine intelligence is carried out. Combined with SPSS statistical analysis software, the statistical characteristics of the ecological construction information of integration of production and education in practice and engineering education under the background of advanced modern man–machine intelligence are analyzed. The distribution of statistical data is shown in Fig. 5.

On this basis, the data analysis of the integration ecosystem of production and education in practice and engineering education under the background of advanced modern man–machine intelligence is carried out, and the descriptive statistical analysis results are shown in Table 3.

From the analysis of Table 3, it is concluded that from the perspective of enterprises, participating in the combination of industry and teaching practice. will lead to an increase in the company's operating costs, which in turn will affect the production efficiency.

At the same time, in the process of training talents with integrated production and education, it is necessary to strengthen the investment in higher practice and

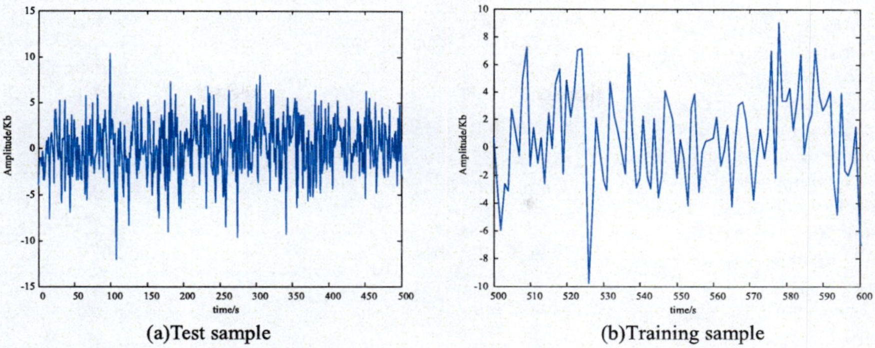

Fig. 5 Distribution of statistical information

engineering education and establish special fund support for enterprises with integrated production and education, so as to enhance the possibility, enthusiasm and sustainable development of all parties involved in the training system of talents with integrated production and education. According to the descriptive statistical analysis results in Table 3, data regression analysis is carried out, and the regression analysis results are shown in Fig. 6.

According to the analysis of Fig. 6, the system designed in this paper has a good quantitative analysis ability for the automatic correlation of the ecological construction of the combination of industry and teaching practice. The key to deepening the combination of industry and teaching practice. in higher vocational colleges lies in the establishment of a mutually beneficial and win–win cooperation concept between schools and enterprises, active cooperation and joint education of talents. In recent years, the state actively advocates the transformation and upgrading of higher vocational colleges, and gives strong support in terms of funds and policies. Schools and enterprises have taken the "first step" in the practice of integrating production with education. Next, how to make schools and enterprises establish a mutually beneficial and win–win talent training concept is very important. First of all, enterprises should be guided to change their misconceptions. Turn passivity into initiative, actively participate in the combination of industry and teaching practice, and establish the concept of mutual benefit and win–win talent cultivation, so that enterprises can fundamentally realize their dominant position in the combination of industry and teaching practice. Deeply realize that enterprises are also one of the main bodies of the integration of production and learning. Secondly, it is necessary for higher vocational colleges to change their traditional concepts and deepen exchanges and cooperation with enterprises. By holding seminars, visiting enterprises, publicizing and promoting schools and other measures, help enterprises to eliminate their concerns. Finally, the government and other relevant departments should take active measures. It is pointed out that the win–win idea of the combination of industry and teaching practice. has brought many benefits to enterprises and schools, as well as tangible benefits. Assist school and enterprise to deepen understanding. The integration of

Table 3 Descriptive statistical analysis results of ecological construction information of practice and engineering education integration of production and education under the background of advanced modern man–machine intelligence

Statistical variable	Distribution of association rules	Earnings management degree	Entropy characteristic quantity
Sample bill 1	0.195	0.267	0.372
Sample bill 2	0.117	0.264	0.363
Sample bill 3	0.112	0.209	0.367
Sample bill 4	0.159	0.244	0.310
Sample bill 5	0.101	0.208	0.363
Sample bill 6	0.105	0.276	0.331
Sample bill 7	0.177	0.271	0.372
Sample bill 8	0.107	0.275	0.340
Sample bill 9	0.168	0.297	0.366
Sample bill 10	0.147	0.245	0.377
Sample bill 11	0.121	0.218	0.331
Sample bill 12	0.165	0.265	0.383
Sample bill 13	0.171	0.229	0.352
Sample bill 14	0.195	0.283	0.305
Sample bill 15	0.130	0.279	0.372

industry and education should adhere to the government's macro-control. Schools and enterprises should pay full attention to the role of talent training in higher vocational colleges in China's social and economic development, integrate high-quality resources, and create favorable conditions for the cooperation between enterprises and schools.

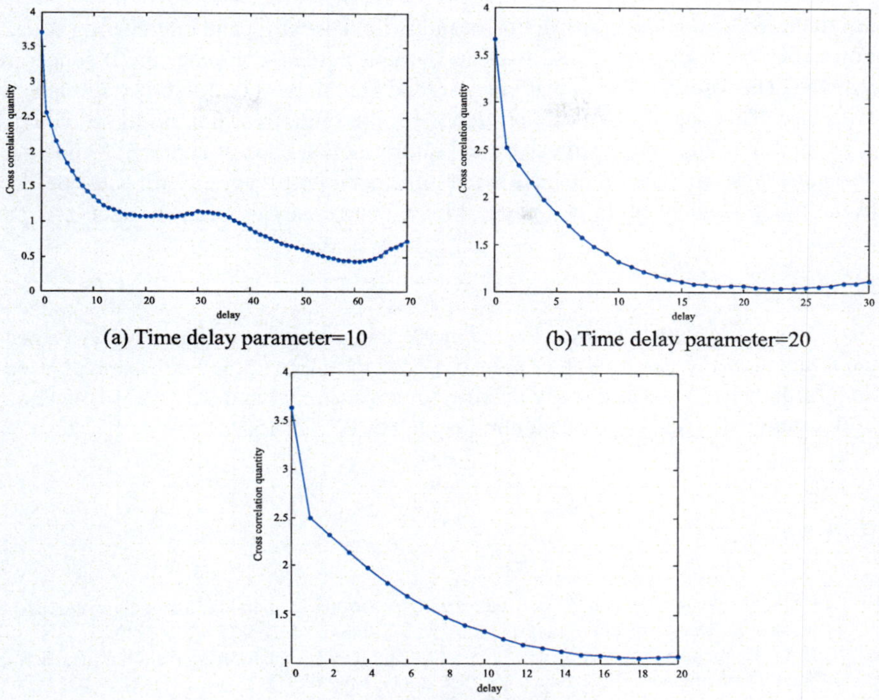

(a) Time delay parameter=10 (b) Time delay parameter=20

(c) Time delay parameter=30

Fig. 6 Regression analysis results of integration of production and education

6 Conclusions

In order to improve the ecological construction ability of the integrated ecosystem of production education practice and engineering education under the background of advanced modern human–computer intelligence, a design method of the integrated ecosystem of production education practice and engineering education based on association rules was proposed. Based on the JBPM workflow engine framework, the integrated ecosystem of production, education, practice and engineering education is designed, and the system integration structure of advanced modern human–computer intelligence is realized by adopting the construction method of comprehensive practice system. Combined with REST, AJAX, JSON and other methods to analyze the characteristics of web data structure, a fuzzy relationship distribution model for information management of industry university research integration ecological construction is constructed, and an association rule mining model for industry university integration practice and engineering education ecological construction under the background of advanced modern human–computer intelligence is established. According to the results of information mining, the algorithm design and system software design of integrated ecological construction of industrial education and engineering education under the background of advanced modern human–computer intelligence are realized. The research shows that the designed system has good stability and low bit error rate. This paper provides a reference for the future research on the ecological construction of the integrated ecosystem of production education practice and engineering education. Due to the insufficient time economy, there are still some defects in the functional system of this paper. I hope that researchers can conduct in-depth research.

Fund Project

2022 Education and Scientific Research Planning Project of Education Advancement Shandong "Construction of industry–education Integrated Ecosystem of Vocational Education, –A Case study of Shandong Vocational and Technical University of International Studies", Project number: JCHKT2022015.

References

1. Z.F. Qian, Construction of personalized learning platform based on intelligent algorithm in the context of industry education integration. Adv. Multimed. **16**(1), 275–315 (2022)
2. G.C. Li, Reform and innovation of Chinese film and television education from the perspective of media convergence. High. Educ. Soc. Sci. **23**(1), 19–27 (2022)
3. J. Tian, The construction of college innovation and entrepreneurship education ecosystem model based on vocational education. Contin. Educ. Res. (5), 5 (2022)
4. X.Zhang, The construction of intelligent education ecosystem of vocational education in the era of artificial intelligence. Invent. Innov. Educ. Inf. **000**(010), 63 (2019)
5. H. Yu, Research on the training path of the integration of production and education to cultivate sports to promote health and applied talents. Curric. Teach. Methodol. **49**(01), 195–213 (2022)
6. J.M. You, W. Zhang, How heterogeneous technological progress promotes industrial structure upgrading and industrial carbon efficiency? Evidence from China's industries. Energy **247**(33), 189–193 (2022)
7. Y.L. Zhao, F. He, Y. Feng, Research on the industrial structure upgrading effect of the employment mobility of graduates from china's "Double First-Class" colleges and universities. Sustainability **14**(4), 2353–3235 (2022)
8. D.Z. Xu, T.Y. Tu, X.Y. Xiao, Talking about the innovative application of big data in enterprise human resources performance management. Math. Probl. Eng. **39**(05), 122–126 (2022)
9. S.Haerani, et al., Structural model of developing human resources performance: empirical study of Indonesia states owned enterprises. J. Asian Financ. Econ. Bus. (JAFEB) **7**(3), 211–221 (2020)
10. Q. Pei et al., Automatic Chinese multiple-choice question generation for human resource performance appraisal. Procedia Comput. Sci. **139**, 165–172 (2018)

Evaluation of College Students' Entrepreneurial Quality Under Internet Environment Based on BP Neural Network

Ping Ye

Abstract In the context of the new era, promoting employment based on entrepreneurship is the main topic of social and economic construction and development, encouraging and supporting college students to entrepreneurship is an effective way to solve the current employment dilemma. From the perspective of the learning and growth trend of college students, in order to build a good entrepreneurial environment in the increasingly complex social environment, it is necessary to not only transfer the basic knowledge and skills needed for entrepreneurship and employment to college students, but also strengthen their comprehensive quality. Therefore, it is very important to comprehensively evaluate the entrepreneurial quality of college students. From the perspective of college students' growth under the Internet environment, this paper constructs the evaluation index system of college students' entrepreneurial quality, and uses BP neural network to conduct empirical research. The final results show that the evaluation of entrepreneurial quality of college students based on BP neural network plays a positive role in the Internet environment.

Keywords BP neural network · College students · Internet · The entrepreneurial qualities

1 Introduction

Under the influence of the financial crisis, college students face great pressure in employment, and entrepreneurship is one of the best ways to solve this dilemma. Although not all college students are suitable for starting a business, the actual survey results show that the success rate of college students' starting a business is only 0.01%, which is far lower than the average success rate. From the actual situation of college students' entrepreneurship and employment, the main factor affecting the probability

P. Ye (✉)
Guangzhou City Construction College, Guangzhou, China
e-mail: y_p128763@163.com

S. Patnaik and F. Paas (eds.), *Recent Trends in Educational Technology and Administration*, Learning and Analytics in Intelligent Systems 31,
https://doi.org/10.1007/978-3-031-29016-9_39

of success is the entrepreneurial quality of college students [1, 2]. Therefore, in order to alleviate the employment pressure of graduates in college students' entrepreneurship, we should focus on improving the entrepreneurial quality of college students in the network environment, and comprehensively evaluate the entrepreneurial quality of professional students. According to xi Jinping thought on building socialism with Chinese Characteristics for a New Era, which has been fully implemented by the Chinese government, and innovation-driven development strategy has been continuously implemented. While actively responding to the call of the state, colleges and universities around the country have begun to strengthen the entrepreneurial quality of college students, so as to enhance the development potential of college students' entrepreneurship in the Internet environment. From the perspective of the Internet, college students' entrepreneurship has the following advantages: First of all, it can alleviate the employment dilemma, have a stronger sense of innovation and entrepreneurship conditions, so as to create more jobs for social development; Secondly, it can guide the high-quality development of social economy, encourage and support college students in innovation and entrepreneurship, improve the level of international competition of social economy, comprehensively upgrade China's industrial structure, and become the backbone of urban construction and development. Finally, it can guide the diversified development of college students, provide a broad platform for college students to show their talents, change their thinking mode, dig into their potential ability, and optimize the comprehensive level of entrepreneurship of graduates. In our country, the full implementation of innovative national construction strategy on the basis of the general public in a positive response to the call for innovative undertaking at the same time, entrepreneurship for college students under the Internet environment quality put forward higher request, this will require social and school combined with Internet technology to develop more education means, thus cultivating all-round talents, provide more opportunities for college students' innovative entrepreneurial activity. Domestic researchers have put forward a variety of methods to evaluate the entrepreneurial quality of college students, such as fuzzy comprehensive evaluation method and analytic hierarchy process. As there are many factors affecting entrepreneurial quality of college students and the content of constructing evaluation index system is complicated, the empirical results lack comparability. And the application of artificial neural network, according to the physiological structure of human brain research human intelligence behavior, simulation analysis of human brain information processing function, with adaptability, learning, parallel processing and other advantages, in the field of system control, prediction and recognition has been widely promoted. In this paper, after constructing the evaluation index system of college students' entrepreneurial quality, BP neural network is used to construct the corresponding evaluation model. The final research results show that the model can enhance the effectiveness of actual evaluation results after mastering the nonlinear mapping between input and output data [3–5].

2 Methods

2.1 BP Neural Network

According to the structural diagram analysis shown in Fig. 1, BP network contains the following contents: First, input layer. All neurons at this level receive information from the outside world and transmit it to the hidden layer; Second, hidden layer. This layer will be transmitted to the input layer. After sorting and analysis, the information of all kinds of neurons will be transmitted forward after completing a learning. Third, the output layer. When the output result is inconsistent with the expected value, it will enter the stage of error back propagation. Forward propagation and error back propagation continue, and the weight values will be adjusted accordingly [6–8].

According to the flow chart analysis shown in Fig. 2, the description of BP neural network algorithm can be divided into the following points: First, the weight value Wji between neuron I and neuron J and the threshold θj of neuron J are initialized; Second, input set I XPL I and output set I ypL I of sample data are defined, where P represents the number of samples and L represents the number of input vectors. Third, calculate Opi of each layer accurately. In this process, the input and output results of the input layer are consistent, while the output calculation results of the neurons of the hidden layer and output layer are shown as follows:

$$O_{pj} = f\left(\sum \left(w_{ji} O_{pi} - \theta_j\right)\right) = f\left(net_{pj}\right)$$

Fourthly, the error signal δ PI of neurons in each layer should be clarified.

Fifth, correct the weight value in the error back propagation, and the specific formula is as follows:

$$w_{ij}(t+1) = w_{ij}(t) + \alpha \delta_{pi} O_{pj}$$

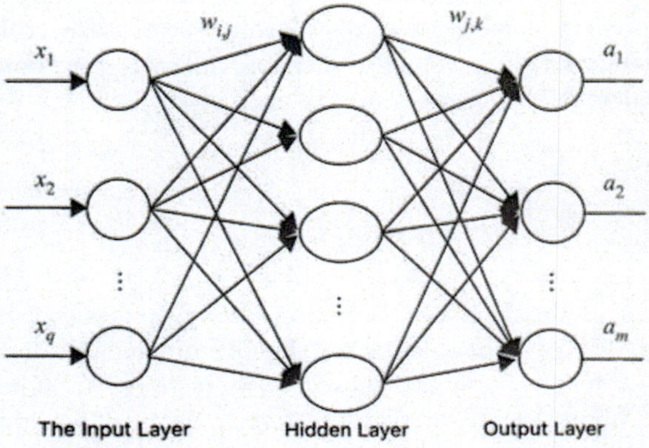

Fig. 1 Structure diagram of BP neural network

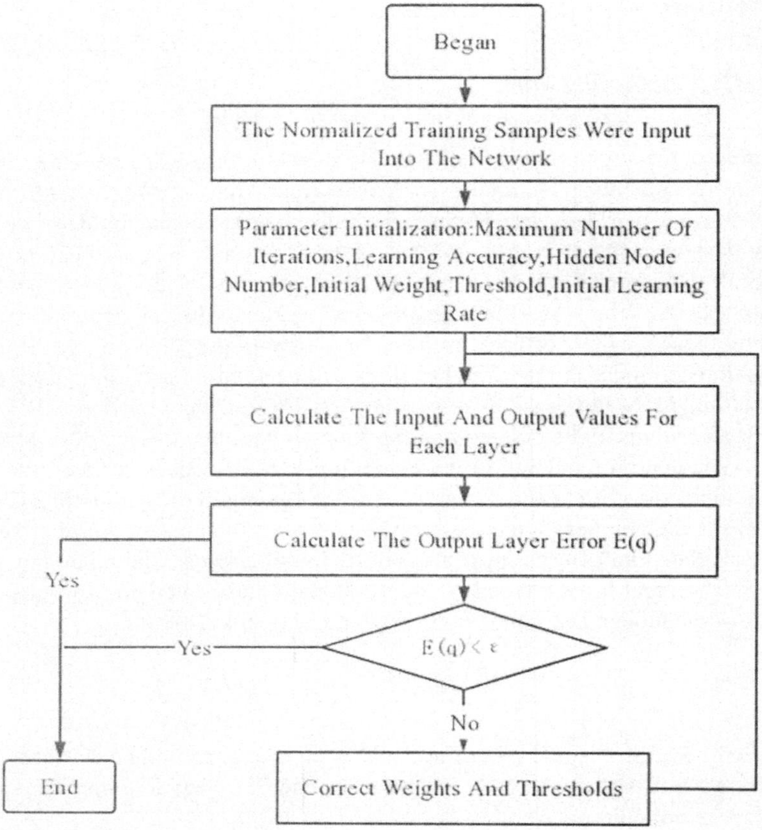

Fig. 2 Flow chart of BP neural network

In the above formula, a stands for learning speed.

Sixth, calculation error. When the error is lower than the given fitting error, the network learning should be ended; otherwise, the third step is turned to continue to calculate the output results of each layer.

2.2 Evaluation Model

The establishment of evaluation index system based on the change of entrepreneurial quality of college students should comply with the following principles: firstly, it refers to orientation, secondly, it refers to timeliness, and finally, it refers to comprehensiveness. Therefore, it is clear that the evaluation system includes entrepreneurial consciousness, entrepreneurial knowledge, entrepreneurial environment, entrepreneurial ability and other contents. The evaluation model structure constructed in this paper is mainly divided into three layers: first, it refers to the input layer. Based on the clear evaluation index system of college students' entrepreneurial quality, the lowest index value is regarded as the number of neurons in the input layer; Secondly, it refers to the hidden layer. When selecting the number of hidden layer neurons, the actual value will determine the learning efficiency and accuracy of the whole network. At present, when determining the number of neurons in hidden layer, no corresponding guiding principle has been put forward, so the specific number should be determined according to the following formula in this study:

$$n = \sqrt{n_1 + n_0} + \alpha$$

In the above formula, n represents the number of neurons in the hidden layer, N1 represents the input node, N0 represents the output node, and A represents the constant term, which meets the condition of $\alpha \in [1, 10]$. Finally, it refers to the output layer. In this paper, the number of output neurons is set as 1. The actual evaluation set contains five levels, and the evaluation principles include the following: $O \geq 0.8$, the evaluation result is good; $0.6 \leq O < 0.8$, the evaluation result is good; When $0.4 \leq 0 < 0.6$, the final evaluation result is average; In the case of 0.2 In the case of $O < 0.2$, the evaluation result is poor.

2.3 Evaluation Procedure

BP neural network is used to evaluate and analyze the entrepreneurial quality of college students in the Internet environment. On the basis of constructing the corresponding model, the actual operation is divided into the following points: Firstly, BP network training analysis should be done well. Combined with the college students' entrepreneurial quality evaluation index system constructed in the Internet environment, master the comprehensive evaluation index value of college students, and do a good job in data pretreatment, so as to obtain the corresponding input set and output set, and input them into the neural network, and then complete the network training according to the BP algorithm. Secondly, search the sample index values that need to be evaluated, and then standardize the relevant index values; Thirdly, the sample values after standardized processing should be input into the trained neural network, and then the output results should be obtained in the calculation and research. Finally, on the basis of clear output results, according to the design principle of evaluation model, the change of entrepreneurial quality of college students is summarized and analyzed, and appropriate solutions are put forward.

3 Result Analysis

In this paper, 20 college students in a certain place were evaluated and screened for their entrepreneurial qualities. Ten college students were selected as representative samples, among which six students were selected as training samples and the remaining four were selected as test samples. Combined with the evaluation index system constructed in the above study, a BP neural network model with different numbers of neurons was constructed, in which the input layer was 20, the hidden layer was 10, and the input layer was 1.

The BP neural network algorithm is programmed by using MATLAB7.0 software, and a three-layer network structure is constructed with the evaluation of entrepreneurial quality of college students as the core. Training samples are selected for network analysis, in which the input sample matrix P is a matrix with 20 rows and 6 columns, and the training output sample matrix T is a matrix with 1 row and 6 columns. After 1457 steps of training, accurate values set as expected could be achieved, with specific results as shown in Fig. 3.

Calculation and analysis are conducted based on the selected test samples, as shown below:

$$\alpha = sim(Net, Ptest)\%Network\ test$$

In the above test analysis, Ptest represents the index value of the test sample, which belongs to the matrix of 20 rows and 1 column. The final test simulation results are shown in Table 1, and the index data evaluation of the test sample is shown in Table 2.

The results show that in the four test samples, there are three college students' entrepreneurial quality is too low, and one college student's entrepreneurial quality is average. According to the data obtained in the above table, it is found that the

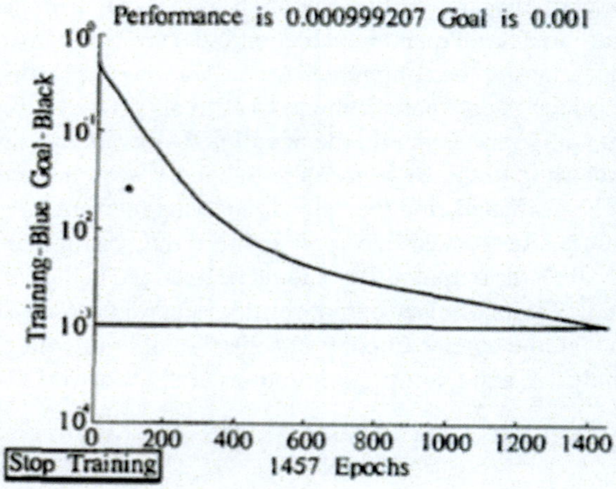

Fig. 3 Error curve of network training

Table 1 Test sample results and error analysis

Test sample number	1	2	3	4
Simulation results (output)	0.3249	0.3846	0.5931	0.3204
Expert evaluation value (actual value)	0.3231	0.3895	0.5974	0.317
Absolute error	0.0018	0.0049	0.0043	0.0034
The relative error	0.56%	1.26%	0.72%	1.07%

Table 2 Index data evaluation results of test samples

Indicators The serial number	Knowledge	Skills	Social role	Self concept	Trait	Motivation
001	85	89	90	91	88	81
002	87	89	86	84	84	90
003	92	87	73	70	69	67
004	88	87	87	85	90	86
005	83	83	69	74	65	70
006	65	77	81	79	80	83
007	57	58	73	76	70	72
008	73	72	68	75	80	77
009	80	89	85	87	89	76
010	53	51	62	63	60	61

gap between the research model constructed by using BP neural network algorithm and the expert evaluation value is still controlled at about 3%, which proves that the research model constructed in this paper has application value. What needs to be noted is that the research in this paper on the evaluation results of entrepreneurial quality of college students with BP neural network as the core is not perfect. The selected training sample values are too few to guarantee the stability of the application of BP neural network, and the actual training results vary to varying degrees. However, the results of computational evaluation can accurately evaluate the entrepreneurial quality of college students, which proves that BP neural network has application value in the index evaluation under the network environment [9].

4 Conclusion

To sum up, this study chooses BP neural network belongs to the artificial neural network method, in the study of artificial intelligence technology with nonlinear mapping function, compared with the traditional sense of the grey clustering method, the fuzzy comprehensive evaluation, etc., get rid of the influence of subjective consciousness, will be processed data information into the network platform, and then constitute the evaluation results in the calculation and analysis. In this process, researchers do not need to specify the weight value, which is helpful to control the artificial influence factors during the evaluation, improve the effectiveness of the evaluation of entrepreneurial quality of college students in the Internet environment, and thus obtain more objective and perfect data information. It is important to note that the algorithm is also wrong, there are some defects in such applications mainly in two aspects: on the one hand, the BP neural network model must have the learning samples, the quality and quantity of the selected samples directly determine the performance model of learning, which requires scientific research scholar in practice to explore the suitable study sample size; On the other hand, the choice of network layers and number of neurons also determines the learning efficiency of the overall network model. Due to the current scientific research has not put forward a clear guiding principle, so in the process of designing the number of slave neurons, the application performance of BP neural network algorithm will be restricted by human factors. Therefore, in order to continue to discuss the evaluation results of entrepreneurial quality of college students in the Internet environment, it is necessary to continue to explore the application performance of BP neural network according to the existing research results, so as to provide technical support for the evaluation of entrepreneurial quality of college students.

References

1. W. Xing, *Research on The Evaluation of College Students' Entrepreneurial Ability and Quality* (Jilin University, 2011)
2. W. Sun, H.E. Yunjing, Research on grey and fuzzy comprehensive evaluation of entrepreneurial quality of college students in china under the background of "Internet +". Educ. Theory Pract. **38**(24), 14–17 (2018)
3. X. Yang, W. Li, J. North China Univ. Water Resour. Electr. Power (Social Science Edition) **26**(01), 90–92 (2010)
4. S. Yinan, Y. Zhu, Y. Chen, The grey fuzzy comprehensive evaluation of college students' entrepreneurial quality. Technoecon. Manag. Res. **02**, 60–62 (2010)
5. L. Peng, X. Zhang, W. Jianxin, L. Wang, A preliminary study on the index system of college students' entrepreneurial quality. Sci. Technol. Entrep. Mon. **02**, 26–28 (2008)
6. M. Zhang, X. Guan, Preliminary study on the structure and evaluation system of college students' entrepreneurial quality. Bus. Times (16), 62+70 (2008)
7. Y.U. Kefa, Structural dimensions and comprehensive evaluation of college students' entrepreneurial quality–an exploratory study from some college students in Jiangxi province. J. Changchun Univ. Technol. (Higher Education Research Edition) **03**, 78–82 (2008)
8. L. Wang, Research on innovation of college students' entrepreneurial quality evaluation mechanism from the perspective of big data. Curric. Educ. Res. (47), 234–235 (2017)
9. B. Jiang, *Research on Cultivation of Entrepreneurial Quality of Contemporary College Students* (Wuhan University of Technology, 2018)

Design and Implementation of Intelligent Voice Answer System for Virtual Volunteer Teachers

Fangyuan Li, Xinhui Tu, Qingyu Cai, and Shijue Zhen

Abstract As the earliest technology developed in the field of artificial intelligence, intelligent speech has greatly improved its speech recognition accuracy with the maturity of deep learning algorithms in recent years, driving a wave of industrial boom. Being able to communicate with machines naturally is something that human beings have been looking forward to, and it is also the development direction of today's science and technology. Through the introduction of intelligent voice technology, the IFLYTEK artificial intelligence open platform is embedded in the system, with remote mountain village education as the background, for students in remote areas, designed a voice Q&A library that meets the teaching scenarios of support teachers, realized the intelligent voice answer system of virtual support teachers on the web page and mobile application side, and promoted the development of education in remote mountain villages.

Keywords Virtual volunteer teachers · Conversation · Intelligent answer · Speech database

1 Introduction

With the continuous development of artificial intelligence, voice technology is gradually becoming mature. People are more likely to use natural language to communicate with machines than the traditional way of interacting with machines through words. Through this more human way of communication, people can increase the level of interest in information. Poverty alleviation through education is a major part of the country's poverty alleviation work [1]. To comprehensively promote rural revitalization and improve the new-type urbanization strategy is the main direction of China in 2021. It is not only the goal of education reform, but also an important task of poverty

F. Li · X. Tu · Q. Cai · S. Zhen (✉)
School of Computer Science, Central China Normal University, Wuhan, China
e-mail: 1848698893@qq.com

S. Patnaik and F. Paas (eds.), *Recent Trends in Educational Technology and Administration*, Learning and Analytics in Intelligent Systems 31,
https://doi.org/10.1007/978-3-031-29016-9_40

alleviation to cut off the intergenerational transmission of poverty through education equity. Behind poverty lies the gap in skills, education and balanced development. Without a good education, there is no way to accumulate knowledge and skills to change our destiny through knowledge. More than 50% of the country's registered poor have less than a primary school education [2]. 22.3% said they were unable to escape poverty due to lack of skills. Many poor families thus fall into a vicious circle of "the poorer they are, the less educated they are, and the less educated they are, the poorer they are". At present, the education support activities in China are mainly carried out by various organizations and groups organizing the teams to the teaching sites [3]. In this process, we face all kinds of problems and difficulties, for example, the quality of support teachers cannot meet the needs of volunteer teaching, and the "migratory bird nature" of volunteer teaching makes it difficult to maintain the emotional relationship between teachers and students [4].

Human researches on related technologies in the field of speech recognition have begun since the early 1950s, when researchers made relevant studies on the factors and characteristics of speech pronunciation. In 1952, Bell LABS of the United States took the lead in the study of specific speaker independent numbers, and developed the first speech recognition system that can recognize 10 English numbers. In 1959, Fry and Denes extracted the features of speech through spectrum analysis, and then adopted the method of pattern matching to realize the system of vowel and consonant recognition, which greatly improved the efficiency and accuracy of speech recognition [5]. Since then, researchers of various countries have actively engaged in the work of speech recognition. At present, the research level of Speech recognition technology in China is basically synchronized with foreign countries. As the largest intelligent speech technology provider in China, IFLYTEK has a long-term research accumulation in the field of intelligent speech technology, and has internationally leading achievements in Chinese speech synthesis, speech recognition, oral evaluation and other technologies. At the same time, China's virtual character industry has risen rapidly. Major network platforms have launched their own virtual people, and our college has also launched virtual counselors to guide students' daily learning and life, which provides development experience and opportunities for the system [6].

Through the study of intelligent voice technology, this paper embedded the IFLYTEK artificial intelligence open platform with voice interaction as the core, built an intelligent voice solution system, and applied it to today's village support education, broke the traditional human–computer interaction mode, and realized the virtual support education teacher intelligent voice solution system. Through real-time voice communication with extremely realistic volunteer teachers, students' confusion in knowledge can be solved in time, their interest in teaching knowledge can be improved, and the development of volunteer education can be effectively promoted.

2 Related Technology

2.1 The Voice API of IFLYTEK

As a well-known listed intelligent voice and artificial intelligence enterprise in the Asia–pacific region, IFLYTEK has been engaged in the research of core technologies such as speech and language, natural language understanding, machine learning reasoning and autonomous learning for a long time since its establishment, and has maintained the international cutting-edge technology level. In order to facilitate the public development of speech recognition system, IFLYTEK provides a series of development interfaces and application tools, including speech recognition interface, grammar interface, endpoint detection interface, audio input interface, management and maintenance interface, so as to make the development of application system more efficient [7].

The system selects IFLYTEK API as the technical support of the voice part of the system. By calling its API database, the system can complete the speech recognition and synthesis functions, so that the virtual volunteer teachers can make immediate response to the voice questions of students. At the same time, IFLYTEK API has a good programming interface, so that in the actual development process of the project, the cohesion of each part is greatly improved, the coupling degree is greatly reduced, and the coding difficulty is reduced [8].

2.2 Construction of Three-Dimensional Virtual Volunteer Teacher Image

The virtual volunteer teacher image of this system adopts a more realistic 3D character model to improve students' interest in learning the course through vision. A 3D mannequin reconstruction algorithm is used to generate a 3D model with a personalized theme, including the outline shape of the body, hair and clothes, the personalized texture of the body model, and an underlying bone structure, by inputting a single single-eye RGB video. On this basis, 3Dmax is used to optimize and beautify the artificial details of the human body model generated by the algorithm. Finally, the virtual teacher character image with both human body and exquisite modeling appearance is derived, which overcomes the disadvantages of long cycle and low precision of single manual modeling [9].

In order to make the virtual volunteer teachers have natural and smooth body movements and voice interaction in teaching courses, the system adopts 3D human skeleton movement recognition algorithm based on feature vector describing the geometric features of the skeleton. The camera coordinates are transformed into human body coordinates through data preprocessing, and then feature extraction is carried out on the data, which is input into the time selection LSTM network for

Fig. 1 Schematic diagram of 3D model feature points

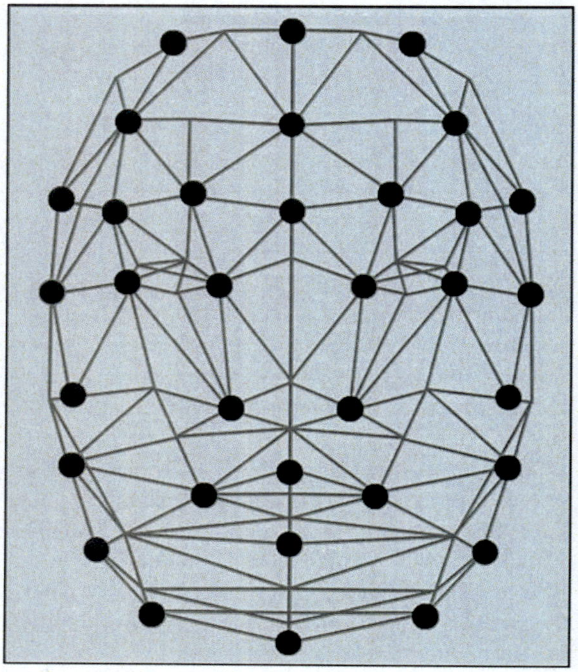

training, so as to generate flexible and fluent virtual teacher lecturing and answering question body action files [10].

In the virtual teacher face modeling method, the use of real face video data intelligent generation of face 3D model, which makes the 3D virtual teacher image transformation more convenient and fast. In order to better match the 3D face model with the input human face data, we first extracted the feature points on the real human face, and then selected 34 feature points (as shown in Fig. 1) on the basis of the reference model of the front face grid, mouth grid, eyeball grid and other information, and integrated the extracted feature points to build the corresponding 3D face model.

2.3 Course Knowledge Automatic Question Bank

After understanding the specific needs of children in mountainous areas for volunteer teaching knowledge content through demand analysis, crawlers were used to crawl primary school Chinese, mathematics and other related curriculum content, while collecting relevant test materials and integrating data. Edit and adjust the format of the collected course knowledge materials to ensure the quality of knowledge materials; Collect all child knowledge base, build teaching curriculum knowledge automatically answer question bank, for the students to ask questions can find the corresponding answer to every question, in the subsequent use continuously add new valuable knowledge in the process, and an orderly, standard, the classification model based on knowledge, improve the automation of information processing.

First in the process of system operation for the user to enter the voice file preprocessing and transformed into the corresponding text information, and then combined with IFLYTEK semantic understanding of language model, the model can determine the statement of intent, finding keywords as a parameter in the statement, parameters and intent constitutes the user's voice commands to convey information, The captured parameters and intentions are matched with the text similarity in the established course knowledge automatic question answering database, and the questions with the highest similarity are recorded. The fragmented knowledge is organized into structured answers. Finally, the text answers are synthesized into voice information and returned to users. In the process of searching questions, inversion list and correlation algorithm are used to speed up the search, and word meaning matching and spelling correction are added to make the results more accurate.

3 Overall System Design

3.1 Overall Framework Design of the System

The system will serve as the basis for the development of the virtual voluntary teaching teacher teaching service platform, which is an educational software product. The platform displays virtual teachers in 2D and 3D dimensions, quadratic and three-dimensional dimensions through client web pages and mobile apps. Users can watch teaching videos of virtual teachers and enjoy 24 h cloud Q&A service of virtual teachers.

When the user starts the application, the external knowledge base will import the course knowledge information into the teaching module, and the virtual teacher image library will import the made 3D virtual teacher model into the teaching module to start teaching for the user. When the user raises a question to the virtual teacher using the APP, the input voice information is sent to the user question processing module, which transmits the voice information to IFLYTEK natural language processing system. The system converts the voice information into text information and imports it to the user question processing module, which then sends the text information to the text speech matching module. The voice text information is imported into the text speech matching module, and the voice corpus is used for matching. The text speech matching module transmits the successfully matched voice information to the user processing module, and the user processing module sends the voice information to the user. The overall framework and workflow of the system are shown in Fig. 2.

The system adopts multi-terminal cross-platform to provide virtual teaching support teacher teaching Q&A service for users. Users can log in to the web page on PC or start mobile APP to switch to the virtual teacher page to realize intelligent

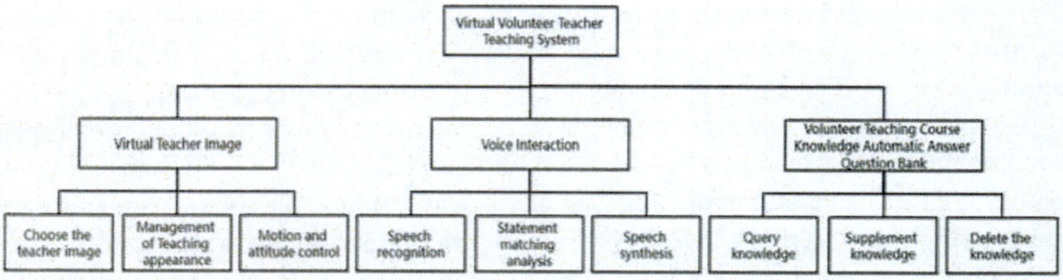

Fig. 2 Overall framework and workflow of the system

voice Q&A,as show in Figs. 3 and 4. When the user starts voice input through the microphone, the system calls the IFLYTEK platform to convert the voice information into text information and send it back to the system background. BM algorithm is used to detect and match the transformed text: if there are keywords in the query database that we have set and the matching index P is greater than a certain preset value, it will be judged as the course teaching, and the audio and video output will be carried out after the query results are obtained. Otherwise, it is regarded as requiring virtual volunteer teachers to answer questions. After judging that the problem is solved, the answer is based on the knowledge base, the speech system architecture design is shown in Fig. 5.

Fig. 3 PC web page Voice chat page

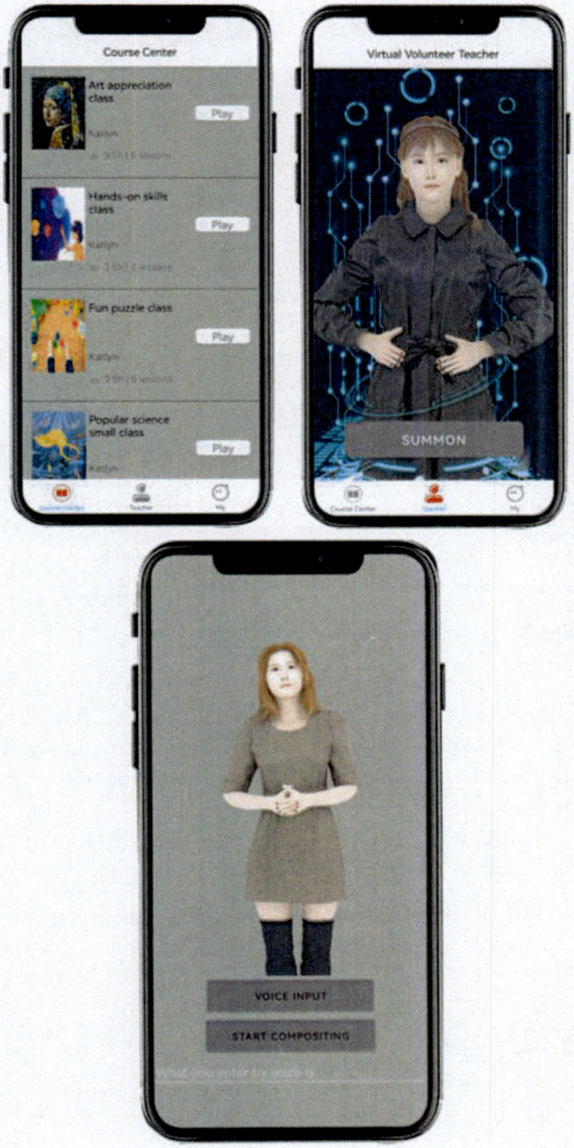

Fig. 4 Mobile APP system

3.2 Speech Database and System Design

The speech dialogue database is designed and constructed for students in remote areas. Before the establishment of relevant data collection activities, integration and

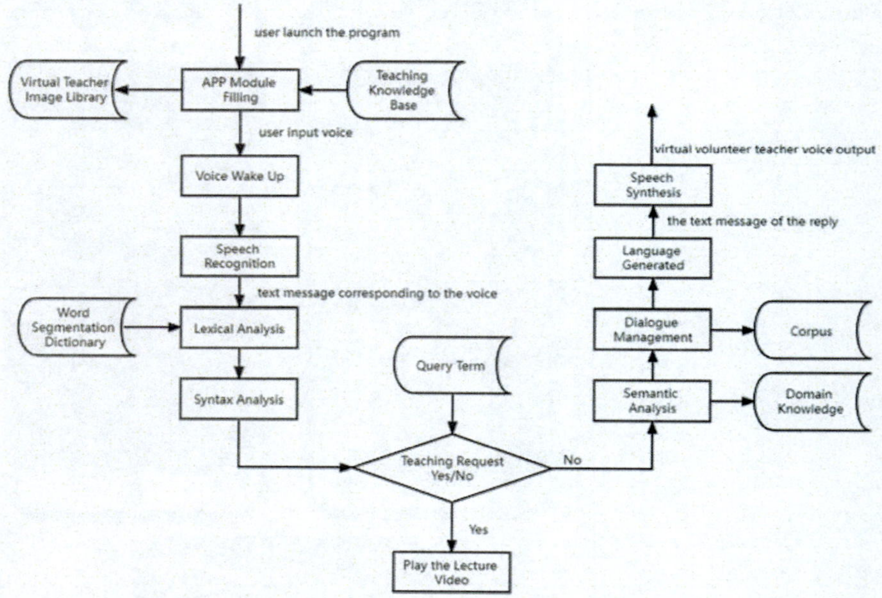

Fig. 5 Speech system architecture design drawing

induction of the corresponding data model. After the steps of requirement analysis, conceptual structure design, logical structure design, database physical design, database implementation, database operation and maintenance, the E-R relationship of the voice database in Fig. 6 is designed.

Through the analysis of the established E-R relationship diagram, the following relationship model is transformed:

VirtualVolunteerTeacher (teacherID, teacherName, teachingCourse)
Student (studentID, studentName, studentGender, studentAge)

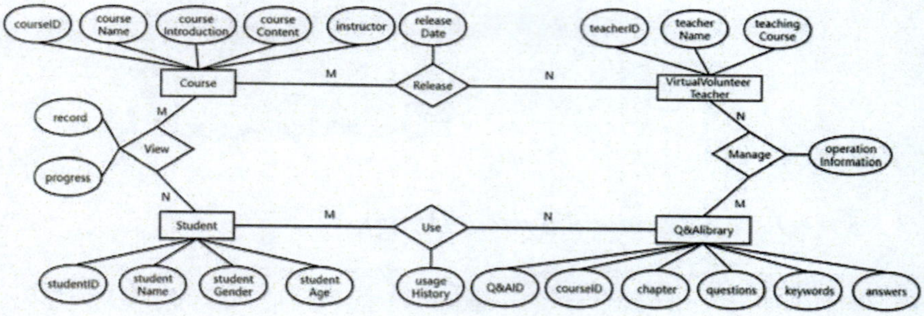

Fig. 6 E-R diagram of voice database

DESKTOP-J4I5G0E...dbo.Q&Alibrary*

列名	数据类型	允许 Null 值
[Q&AID]	int	☐
courseID	nchar(20)	☐
chapter	nchar(20)	☐
questions	nchar(50)	☐
keywords	nchar(25)	☑
answers	nchar(200)	☑

Fig. 7 Volunteer teaching knowledge Q&A database form

Course (courseID, courseName, courseIntroduction, courseContent, instructor, releaseDate)
Q&Alibrary (Q&AID, courseID, chapter, questions, keywords, answers)
Manage (teacherID, Q&AID, operationInformation)
Use (studentID, Q&AID, usageHistory)
View (studentID, courseID, record, progress).

Figure 7 defines fields in a record, such as field name, field type, length, primary/foreign key, and field value constraint, as shown in the table. Other lists are not listed one by one due to space limitations.

3.3 *Voice Call and Implementation*

The voice information is input into the computer through the microphone, and the computer converts the original analog signal data into digital terminal signal. Then, the speech signal is preprocessed, including pre-weighting, framing, windowing, endpoint detection and so on. Then feature parameters are extracted from the preprocessed speech data to obtain the entry model. At the same time, the processed template generated by the speech signal is tested and matched, and the final text is obtained for speech recognition. The schematic diagram of speech recognition is shown in Fig. 8.

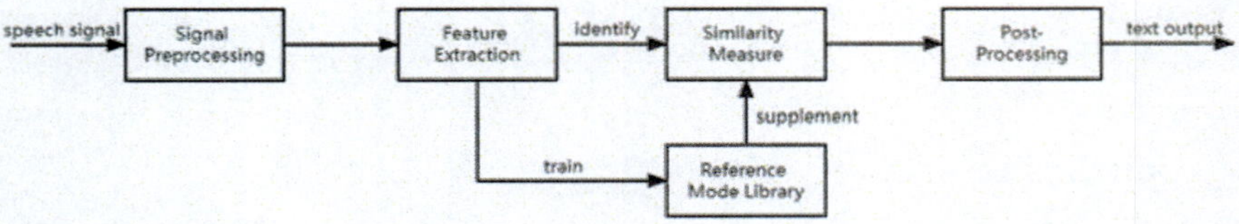

Fig. 8 Speech recognition schematic diagram

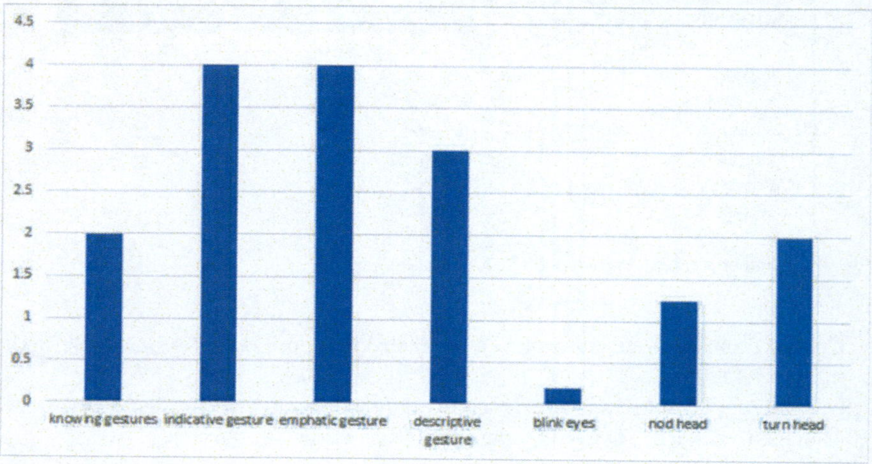

Fig. 9 Expected technical indicators

4 System Test

In order to make the teaching process of teachers have a sense of science and technology and a good sense of spatial picture, and enhance the immersion of students in class, we hope to build a virtual volunteer teacher close to the image of people, and at the same time ensure that the virtual volunteer teacher in the teaching and answering process presents a strong sense of reality. Some of the expected technical indicators are shown in Fig. 9. (The abscissa represents motion and attitude; The ordinate is the time, in seconds, it takes to complete the pose.)

The input voice volume is above 45 dB, the environmental noise is below 15 dB, the accent is relatively standard Mandarin, the normal speed of speech, and the bandwidth is 10Mbps. The test results of the system's speech recognition and response time are shown in Fig. 10. (The abscissa represents the number of experimental groups, unit: group; The ordinate is recognition or response time, in seconds.)

According to the ten groups of data, the average speech response time is 1.59728 s, and the average speech recognition time is 0.73209 s. Basically achieve the expected design effect.

5 Conclusion

With the continuous development of science and technology, the future era will be an era of intelligence. Deep integration of ARTIFICIAL intelligence into education can build a high-quality education system. By applying intelligent voice technology, 3D character modeling technology and 3D character pose generation technology,

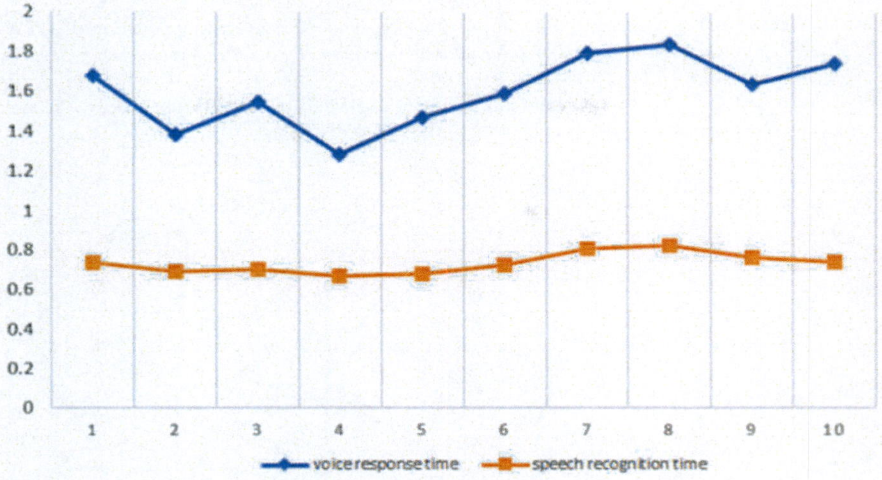

Fig. 10 Speech recognition and speech response time

embedded in IFLYTEK API interface, this paper constructs virtual volunteer teachers with vivid images and real-time voice answers to relevant questions, realizing the intelligent voice solution system for volunteer teachers, providing new ideas for volunteer teaching in remote areas. To some extent, students' participation in class will be enhanced, the teaching quality of volunteer teaching will be improved, and the Gospel will be brought to children in mountainous areas.

Acknowledgements The work was funded by the National Nature Fund, No. 61572223.

References

1. F. Xiong, *Research and Implementation of Embedded Speech Recognition Algorithm* (Taiyuan University of Technology, 2008), pp. 1–2
2. L. Yu, *Implementation of Simple Speech Dialogue System based on Google Speech-API* (South China University of Technology, 2012), pp. 1–2
3. Z.G. Pan, H.W. Yang, Z. Liu, J. Comput. Aided Des. Comput. Gr. **19**(12), 1509–1516 (2007)
4. S.Y. Su, J.M. Chen, J.G. Pan, Research on behavior modeling in virtual environment. Comput. Sci. **34**(2), 270–273 (2007)
5. G. Luo, C.Y. Hao, W. Zhang, Y.Y. Fan, Review of virtual human technology. Comput. Eng. **18**, 7-9:7-8 (2005)
6. Z.L. Wang, *Artificial emotion* (China Machine Press, Beijing, 2009), pp. 42–42
7. Y.Y. Yu, *Design and Development of Campus Service Dialogue System based on Android* (Fudan University, 2013), pp. 12–13
8. X.Z. Li, *Research on Speech Recognition Algorithm and Application Technology* (Chongqing University, 2010), pp. 1–3

9. C. Wan, L.L. Liu, Research and improvement on embedded system application of DTW-based speech recognition, in *2nd International Conference on Anti-counterfeiting, Security and Identification, ASID* (2008), pp. 401–404
10. L.Y. Min, T.T. Zhao, Research and improvement of BM algorithm. J. Wuhan Univ. Technol. (Transportation Science and Engineering) (03), 528–530 (2006)

Research on Network Psychological Education Model Based on Cloud Computing

Lina Liu

Abstract In order to construct the effective network psychological education model, the cloud computing is applied for it. Firstly, current situation of mental health education is summarized. Secondly, development goal of constructing network mental health education system is analyzed. Thirdly, application of cloud computing technology in college students' network mental health education is studied. The cloud computing model is established. Finally, the network mental education system is established based on cloud computing, and the performance of the system is analyzed. Results show that the proposed system has better performance, which can provide the stability of network mental education system.

Keywords Cloud computing · Network mental education system · System

1 Introduction

In recent years, various campus malignant events appear in the news from time to time, which makes college students' mental health and psychological crisis gradually become the focus of social attention, which is a problem that colleges and universities need to solve and pay attention to. At present, Chinese college students look happy and worried. Fortunately, they have strong independence and the ability of self-management, learning and development. The worry is that their psychological status is closely related to the mental health status of Chinese people under the background of the whole society. Various ethos in the society continue to affect the original pure university campus. The campus is no longer a pure land, and the psychological problems of college students have been very concentrated and prominent [1].

With the rapid development of network technology, in the training and educational activities for students, it is necessary for colleges and universities to innovate the

L. Liu (✉)
Morden Education College of Changchun Guanghua University, Changchun, China
e-mail: liulina1@ghu.edu.cn

S. Patnaik and F. Paas (eds.), *Recent Trends in Educational Technology and Administration*, Learning and Analytics in Intelligent Systems 31,
https://doi.org/10.1007/978-3-031-29016-9_41

traditional mental health education model, build a network mental health education system and adapt to the development trend of the times. Combined with the advantages of network mental health education, make use of the functions of the network in mental health education, combined with the mental health needs and growth law of college students, build a fully functional network mental health education system, make the mental health education work advance with the times and improve the level of modernization and networking. The construction of mental health service system can not only enhance the people's sense of access, happiness and security, but also enrich the content of national social governance system, improve social governance capacity and help the modernization of national governance capacity [2, 3].

Network mental health education provides a new mental health education model to support and make up for the traditional mental health education in Colleges and universities. From the perspective of development, the "post-00 s" who grew up with the Internet have gradually become the core group of university campuses. They pursue individual value, networked entertainment life and independent learning style. It can be seen that network mental health education will become the basic trend of mental health education in Colleges and universities in the twenty-first century. In recent years, the application of cloud computing technology indicates a new revolution in network technology. The arrival of the "cloud era" has brought great changes to social production and life. It is of great practical significance and operational value to explore how the network mental health education in the "cloud era" is constructed and carried out. With the help of modern information technology, this paper discusses the construction of an information platform for real-time monitoring of College Students' mental health in Colleges and universities, which can be accessed from the client by using mobile personal devices such as smart phone pad, e-book bag and computer. Users at all levels can also report the psychological crisis status of individual or group students at any time through the user terminal, and the platform system intelligently pushes it to relevant mental health managers or psychological crisis intervention personnel, so as to prevent the psychological crisis of early warning objects at the first time.

The greatest advantage of cloud computing lies in the integration of resources, which can facilitate the coexistence of heterogeneous mental health education resources, and can maximize the integration of resources without too much change to the existing data operation platform. Using cloud computing to integrate high-quality resources of mental health education into a "cloud", students can build and share each other's high-quality resources to achieve their own strengths. The mental health education resource "cloud" is an opportunity set of high-quality mental health education resources. It consists of two parts: first, the basic classroom "cloud", which provides online compulsory classes. Students can choose teachers and learning progress, and master the basic knowledge and theory of mental health through interactive teaching with psychological teachers; The second is the special lecture "Cloud", where students choose the urgently needed special knowledge and theory of mental health according to their needs, and deeply master some aspects of psychological adjustment skills. Teachers can recommend learning content to students according to their information. These resource "clouds" require the least equipment for both schools and teachers and students. Users can access and use the resource "cloud" at any time and place in the form they are accustomed to, as long as they have terminal devices that can access the Internet.

2 Current Situation of Mental Health Education

Many colleges and universities have done a lot of work in college students' mental health education, such as establishing college students' mental health education center and psychological counseling center. Many colleges and universities regularly carry out the general survey of College Students' mental health, interview and coach the objects with psychological crisis tendency, and usually hold some mental health education related activities to publicize the importance of College Students' mental health. With the advent of the Internet age, some people put forward the "dual structure" of College Students' psychological counseling online and offline, that is, online psychological counseling and practical psychological counseling. Some colleges and universities have established three or four levels of mental health education networks, such as school level college students' mental health education consulting centers, college and department mental health education workstations, college students' psychological mutual aid groups and so on. For example, the dormitory psychological liaison is taken as the first level alone, and the class psychological Committee and psychological community are taken as the first level to form a four-level network structure, and even the medical system or relatives and friends are taken into account. Some colleges and universities have also established a four-level network form from dormitory psychological informants, class psychological health care workers, college and department mental health education groups to mental health education and counseling centers [4, 5].

The establishment of the existing "three-level network" or "four-level network" and other educational monitoring and early warning models for college students' mental health education. This multiple level mental health education combining educational guidance and consultation self-help is conducive to college mental health education and psychological crisis intervention, but there are also some problems, The above models need to judge the psychological abnormality or crisis of college students through the observation of other relevant mental health personnel outside. After the crisis is determined, the crisis warning and report will be issued. In this process, the effective time of psychological intervention will be delayed. Therefore, the psychological crisis can not be uploaded and released in time and effectively, can not achieve real-time monitoring and early warning, and lack of effective interaction. People's psychology changes dynamically. Some individual hidden psychological problems can only be stimulated or manifested in a specific stress environment. The screening data obtained by psychological evaluation and other methods are lagging, and the psychological change process can not be obtained in time, which will reduce the timeliness of College Students' psychological crisis identification and early warning [6].

3 Development Goal of Constructing Network Mental Health Education System

(1) First, build a three-dimensional, multi-dimensional and interactive network mental health education system and build a brand of network mental education. Online psychological consultation, psychological knowledge publicity and crisis investigation can be carried out through the official website, psychological teachers group, psychological association group, questionnaire star, wechat public platform and Chinese college students' mental health evaluation system, so as to improve the efficiency of online mental health education [7].
(2) Create a positive, happy and harmonious network mental health education environment and cultivate students' positive psychological quality. The network mental health education system is operated and managed by a professional network team of teachers and students, regularly updates the website content, timely publishes psychological network information, provides excellent network cultural products to teachers and students, spreads positive psychological energy, creates a positive campus psychological atmosphere, enriches students' campus network cultural life, and cultivates students' good psychological quality with a good campus network psychological education environment.
(3) Innovate the psychological education mode of online, offline and two pronged, and improve the effectiveness of College Students' Ideological and political education. As one of the important contents of Ideological and political work in Colleges and universities, college students' mental health education is also an important part of talent training in Colleges and universities. Healthy psychology is the key basis for students to grow up healthily. The school innovates the online and offline mental health education mode, expands the ways of mental health education, innovates working methods, and takes a two pronged approach to effectively improve the effectiveness of College Students' mental health education, so as to ensure the healthy growth of students, so as to provide enterprises and society with talents with physical and mental health, integrity and ability [8].

4 Application of Cloud Computing Technology in College Students' Network Mental Health Education

(1) Requirement analysis

The mental health education auxiliary service system based on cloud computing is aimed at customers of all ages across the country. It aims to build a professional psychologist serving the common people. While helping customers diagnose, it provides corresponding counseling services according to the data analysis of the cloud computing center, and at the same time provides one-to-one services, which not only ensures the privacy of patients, but also reduces the difficulty of patients seeking medical care.

The mental health education auxiliary service system based on cloud computing technology consists of four modules, which are respectively cloud computing health service system, physical health detection system, mental health management system and mental health consulting system. The mental health education auxiliary service system based on cloud computing technology consists of four modules, which are respectively cloud computing health service system, physical health detection system, mental health management system and mental health consulting system. The physiological health detection system is a physical connection with customers. It mainly detects various physiological signs of customers and records them in real time according to the detection information; The mental health management system is a comprehensive storage database, which will record each customer's information and adjust the counseling strategy in real time according to the information detected by the customer; The last is the mental health consulting system, which is an interactive system between the system and customers. Customers interact with the cloud computing health service system through this interactive system, and check their mental health level by answering test questions. If the results automatically generated by the system are not targeted to customers, online one-to-one consulting can be conducted through this system.

Due to the real-time information transmission needs between customers and the system, the system belongs to online consulting, so it needs to get rid of the constraints of time and space. The Internet platform can effectively solve this problem, but the bridge connecting each other needs a safe and stable information channel. After the customer wears the supporting facilities, they will enter the service interface. When the customer chooses to do some basic tests, they will focus on the detection of customer data information according to the content selection, and make a comprehensive judgment on the customer through different data feedback. The system will provide customers with text communication and voice communication at the initial stage, and one-to-one video communication will also be provided in the later stage with the upgrading of consultation. Of course, whether to show up or not will be selected according to the needs of customers. However, any of the above communication needs stable and timely data exchange.

The system needs good compatibility and integrity. At present, customers choose many channels to access the Internet, including mobile terminal and PC terminal. The mobile terminal is mainly divided into Android terminal and IOS terminal. The two are completely different logical architectures, so they need to develop mobile terminals separately; In order to give customers the best experience, the PC side not only needs to develop a separate client but also needs to develop a web side. In the web side, customers can not only log in through computers, but also enter through the mobile side through the web side, but in the background, all the ports above need to achieve data interoperability.

Progressiveness of the system. The system adopts a multi port structure, which is convenient for customers to the greatest extent. It integrates all ports on the PC end and mobile end, achieves data exchange in the background, which is convenient for customers to log in to the system through any channel. Moreover, various personalized settings are displayed, and all customer information will be loaded.

The efficiency and accuracy of the system. With cloud computing as the core, the system can collect massive mental health data, analyze and correlate the data through the data warehouse, match the corresponding information according to the customer's mental health test, which can accurately judge the customer's mental health, and also output the best counseling plan.

(2) Basic model of cloud computing

"Cloud" is a virtual computing resource, which can be self maintained and managed. It can centralize computing resources and realize automatic management through special software without human participation. It is generally composed of clusters of some large servers, including computing servers, storage servers and broadband resources. Cloud computing provides cheap distributed computing power through the network. The basic architecture of cloud computing platform is shown in Fig. 1.

Cloud computing has become one of the current development hotspots and important trends. Many information giants have participated in the research and development of cloud computing. At present, commercial cloud computing platforms include Microsoft, Google, Oracle, Amazon, salesforce, Wangtian cloud services, EMC and China Mobile. The amount of mental health data of college students is so huge that it needs a safe, effective and high-capacity storage device, and cloud computing technology can solve the problems of storage and operation of these massive data. With the rapid growth of educational digital resources, one of the hot issues in the development of educational informatization is to solve the resource integration management

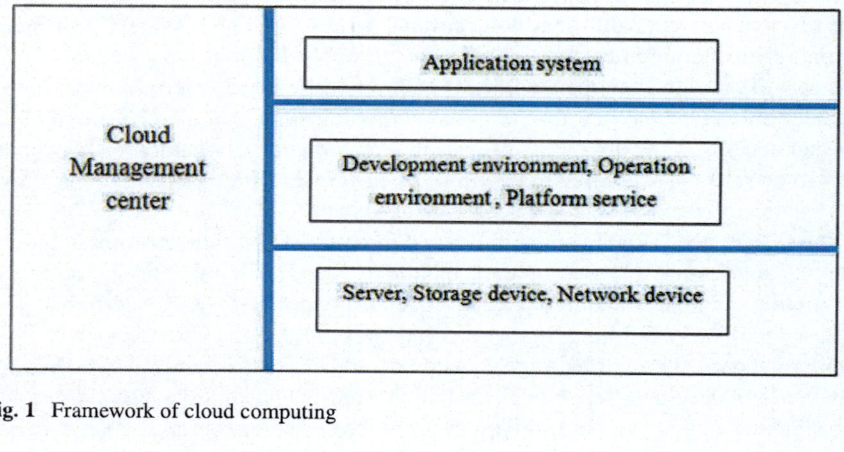

Fig. 1 Framework of cloud computing

and data center construction based on cloud services, so that various data resources can run in personal devices such as smart phones, pads, e-book bags and so on in the form of cloud services. At present, there are many researches on various information platforms in the cloud computing environment, such as e-government platform, educational resource platform, trading platform, talent market and employment platform, but there are few researches on the monitoring and early warning of College Students' Psychological Education in the cloud environment at home and abroad. Although the existing methods realize the optimization of virtual machine migration to a certain extent, they are not suitable for cross data virtual machine migration. Therefore, the cloud computing virtual machine migration model based on deep hash algorithm is studied to reduce the bandwidth overhead of cross data virtual machine migration [9].

The depth hash algorithm is used to extract the similarity of the basic image, and the hash map is constructed according to the similarity of the basic image. This method mainly includes unsupervised stage and supervised stage. The trestle automatic coding neural network is used to learn the initial parameters of the deep convolution neural network. The trestle automatic coding neural network includes several layers of sparse encoders, in which the input of the next layer is the output of the previous layer, and the coding operation is to run the self encoders of each layer according to the upper and lower order; The decoding operation is to expand the decoding according to the opposite direction of the above operation.

After the trestle automatic coding network training, the samples with labels are used to supervise and adjust the trestle automatic coding, and the adjustment range is small, so as to ensure that the basic image still has a high degree of similarity in Hamming space.

Define sample $(x_1, x_2) \in X$, $A \in [0, 1]\ [0, 1]$, training basic image with label is defined as X, the training label is defined as Y, and the Euclidean distance of sample pair mapping space is calculated by [10]

$$D_E(\vec{x}_1, \vec{x}_2) = \|G_E(\vec{x}_1) - G_E(\vec{x}_2)\| \tag{1}$$

where G_E denotes the mapping function.

Loss function in supervised stage is calculated by

$$\rho(R) = \sum_{i=1}^{P} L(R, (A, \vec{x}_1, \vec{x}_2)^i) \tag{2}$$

$$L(R, (A, \vec{x}_1, \vec{x}_2)^i) = (1 - A)L_s(D_R^i) + AL_E(D_R^i) \tag{3}$$

where, the parameter in the supervised stage is W; The matching label of the i th sample pair is $(A, \vec{x}_1, \vec{x}_2)^i$, and the partial loss functions between similar point pairs and non similar point pairs are L_s and L_D respectively; The number of training sample pairs is P; Let the number of samples be N, $P = N^2$.

The steps of the deep hash algorithm are:

Step 1: input the original basic image data set in the DCNN network to obtain the characteristics of the basic image;

Step 2: train stacked auto-encoder to obtain binary hash code through threshold method;

Step 3: obtain the basic image data set with similar attributes according to the binary hash code;

Step 4: semantically analyze the label of the basic image data set with similar attributes to obtain the approximate weight of the label;

Step 5: change the stacked auto-encoder parameters according to the semantic similarity of the basic image pair;

Step 6: solve the Hamming distance to obtain the similarity of the basic image.

5 Case Study

Select a university as the research object to build an information platform for monitoring and early warning of College Students' mental health education based on cloud computing. Through the deployment of cloud services, cloud storage, computing and services in the data center are realized, and access and information push on personal hand held terminal devices such as mobile phones and personal computers are realized, so as to realize the real-time early warning function of psychological crisis. Establish a psychological early warning information database for college students. The platform calculates the early warning objects according to the set psychological crisis index system, timely pushes information to relevant personnel, and issues crisis warnings, so that managers can intervene and treat psychological crisis at the first time, so as to prevent abnormal or even dangerous situations. The platform model is conducive to integrate resources, mobilize the enthusiasm of participants at all levels, improve the effectiveness and real-time of psychological crisis intervention, and make the psychological crisis prevention work achieve early prevention, early detection and early treatment.

Select the cloud data center genetic algorithm migration model as the comparison model of this model. Use the two models to migrate 60 virtual machines at the same time under different similar degrees and different loads. Compare and analyze the total migration time and network traffic in the migration process of the two models. The number and meaning of different loads are shown in Table 1, and the comparison and analysis results are shown in Table 2.

According to the results in Table 2, in different cases, the network traffic of the model in this paper is the lowest, indicating that the bandwidth overhead of the method in this paper is the lowest. When migrating multiple virtual machines at the same time, this model has the minimum total migration time, the lowest network

Table 1 Number and meaning of different loads

Number	Meaning
I	Virtual machine unused state
II	The virtual machine contains a static web application, which is used by the operator to download iterative files until the migration is completed, so as to ensure the uninterrupted operation of the program during the migration
III	Virtual machine contains dynamic TPC-W application, which is responsible for solving server logic and occupies more resources than static web

Table 2 Comparative analysis results

Load number		I	II	III
Total migration time of this model/s	Similarity (30%, 30%)	19.5	30.6	116.3
	Similarity (65%, 65%)	19.2	29.6	115.6
Total migration time of model based on genetic algorithm/s	Similarity (30%, 30%)	22.5	45.7	124.7
	Similarity (65%, 65%)	21.2	44.1	123.4
Network flow of this model/MB	Similarity (30%, 30%)	1095	2103	6682
	Similarity (30%, 30%)	832	1754	5439
Network flow network flow/MB	Similarity (30%, 30%)	1392	2496	6754
	Similarity (30%, 30%)	854	1844	5743

traffic, that is, the minimum bandwidth overhead, and has better cross data center virtual machine migration performance.

The cloud platform consists of hardware and software facilities, including data center and application services. The data center realizes the cloud computing function, which is composed of servers and storage devices. Each user is assigned virtual computers, which are distributed on one or more physical devices to realize distributed virtual cloud computing. Through network terminal access, users can connect to the cloud computing platform and log in to their virtual machine anytime and anywhere, which is not different from ordinary computers.

6 Conclusions

With the continuous application of mental health education system based on cloud computing, college students will enjoy more psychological services, and the effect of resource utilization and integration will be further improved. In addition, college mental health educators and users can use the data traces left by college students when choosing services in the "mental health education cloud" as the basis for mental health education. Judge the mental health problems that college students pay attention to according to the most visited contents, or improve the electronic archives of College Students' mental health according to the results of College Students' online measurement, master the situation of College Students' psychological development and changes, and achieve prevention and intervention in advance on the basis of constant monitoring, so as to effectively improve the effect of mental health education.

This work is suported by the Subject of Higher Education Teaching Reform in Jilin Province (Project No.: JLJY202135841095).

References

1. N. Zhang, C. Zhang, D. Wu, Construction of a smart management system for physical health based on IoT and cloud computing with big data. Comput. Commun., **179**(11), 183-194 (2021)
2. H.-C. Lin, Y.-C. Kuo, M.-Y. Liu, A health informatics transformation model based on intelligent cloud computing-exemplified by type 2 diabetes mellitus with related cardiovascular diseases. Comput. Methods Programs Biomed. **191**(7), 105409 (2020)
3. A. Kaginalkar, S. Kumar, P. Gargav, D. Niyogid, Review of urban computing in air quality management as smart city service: An integrated IoT. AI, and cloud technology perspective, Urban Climate **39**(9), 100972 (2021)
4. G. Ortiz, M. Zouai, O. Kazar, A. Garcia-de-Prado, J. Boubeta-Puig, Atmosphere: Context and situational-aware collaborative IoT architecture for edge-fog-cloud computing. Computer Standards & Interfaces **79**(1), 103550 (2022)
5. W. Jun, Z. Zhiqiang, T. Rui, L. Dong, L. Tied, Research on recognition of ice and snow athletes based on feature extraction and cloud computing platform. Microprocess. Microsyst., **80**(2), 103338 (2021)
6. P. Stefande, Carvalho Julio Cezar, Mairesse Siluk Jones, Luís Schaefer, Mapping of regulatory actors and processes related to cloud-based energy management environments using the Apriori algorithm. Sustain. Cities Soc. **80**(5), 103762 (2022)
7. O. Sohaib, M. Naderpour, W. Hussain, L. Martinez, Cloud computing model selection for e-commerce enterprises using a new 2-tuple fuzzy linguistic decision-making method. Comput. Ind. Eng. **132**(6), 47–58 (2019)
8. H. Elahi, G. Wang, Y. Xu, A. Castiglione, Q. Yan, M.N. Shehzad, On the characterization and risk assessment of ai-powered mobile cloud applications. Comput. Stand. & Interfaces, **78**(10), 103538 (2021)
9. H. Hassan, H. Mohamad, Hsbollah Rosli Mohamad, Examining the interlink of social media use, purchase behavior, and mental health. Procedia Computer Science **196**(1), 85–92 (2022)
10. C. Goetz, R. Bavaresco, R. Kunst, J. Barbosa, Industrial intelligence in the care of workers' mental health: A review of status and challenges. Int. J. Ind. Ergon. **87**(1), 103234 (2022)

Research on Rural Innovation and Entrepreneurship Platform Based on Cloud Computing

Qiuju Liu and Chengcai Tan

Abstract In order to establish the reliable rural innovation and entrepreneurship platform, the cloud computing is used to design it. Firstly, development mechanism of rural innovation and entrepreneurship in the new era is analyzed. Secondly, construction of "internet +" rural innovation and entrepreneurship support platform based on cloud computing technology. Thirdly, the performance optimization model of rural innovation and entrepreneurship platform based on cloud computing is established based on ant colony algorithm, and simulation analysis is carried out, results show that the proposed algorithm can improve the performance of rural innovation and entrepreneurship platform.

Keywords Rural · Innovation and entrepreneurship · Cloud computing

1 Introduction

In recent years, China has opened an era of innovation and entrepreneurship, and rural innovation and entrepreneurship have also flourished. China will implement the Rural Revitalization Strategy, support and encourage farmers to find jobs and start businesses, and broaden channels for increasing income. China encourages migrant workers, college graduates, veterans and urban talents to return to the countryside for innovation and entrepreneurship, supports the establishment of various forms of entrepreneurship support service platforms, and improves the rural innovation and entrepreneurship support service system. China has established a rural innovation and entrepreneurship and incubation training base, strengthened the training of rural craftsmen, cultural talents, craftsmen, management talents and other innovation and entrepreneurship subjects, and improved entrepreneurial skills. The policies mentioned above have played a positive role in promoting the in-depth development

Q. Liu · C. Tan (✉)
Marxist College of Changchun Guanghua University, Changchun, China
e-mail: tanchengcai@126.com

S. Patnaik and F. Paas (eds.), *Recent Trends in Educational Technology and Administration*, Learning and Analytics in Intelligent Systems 31,
https://doi.org/10.1007/978-3-031-29016-9_42

of rural innovation and entrepreneurship. However, with the increasing downward pressure on the national macro-economy, overcapacity in non-agricultural industries, the deep promotion of agricultural supply side structural reform, rural mass entrepreneurship and innovation still faces many challenges. Although the government has provided strong support in all aspects to promote the development of rural innovation and entrepreneurship in recent years, the process of rural innovation and entrepreneurship is still hindered by the lack of rich entrepreneurial experience and professional guidance, and the lack of applicable innovation and entrepreneurship platform [1, 2]. China should accelerate efforts to make up for the shortcomings of rural infrastructure and promote the interconnection of urban and rural infrastructure. Village road construction should not only improve the hardening proportion of natural village roads, but also solve the problem of too narrow village roads, and speed up the improvement of the long-term mechanism of rural road management and maintenance. Rural water supply projects and pipe networks should be upgraded at a faster pace, with emphasis on improving the construction of sewage treatment supporting facilities.

2 Development Mechanism of Rural Innovation and Entrepreneurship in the New Era

In recent years, the increasing number of rural innovation and entrepreneurship personnel has injected new elements, added new vitality and new impetus into the development of agriculture and rural areas, and created favorable conditions for the high-quality development of rural economy [3].

In recent years, the speed of China's agricultural and rural transformation and development has been accelerating. On the one hand, profound structural changes are taking place in agricultural development. The modern agricultural structure of overall planning of grain, economy and feeding, integration of planting, breeding and fishery, and combination of agriculture, animal husbandry and fishery is gradually taking shape. The direction of high-value, efficient, green, safe, personalized and multifunctional agricultural structure adjustment has opened up more new industrial prospects for Rural Entrepreneurship and innovation, and cultivated a number of new agricultural entrepreneurship and innovation subjects. On the other hand, after 15 years of new rural construction, rural development has greatly improved various infrastructure conditions, and the rural industrial form has rapidly shifted from the primary industry to the integration of primary, secondary and tertiary industries. Farmers' consumption level, living environment and living convenience have been greatly improved, which has profoundly changed the soft and hard entrepreneurial environment and entrepreneurial resources of rural mass entrepreneurship and innovation. The changes in all aspects of "agriculture, rural areas and farmers" have expanded the original rural market, created new business opportunities, improved entrepreneurial willingness and enthusiasm, and increased the possibility of entrepreneurial success.

More and more migrant workers, graduates of colleges and universities, retired soldiers and scientific and technological personnel, as well as many business owners who love agriculture, understand operation and good management, returned overseas students and urban personnel wait for rural innovation and entrepreneurship. Among them, the primary factor for about 80% of the mass entrepreneurship and innovation entities to choose to return home or start a business in the countryside is related to their "Hometown" or the positive changes in the rural production and living environment [4]. Establish rural financial service products for rural innovation and entrepreneurship subjects. Explore and deepen the cooperation mechanism of "bank, government and insurance", set up special government funds for rural innovation and entrepreneurship entities, and encourage financial institutions to actively explore the business of farmers' microcredit loans, policy pledge loans, agricultural machinery and greenhouse facilities mortgage loans through government guarantee and "government and insurance cooperation". Local governments should actively explore the establishment of integrity accounts and information bases for rural innovation and entrepreneurship personnel, give full play to the role of third-party credit service institutions, and promote the establishment of credit system rating standards for innovation and entrepreneurship subjects. Guide financial institutions to lend in combination with the credit status of rural innovation and entrepreneurship personnel and enterprise financial information, and issue pure credit loans to rural innovation and entrepreneurship personnel with normal operation and good reputation.

In recent years, China's industrialization and urbanization have provided important material basis, technical conditions and market space for rural innovation and entrepreneurship, while the rapid development of informatization represented by the Internet has provided impetus for rural innovation and entrepreneurship. "Internet plus" has the important significance of promoting the transformation of traditional industries, cultivating new economic growth points, promoting the transformation of various links in commodity production, circulation and consumption, promoting the innovation of business mode, promoting the transformation of personal thinking mode and social transformation, and promoting the development of entrepreneurship wave. The joint action of industrialization, urbanization and informatization on rural mass entrepreneurship and innovation will greatly change the traditional limitations of human capital, social capital, team heterogeneity and other factors of mass entrepreneurship and innovation on the identification of mass entrepreneurship and innovation opportunities and the improvement of mass entrepreneurship and innovation performance, and create the possibility for relatively isolated villages to realize all-round opening and integrate into the market. Through rural innovation and entrepreneurship, not only all kinds of agricultural products and local specialties can effectively connect with consumers and realize higher market premium, but also rural characteristic natural and human resources can realize unique economic value through relevant creative products and marketing methods; Moreover, rural areas with relatively abundant labor force and low factor cost can also become an important carrier of urban secondary and tertiary industry transfer and expand rural employment capacity [5].

In the context of rural revitalization, agricultural departments at all levels earnestly implement the decisions and arrangements of the Party Central Committee and the State Council, and promote the support of rural innovation and entrepreneurship as an important task of agricultural and rural work and an important starting point for deepening the structural reform of the agricultural supply side. All localities have issued Implementation Opinions on the implementation of the opinions of the state office. Many provinces have established a coordinated promotion mechanism of rural mass entrepreneurship and innovation with the participation of multiple departments, and formulated special supporting policies in terms of financial support, financial guarantee and tax preference. Most provinces have given strong support in many aspects, such as rural mass entrepreneurship and innovation incubation base, park platform construction, mass entrepreneurship and innovation talent training, commercial and administrative services, setting up models, publicity and promotion [6].

Rural areas are the beautiful homes of farmers and even the people of the whole country. It is the right of the people to engage in mass entrepreneurship and innovation in rural areas. It is also a major opportunity for all urban and rural residents, including farmers, to share development opportunities and achievements. In the future, the rural mass entrepreneurship and innovation environment will be more superior, and the government will further increase its support for mass entrepreneurship and innovation. In the near future, all kinds of mass entrepreneurship and innovation will become popular in rural areas and become the engine of urban and rural economic development.

3 Construction of "Internet +" Rural Innovation and Entrepreneurship Support Platform Based on Cloud Computing Technology

By integrating innovation and entrepreneurship resources such as rural governments, universities and enterprises, we will build a comprehensive cloud service platform for rural innovation and entrepreneurship, provide innovation and entrepreneurship services such as policy consultation, talent information, supply and demand docking, achievement transformation and incubation investment for rural innovation and entrepreneurship entities, build a supply and demand docking bridge between innovation and entrepreneurship resources and innovation and entrepreneurship projects, create a strong atmosphere for innovation and entrepreneurship, and reduce innovation and entrepreneurship costs, Promote innovation driven development of rural economy. The rural innovation and entrepreneurship support platform can provide the following services:

(1) The platform can integrate resources such as innovation and entrepreneurship talents, technological achievements and scientific and technological services, promote the close integration of talent chain with industrial chain, innovation

chain and capital chain, and create a good environment for innovation and entrepreneurship services;

(2) The platform can provide convenient and efficient "one-stop" introduction and training of high-level talents through scientific and technological innovation and entrepreneurship talent network management, information personality recommendation, and directional push of innovative results (supply and demand information);

(3) It can provide accurate matching and docking services between innovation and entrepreneurship projects and innovation and entrepreneurship resources through the combination of online and offline operations.

The main goal of the Internet plus rural innovation and entrepreneurship support platform is to provide Internet technology support and services for rural innovative entrepreneurs in county. In order to reduce the threshold of Internet technology requirements for entrepreneurs, and reduce the cost of innovation and entrepreneurship activities, the Docker Engine cloud computing technology is used to build an Internet technology PaaS platform serving rural innovation and entrepreneurship. As shown in the Fig. 1, entrepreneurs can directly use the infrastructure and innovation services of the rural entrepreneurship environment without the support of the underlying software development and innovation management environment [7].

According to the Internet technology application needs of rural innovation and entrepreneurship activities collected and combed, we classify and package the innovation and entrepreneurship environment required by rural innovation and entrepreneurship into template images, such as "rural e-commerce service" template,

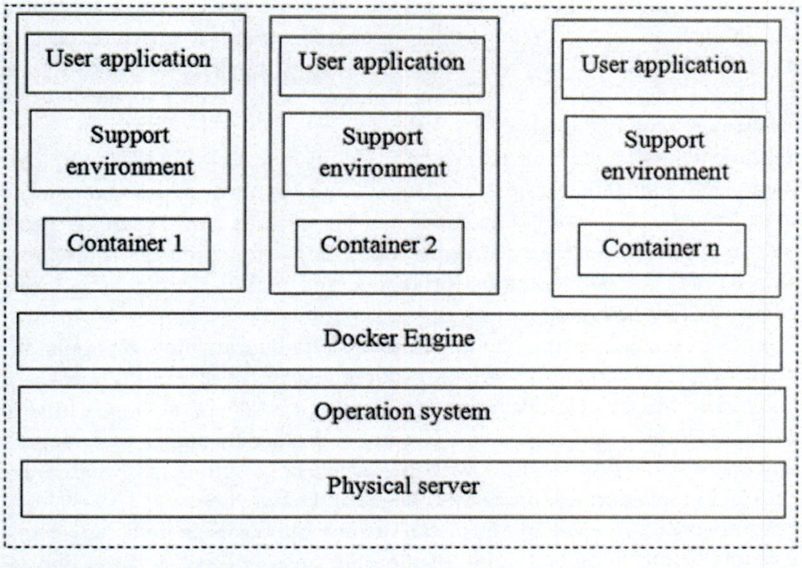

Fig. 1 Architecture of lightweight virtualization technology

"rural characteristic tourism service" template, etc., allocate corresponding resources according to the specific needs of rural innovation and entrepreneurship, and generate an operation container that can meet the needs of innovation and entrepreneurship, Provide it for innovation entrepreneurs to support the development of innovation and entrepreneurship activities of rural innovation entrepreneurs; In addition, the image warehouse supports dynamic management, and the template image can be adjusted dynamically according to the actual needs [8].

As a new network architecture, cloud computing can recalculate and reasonably allocate network resources, meet the different needs of clients, continuously improve and improve the quality of network services, and enable rural innovation and entrepreneurship platforms to obtain the largest network information resources. The requirements for cloud computing scheduling resource technology and strategy are also constantly raised. Due to the lack of necessary accuracy and security of traditional computing methods, a cloud computing scheduling resource model based on ant colony optimization is proposed. In this model, ant colony uses adaptive search, which has high robustness, collaboration, security, easy scalability and parallelism. It can make the ant colony in cloud computing environment achieve the best optimization ability and improve efficiency.

The rural innovation and entrepreneurship cloud platform concludes following functions:

(1) Policy services. The cloud platform can reprint and release the latest policy information on mass entrepreneurship and innovation, provide online policy consulting services, and integrate links to government websites and innovation and entrepreneurship platforms at all levels.
(2) Innovation and entrepreneurship services. The cloud platform is supported by the innovation and entrepreneurship resource library, it provides users with classified navigation, keyword query and supply and demand docking services for various resources such as talents, maker spaces, entrepreneurial services, investment and financing.
(3) Technical services. The platform can release the scientific research achievements and intellectual property rights obtained by universities and enterprises in the region, release the demand for scientific research crowd sourcing, promote industry university research cooperation, help innovation and entrepreneurship entities to solve technical problems, and provide technical services for enterprises in the region.
(4) User Center. The platform can provide registered users with the population for demand release, project undertaking and management of various information.
(5) University library. The platform can display the talents, achievements, mass entrepreneurship and innovation carriers and other resources of colleges and universities, and provide users with the service of searching mass entrepreneurship and innovation resources by colleges and universities.
(6) Enterprise Library. The platform can display the business field, achievements and intellectual property rights of the enterprise, project demand and talent

demand information of the enterprise, and provide users with the service of searching mass entrepreneurship and innovation resources by enterprise.

(7) Background management. The cloud platform can provide the platform administrator with functions such as resource review, resource library maintenance and management, prompt notification of platform supply and demand docking messages, and statistics and maintenance of various information to ensure the smooth operation of the platform.

In cloud computing scheduling resources, its scheduling resources are an important part. In general, in cloud computing, its scheduling resources are effectively decomposed into multiple subtasks. According to the size of subtask data, reasonably schedule and allocate resources to effectively improve the computing efficiency. Its function can be shown in formula (1) [9].

$$P(t_f) = \sum_{j=1}^{d} R_{jf} T(U(W_b) + (C(d_j k_{ju}, W_b))) \tag{1}$$

where $P(t_f)$ is the total time of scheduling and allocating resources, W_b is the virtual machine set by cloud computing, and its value space is $W = \{W_1, W_2, \cdots, W_b\}$. U is the service information point, and its value space is $U = \{U_1, U_2, \cdots, U_b\}$. C is the working time of subtask d_j on virtual machine W_b. j is the workload of cloud computing operator, and its value space is $j = \{ j_1, j_2, \cdots, j_b\}$. k_{ju} is the mapping of scheduling resources. R_{jf} is the scheduling relationship of subtasks on the service information point W_b.

Cloud computing scheduling resources use ant colony MATLAB optimization algorithm, which is to simplify, distribute, parallel, simplify and map a large number of networks, servers, clients and shared resources, effectively decompose a large amount of information and resources, generate a multiple task system, label each task system, and randomly obtain shared information service point resources. The model is shown in formula (2) [10].

$$Q_{mn}^p = \begin{cases} \dfrac{F_{mn}^g(i)[E_{mn}(i)]H_{mn}(i)}{\sum\limits_{m,n \in S} [E_{mn}(i)]H_{mn}(i)} \\ 0, m, n \notin S \end{cases} \tag{2}$$

where Q_{mn}^p is the resource scheduling function of ant colony p from foraging starting point m to ending point n, and i is the time used from starting point m to ending point n, $F_{mn}^g(i)$ is the amount of shared information resources obtained by the current information service point after the ant colony p runs i time, and g is the capacity of the shared information resources of the information service point. $H_{mn}(i)$ represents the amount of shared resources remaining after ant colony P passes through the information service point, s represents the path table that ant colony p can pass through, and $E_{mn}(i)$ represents the network traffic that ant colony p needs to support from starting point m to ending point n.

The algorithm procedure based on expression (2) is listed as follows:

Step 1: The ant colony starts from the starting point m of foraging.

Step 2: Obtain relevant initial foraging parameters.

Step 3: Judge whether it is in the traversable path tables.

Step 4: If the service point is running, the relevant information can be collected in the service point.

Step 5: If it is not in the runnable path, the ant colony returns to the starting point m, re obtains the initial parameters and re runs the judgment.

Step 6: After collecting the information resources of all service points in the current path, the ant colony determines whether to complete the collection. If it is completed, go to the foraging end point n and output the results. If the collection task is not completed, jump back to the information service point and re obtain the remaining shared resources.

4 Case Study

The simulation experiment uses 20 PCs to build a simulated LAN in the cloud computing environment. The PC hardware configuration is shown in Table 1.

One of them is built as a cloud service platform, which is mainly used to set up service information points and provide relevant information sharing resources. It has the functions of data mining, storage and so on. In order to make the platform have the simulation environment of cloud computing, the scheduling resources are supported by C language, and the result analysis and calculation are realized by Matlab platform with distributed architecture. This platform can provide efficient data acquisition function. In the simulation experiment, each client is set to have 6 service information points as data collection points, and the communication association is established between each service information point.

Table 1 Hardware of PC

Name	Type
CPU	G7400
CPU interface	LGA 1700
Core number	Double
Memory	8G
Operating system	Windows 10
Hard disk capacity	500G

Table 2 Time to complete task

p	Time to complete task/ms	
	Ant colony algorithm	Traditional method
200	431	354
400	647	560
600	869	715
800	1238	1105
1000	1457	1328

Conduct 10 simulation experiments and count the average value. The completion time of tasks of different sizes is shown in Table 2.

With the increase of the number of tasks submitted by the traditional ant colony algorithm, the optimal task completion time of the two algorithms is the shortest, which shows that the optimal task completion time of the traditional ant colony algorithm can also increase.

Cloud computing technology is used to build a rural innovation and entrepreneurship platform. From the rural e-commerce services, rural characteristic tourism website services, WeChat official account management services, data management as a service, development platform as a service, and operation environment as services, the Internet plus innovation platform is constructed to provide the Internet technology support and service for rural entrepreneurs.

Rely on government departments or industry associations, special operation teams and institutions are established to be responsible for platform promotion and operation.

(1) Online promotion

Online promotion is carried out through paid and free channels such as WeChat official account, forum, search engine website, email, etc. to improve the visibility of rural innovation and entrepreneurship cloud platform.

(2) Offline operation

With the support of the regional government departments, relying on the policy support and guidance, it has attracted university resources, enterprises and innovation and entrepreneurship service institutions in the region to settle on the platform. Relying on the school's mass entrepreneurship and innovation base, attract talents to settle on the platform through special training or lectures, forums, seminars, research meetings, etc., and develop a large number of new users. Carry out interactive gathering among members of the platform and spread among members. For example, holding member gatherings, tea parties, outings and other activities to promote mutual exchanges between talents and promote mutual cooperation between innovative and entrepreneurial talents and enterprises. Cooperate with rural governments, enterprises and institutions, participate in various industrial and capital promotion conferences held by governments, enterprises and institutions, promote the platform in the activities, and attract more talents.

5 Conclusions

This paper proposes a rural innovation and entrepreneurship platform based on cloud computing technology, which can provide a strong basic guarantee for rural innovation and entrepreneurship. Targeted feedback and Optimization Based on the subsequent application of the system platform can provide more comprehensive and targeted platform services for cultivating rural innovation and entrepreneurship.

The rural innovation and entrepreneurship platform based on cloud computing is not only the coordination organization of government service departments, but also the communication network of regional rural entrepreneurs and entrepreneurs, but also the link organization for the coordinated development of industry, University and research. Its main purpose is to reduce administrative execution costs and market transaction costs. We should not only improve the diversified service functions in land management, ecological environment management and infrastructure management, but also give full play to the service role of industry associations composed of entrepreneurs, professionals and rural capable people, coordinate operation management, collectively respond to the crisis, and do a good job in medical insurance, children's education and other services for themselves and their employees.

Acknowledgements This paper is supported by the scientific research planning project of Jilin Provincial Department of Education in 2022 "Research on the Construction and Practice of Innovation and Entrepreneurship Education System in Private Colleges and Universities from the Perspective of Maker Space" (No.: JJKH20212888JY)

References

1. D. Zhou, M. Kautonen, W. Dai, H. Zhang, Exploring how digitalization influences incumbents in financial services: The role of entrepreneurial orientation, firm assets, and organizational legitimacy. Technol. Forecast. Soc. Chang. **173**(12), 121120 (2022)
2. D. Reuschke, C. Mason, S. Syrett, Digital futures of small businesses and entrepreneurial opportunity. Futures **135**(1), 102877 (2022)
3. Z. Xie, X. Wang, L. Xie, K. Duan, Entrepreneurial ecosystem and the quality and quantity of regional entrepreneurship: A configurational approach. J. Bus. Res. **128**(5), 499–509 (2021)
4. S. Safdar, M. Ren, M.A.Z. Chudhery, J. Huo, H.-U. Rehman, R. Rafique, Using cloud-based virtual learning environments to mitigate increasing disparity in urban-rural academic competence. Technol. Forecast. Soc. Chang. **176**(3), 121468 (2022)
5. Y. Tan, X. Li, The impact of internet on entrepreneurship. Int. Rev. Econ. Financ. **77**(1), 135–142 (2022)
6. I. Ali, M. Balta, T. Papadopoulos, Social media platforms and social enterprise: Bibliometric analysis and systematic review. Int. J. Inf. Manag., 102510 (2022). in press
7. M. Rodrigues, M. Franco, Digital entrepreneurship in local government: Case study in Municipality of Fundão. Portugal, Sustainable Cities and Society **73**(10), 103115 (2021)
8. P. Gregori, P. Holzmann, Digital sustainable entrepreneurship: A business model perspective on embedding digital technologies for social and environmental value creation. J. Clean. Prod. **272**(11), 122817 (2020)

9. M. Shamsul, Karima Sharmin Nahar, Mehmet Demirbag, Resource-based perspective on ict use and firm performance: A Meta-analysis Investigating the Moderating Role of Cross-Country ICT Development Status. Technol. Forecast. Soc. Chang. **179**(6), 121626 (2022)
10. F. Khod, Parast Chandni Sindhav, Seema Nikam Hadiseh, Izadi Yekt, Cloud computing security: A survey of service-based models, Technological Forecasting and Social Change. Comput. Secur. **114**(3), 102580 (2022)

The Health and Sustainability Evaluation of National Higher System via the CIPP Method and Its Application

AoQi Tan and Xiang Xie

Abstract The higher education system is the essential to further improve the level of civic education in all countries around the world. In this paper, we construct a three-tier indicator system to evaluate different countries' higher education systems called CIPP, which is composed of context evaluation, input evaluation, process evaluation and products evaluation as the first class indicators and have 9 s class indicators, 20 third class indicators. Meanwhile, the standard of system health and sustainability is obtained by K-Means algorithm, and specifically classified into normal, good, excellent three states. Next, we calculate the scores of CIPP for the 178 countries collected and analyze three countries. We find that China's higher education system is relatively weak compared to the other two countries. Then, we analyze the strengths and barriers of its current context and situation and set an ultimate improvement goal. Further, we use principal component analysis to determine the main indicators of each first class indicator. Integrating the above information, we propose 9 targeted policies for it. And a schedule for implementing this policy under different stages in the future. To assess the effectiveness of the policies, we use GM (1,1) to make predictions about the state of the country's higher education system before and after the policy adjustment. The results show that our policies will have an increasingly positive impact on the health and sustainability of the country's higher education system over time, and that the country will reach excellent state by 2043.

Keywords Higher education · CIPP · Evaluation · K-Means · Policy

1 Introduction

The higher education system is an important component of countries around the world to further improve the education of their citizens. It not only plays the function

A. Tan (✉) · X. Xie
Beijing Jiaotong University, Beijing 100044, China
e-mail: tanaoqi2000@163.com

S. Patnaik and F. Paas (eds.), *Recent Trends in Educational Technology and Administration*, Learning and Analytics in Intelligent Systems 31,
https://doi.org/10.1007/978-3-031-29016-9_43

of educational governance, but also influences the economic and civil development of the country. Looking around the world, we see that each country has its different approach to higher education.

However, diverse countries' higher education systems have their strengths and weaknesses, so how to build a system of indicators to evaluate the health and sustainability of higher education? What constitutes an effective and self-renewing system? What policies and measures can be taken to bring the country to the state we want? Built on the above considerations about the higher education system, we will design a comprehensive and feasible indicator system to evaluate the higher education system and address the above questions through mathematical models.

2 Evaluation Model of Higher Education System

The existing literature shows that the evaluation of the higher education system is a complex concept with different methods of analysis and perspectives of interpretation [1]. The framework of educational inputs, processes and outputs is by far the most commonly used analytical process in the construction of education indicator systems at the national level [2]. On the basis of this framework, the CIPP model of educational evaluation proposed by the renowned educational evaluation expert Stuffleam, L.D., is more applicable to the evaluation of the higher education system [3]. CIPP consists of four indicators that represent distinctive aspects of Context Evaluation, Input Evaluation, Process Evaluation, and Product Evaluation.

The CIPP model can assist in the definition of programme objectives, the revision of research plans, the realization. Therefore, in this process of developing an evaluation system for higher education systems, we followed the above model and created an independent mathematical model for evaluating the health and sustainability of higher education systems in each country.

Here we start to analyze the indicators which can imply the four aspects of the CIPP model. We have built an indicator system called CIPP to measure the health and sustainability of the country's higher education system. A larger CIPP implies a better development situation in aspects of the background, input, process and results. To construct the CIPP, we successively construct metric of CTE, IPE, PCE and PDE, Separately measures the context evaluation, the input evaluation, the process evaluation and the product evaluation of higher education.

2.1 *Basic Framework of the Indicator System*

Hongyi Xu et al. followed the analytical framework of context, input, process and outcome as the first class indicators of the evaluation system [2]. After that

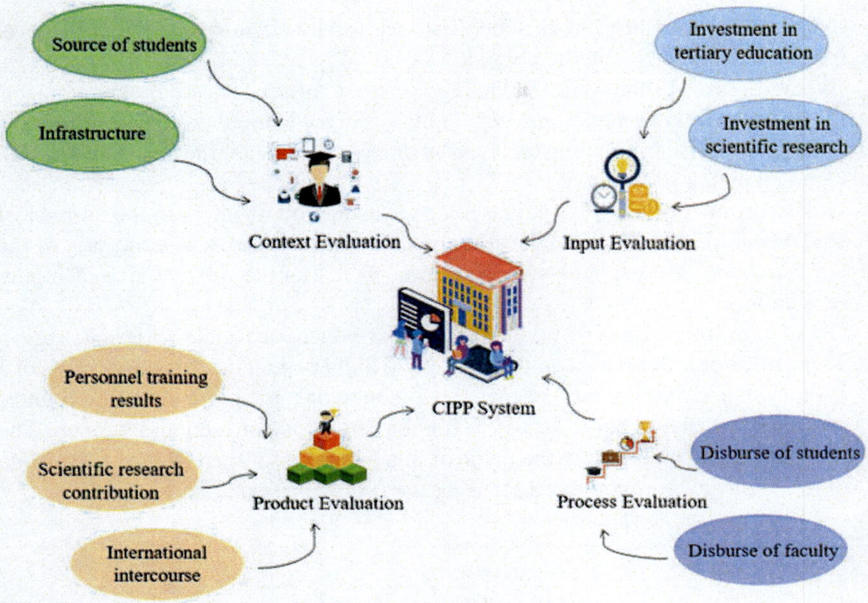

Fig. 1 The second class indicators for CIPP system

its identified twelve second class indicators with diagnostic and orientation functions. The construction of the indicator system follows the analytical framework of input-process-outcome and there is a feedback loop in the system.

Therefore, we used this system as a reference to determine 9 s class indicators of Source of students, Infrastructure, Investment in higher education, Investment in scientific research, Disburse of students, Disburse of faculty, Personnel training results, Scientific research contribution, International intercourse. Figure 1 shows the second class indicators.

2.2 *Metric of CTE*

Selection of indicators

- Source of students.

Demographic statistics are inseparable from the evaluation of a country's educational level, which is also an important element for higher education.

- Duration of compulsory education: The number of years of schooling per capita reflects the overall level of education received by a country's population; The longer the number of years of schooling received by citizens, the higher the level of education received by the population.

- Average years of schooling: The number of years of schooling per capita reflects the overall level of education received by a country's population; The longer the number of years of schooling received by citizens, the higher the level of education received by the population.
- Gross enrolment ratio of population higher education: it refers to the number of students enrolled in higher education in an academic year as a proportion of the corresponding total school-age population. It marks the relative size of higher education.
- Gross enrollment ratio of population higher education (male to female ratio): This indicator reflects the equity of access to higher education for the citizens of a country. The closer the ratio of men to women to one, the more effort the country is making to ensure equal access to higher education for men and women. The indicator is calculated from the value of this indicator for men (R1) and the value of this indicator for women (R2), using the following formula:

$$CET_{14} = \frac{R_1}{R_2} \tag{1}$$

- Infrastructure

Personal computers not only provide new channels for academic communication for people deep in higher education, but also provide a way for university and social resources to cooperate with each other. Therefore, we choose the computer penetration rate of school students to represent the capacity of national infrastructure.

The CTE Metric system

Based on the above analysis, we developed a method, called Context Evaluation (CTE), to assess the foundations of higher education. The indicator includes two second class indicators and five third class indicators. We show the metrics of CTE in Table 1.

2.3 Metric of IPE

Selection of indicators

- Investment in Higher education.

Table 1 The indicators used in CTE Metric system

CTE	Indicators	Notation	Indicators	Notation
	Source of students	CTE_1	Duration of compulsory education	CET_{11}
			Average years of schooling	CET_{12}
			Gross enrolment ratio of population higher education	CET_{13}
			Gross enrollment ratio of population higher education (male to female ratio)	CET_{14}
	Infrastructure	CTE_2	Computer penetration rate of college students	CET_{21}

The metric of CTE can be finally expressed by a function of CTE_1 and CTE_2 as $CTE = f(CTE_1, CTE_2)$.

Among the various indicators that reflect investment in higher education, the level of government financial support for higher education has a direct impact on what level of higher education can be achieved.

- Higher education spending as % of GDP: This indicator directly reflects the total amount of a country's investment in higher education and is the main indicator of the extent of the national government's efforts in education. An increase in financial investment in education can provide strong support for education reform and development.
- Higher education expenditure as % of government expenditure: This indicator reflects how a country allocates its overall financial resources to all levels of education. The higher a country spends on higher education, the better funded it is and the more sustainable it is.
- Staff compensation as % of total expenditure: Teachers and students are the two main bodies of the university. Among them, teachers are the relatively stable group in the university. We examined the expenditure structure of universities and found that the core indicator is the percentage of all staff salaries to the total expenditure on higher education institutions. We have therefore selected this indicator.

• Investment in scientific research.

"Research and development expenditure" is an important support for the research progress of the higher education system. "Research and development expenditure" reflects the level of investment in research and development by companies, governments, private non-profit organisations and the higher education sector in a country.

The IPE Metric system

Based on the above analysis, we developed a method, called Input Evaluation (IPE), to assess the input of higher education. The indicator includes two second class indicators and four third class indicators. We show the metrics of IPE in Table 2.

Table 2 The indicators used in IPE Metric system

IPE	Indicators	Notation	Indicators	Notation
	Investment in Higher education	IPE_1	Higher education spending as % of GDP	IPE_{11}
			Higher education expenditure as % of government expenditure	IPE_{12}
			Staff compensation as % of total Expenditure	IPE_{13}
	Investment in scientific research	IPE_2	Research and development expenditure	IPE_{21}

The metric of IPE can be finally expressed by a function of IPE_1 and IPE_2 as $IPE = f(IPE_1, IPE_2)$.

2.4 *Metric of PCE*

Selection of indicators

- Disburse of students

  Students, postgraduates and PhD are the main body of study in higher education, and the effectiveness of their studies depends not only on the teaching arrangements of the university, but also on the degree of importance they attach to their studies.

- Gross attendance rate: Attendance reflects the motivation and self-discipline of students in a country's higher education institutions and has a direct impact on the effectiveness of classroom teaching, which has an impact on the quality and competence of students.
- School life expectancy: This indicator reflects the number of years that students who have attended higher education have been in the institution and reflects the overall level of access to education for this group in the country in terms of the time dimension.

- Disburse of faculty

The number of teachers in higher education is the basis for the development of higher education. The number of teachers in higher education, and the level of teachers in higher education, reflects the quality of higher education to a certain extent. Therefore, the indicator 'Teachers in higher education programmes is chosen to measure the input of teachers in the national education system.

Table 3 The indicators used in PCE metric system

PCE	Indicators	Notation	Indicators	Notation
	Disburse of students	PCE_1	Gross attendance rate	PCE_{11}
			School life expectancy	PCE_{12}
	Disburse of faculty	PCE_2	Teachers in higher education programmes	PCE_{21}

The metric of PCE can be finally expressed by a function of PCE_1 and PCE_2 as $PCE = f(PCE_1, PCE_2)$.

The PCE Metric system

Based on the above analysis, we developed a method, called Process Evaluation (PCE), to assess the process of higher education. The indicator includes two second class indicators and three third class indicators. We show the metrics of PCE in Table 3.

2.5 *Metric of PDE*

Selection of indicators

- Personnel training results

  This second class indicator focuses on the quality of graduates from higher education and the extent to which they contribute to society.

- Gross graduation ratio from first degree: The overall graduation rate reflects the ability of universities to train people for higher education, as well as the level of teaching and the ability of students to access higher education.
- Programmes Percentage of population age 15 + with tertiary schooling: This indicator reflects the number of people trained in higher education. The higher the share of this indicator in a country, the stronger the academic quality of its citizens.
- Labor force with advanced education: The proportion of the group with higher education in a country's main workforce can reflect the contribution that a country's higher education sector makes to the country and reflects the results of talent development.
- Percentage of population age 25 + with a doctoral degree: This indicator reflects the proportion of PhD in the corresponding age group of citizens and reflects the density of highly knowledgeable people in a country.

- Scientific research contribution

In the national development strategy, science, technology and innovation capacity are considered to be an important part of the improvement of comprehensive national power. Therefore, we have selected two indicators: 'Number of patent applications per capita in universities', 'Number of published SCI and EI per capita in universities'.

- International intercourse

The advent of the knowledge economy has brought new challenges and opportunities for higher education [4]. Under this second class indicator, we have selected 'Inbound mobility rate' and 'Percentage of students going abroad', which reflect the ratio of international students to national higher education students in a country and the ratio of students going abroad to national higher education students in a country respectively. These two indicators reflect the intensity of the resources that a country's higher education system receives from other systems [5].

The PDE Metric system

Based on the above analysis, we developed a method, called Product Evaluation (PDE), to assess the development performance of higher education. The indicator includes three second class indicators and eight third class indicators. We show the metrics of PDE in Table 4.

Table 4 The indicators used in PDE metric system

PDE	Indicators	Notation	Indicators	Notation
	Personnel training results	PDE_1	Gross graduation ratio from first degree programmes	PDE_{11}
			Percentage of population age 15 + with tertiary schooling	PDE_{12}
			Labor force with advanced education	PDE_{13}
			Percentage of population age 25 + with a doctoral degree	PDE_{14}
	Scientific research contribution	PDE_2	Number of patent applications per capita in universities	PDE_{21}
			Number of published SCI and EI per capita in universities	PDE_{22}
	International intercourse	PDE_3	Inbound mobility rate	PDE_{31}
			Percentage of students going abroad	PDE_{32}

The metric of PDE can be finally expressed by a function of PDE_1, PDE_2 and PDE_3 as $PDE = f(PDE_1, PDE_2, PDE_3)$.

3 Analysis for National Higher Education System

3.1 Data Preprocessing

Data collection

Collecting sufficient data is the basis of establishing a complete index system. We collected data mainly through the official website of the World Bank Group and found 13 indicators from 178 countries. On this basis, we also introduce a total of 7 indicators from knoema, Publons and other platforms.

Data padding

The availability of data is an issue that can not be ignored. If unreliable or untruthful data is used, no matter what measure it is, no matter how valuable it is, it can not provide an efficient assessment. Thus it can be seen that it is particularly important to ensure the authenticity and continuity of the data. However, not all data can be collected. In order to improve this situation, we propose the following four ways to improve the data:

- If the data value of the indicators are smooth, the previous data can be replaced instead.
- The average value can be considered as missing.
- If two groups of data are similar, we replace the missing data in the other group with the data in the same position in one group.
- The data can be fitted by interpolation.

3.2 Weight of Indicators

Weighting model based on information entropy and coefficient of variation (IECV)

In this section, with the evaluation indicators defined above, we further determine the weights of these 20 indicators. We first use the data normalization to eliminate the data incommensurability caused by data scale inconsistency, based on the attribute type of the original indicators, we use the standard 0–1 transformation and the given optimal interval method to normalize the dimension. Then we obtain the weight of each indicator by combining the information entropy and coefficient of variation. The detailed algorithm is shown as follows.

Step 1: Data normalization.

Let x_{ij} denote the $i-th$ indicator of $j-th$ country.

For the benefit-type indicator:

$$x_{ij} = \frac{x_{ij} - x_i^{min}}{x_i^{max} - x_i^{min}} j = 1, \cdots, n \tag{2}$$

where x_i^{max} is the largest value of the ith indicator and x_i^{min} is the smallest value:

$$x_i^{max} = max\{x_{i1}, x_{i2}, \cdots, x_{in}\} \tag{3}$$

$$x_i^{min} = min\{x_{i1}, x_{i2}, \cdots, x_{in}\} \tag{4}$$

For the interval-type indicator:

$$x_{ij} = \begin{cases} \frac{x_{ij}}{a} & x_{ij} \leq a \\ 1 & a \leq x_{ij} \leq b \\ 1 - \frac{x_{ij} - b}{x_i^{max} - b} & x_{ij} > b \end{cases} \tag{5}$$

where [a, b] is the best value range of the indicator.

Step 2: Calculation of information entropy.

Let q_{ij} denote the ratio of each indicator, it can be calculated by the following formula:

$$q_{ij} = \frac{x_{ij}}{\sum_{j=1}^{n} x_{ij}} \tag{6}$$

then the entropy value e_i is obtained by the following formula:

$$e_i = -K \sum_{j=1}^{n} q_{ij} \ln q_{ij} \; i = 1, 2, \ldots, m \tag{7}$$

where the constant value K equal to $1/\ln n$.

Step 3: Calculation of coefficient of variation.

The entropy value v_i is obtained by the following formula:

$$\mathrm{v}_i = \frac{s_i}{\overline{x}_i}, i = 1, 2, \ldots, m \tag{8}$$

where the q_i and $\overline{x}_i$ is the mean and standard deviation of $i - th$ indicator:

$$\begin{cases} \overline{x}_i = \frac{1}{n} \sum_{j=1}^{n} x_{ij} \\ s_i = \sqrt{\frac{\sum_{j=1}^{n} (x_{ij} - \overline{x}_j)}{n-1}} \end{cases} \tag{9}$$

Step 4: Obtain the Weights.

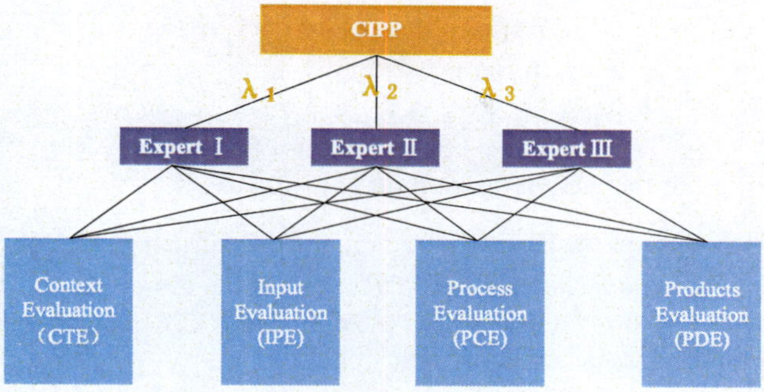

Fig. 2 The flow of calculation weights of indicators using GDM

We get the weight of indicators by calculating the weighted average of the results calculated above and define the equation as follows:

$$\mathrm{w}_i = \xi \cdot \frac{1 - e_i}{m - \sum_{i=1}^{m} e_i} + (1 - \xi) \cdot \frac{\mathrm{v}_i}{\sum_{i=1}^{m} \mathrm{v}_i}, \; i = 1, 2, \ldots, m \tag{10}$$

where we set the value of ξ is 0.8.

Weighting model based on Group Decision Making Method (GDM)

In order to get the score of our CIPP system, we use consistent judgment matrix for each metric for weighting. The following Fig. 2 shows how the GDM work.

For each expert, we obtain a positive reciprocal judgment matrix, which is scored by expert according to their experience, and indicators are scored according to the scoring table. Matrixes are shown as below:

$$E_I = \begin{matrix} C_1 \\ C_2 \\ C_3 \\ C_4 \end{matrix} \overset{\begin{matrix} C_1 & C_2 & C_3 & C_4 \end{matrix}}{\begin{pmatrix} 1 & 5 & \frac{1}{3} & \frac{1}{7} \\ 5 & 1 & 2 & \frac{1}{2} \\ 3 & \frac{1}{2} & 1 & \frac{1}{3} \\ 7 & 2 & 3 & 1 \end{pmatrix}} \quad E_{II} = \begin{matrix} C_1 \\ C_2 \\ C_3 \\ C_4 \end{matrix} \overset{\begin{matrix} C_1 & C_2 & C_3 & C_4 \end{matrix}}{\begin{pmatrix} 1 & \frac{1}{3} & \frac{1}{5} & \frac{1}{2} \\ 3 & 1 & \frac{1}{2} & 2 \\ 5 & 2 & 1 & 3 \\ 2 & \frac{1}{2} & \frac{1}{3} & 1 \end{pmatrix}}$$

$$E_{II} = \begin{matrix} C_1 \\ C_2 \\ C_3 \\ C_4 \end{matrix} \overset{\begin{matrix} C_1 & C_2 & C_3 & C_4 \end{matrix}}{\begin{pmatrix} 1 & \frac{1}{3} & \frac{1}{2} & \frac{1}{4} \\ 3 & 1 & 2 & \frac{1}{2} \\ 2 & \frac{1}{2} & 1 & \frac{1}{2} \\ 4 & 2 & 2 & 1 \end{pmatrix}} \tag{11}$$

By solving the eigenvalues and eigenvectors of the above matrix, the weights obtained by the three experts are as follows:

$$\begin{cases} W_{\text{I}} = [0.0601, 0.2878, 0.1615, 0.4905] \\ W_{\text{II}} = [0.0882, 0.2720, 0.4829, 0.1570] \\ W_{\text{II}} = [0.0931, 0.2729, 0.1753, 0.4587] \end{cases} \quad (12)$$

The reliability of the weight is examined as follows:

- Calculate the maximum of eigenvalue λ_{max} and the consistency index $CI = \frac{\lambda_{max}-n}{n-1}$.
- Then calculate the consistency ratio $CR = \frac{CI}{RI}$,as the dimension of the judgment matrix is four, so the value of RI is 0.89. Because CR is smaller than 0.1, the judgment matrix passes the consistency check.

Further we calculate the final weight by the following equation:

$$W_{\text{final}} = \lambda_1 W_1 + \lambda_2 W_{11} + \lambda_3 W_{ll}$$

Now we can calculate the CIPP score for each country:

$$CIPP = \text{w}_1 \cdot CTE + \text{w}_2 \cdot IPE + \text{w}_3 \cdot PCE + \text{w}_4 \cdot PDE$$

where the w_i is given in $W_{\text{final}} = [0.0784, 0.2786, 0.2621, 0.3809]$.

3.3 The Standard of CIPP

In Sect. 2, we have established a three-tier of evaluation. However, an appropriate standard is still needed to assess the healthy and sustainability of the higher education system. Therefore, we adopted K-means Clustering Algorithm to establish a reasonable standard, which is used to determine the status of higher education systems in different countries.

In this algorithm, for each first class indicator, we clustered data from 178 countries to calculate three cluster centers. Then, the average value of the center of the index is used as the standard boundary. As shown in Fig. 3, we assign different colors to the different higher education system's levels of weak, moderate and strong, so as to intuitively display the state of the higher education system.

Normal: the national higher education system still has a lot of room for improvement.

Good: there is some room for improvement in the national higher education system.

Excellent: the national higher education system is among the best in the world and is successful.

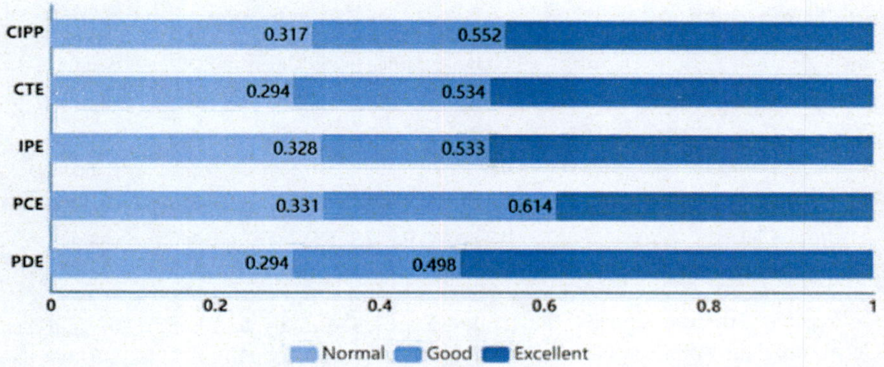

Fig. 3 The Standard of CIPP, CTE, IPE, PCE and PDE

3.4 Analysis of Different Higher Education Systems

Based on our model, we obtain the CIPP value of each country and the value of each indicator. As shown in Fig. 4, we visually display the results on the map. The darker the color of a country, the higher the CIPP value of the country, and the healthier and more sustainable the national higher education system.

We ranked the selected 178 countries based on the CIPP value. And select some countries for display, as shown in the Table 5.

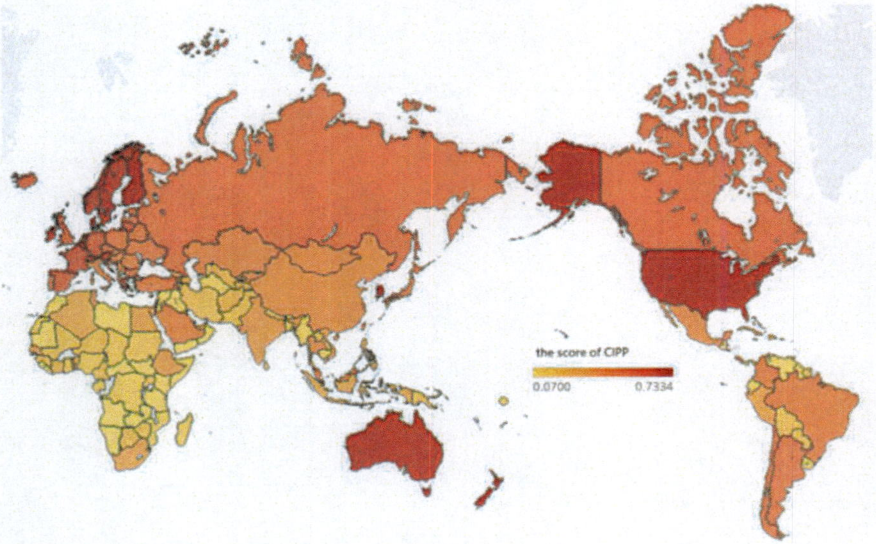

Fig. 4 World map for CIPP score

Table 5 Specific scores and states of some countries

Rank	Country	CIPP	CTE	IPE	PCE	PDE	State
1	Denmark	0.733	0.702	0.718	0.891	0.642	Excellent
3	United States	0.696	0.803	0.629	0.775	0.669	Excellent
17	Germany	0.581	0.675	0.509	0.516	0.660	Excellent
28	Japan	0.504	0.516	0.418	0.433	0.612	Good
58	China	0.359	0.239	0.337	0.295	0.445	Good

We select three countries, the United States, Germany and China for analysis. Combining the four second class indicators. As shown in Figs. 5 and 6, we use the radar chart to reflect the gap between the indicators of the selected countries intuitively, and show the comparison of the CIPP of the three countries by a line chart.

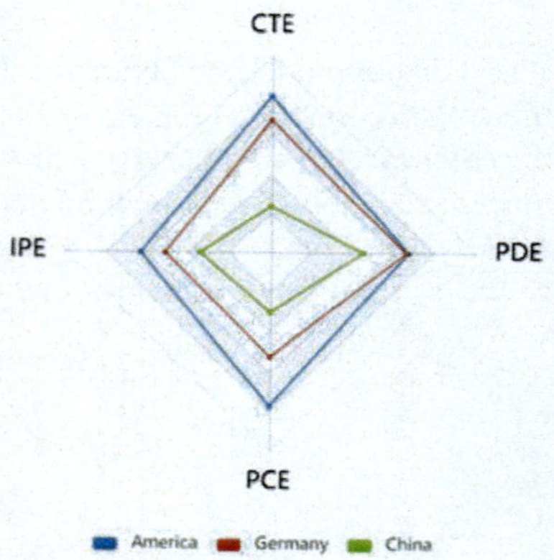

Fig. 5 Comparison of the first class indicators in three countries

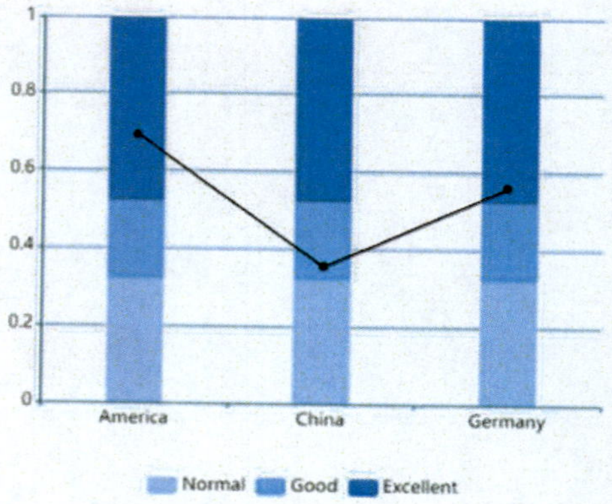

Fig. 6 Comparison of CIPP in three countries

From the above analysis, we can see that the score of CIPP of China's higher education system is 0.359, and the CTE, IPE, PCE and PDE are 0.239, 0.337, 0.295, and 0.445 respectively, all of which have not reached an excellent state. It shows that there is still a lot of room for improvement in the system. Therefore, we will choose China's higher education system as the research object and analyze it in detail in the next section.

4 The Migration Plan for Chinese Higher Education System

The above analysis shows that there is still much place for improvement in the area of China's higher education system. To improve the existing measures to help the system be able to reach the excellent state, we develop an improvement plan for them according to the following steps:

- Set the goal of the final state.
- Identify the status of the system, analyze the advantages and disadvantages.
- Formulate a range of policies for goals and status.
- Formulate development plans for different states.
- Assess the feasibility of the policy.

4.1 Set the Goal

Based on the analysis in Sect. 3, China's higher education system does not achieve excellent state in any of the four dimensions involved in the CIPP model. Considering that any institutional reform needed in the system requires policies to be implemented over a longer period of time in order to be effective, the goal we set for the country is that in 25 years the country's CIPP score will reach excellent state and the gap between the four areas will be reduced to about 50% of the current level.

4.2 Identify the Status of the System

In this section, we will analyze the strengths, obstacles and challenges of China's higher education system in its current development path.

- Strengths

In terms of the stage of higher education development, China's higher education has entered the stage of massification and ranks first in the world in terms of the number of universities. China has formed a new pattern of diversified forms and multi-level types of schooling and common development of public and private education [6]. The government and society have continuously increased their support and investment in the development of higher education, and the deepening of higher education reform has made China achieve historic achievements in recent years.

- Barriers

Compared with countries in an excellent state, China's higher education system does not have enough input of human, material and financial resources, and there is uncoordinated between them, such as the inconsistency between Student education and social needs, and the unreasonable structure of faculty, etc. These contradictions are the outstanding problems that affect the sustainable development of the higher education system [7].

4.3 *Formulate a Range of Policies*

We selected a total of 20 third class indicators in the evaluation index system of higher education system. Based on this, we use principal component analysis (PCA) to analyze the main indicators affecting the two first class indicators, CTE and PDE, in order to make recommendations for the current higher education system in China more specifically. For other two metrics, IPE and PCE, we subjectively determine their main indicators. Table 6 shows the core indicators corresponding to different policies.

We analyze each metric and in relation to the main indicators we propose the following policies for China's higher education system.

- CTE: China's current per capita GDP belongs to the third echelon in the world, and the economy can hardly support too large scale of higher education. China has been improving the academic infrastructure of higher education in recent years, but there is still a big gap in the level of facilities with the countries in the excellent

Table 6 Core indicators corresponding to different policies

Metric	policy	Indicator components
CTE	W_1	CTE_{12}, CTE_{14}
	W_2	CTE_{21}
IPE	W_3	IPE_{11}
	W_4	IPE_{13}, IPE_{21}
PCE	W_5	PCE_{11}, PCE_{12}
	W_6	PCE_{21}
PDE	W_7	PDE_{11}, PDE_{12}
	W_8	PDE_{21}, PDE_{22}
	W_9	PDE_{31}, PDE_{32}

state. In addition, the development of higher education in China is relatively late compared to some developed countries in the West, which leads to a relatively low level of student source in China. Combined with the analysis of the actual national situation, we believe that we should not be overly optimistic about the expansion rate of the scale, but should focus on the quality of school operation, and determine the development plan realistically [7]. Two requirements for the development of relevant policies are as follows:

- W1: Control the size of enrollment rationally
- W2: Increasing academic infrastructure at a steady rate

- IPE: The government's investments, distribution and utilization of China's higher education system are not reasonable [8]. This situation will make the gap between universities increasing. As a result, we have developed an adjustment-type policy as follows:
  - W3: The government should increase the overall revenue appropriately
  - W4: Optimize the financial allocation ratio, narrow the gap in funding allocation among universities of different levels, and improve the effectiveness of resource use [9]
- PCE: For the higher education system, the participation and motivation of students and teachers directly affect the output of the system. Therefore, we develop the policies to improve the process of higher education system in China as follows:
  - W5: Establish effective course control mechanisms to improve the quality of the lessons for students and teachers
  - W6: Establish a reasonable evaluation mechanism for teachers and students to reflect the problems objectively and truly, so that the problems can be corrected accurately [10]
- PDE: China's higher education suffers from unreasonable structure of disciplines and specialties, insufficient awareness of industry-university-research cooperation, and weak effectiveness of innovative talent cultivation. For this reason, we propose the policy recommendations below as follows:
  - W7: Evaluate the education of colleges and universities to enhance the ability to proactively adapt to society, thus improving the quality of education
  - W8: Promote the integration of industry-university-research education, strengthen the combination of universities and enterprises, and boost the development of research output
  - W9: Strengthen academic exchanges with international universities in proactive way to enhance the international competitiveness of higher education

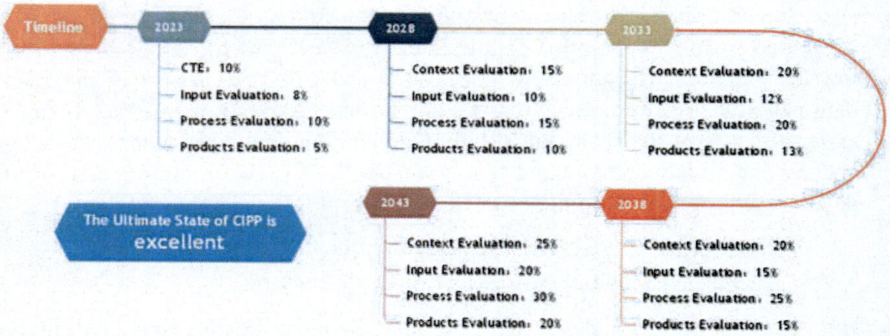

Fig. 7 The timeline for China's higher education system

4.4 Formulate Development Plans

In this section, we set the goal with CIPP's four dimensions of comprehensive and coordinated development and give specific timelines in Fig. 7.

The ratios in Fig. 7 indicate the percentage increase in the scores of the first indicators after the policy adjustment compared to the current state at the time of the plan year.

4.5 Assess the Feasibility of the Policy

In this section, under the evaluation model we developed, we use a gray prediction model to predict the effects of the policies proposed at different time periods. If a set of time series data has obvious trend, Grey Forecasting Model in the Grey System Theory can give a precise prediction even with few data. We apply GM(1,1) model, the most widely used Grey Forecasting Model, to forecast the score of CIPP for China's higher education system in 2023, 2028, 2032, 2038 and 2043 under the policies we developed.

Through the Fig. 8, we can intuitively observe that, our plans demonstrate increasing positive effects on the CIPP, that's to say our plans will exert better influence on the healthy and sustainable of the system as time goes by, from now to 2043.

Next, we use the same projection method to obtain the score of CIPP, CTE, IPE, PCE, and PDE in 2043 after and before the policy adjustment respectively, and compare them with the current state.

As shown in Fig. 9, by 2043, without policy adjustments, the score of CIPP would be 0.47, which is in food state. There is still a gap between being an excellent healthy sustainable system. After implementing our proposed policy, the CIPP value will reach the EXCELLENT state, which means that our policy is effective.

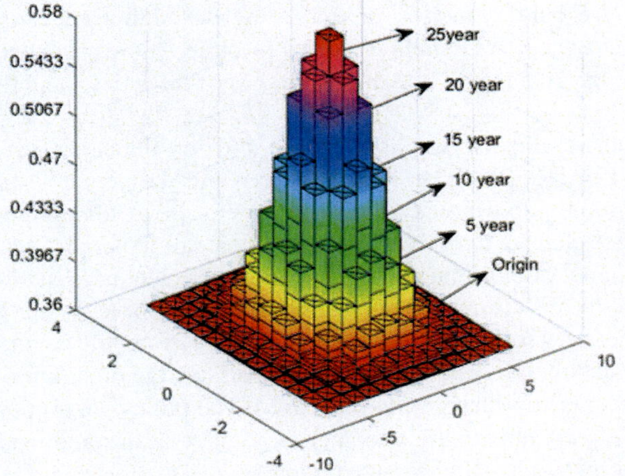

Fig. 8 The predicted value of CIPP in five time periods

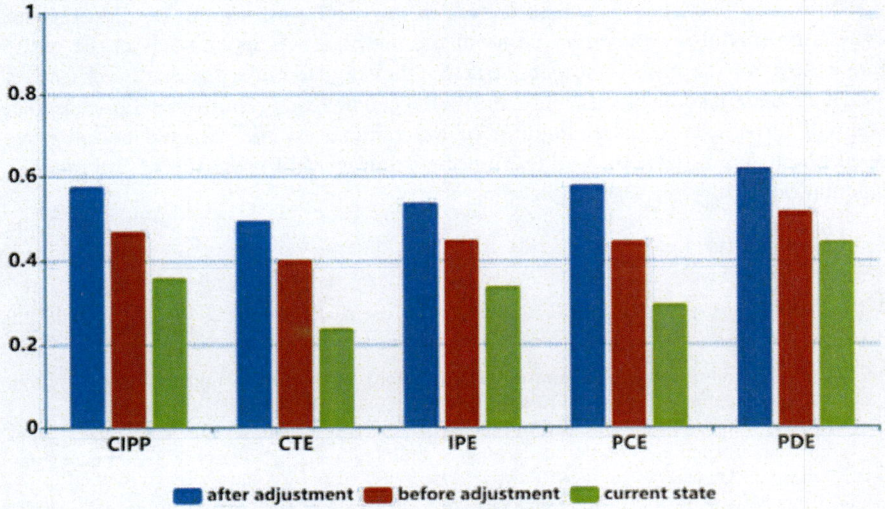

Fig. 9 The end value of CIPP, CTE, IPE, PCE, PDE

5 Conclusion

In this paper, we build a model for evaluating the higher education system and propose a series of policies to help it reach a healthier, more sustainable state. Next, we summarize the advantages and disadvantages of the article, and further research directions.

In term of the strengths, first of all, the evaluation indicator system is comprehensive and effective. Based on the references and the actual situation, we have developed a three-tier indicator system. The system allows for a more intuitive analysis of all aspects and we have described each indicator in detail. Secondly, we use a combination model of information entropy, and coefficient of variation, grey prediction models, K-means clustering and principal component analysis. These methods make our model very robust and convincing. What's more, the countries chosen are representative. China is a developing country with a large population and its higher education system is constantly being improved. The policy we propose for China is beneficial to help countries with similarly large populations and inadequate higher education systems, such as India, for example.

However, there still some weaknesses. For example, there are biases in the data. We collect data on multiple websites. As the statistical criteria adopted by different websites may vary, this may lead to bias in the results we obtain. In addition, our evaluation model is subjective. Some of the methods we apply, such as the group power decision method, are subjective in that they tend to make the results subjective.

There are still some places where the model can be further improved. For example, we will have a deep understanding of the policies of the selected country, we need obtain more data, thus enhancing the accuracy and precision of the model's calculations.

References

1. C. Sun, X. Wang, X. Xu, Evaluation of the health status of the national higher education system based on mathematical model. Front. Educ. Res. **4**(3), 89–97(2021)
2. H. Xu, Q. Zhou, The construction of the evaluation index system for the quality and level of higher education: based on the perspective of building a strong country in higher education. Mod. Educ. Manag. (05), 43–46(2010)
3. Y. Peng, P. Jiang, H. Li, Health insurance of higher education. Trans. Comp. Educ. **3**(2), (2021)
4. M. Hong, A comparative study of the internationalization of higher education policy in Australia and China (2008–2015). Stud. High. Educ. **45**(4), 1–12(2020)
5. Y. Liu, N. Li.: Construction of comprehensive evaluation index system for internationalization ability of higher education. J. High. Educ. Manag. **13**(05), 52–60(2019)
6. Z. Jiang, Z. Sheng, On the sustainable development of the higher education system. J. Hunan Univ. Sci. Technol. (Soc. Sci. Ed.) **13**(06), 149–151(2010)
7. S. Han, B. Zhang, Optimization strategy of resource allocation in China's colleges and universities in the stage of popularization of higher education. High. Educ. Explor. (12), 14–20(2021)

8. W. Zhang, J. Mei, L. Wang, Analysis of regional differences and influencing factors of public investment performance in higher education in China. Heilongjiang Researches on Higher Education 39(01), 66–72(2021).
9. L. Wang, Research on the optimal allocation of educational resources within colleges and universities under the background of supply-side reform. Jiangsu High. Educ. (04), 57–61(2021)
10. W. Li, L. Shi, J. Han, J. shouJun, Research on the satisfaction evaluation system of teachers and students of service-oriented party organizations in colleges and universitie. Time Report(05), 73–75(2022)

Research on the Current Situation and Implementation Strategies of Artificial Intelligence (AI) Education in K-12 Schools

Ju Pan and Xiaohong Lan

Abstract Carrying out artificial intelligence (AI) education in K-12 schools is of great significance to cultivating students' higher-order thinking and improving AI literacy. Firstly, the research makes a visual analysis of AI education in K-12 schools in China, and then analyses and summarizes its existing problems. These problems include the lack of unified curriculum standards and fuzzy curriculum positioning; the lack of appropriate course materials and fragmented teaching content; the lack of professional course teachers and poor teaching effect; the lack of a good course atmosphere and difficult course implementation, etc. Finally, targeted implementation strategies for AI education in K-12 schools are proposed. It provides a reference for future development in this field.

Keywords Artificial intelligence (AI) education · K-12 schools · Current situation · Implementation strategy

1 Introduction

In recent years, China has vigorously developed artificial intelligence (AI) education and cultivated professional talents in the field of AI. In 2017, the State Council issued the Development Plan for A New Generation of Artificial Intelligence. It pointed out that AI has become a new focus of international competition, and that China should gradually carry out an intelligent education program for all, set up AI-related courses at the K-12 school levels, gradually promote programming education, and build an AI discipline [1]. In 2019, clear requirements for AI education and information literacy cultivation in K-12 schools were put forward in the China Education Modernization 2035 [2]. With the promulgation and introduction of national policies and the development and progress of AI technology, AI education in K-12 schools has shown new

J. Pan (✉) · X. Lan
College of Computer and Information Science, Chongqing Normal University, Chongqing, China
e-mail: panju9612@163.com

S. Patnaik and F. Paas (eds.), *Recent Trends in Educational Technology and Administration*, Learning and Analytics in Intelligent Systems 31,
https://doi.org/10.1007/978-3-031-29016-9_44

changes. Therefore, based on analyzing and summarizing the existing problems of AI education in K-12 schools, this research proposes targeted implementation strategies to better promote the ground implementation of it and accelerate the systematic development of AI.

2 Visual Analysis of AI Education in K-12

To analyze the current situation and trends of AI education research in K-12 schools in China, the research firstly searched academic journal literature in China National Knowledge Infrastructure (CNKI) database with the subject terms “AI education in K-12 schools” “AI teaching in K-12 schools” “AI curriculum in K-12 schools” and “AI subjects in K-12 schools”. The search spanned from 2003 to 2021 and selected 220 articles as research samples by excluding those that did not match the subject.

The chronological change in the number of journal literature publications is an important indicator to measure the development history of the field and reflects how active the field has been in various periods [3]. As shown in Fig. 1, scholars in China conducted research on related topics in 2003. Since 2017, AI education in K-12 schools has received increasing attention from researchers, and the scope of research has gradually expanded, with an increasing trend in the number of publications.

Keyword co-occurrence mapping can reflect the current research hotspots in a field and which hotspots have appeared in the past. From Fig. 2, it can be seen that AI, K-12 education, intelligent education, information technology, robotics, smart education, teaching, programming education, machine learning, and talent training are the research centers and hotspots in China, mainly in K-12 AI education.

The timeline view facilitates the identification of changes in the scope of a research topic, the relevance and legacy of research, and the evolutionary trajectory of the

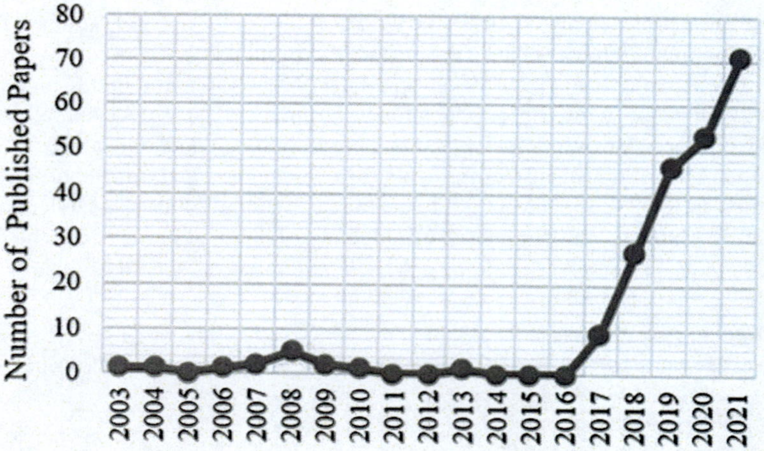

Fig. 1 Literature statistics of AI education in K-12 schools

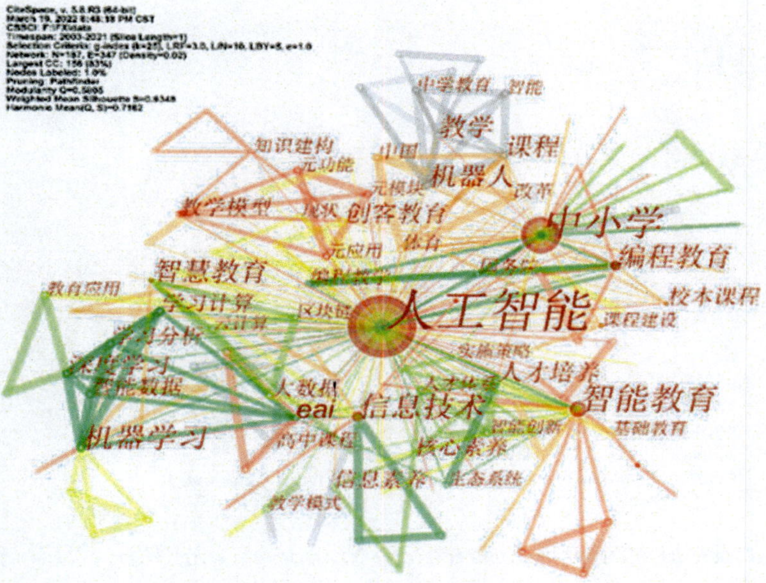

Fig. 2 Keyword co-occurrence mapping of AI education in K-12 schools

research focus. In this paper, keyword timeline mapping is done through CiteSpace software, as shown in Fig. 3, to analyze the development of AI education in K-12 schools in China over different periods. Cluster #0 K-12 schools, which has been a research hotspot since the first keyword appeared in 2003 until 2021, mainly explores the construction of AI courses in K-12 schools. Cluster #5 teaching, received more attention from 2003 to 2020. Cluster #7 teaching models, began to receive attention in 2019. After the first keyword of cluster #1 programming education, #2 machine learning, and #4 intelligent education appeared, the research popularity maintained from 2017 to 2021.

3 Exsisting Problems of AI Education in K-12 Schools

Since AlphaGo defeated the world champions Lee Sedol and Ke Jie, AI has once again triggered a new wave of upsurge. There are many industries that have moved closer to AI. It is worth noting that the implementation of AI education in K-12 schools has always received great attention and importance from all sides. It occupies a very important position in education and is a basic project for talent training. At present, AI education in K-12 schools in China has been promoted in terms of curriculum planning, curriculum positioning, educational equipment, and resource construction, but the following problems still exist in the implementation process.

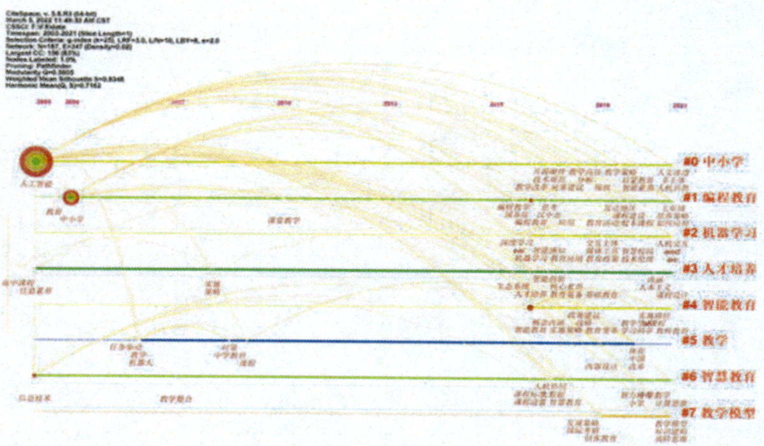

Fig. 3 Timeline mapping of keywords for AI education in K-12 schools

3.1 *Lack of Uniform Curriculum Standards and Fuzzy Curriculum Positioning*

Clarifying the curriculum standards of AI courses at all stages of K-12 schools is the basis for effectively promoting AI education in K-12 schools. However, currently, only the Curriculum Standard of Information Technology in Senior High School (2017 Edition, 2020 Revision) has clearly defined the high school curriculum and has clear academic requirements. However, due to age differences and different foundations, this curriculum standard does not apply to the development of AI courses at the primary and junior high school levels. The lack of unified and comprehensive curriculum standards has led to a more ambiguous positioning of AI courses with rich content and extensive knowledge. It is easy to confuse society, schools, individuals, etc., making it even more ambiguous with programming education, maker education, robotics education, and other segments. Although some schools have set up AI courses, it is just old wine in a new bottle. The previous programming courses and maker education courses are slightly “packaged”, and then continuing to use them. Therefore, the development of AI education in K-12 schools in various places is not good, and the content is mostly scattered and fragmented, which is not a system. It is difficult to achieve the goal of AI education.

3.2 *Lack of Appropriate Course Materials and Fragmented Teaching Content*

At present, there are some AI textbooks for K-12 schools on the market, and the textbooks published in the past three years account for a relatively high proportion. The contents of the textbooks mainly cover the basic knowledge of AI, key technologies, and prospects. Through sorting out, it is found that the forms of AI textbooks for K-12 schools include single textbooks and series of textbooks. The series textbooks are dominant, and the number of textbooks for primary school is the largest [4]. However, the quality of teaching materials is uneven, the core content and focus are different, the level gap is large, and there is a lack of authoritative guidance. Some textbooks do not grasp the relationship between the knowledge points of AI, resulting in a weak system, some concepts that should be involved in general courses are not available, and some more professional concepts are presented too much. Some textbooks are biased, which can easily make readers mistakenly think that it is the whole of AI, resulting in misunderstandings in knowledge. It is difficult to judge whether the existing teaching materials meet the teaching objectives of the curriculum, whether they are consistent with the students' learning rules, cognitive characteristics, and physical and mental development, and whether they contain sufficient ethical and moral knowledge. When it is difficult to select suitable teaching materials for AI courses, some schools do not use relevant teaching materials, but teachers prepare and teach by themselves. Therefore, there are many deviations in knowledge, and the content is complex and differentiated.

3.3 Lack of Professional Course Teachers and Poor Teaching Effect

There is a large shortage of AI talents in China, which leads to the increasing contradiction between the rapidly growing demand for AI talents and the supply of backward talents. The Beijing-Tianjin-Hebei region, the Yangtze River Delta, and the Guangdong-Hong Kong-Macao Greater Bay Area are the main gathering places for AI talents in China at this stage, while some underdeveloped areas lack talents. At the same time, the vast majority of practitioners in the field of AI will choose companies with significantly higher salary levels, rather than school teachers. It is difficult for K-12 schools to recruit suitable course teachers, and the existing professional teachers are very few. It is difficult to be effectively supplemented in a relatively short period. Therefore, most schools take the approach of teaching part-time to teachers of information and other related disciplines. Due to the characteristics of a large comprehensive AI system, deep cross-content, and rapid update and iteration, coupled with the lack of AI knowledge background and disciplinary literacy of these teachers, many of them do not know where to start teaching [5]. Only the course name was changed, but the actual teaching content did not meet the requirements of the subject so students lacked interest in learning, their enthusiasm was not high, and the final teaching effect was not good, which hindered the cultivation of students' thinking and ability development [6].

3.4 Lack of Good Course Atmosphere and Difficult Course Implementation

Although AI education has emerged in K-12 schools, setting off a learning boom, there are still many problems in the specific implementation of each school. The reason is that, first of all, the school has not carried out the top-level design for AI courses, lack of integration, and the implementation effect is not good. Secondly, the class hours of such courses are few and cannot be guaranteed, and some schools have the phenomenon of coping with inspections and muddling through. Even if some K-12 schools have set up corresponding AI courses, some other things, such as Chinese, mathematics, and other major examination subjects, will occupy the time of the planned AI courses. Finally, parents think that their children are wasting their time and energy by studying AI now. They can learn it later if they want to pursue a related career. It can be seen that the root is mainly due to the drawbacks left by exam-oriented education. People's ideas have not been changed, and it is useless to simply think that those who do not take the entrance exam are useless. Therefore, AI education has not been fully supported, recognized, and cooperated by schools and parents in the implementation process, and the course atmosphere is poor, which seriously affects the enthusiasm of students to learn and hinders the development and promotion of related work [7].

4 Implementation Strategy of AI Education in K-12 Schools

In order to solve the above problems in a timely and effective manner and promote the development of AI education in K-12 schools, after fully investigating the current situation of AI education development in K-12 schools and analyzing the existing problems, this research mainly explores the implementation strategies in this field from the following aspects, as shown in Fig. 4.

4.1 Formulate Uniform Curriculum Standards

The formulation of unified AI curriculum standards is a necessary prerequisite for K-12 schools to carry out AI education. After determining the basic elements such as teaching objects, course objectives, course forms, and course concepts, combined with the existing high school AI curriculum standards, clarify the AI teaching boundaries at different stages of primary school, junior high school, and high school, and formulate a segmented, vertically integrated unified AI curriculum standard for K12 [8]. When formulating standards, experts in the field of AI and education can be invited to discuss and revise together. According to the psychological and cognitive level and cognitive characteristics of students, the curriculum standards can be

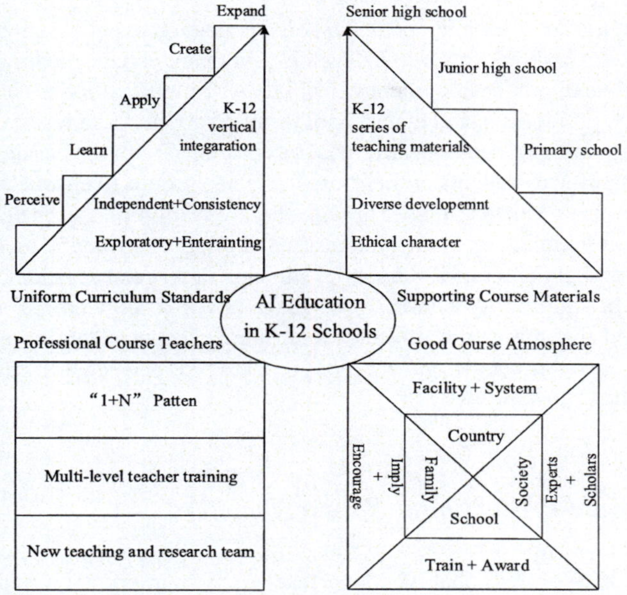

Fig. 4 Implementation strategies of AI education in K-12 schools

advanced according to the five levels of perception, learning, application, creation, and expansion. With the continuous improvement of students' grades, information awareness, computational thinking, digital practice ability, etc., higher-level learning will be carried out [9]. In addition, the curriculum standards must not only ensure the independence of the knowledge modules at each stage but also ensure the continuity of the advanced level, so that students can be more interesting and challenging when learning. Only when a standardized curriculum standard system is introduced can it play a guiding role for schools, teachers, and students. Teachers can continuously improve the content and teaching form according to the outline, and finally, form a high-quality course for AI education in K-12 schools.

4.2 *Develop Supporting Course Materials*

The development of supporting AI course materials is a key link in the development of AI education in K-12 schools. Textbooks are reference books for students to learn and reference books for teachers to teach. Therefore, a professional, comprehensive and excellent AI course material is particularly important. AI is a relatively new field and covers a lot of knowledge content, which also puts forward higher requirements for the teaching material development team. Given this situation, a teaching material development team consisting of multiple parties and absorbing a variety of personnel is proposed, such as academicians, AI experts, education experts, school e teachers, technicians, etc., to create a diverse development team, enhance team strength, and do a good job textbook development work. The AI course textbooks can correspond to elementary, junior high, and high school respectively, and a series of textbooks for the K-12 stage must be developed. It must be combined with the learning characteristics and interests of students of their respective age groups to ensure the coherence and scientific nature of learning. Through the cultivation of the basic knowledge, awareness and thinking, technology and skills of AI, the self-confidence, innovative spirit and practical ability of K-12 school students in AI-related fields can be established [10]. In addition, typical cases of AI can be incorporated into the teaching materials, and the cultivation of ethics and moral character can be reflected. The relationship between people and AI must be clarified, and legal norms and social ethics must be followed.

4.3 Cultivate Professional Course Teachers

The cultivation of professional AI course teachers is an important measure for K-12 schools to carry out AI education. It is very important and necessary to train teachers who teach AI courses. This research proposes the cultivation path of professional course teachers, as shown in Fig. 5. On the one hand, the "1 + N" pattern is implemented, that is, an AI teacher drives multiple teachers of related disciplines, transfers knowledge, shares experience, and establishes a lifelong learning mechanism [11]. The relevant disciplines include information technology, general technology, science, and other information disciplines. Considering the lack of AI teachers, this pattern can be constructed in units of provinces and cities, districts, and counties, which is not only conducive to the expansion and training of teachers but also promotes the exchange and integration of disciplines. On the other hand, carry out multi-level teacher training. Various provinces and cities, districts, and counties can divide AI teachers into different groups according to the actual development of local the course. Then carry out various types of training that are highly targeted, focused, and effective. For example, teachers with weak knowledge in the field of AI can first carry out universal type training. Teachers who already have a certain foundation carry out professional type training, while professional teachers carry out advanced type training. All types of training are advanced layer by layer. In addition, joint enterprise experts set up a new teaching and research team to explore innovative teaching models. The team includes enterprise experts in the field of AI, as well as school teachers who are familiar with the laws of students' physical and mental development and learning characteristics, to speed up the cultivation and training of professional teachers.

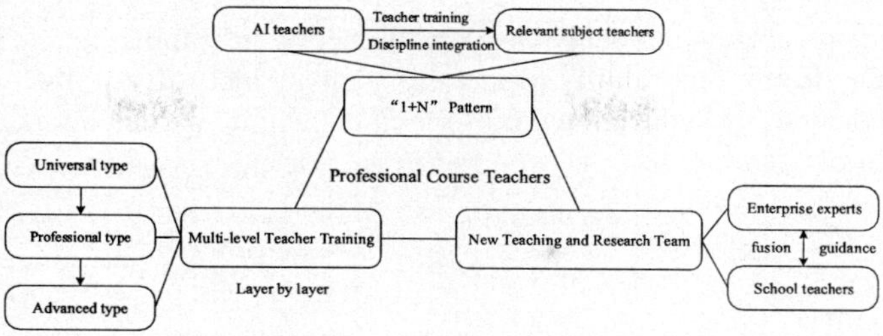

Fig. 5 Path of professional course teacher cultivation

4.4 *Create a Good Course Atmosphere*

Creating a good atmosphere for AI courses is an objective requirement for K-12 schools to carry out AI education. Encourage multiple forces to jointly build an AI education ecosystem for K-12 schools. At the national level, build and improve AI popular science infrastructure, encourage AI enterprises and scientific research institutions to build open-source platforms, and open AI pavilions to the public, playing an active leading role. In addition, the introduction of appropriate policy mechanisms and the formulation of a teaching and assessment system to ensure the effective implementation of the AI course and to provide action guidelines for the specific implementation of schools. At the social level, AI experts, scholars, and enthusiasts in this field are called on to join the school's teaching and research team to jointly create a good educational atmosphere and work together. At the school level, organize relevant teacher training, and set up a certain reward mechanism to continuously stimulate teachers to improve and expand in this field. At the same time, the school can carry out a series of AI competitions to fully demonstrate the school's characteristics. At the family level, parents change the traditional concept of exam-oriented education, give students positive psychological hints and encouragement to learn AI, and create a positive family learning atmosphere.

5 Research Summary

Talent is one of the most critical factors that determine whether a country can lead in the field of AI in the future. One of the strategic goals and key tasks of China's AI development is talents training. The establishment of the AI course in K-12 schools is an important guarantee [4]. The fundamental purpose of AI education in K-12 schools is not to train every student to become an industry expert, but to nurture people, i.e. to follow the logic of progressive cognition and knowledgeability of different ages under the five major concepts of AI education (perception, representation and reasoning, machine learning, human–computer interaction, and social impact), and to encourage innovation, practice, inquiry, and collaboration, so that students can build their interest and confidence in AI, train their logical thinking and innovation ability, and cultivate their ability to identify, analyze and solve problems. How to fully implement the AI education policy requires the joint participation of the state, society, schools, families, etc., as well as further research and practice.

References

1. The State Council, Notice of the State council on printing and distributing the development plan of the new generation of artificial intelligence. [EB/OL]. (2017). http://www.gov.cn/zhengce/content/2017-07/20/content_5211996.htm
2. The CPC Central Committee, the State Council, China's Education Modernization 2035. [EB/OL]. (2019). http://www.moe.gov.cn/jyb_xwfb/s6052/moe_838/201902/t20190223_370857.html
3. Y. Chen, C.M. Chen, Z.G. Hu, *Principles and applications of analysing a citation space* (Science Press, 2014)
4. D.L. Wang, D.Q. Zhou, Y.R. Wang, et al., Review of artificial intelligence teaching materials for primary and secondary schools—analysis based on 45 published textbooks. Mod. Educ. Technol. **31**(02), 19–25 (2021)
5. H.Y. Gao, Y.S. Li, S. Dai, et al., How programming education can better promote the development of computational thinking in early childhood—A systematic review based on international empirical studies. e-Educ. Res. **42**(11), 121–128 (2021)
6. Z.X. Zhang, H. Du, L. Gao, et al., Current situation, problems and countermeasures of artificial intelligence curriculum construction in primary and secondary schools in developed areas. China Educ. Technol. **09**, 40–49 (2020)
7. R. Xiao, H.M. Shang, J.J. Shang, Artificial intelligence and educational reform: Prospects, difficulties, and strategies. China Educ. Technol. **04**, 75–86 (2020)
8. D. Sun, Y. Li, The development status, research hotspots and enlightenment of youth programming education in China and foreign countries: Also discuss the implementation strategy of programming education for China in the age of intelligence. J. Distance Educ. **37**(03), 47–60 (2019)
9. D. Zhang, G.Z. Cui, Research on the artificial intelligence education in primary and secondary schools. Mod. Educ. Technol. **30**(01), 39–44 (2020)
10. Y. Yu, P. Xu, W.Y. Liu, Current situation and enlightenment of the curriculum system of artificial intelligence education in primary and secondary schools in Japan. China Educ. Technol. **08**, 93–99 (2020)
11. Y. Lu, X.Y. Tang, J.C. Song, et al., Artificial intelligence education in K-12 schools in the intelligent era: strategic positioning and core content domains. Chin. J. Distance Educ. **556**(05), 22–31+77 (2021)

A RIO-Based Cybersecurity System Construction Method for Campus Network-Using Economic Concepts to Build a Network Security Protection System

Lanjun Li and Zhongyi Liang

Abstract By introducing RIO into the construction of campus cybersecurity, quantitative analysis of input and output is carried out to solve the problem of "bottomless investment and invisible results" of cybersecurity. This paper innovatively adopts RIO in the PPDRRF (Policy, Protection, Detection, Response, Recovery and Forensic) model, and quantitatively analyzes some key technologies of campus cybersecurity system construction, proposes AMTC Cybersecurity System Architecture (ACSA) and ensures the smooth progress of cybersecurity construction under resource constraints.

Keywords Cybersecurity · Rate of Input and Output · Return on investment · Classified protection 2.0 · Campus network · ACSA

1 Introduction

With the enforcement of Cybersecurity Law of the People's Republic of China [1] (referred to simply as the Security Law), as the network provider of the campus network, universities need to take the responsibility of protecting the legitimate use of the campus network. Current campus network is a huge and complex system, which supporting the business operation and development of universities. The construction of campus informatization is the basic required action for universities, which including website system, campus information management system, campus card system, Information Security Management System [2], etc. Cybersecurity threats

L. Li · Z. Liang (✉)
An Hui Broadcasting Movie and Television College (AMTC), Network and Information Center, Hefei, China
e-mail: Lzhy0009@amtc.edu.cn

L. Li
An Hui Broadcasting Movie and Television College, School of Information Engineering, Hefei, China

S. Patnaik and F. Paas (eds.), *Recent Trends in Educational Technology and Administration*, Learning and Analytics in Intelligent Systems 31,
https://doi.org/10.1007/978-3-031-29016-9_45

faced by information systems are also growing, more and more vulnerabilities or weaknesses are found, and information security risks are becoming increasingly prominent, becoming one of the important and urgent problems. On the other side of the coin, universities often fall into the embarrassing situation of shortage of human resources, financial resources and equipment when cybersecurity problems happening.

At present, most of the research on the campus network security protection system stays on the dynamic integrity of the system, that is, it emphasizes the comprehensive protection, and can realize that network security is a dynamic process, and the campus network security protection system must keep pace with the times. The landing of some campus network security protections only emphasizes the application of one or several key technologies, and does not consider the systematization of protection. In the above two directions, because the quantitative problems of resource input and output are not considered, the plan is often unenforceable due to insufficient resources, or the result of serious shortcomings in protection after implementation, they have lost the practical guiding significance of the landing of the program.

This article introduces the concept of ROI (return on investment) from the field of economics and extends it to RIO (Rate of Input and Output on cybersecurity) as the main basis for quantification of cybersecurity construction.

2 RIO: A Feasible Indicators for the Construction of Cybersecurity Systems

2.1 *What is ROI and ROSI?*

ROI is a performance measure used to evaluate the efficiency of an investment or to compare the efficiency of several different investments [3]. To calculate ROI, the return of an investment is divided by the cost of the investment; the result is expressed as a ratio. The return on investment formula:

$$ROI = \frac{Revenue - Total\,Investment}{Total\,Investment} \times 100\% \tag{1}$$

Return on investment (ROI) refers to the economic return of an enterprise from an investment, and it is a comprehensive index to measure the operating efficiency of the enterprise. There are many other definitions of ROI in the literature. Each definition focuses on certain ROI aspects. This definition reflects that the calculation method of return on investment varies from company to company. Nevertheless, the basic idea is the same: the numerator of ROI is the "net income" (return, profit and income) of the project, while the denominator is the cost spent [3]. However, the traditional ROI calculation method is not very suitable for information security investment, and ROI calculation is a very difficult task for investment on cybersecurity, because security investment does not bring very direct benefits and profits to most enterprises. It is usually to prevent the loss of company property, in other words, investment on cybersecurity do not expect a profit return.

Therefore, the calculation of the ROI of cybersecurity should be calculated to calculate how much loss was avoided through the security investment, also known as ROSI (Return On Security Investment), ROSI is calculated as follows:

$$ROSI = \frac{\text{Reduced loss} - \text{Cost of security measures}}{\text{Cost of security measures}} \times 100\% \tag{2}$$

In the calculation process of ROSI, there is the most important problem: How to obtain the reduced loss? A relatively accepted model of risk expected loss in the information security industry is calculated as follows:

$$Reduced\,loss = \sum \begin{pmatrix} PreImplementation\ Risk\ Loss\ Expectancy \\ -\ PostImplementation\ Risk\ Loss\ Expectancy \end{pmatrix} \tag{3}$$

At present, there is no uniform standard for risk loss expectancy. ALE (Annual Loss Expectancy) was introduced, which considering Asset value, level of damage, ARO (Annual Rate of Occurrence). ALE is the loss caused by the annual cyber-attack, which refers to the potential loss that cyber-attack may cause to the company or organization in one year, calculated as ALE = ARO * SLE. ARO (Annual Rate of Occurrence) is the probability that a cybersecurity incident will occur in a year, that is, the probability and number of times a certain cyber threat may occur in a year. SLE (Single Loss Expectancy) is loss expectancy caused by the occurrence of a cyber-attack alone. However, the particularity of cyber threats is relatively complicated when calculating losses, such as a notebook lost, regardless of the value of the notebook, but also to add the cost of purchase, IT support, loss of productivity, reputation, intellectual property loss and so on. ROSI calculation relies on the monetization calculation of losses, ROI calculation relies on the monetization calculation of income, and the monetization calculation of network security income or loss often depends on the definition of income and loss of specific individuals, this assessment is based on the specific network environment and security protection measures, but also based on experience to obtain an estimate, so the implementation process often causes the calculation results to be inconsistent [5].

Therefore, we need a new feasible method to calculate return on cybersecurity system, which contains all factors above, such as protected asset value, potential loss, rate of occurrence, reduced loss, etc.

2.2 What is RIO?

According to GB/T 22,239–2019 [6], The core of classified protection of cybersecurity (also known as classified protection 2.0) is using key information infrastructure to protect the security of the cyberspace, leveled by protected asset value and the destruction consequences of cyber-attacks. When a level 2 system is evaluated by classified protection 2.0, for example, it is assigned a score according to the implementation of security countermeasures. The score is a feasible indicator of "net gain" (return, profit, benefit) on cybersecurity investment, we call the score CPS (Classified Protection Score) in this paper:

$$CPS = \begin{cases} INT(score), score \geq 70 \\ 0, score < 70 \end{cases} \tag{4}$$

The level of classified protection 2.0 is divided according to the importance of the evaluated system and the destruction consequences, and the qualifying score is 70 points (out of 100) [6, 7]. In other words, the calculation factor of CPS already includes protected asset value, potential loss of asset, rate of occurrence, reduced loss. CPS is the output of cybersecurity system construction.

We propose the concept of RIO (Rate of Input and Output on cybersecurity). Input on cybersecurity is cost of countermeasures, and output on cybersecurity is CPS. The RIO proposed calculation formula is:

$$RIO = \frac{Cost\ of\ countermeasures}{CPS} \tag{5}$$

Cost of countermeasures is obtained from median of security vendor quotes corresponding security measures, CPS is calculated according to formula 4. RIO is combined with the advantages of ROI in formula 1 and ROSI in formula 2. The smaller the RIO, the better the security countermeasures.

As a key information infrastructure of classified protection 2.0, we take malicious code prevention as example, the calculation of RIO is shown in Table 1.

As a level 2 evaluation object of classified protection 2.0, website "www.amtc.edu.cn" was found 28 cybersecurity issues. A total of 135 evaluation indicators were selected. The overall score for the evaluation was 77.97. The security measures we

Table 1 An example of RIO

Category	RIO = Cost/CPS			
	Countermeasures	Cost(¥)	CPS	RIO
Malicious code prevention	Free antivirus software	0	0	∞
	Centralized antivirus software	20,000	76	253.16
	Antivirus firewall	25,000	77	324.68
	Antivirus software & firewall	45,000	78	576.92

used in the first evaluation were antivirus firewall plus centralized antivirus software. Now we use alternative countermeasures in the malicious code prevention category and bring it into the scoring system. Finally, the score is rounded or zeroed to get the CPS. The calculation results are shown in Table 1.

In Table 1, RIO minimum value 253.16 corresponds to "Centralized antivirus software", so the most cost-effective countermeasure in the malicious code prevention category is centralized antivirus software. In a similar way, we got the most cost-effective countermeasures in the key categories of classified protection 2.0.

According to this idea, we can design our cybersecurity system architecture.

3 Amtc Cybersecurity System Architecture

3.1 ACSA

As mentioned above, we identify the most critical layers of cybersecurity protection and proposed AMTC Cybersecurity System Architecture (ACSA) drawing on adaptive network security model (ANSM [8], also known as P2DR model), shown as Fig. 1.

In ACSA, we followed the basic idea of classified protection 2.0, which is featured by "one center and three lines of defense" [6]. The protective measures taken can meet the requirements of the P2DR model. In the late 1990s, the American ISS proposed a time-based security model-the adaptive network security model (ANSM),which is also called P2DR (policy protection detection response) model [8], Based on the best practices and P2DR model, the NSA released the Community Gold Standard v2.0 (CGS2.0) in June 2014 [9]. The CGS2.0 standard framework emphasizes the

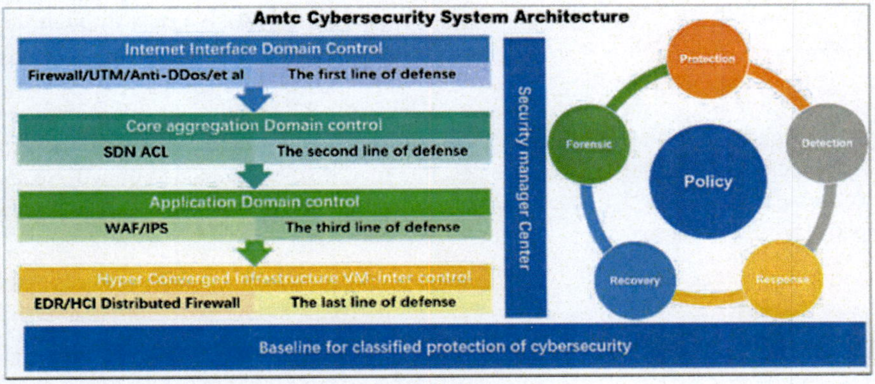

Fig. 1 AMTC cybersecurity system architecture (ACSA)

four overall functions of cyberspace security: govern, protect, detect, respond and recover.

Based on the baseline for classified protection of cybersecurity, ACSA has four lines of cybersecurity defense and security manager center, combining the PPDRRF ((Policy, Protection, Detection, Response, Recovery and Forensic)) model.

3.2 *Baseline for Classified Protection of Cybersecurity*

GB/T 22,239–2019 [6] divides cybersecurity into technical and management requirements. Managerial requirements include security management policy, security management organization, security management personnel, security construction management and security system maintenance. Technical requirements include secure physical environment, secure communication network, secure boundaries, secure computing environment and secure management center.

3.3 *PPDRRF*

In the PPDRRF model, protection, detection, response, recovery, and forensics constitute a complete and dynamic security cycle, which together realize security assurance under the guidance of security policies.

At the policy level, we proposed idea of "three stages—four levels—one penetration", in which "three stages" refer to the design and development stage, deployment and implementation stage, and management and maintenance stage; "four levels" refer to the Application-level security, system-level security, network-level security and physical-level security; "one penetration" means that security management runs through the life cycle of the entire information system, and also runs through the "three stages" and "four levels".

We proposed four lines of defense from baseline for classified protection of cybersecurity. For maximum cost performance, we use the RIO method to select countermeasures. The four lines of defense and security manager center are constructed to implement ACSA model, meeting the baseline for classified protection of cybersecurity and realizing PPDRRF model, which includes Protection, Detection, Response, Recovery, and Forensic.

3.4 *Lines of Defense and Security Manager Center*

- The first line of defense: The first line of defense mainly realizes internet interface domain control. We deploy a next-generation firewall at the entrance of the campus network as the first line of defense for cybersecurity. The firewall explicitly deny access to high-risk ports, such as 1433, 445, 135–139, etc. in the access control list, and set highest priority to avoid unintentional accept in the later maintenance process. The list of high-risk ports changes dynamically. When a new worm breaks out, the ports will be dynamically added to the list. The firewall policy is set to deny by default, allowing only necessary ports such as websites and campus apps. The next-generation firewall integrates anti-distributed-denial-of-service (DDos) implementing the functions of protection, detection, and response in the PPDRRF model.

- The second line of defense: The second line of defense mainly realizes core aggregation domain control. With the help of SDN technology, we add access control list (ACL) to the core switch to control traffic between VLANs. According to statistics, 80% of network attacks originate from the internal network [8]. Therefore, it is necessary to strengthen the security control and prevention of the internal network. In addition to the first line of defense, the remaining lines of defense are both internal and external. Departments with different VLANs are divided into security areas. Traffic between different VLANs must pass through the core switch, so that ACLs applied to different VLANs can block high-risk behaviors such as virus outbreaks and hacker jumping. Even if some terminal poisoning occurs, the virus can be controlled within a certain range, providing time for the next step to quickly determine, isolate and deal. In the core switch, by mirroring the traffic to the behavior audit equipment, the functions of protection, response and forensics in the PPDRRF model are realized.
- The third line of defense: The third line of defense mainly realizes application domain control. We deploy a web application firewall (WAF) integrated Intrusion Protection System (IPS) function at the boundary of the server area. Through the filtering of the first and second lines of defense, the traffic entering the server area is mainly business traffic, but application layer attacks such as SQL injection and cross-site scripting (XSS) also follow. WAF can go deep into application layers to detect and block the attacks. Because it is deployed at the boundary of the server area, in addition to external attacks from the Internet, WAF can also block application layer attacks from inner area, and realize the functions of protection, detection, response and forensics in the PPDRRF model.
- The last line of defense: The last line of defense mainly realizes Hyper Converged Infrastructure (HCI) VM-inter control. With the advancement of virtualization, the server area of the campus network has been moved into the private cloud, and the servers communicate through the virtualized network. Server-to-server communication traffic inside a private cloud, we call it east–west traffic. If a server is compromised, the entire server zone will be compromised. In order to avoid this situation, we enable distributed firewall in the virtualized network. and installed centralized antivirus software "Endpoint Detection and Response (EDR)" for antivirus, OS security configuration and system patch updates. EDR and HCI distributed firewall block the transmission of VM-inner and VM-inter malicious traffic, and realize the functions of protection, detection, response, recovery and forensics in the PPDRRF model.

- Security manager center: Security manager center mainly realizes security management. Through security management, we can regularly analyze the security logs generated by each line of defense, scan security vulnerabilities, dynamically adjust the strategies of related security devices in the lines of defense, implement the security management system formed by the guidance of security strategies, and protect against new attacks such as APT.

4 Conclusion and Future Work

In this paper, we aimed to construct a reasonable cost-effective cybersecurity system. We introduced the concept of ROI (return on investment) from the field of economics and extended it to RIO (Rate of Input and Output on cybersecurity). The smaller the RIO, the more cost-effective the security countermeasures. According to this idea, we designed AMTC Cybersecurity System Architecture (ACSA). Based on the baseline for classified protection of cybersecurity, ACSA has four lines of cybersecurity defense and security manager center, combining the PPDRRF (Policy, Protection, Detection, Response, Recovery and Forensic) model. Finally, we constructed four lines of defense and security manager center to implement ACSA.

However, the calculation of RIO may be inconsistent, cost of countermeasures in formula 5 is obtained from median of security vendor quotes, which fluctuates greatly from vendor to vendor. Therefore, in the future we will focus on optimizing the calculation of cost of countermeasures, one of the directions is to obtain data from bidding online for big data analysis.

Acknowledgements This paper is supported by Anhui Province Higher Education Natural Science Research Project (KJ2019A1145) and Teaching Team of Computer Network Technology Project (2020jxtd050). The completion of the paper depends on the AMTC's SDN network and HCI platform, which designed by professor He Xiaodong, here to express my thanks to him, and also thanks for many ideas and help from professor Ling Baohong.

References

1. Cybersecurity Law of the People's Republic of China. http://www.npc.gov.cn/npc/, 2016–11–7
2. ISO/IEC 27000:2013, *Information technology- Security techniques- Information security management systems--Overview and vocabulary*, 2013
3. P. Andru, A. Botchkarev, ROI for technology projects: measuring and delivering value—D. Brian Roulstone and Jack J. Phillips (Woburn, MA: Butterworth Heinemann, 2008, pp. 343, ISBN-10: 0750685883; ISBN-13: 978–0750685887). IEEE Trans. Engineering Management, **59**(4), 766–770(2012)
4. W. Sonnenreich, J. Albanese, B. Stout, Return on security investment (ROSI): A practical quantitative model. J. Res. Pract. Inf. Technol. **38**(1), 45–56 (2005)
5. Daniel Schatz and Rabih Bashroush, Economic valuation for information security investment: a systematic literature review. Inf. Syst. Front. **19**(5), 1205–1228 (2017)

6. GB/T 22239–2019, *Information security technology—baseline for classified protection of cybersecurity*. (Standards Press of China, Beijing, 2019)
7. GB/T 28448–2019, *Information security technology—Evaluation requirement for classified protection of cybersecurity*. (Standards Press of China, Beijing, 2019)
8. M.E.N.G. Xuejun, S.H.I. Gang, Network security architecture based on P2DR model. Comput. Eng. **04**, 99–101 (2004)
9. NAS/CCS. The community gold standard framework version 2.0 [EB/OL]. [2018–06–10]. https://apps.nsa.gov/iaarchive/library/ia-guidance/ia-standards/cgs/community-gold-standard-framework.cfm
10. L. Ma, G. Zhu, L. Lu, Baseline for classified protection of cybersecurity (GB/T 22239–2019) standard interpretation. Netinfo Secur. **19**(2), 77–84(2019)
11. G.U.O. Qiquan et al., *Training course on cybersecurity law and classified protection of cybersecurity* (Publishing House of Electronics Industry, Beijing, 2018)

Quantitative Analysis of Student Emotion Based on Face Recognition Technology

Ji Hongjian and Sheng Jing

Abstract In order to solve the problem that the classroom learning state is difficult to quantify and evaluate, a quantitative analysis method of student emotion based on face recognition technology is proposed. In this paper, the images of student faces in the natural state are collected in a non-intrusive way, and the real-time expression characteristics of students are obtained by using cascade regression tree and hog algorithm. Meanwhile, a quantitative emotion model is established to evaluate the emotion score of students, and active intervention strategies is used to improve the teaching environment. The results show that after the intervention, the learning state of students increases significantly, which proves that appropriate intervention strategies can effectively improve the quality of education. The research content and results provide data and method support for quantifying the situation of classroom learning.

Keywords Face recognition · Machine learning · Quantitative characterization · Teacher intervention

1 Introduction

With the deepening and popularization of intelligent education, the learning situation of student has attracted extensive attention. However, there are great differences in the learning status, learning efficiency and the digestion and absorption of knowledge points of each student, and the impact of the correlation between student learning results and classroom learning status is difficult to quantify, so it is difficult for teachers to define and evaluate them through learning indicators. Many scholars have also investigated and studied in classroom. Wang et al. [1] proposed

J. Hongjian · S. Jing (✉)
School of Mechanical and Automotive Engineering, Xiamen University of Technology, Xiamen, China
e-mail: shengqqqqjing@163.com

S. Patnaik and F. Paas (eds.), *Recent Trends in Educational Technology and Administration*, Learning and Analytics in Intelligent Systems 31,
https://doi.org/10.1007/978-3-031-29016-9_47

an intelligent teaching data collection model based on situational awareness technology, which adopts six types of information representation: user, task, location, time, equipment and infrastructure, according to the characteristics of multi-source heterogeneity, incompleteness and strong relevance of dynamic data collected in the teaching process. Zeng et al. [2] used city space software to visually analyze the relevant literature, summarized the research trend of intelligent learning, and concluded that there are few studies on intelligent learning, the development is uneven, and the regional differences are obvious. Li et al. [3] constructed a seamless and continuous learning environment with the help of flipped classroom model and situational awareness technology. On this basis, they carried out empirical research on College English curriculum. The results show that this model can improve learning attitude and cultivate learning autonomy. With the help of intelligent teaching system, Wang [4] collected the interactive behavior data of teachers and students in the classroom through app, designed the intelligent classroom interaction model, and excavated and revealed the significance of teacher-student interaction for teaching effect. Zhang et al. [5] proposed a new ultra wide regression network model to realize adaptive micro expression recognition. However, the above system design for classroom situation perception is mainly limited to using the ways that students can perceive, including app answer interaction, teacher-student question and answer to collect the interactive information between students and teachers during class time. The real emotion and learning state of students are limited to the choices provided, which can not accurately reflect the state and classroom participation of students [6]. Image recognition algorithms such as CNN [7], OpenFace Framework [8] and LBP [9] can realize emotion recognition in complex environments. But also there is no quantitative analysis method to quantify student learning situation. Therefore, by quantitatively analyzing the facial expressions of middle school students in the teaching process, combined with classroom teaching data and teacher intervention strategies, this paper establishes a quantitative model of learning emotion, and explores the rules of classroom learning, so as to improve the quality of teaching.

2 Implementation Method

2.1 *Face Recognition Algorithm*

The classroom data is mainly composed of face data. This paper uses cascade regression tree technology [10] (gbdt) to realize face alignment, including the alignment of face contour and face features (eyes, eye spacing, nose and mouth). The ultimate purpose is to accurately locate its shape on the known position of face wire-frame. In the regression tree, each tree is established serially, that is, the establishment of the latter tree is based on the previous tree. The residual regression quantity is stored on the leaf node of each tree. When the input falls on the node, the calculated residual will be added to the output to make the shape return continuously, so as to return to the real shape.

The establishment of cascade regression tree needs to set the initial shape as the starting point of regression. All test images are predicted and regressed through the initial shape. In this paper, the average shape is used as the initial shape, and each image adaptively extracts the features of the initial image. Therefore, although the initial shape is the same, the extracted features are different, so each tree can be split next time. Before splitting, establish a feature pool, select two useful points in the feature pool randomly, calculate the pixel value and pixel difference of each image at the above two points, at the same time, randomly generate a splitting threshold, judge the splitting direction of the image according to the value of the threshold and pixel difference, judge the splitting effect based on variance, repeat the splitting step until it splits to the leaf node, and each image will split to different leaf nodes. After calculating the residuals of the current shape and the real shape of each picture, average the residuals of all pictures falling on the same leaf node, save the model of the first regression tree, and construct the second tree through the serial structure, so the latest shape of the regression tree is changed to the initial shape plus the residuals [11]. After repeated iterations, the accurate contour and feature points of the aligned face are finally obtained, The cascade regression algorithm is as

$$\hat{S}^{(t+1)} = \hat{S}^{(t)} + r_t(I, \hat{S}^{(t)}) \tag{1}$$

where (t) is the vector of multiple point coordinates, t is the cascade sequence number, I is the image, and rt is the regressor.

2.2 *Quantitative Method of Student Emotion*

Based on the facial expression data collected in the classroom, the quantitative representation of classroom learning situation is carried out. Among them, the expression types are judged and classified through the point set composed of the facial feature points mentioned above, combined with the facial expression threshold. The eyebrow height is calculated by the ratio of the distance d_1 from the eyebrow to the upper eyelid and the width w of the face wire-frame. The eyebrow inclination is calculated by calculating the inclination d_2 between the eyebrow and the horizontal line l. The eye opening size is calculated by the ratio of the distance d_3 between the upper and lower eyelids and the width w of the face wire-frame. The grinning degree of the mouth is calculated by the ratio of the distance d_4 between the center of the upper lip and the lower lip and the width w of the face wire-frame. Through the continuous test and adjustment of the data, it is concluded that the threshold setting of each feature point in each expression is shown in Table 1. In the table, the proportion between each feature and the recognition wire-frame is used to judge the type of face expression, and the obtained emotions are distracted d, natural n, confused t and surprise r.

Table 1 Emotional threshold setting table

Emotion	d_1/w	d_2/l	d_3/w	d_4/w
Distract	0.1	−0.1	0.03	0.01
Nature	0	0	0.06	0.1
Confuse	0	0.1	0.04	0.2
Surprise	0.3	−0.2	0.1	0.3

Based on the facial expression data collected in the teaching process, it is assumed that the facial expression $T = \{C_1, C_2, \ldots, C_4, \ldots, C_i\}$ of students is collected, and the facial expression of students is weighted numerically to represent the depth of thinking. The depth of thinking determines that students are in the learning state C_j, distracted state C_k or natural state C_h, where $C_j \in \{t, r\}$, $C_k \in d$, $C_h \in n$. When a student is judged to maintain a natural state for a continuous t period of time ($t > 30$ s), And then there was an expression of surprise or confuse, he(she) will be judged as keeping learning (belong to C_j), The system will increase his(her) score by $0.05t_j$. While the students who have been in the natural state for a long time($t > 600$ s) are determined to be in a daze, and the scores decreases, with a decrease of $0.03t_k$. The rest of the students are calculated according to the average value under the natural listening state. The above learning index of student is defined as the emotion score D_i, as shown in Eq. (2)

$$D_i = S_i\left[(\sum_{j=1}^{n} C_j + 0.05t_j) + (\sum_{k=1}^{n-j} C_k - 0.03t_k) + (\sum_{h=1}^{n-j-k} C_h - C_j - C_k)\right] \quad (2)$$

where S_i refers to the proportion of the number of people who look up at time i in the total number of people.

2.3 Design of Emotion Detection System

The construction of the detecting system is shown in Fig. 1 below. The whole system transmits pictures to raspberry pi 4B through the image sensor. The raspberry pi carries out algorithm processing on the face data of each frame, continuously iterates through cascade regression tree prediction and regression, and aligns the face features. Finally, the collected data is saved to raspberry pie, and the data is substituted into the classroom learning quantitative model. The experiment will be divided into control group and experimental group, and the teachers will intervene in the teaching in the experimental group. Finally, the results of learning situation analysis with graphical representation are obtained.

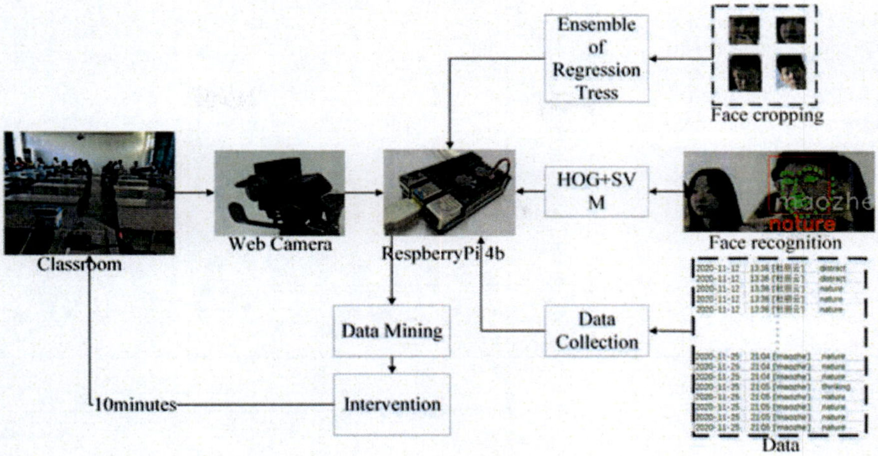

Fig. 1 System working diagram

3 Experiment Results and Analysis

The two experiments were divided into two parts: the control group and the experimental group. The teacher would remind the students every 10 min in the experimental group (only language reminder). We use Wilcoxon rank sum test to test the proportion of people in distracted state in the total number of people in the two groups, so as to test the effect of teacher intervention, as shown in Table 2. In both experiments, we assumed that more students in the experimental group are distracted than those in the control group. We choose the significance level of 0.05, that is, when the p value is less than 0.05, we reject the original assumption and come to the conclusion that the number of people in the distracted state in the control group is greater than that in the experimental group. The experimental results are shown in Fig. 2.

We found that the number of students in the distracted state in the experimental group was less than that in the control group. From the 10th minute of the intervention, the proportion of students in the distracted state in the experimental group was less than 0.05 since the control group (see Table 2), which means that the effect of teachers' active intervention is significantly better than that of students in the natural learning state.

Table 2 Comparison of p-value in two experiments

	Experiment a		Experiment b	
Time	0–10 min	10–45 min	0–10 min	10–45 min
p-value	0.106*	$1.4 \times 10^{-16\#}$	0.192*	$1.4 \times 10^{-16\#}$

*Indicates that the result is consistent with the original assumption
#indicates that the result is inconsistent with the original assumption

Fig. 2 The ratio of the number of students with negative emotions. **a** Comparing the two ratios in experiment a; **b** comparing the two ratios in experiment b

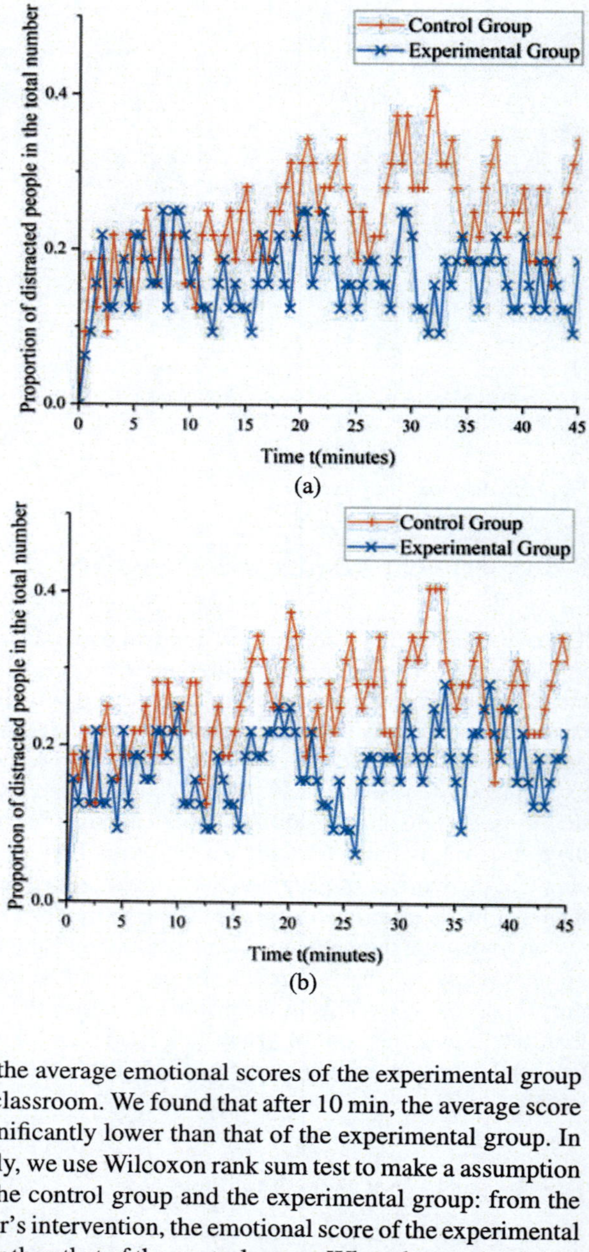

For Fig. 3, we compared the average emotional scores of the experimental group and the control group in the classroom. We found that after 10 min, the average score of the control group was significantly lower than that of the experimental group. In order to confirm this, similarly, we use Wilcoxon rank sum test to make a assumption on the emotional scores of the control group and the experimental group: from the first ten minutes of the teacher's intervention, the emotional score of the experimental group was significantly lower than that of the control group. When the p-value is less than 0.05, the original assumption will be rejected, and the emotional score of the experimental group is higher than that of the control group. The experimental results are shown in Table 3.

Fig. 3 The average value of the emotion scores. **a** Comparing the average values in experiment a; **b** comparing the average values in experiment b

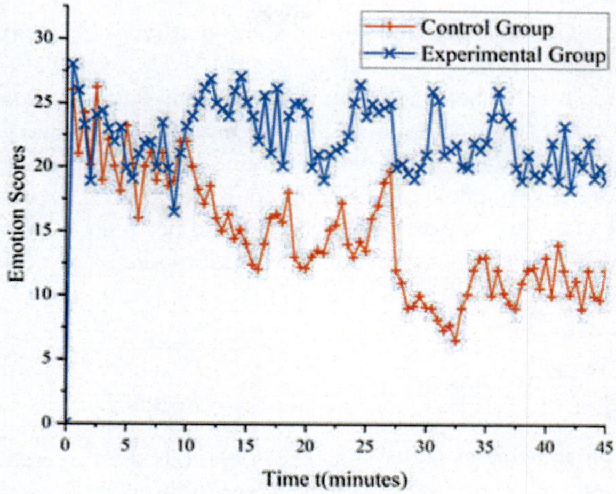

(a)

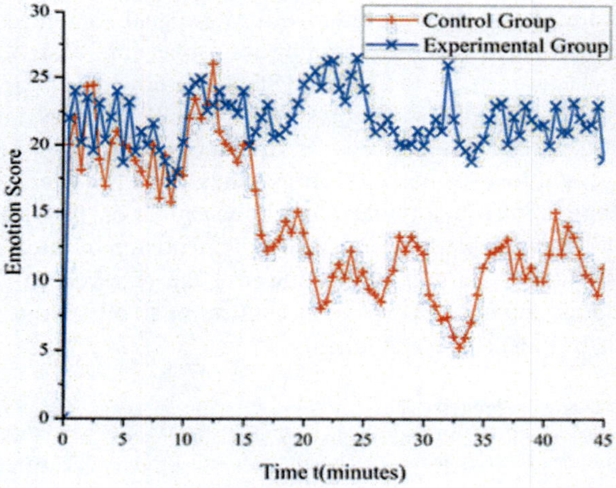

(b)

Table 3 The average score and correlation p-value in two experiments

	Experiment a		Experiment b	
Time	0–10 min	10–45 min	0–10 min	10–45 min
Control group	20.925	12.791	19.29	10.591
Experimental group	22.045	23.481	20.385	22.277
p-value	0.115*	$2.3 \times 10^{-16\#}$	0.102*	$2.3 \times 10^{-16\#}$

*Indicates that the result is consistent with the original assumption
#Indicates that the result is inconsistent with the original assumption

The experimental results show that within the first 10 min, the p-value of the experimental group and the control group is greater than 0.05, while after 10 min, the p-value between them is far less than 0.05. Therefore, the original assumption is rejected, and the emotional score of the experimental group is higher than that of the control group, which verifies the reliability of the algorithm. Although the p-value of Experiment a and Experiment b was greater than 0.05 in the first ten minutes, it can be explained that the teacher did not intervene in the first ten minutes, so this does not conflict with our preset assumption.

4 Conclusion

This paper takes the students in class as the research object, builds a platform for student emotional information recognition in the classroom scene, quantifies the learning state of student through the emotional big data, and analyzes the changes of learning state after the intervention through the active intervention strategy of teacher, which can improve the teaching atmosphere and make the students in a good listening and learning state to a certain extent. It is hoped that this study can provide a new idea for the development and research of intelligent classroom.We believe that there are still deficiencies in this paper. In order to realize high-performance multi person face recognition, we need to further study the CNN (convolutional neural network) for application to emotional information extraction, the performance optimization design of hardware platform and the implementation of intelligent intervention strategy. In the future, we will further improve the research on students' classroom learning status, and hope to expand its application to other aspects of educational activities to help students learn efficiently.

Acknowledgements The authors would like to acknowledge the financial support from Fujian Science and Technology Planning Project (Grant No. 2021H0027), 2021 Fujian Province Undergraduate Education and Teaching Reform Research Project (Grant No. FBJG20210083).

References

1. W. Dongqing, E. Han, Q. Meiling, L. Haiyan, Dynamic generative data collection method and model of smart classroom based on situational perception. Res. Audio Visual Educ. **39**(05), 26–32 (2018)
2. Z. FanMei, W. Yan, Research hotspot and trend analysis of domestic intelligent learning based on CiteSpace. China Educ. Inf. (03), 6–10 (2018)
3. L. Xiaodong, W. Baoyun, Research on college English classroom situational flipping model. For. Lang. Audio Visual Teach. (06), 71–77 (2017)
4. W. Dian, *Design and application of intelligent classroom interaction model based on learning situation analysis.* Yunnan Normal University (2020)
5. X. Zhang, T. Xu, W. Sun, A. Song, Multiple source domain adaptation in micro-expression recognition. J. Amb. Intell. Human. Comput. (2020)

6. J. Li, D. Shi, P. Tumnark, H. Xu, A system for real time intervention in negative emotional contagion in a smart classroom deployed under edge computing service infrastructure. Peer-to-Peer Netw. Appl. (2020)
7. H. Qin, J. Yan, L. Xiu, X. Hu, Joint training of cascaded CNN for face detection, in *2016 IEEE Conference on Computer Vision and Pattern Recognition (CVPR)* (2016)
8. H. Monkaresi, N. Bosch, R.A. Calvo, S.K. D'Mello, Automated detection of engagement using video based estimation of facial expressions and heart rate. IEEE Trans. Affect. Comput. **8**(1), 15–28 (2017)
9. T. Ojala, M. Pietikäinen, D. Harwood, Performance evaluation of texture measures with classification based on Kullback discrimination of distributions. Int. Conf. Pattern Recogn. **3**, 582–585 (1994)
10. V. Kazemi,J. Sullivan, One millisecond face alignment with an ensemble of regression trees, in *IEEE Conference on Computer Vision & Pattern Recognition.* IEEE (2014)
11. Z. Zhang, C. Jung, GBDT-MO: gradient-boosted decision trees for multiple outputs. Inst. Electr. Electron. Eng. (IEEE) (7) (2021)

Research on Student Media Literacy Construction Under the Framework of Network Information Security

Zhenyu Zhang

Abstract Network information security, as the focus of current social development, to be in the development of science and technology innovation, to build with computer technology as the core of the campus network information security framework, the full implementation of the student's media literacy training work, must be in the original, on the basis of network information security issues, from the perspective of the all-round development of students, In view of the requirements of monitoring security audit of network system, the model and implementation process of network information security framework are clarified, so as to lay a foundation for cultivating students' media quality. Therefore, on the basis of understanding the current situation of the construction of the network information security framework, this paper makes an in-depth analysis of the overall framework design and key technologies for the requirements of practical education innovation, so as to clarify how to cultivate students' media literacy under the network information security framework.

Keywords Network · Information security · Students · Media literacy

1 Introduction

More in science and technology innovation, the social economy gradually into the information age, the use of computer technology as the core of information science and other fields is widely used in social economy, not only can strengthen the comprehensive strength of our country, also can further enhance the level of social and economic development, so the content is also seen as evaluating the main index of national economic power. Network information not only changes human life and work mode, but also transmits information data to all parts of the world in a short time, which not only brings convenience to the public, but also becomes the basis

Z. Zhang (✉)
School of Marxism of Hohai University, Hohai, China
e-mail: zhzhyu168@163.com

S. Patnaik and F. Paas (eds.), *Recent Trends in Educational Technology and Administration*, Learning and Analytics in Intelligent Systems 31,
https://doi.org/10.1007/978-3-031-29016-9_48

for the design of network information infrastructure and security architecture. Especially to the field of education, comprehensive promotion in our country's network information platform, the diversification of information transmission channel has changed the traditional teaching mode, the education reform is the most crucial is a comprehensive grasp of the Internet technology theory, such not only can this solve the problem of practice teaching, teaching resources platform can also and all over the world to build good relations of cooperation, It can also carry out online teaching and so on. From the perspective of overall development, the network information security framework built on the basis of the Internet can make education management work more secure and effective. The construction of students' media literacy and cultivate work is under the background of new era, an important part of the promotion of Internet information technology, this not only helps students to actively participate in related construction, can also have more high quality in the new media education environment of knowledge and skills, this is the practice of education reform and to promote the network information security framework of basic requirements. Therefore, in the continuous development of social economy and science and technology, faced with people's increasingly high expectations of application and development, researchers should pay attention to the in-depth discussion of network information security in the field of education, so as to build a good educational guidance environment for students' media literacy. At present, with the popularization of education network application, the security problem of campus network is becoming more and more obvious, and the information security performance is also affected by this, and higher requirements are put forward. Proposed under the background of facing the era of big data and the urgent need of network security audit system monitoring based on the sustainable development of the education operation management goal, further discusses the applicable to domestic media quality education of university students of the network information security framework, system functions, and contains the overall framework, database design, system implementation, this paper analyzed the content of the, Finally, the overall function of the system was tested and studied. It is found that the construction of students' media literacy under the framework of network information security has certain practical significance [1–3].

2 Method

2.1 *Theory and Technology*

As the simplest network management protocol, S N M P protocol is a management standard based on transmission control protocol. It helps network management personnel to deal with problems scientifically in a timely manner and improves the operation efficiency of network management. At the same time, the network manager can also use S N M P to obtain the notification information and event report of the network node, so as to identify the network operation problems. The construction of database system is mainly to meet the storage and retrieval requirements of structured data information. On the one hand, sharing function should be provided; on the other hand, unified voice control should be carried out to ensure that data pairs exist independently.

In the network information security education, the student-centered media literacy education can be regarded as a kind of educational practice with strong purpose. How to propose clear educational objectives is the basic condition for actively carrying out the cultivation of students' media literacy. In ideological and political education in colleges and universities, students' media literacy is an extremely important educational content. Its purpose is to cultivate students' consciousness ability to cope with various media and information in the era of "we media", so that they can have corresponding media literacy and become active "media citizens" in the era of big data. According to the accumulated experience of media literacy education in colleges and universities, metacognition, as the extension and reconstruction of information literacy by technology characteristics in the age of social media, will be regarded as the main basic theory. On the basis of paying attention to multiple abilities such as information retrieval, information acquisition, weekly understanding and information evaluation, More attention should be paid to students' ability to participate in sharing, cooperative innovation and integrated application in social media, mobile technology, open resources, online communities and other environments. Compared with media literacy cultivation teaching in the traditional sense, meta-literacy pays more attention to setting participatory and cooperative learning in interactive space, so as to pay more attention to the development of advanced critical thinking and metacognitive learning while creating and disseminating media information, so as to have the awareness and ability of lifelong learning.

2.2 System Requirements

First of all, it is necessary to make clear the implementation requirements of network construction. Since the campus information network system is extremely comprehensive, it is necessary to combine the work needs of the school and cooperate with professional construction personnel to communicate and design, so as to build diversified network schemes and technical routes. The actual basic structure is shown in Fig. 1: When carrying out media literacy education in college education, we should ensure that the educational content is applicable and popular, and focus on serving students' life and learning ability. Nowadays, when educational scholars pay more attention to the perspective of comprehensive development when studying the curriculum system of cultivating students' media literacy education. The specific contents are as follows.

Secondly, it is necessary to study the security performance of the system. This work is mainly divided into two aspects. On the one hand, it refers to the security threats that the campus network is currently dealing with, and on the other hand, it

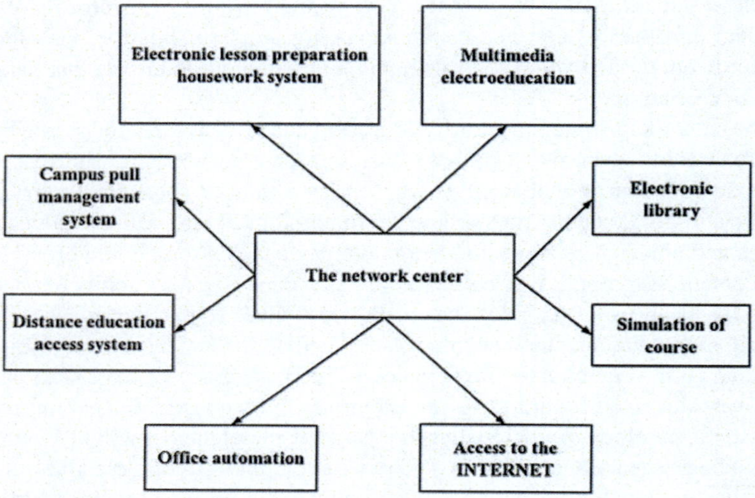

Fig. 1 Basic architecture of campus information network system

refers to the development needs of the network equipment monitoring and management system. For example analysis of campus internal network security problems, found that in order to avoid the hacker attacks, technical destruction and illegal access to bad behavior such as security question consciousness, should be set during practice running perfect management system, pay attention to the internal network systematic regulation measures are put forward according to the campus, only in this way can effectively prevent harmful related issues [4–6].

Again, the functional requirements of the system operation should be clarified. The specific content involves user management, object management, fault management, daily inspection and monitoring management. Taking fault management as an example, monitoring objectives are divided into five states, and the status of the monitoring target is the same as that of the monitoring indicator at the most serious level. If one indicator status is displayed as a failure, then the status of the monitoring target is also a failure. Comprehensive management and control of system faults requires management personnel to scientifically adjust and design fault parameters on the basis of identifying all system users to ensure that all internal functions of the system can cooperate with each other. The specific call method is shown in Fig. 2.

Finally, the non-functional requirements of the system are studied, which refers to the essential functions provided by the operation of the microsystem, including performance requirements, security requirements, reliability requirements, application requirements and other contents.

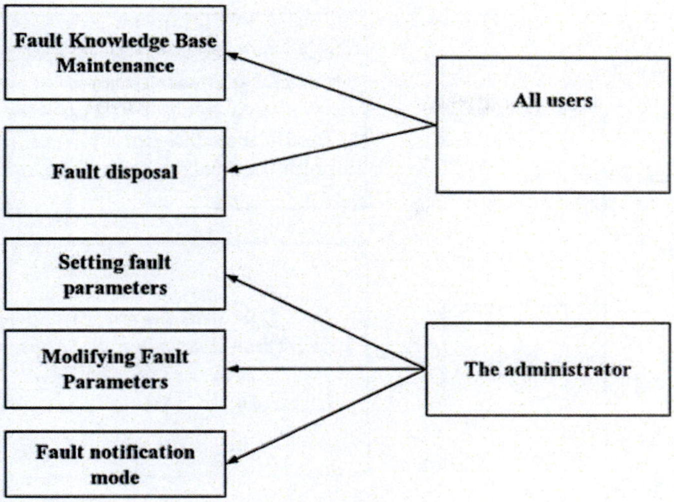

Fig. 2 Invocation process diagram of fault management

2.3 *System Design*

The design and promotion of network information security framework should be combined with the above requirements analysis for the overall design, and focus on the overview of user management, target management, fault management and other content. The system topology diagram is as follows: The educational content mainly starts from three aspects: first, the cultural basis. On the one hand, we should pay attention to the cultural background, on the other hand, we should explore the students' scientific spirit; Secondly, independent development. On the one hand, students should be guided to study independently, have self-reflection and confidence consciousness, on the other hand, they should form a sound personality, have the right concept of self-management and cherish life. Finally, social engagement. On the one hand, students should be guaranteed to take responsibility actively in the network information environment. On the other hand, students should learn to use theory and technology to solve problems and constantly innovate. It should be noted that, when constructing the curriculum system of students' media literacy, we should pay attention to comprehensively analyzing students' behavioral characteristics and learning needs, create curriculum models and teaching strategies consistent with the era of big data, and pay attention to guiding them to establish correct views on media and society, so as to improve the quality of educational guidance.

Under the background of education in the new era, taking media convergence as an example, teachers can guide teaching according to the content as shown in Fig. 3 below. Firstly, teachers can explain the main definitions of relevant theories to students, then guide students to clarify the influence brought by media convergence, and finally evaluate the development trend of media convergence, so that students can

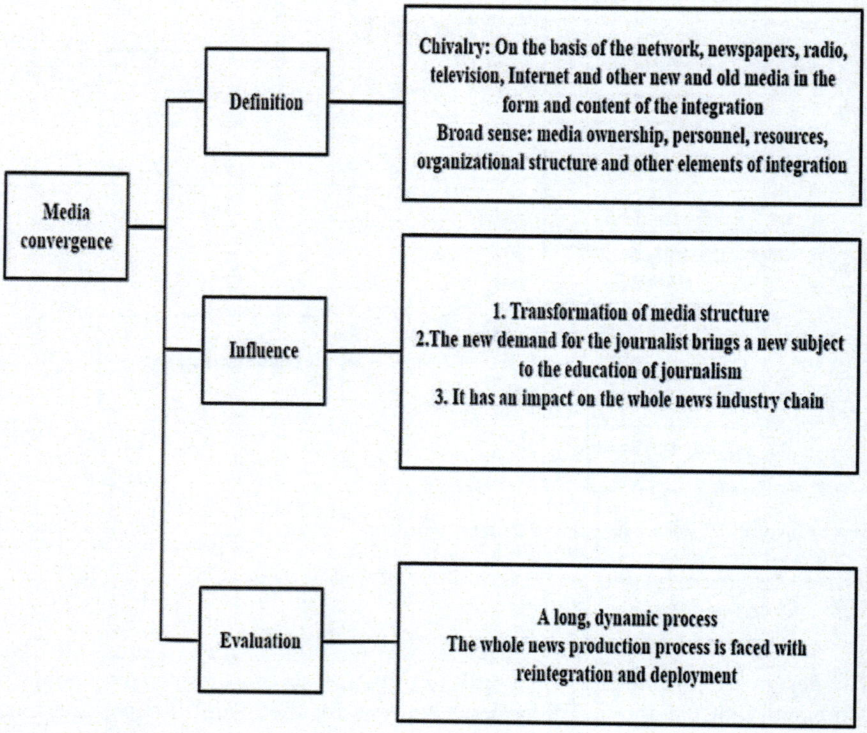

Fig. 3 Topology of the system framework

truly realize the important role of media convergence in the news industry. Change their own ideas and consciousness, and construct a brand new idea of education and learning.

From the perspective of the learning task of "media convergence" course, attention should be paid to guiding students to actively learn the use of different media languages under the background of media convergence, such as information acquisition, information expression, observation thinking, etc., combining with the study of its technical characteristics and development rules, to improve the ability of cross-media sharing and communication, and improve the level of understanding, analysis, evaluation and portable media communication content. On the basis of the basic connotation of correct value information, cultivate the development of students' practical attitude. The actual function realization flow chart is Shown in Fig. 4.

The first is user management, which pays more attention to the setting of user permissions and the relationship between the access to resources. Generally speaking, the design of system permissions should comply with the following principles: first, to use the selected way to predict the number of roles in the system, so as to control access according to the role permissions; Second, users can have multiple roles, thus

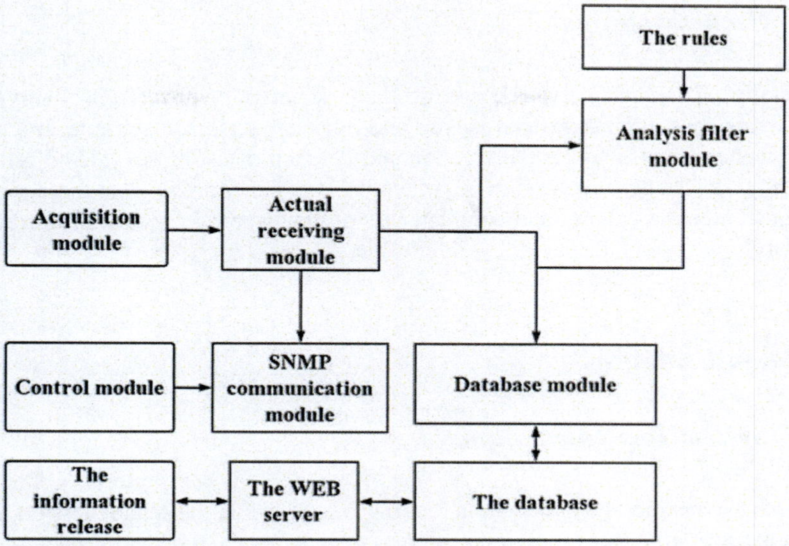

Fig. 4 Realization model diagram of system functions

obtaining the actual permissions of different roles. Thirdly, the user authority design should be realized by using control list, user ID and role mapping table.

This business function is mainly divided into two parts: on the one hand, it refers to the registration and processing of monitoring targets; on the other hand, it refers to the processing of monitoring targets. The former needs to register indicators during monitoring, while the latter needs to record data information during monitoring.

Finally, it refers to fault management. This work is mainly divided into three points: one is to maintain and deal with the fault parameters, the other is to warn the fault information, and the third is to set up the daily inspection scheme for the operation of the campus network information security framework.

The construction and education of students' media literacy under the framework of network security is directly related to the expression of their network demands. From the perspective of time, students' media literacy refers to the ability and quality of receiving and transmitting information by using network technology, which involves their own values, network attitudes, development level and other contents. Since the Internet has both media attributes and social attributes in practice, students' media literacy not only reflects media literacy, but also civic literacy. Combined with the network information security framework to guide students to strengthen their media literacy, can in the new media age become builders of the campus network culture, consumers, producers, fully arouse the enthusiasm of their independent to participate in school affairs, expand their practice in the development of various channels, and have respect and understand others speech right in good quality, Present your concerns and suggestions in a constructive manner [7–9].

2.4 Security Design

Generally speaking, network system data security control from encryption processing, operation log, data backup and other aspects. For example, log operation management involves the creation and deletion of user roles, the adjustment of user permissions, and the start of system backup, etc. Starting from these aspects can guarantee the validity and completeness of data stored in network information security.

3 Result Analysis

3.1 System Implementation

In order to construct the information security framework in the school network platform, the core work is to monitor the interactive network and transfer the data packets from the monitored host to the computer for analysis and filtering. The system uses the technology of S N M P and R N O N to complete data acquisition. The former is a simple network management protocol, while the latter is a special remote monitoring management information base, which can exchange network information data between the network platform and the control system. Under the framework of network information security, the device status monitoring module should be realized. On the one hand, it is necessary to collect the state data information of the equipment, usually combined with the development platform of U C D S N M P to design and implement; On the other hand to monitor network equipment status data, mainly within the collection of the campus network router port outflows and inflows after the number of bytes, a statistical data, the accurate calculating various kinds of key data, and use charts in a web browser, managers can through real-time observation in finding potential problems, and on the analysis of the prediction of mastering the development trend of the network.

3.2 System Test

For the performance test analysis of the system mentioned above, a variety of test scenarios should be constructed to simulate various situations that the system needs to deal with under complex conditions, so as to clarify the processing performance, effectiveness and stability of the system operation. In this paper, servers, switches and routers are regarded as the main monitoring equipment, servers matching the production environment are selected for monitoring, and private networks are used to effectively connect the monitored equipment and the network monitoring system test machine. The specific structure is shown in Fig. 5.

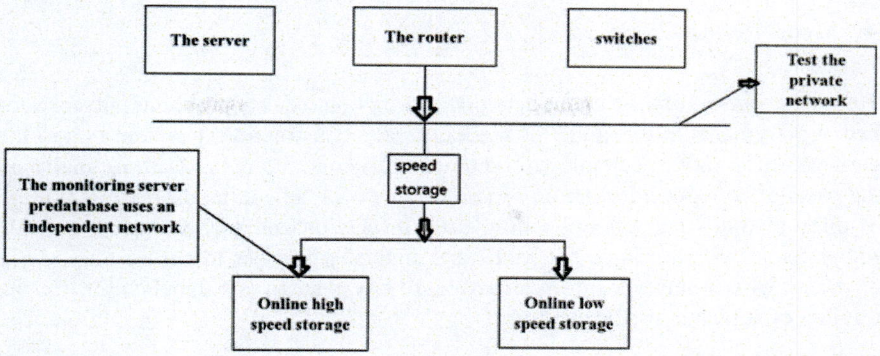

Fig. 5 Test environment design

On the basis of defining benchmark test scenarios, selecting test cases and mastering concurrency, we can use benchmark scenario tests to understand the operating data of the system before optimization. After setting the pressure test scenario, select the same test cases and concurrent quantity as the benchmark test scenario, scientifically adjust the system operation parameters, and continuously optimize the system operation performance. The final results show that the system has a strong concurrent processing capacity, in line with the development goals of system construction. At the same time, the system can be adjusted and optimized according to specific problems in multiple benchmark analysis to ensure that the processing capacity of the system meets the initial requirements [10]. When creating the network information security framework, universities should first use the firewall to control and monitor system service attack, and timely prevent rejection; secondly, use wireless LAN to provide security service function, orderly complete wireless intrusion detection and access control of mobile devices; and finally use domain controller to realize user management service. Among them, the firewall refers to the protection barrier composed of software and hardware equipment, created in the interface between the internal network and the external network, between the private network and the public network, is an effective method to enhance the network security performance. By scanning through the network communication, can effectively filter part of the attack, so as not to be executed in the target computer, at the same time firewall can also close the port, prohibit specific port outflow communication, blocking Trojan horse, finally prohibit access from special site, prevent all communication from unknown invaders. From the perspective of practical application, the network information security framework can provide technical guarantee for the implementation of students' media literacy education, and can avoid excessive security problems during the operation of the system.

4 Conclusion

To sum up, in view of the campus network information security problems puts forward new requirements to the student's media literacy construction, pay attention to the network information technology combined with continuous innovation, finally on the basis of network information security framework, select the user, target, failure of three methods for data collection, daily inspection and monitoring of multiple functions such as management, statistical analysis, In order to ensure the orderly operation of the network information security system, to lay a foundation for the training of students' media quality.

Acknowledgements Project name and No.: Basic Research Fee of Central Universities Project "Deng Xiaoping Ideas and Faith and Connotation Development Research in the New Era", Project No.: 2019B21414.

References

1. L. Sun, J. Zhang, D. Li, Research status and Necessity of media literacy education for medical students. News Sentinel **10**, 117–118 (2021)
2. J. Hu, On online media education in the teaching of morality and rule of law in primary schools—a case study of online media education for grade 4 to 6 students. Digital Teach. Primary Second. Schools **09**, 41–44 (2021)
3. J. Yu, Investigation and countermeasures on mobile phone use in class of vocational college students. Road Success (26), 53–55 (2021)
4. K. Xue, Research on Chinese teaching in higher vocational colleges from the perspective of media literacy education. J. Shaanxi Youth Vocat. College **03**, 38–41 (2021)
5. Y. Li, Let the Internet become the biggest increment to solidify the sense of community of the Chinese nation. China National Daily, 2021–08–10 (006)
6. H. Li, C. Liu, A brief analysis on the cultivation of Students' media literacy. New Wisdom (22), 65–66 (2021)
7. P. Zhang, The application of new media in college students' mental health education in the Era of "Internet+". Teach. Educ. (Higher Education Forum) (21), 52–54 (2021)
8. L. Yang, Discussion on media literacy education of marketing students under the background of new media. Qual. Mark. (14), 37–39 (2021)
9. Q. Jiang, Thinking on integrating media literacy into higher vocational ideological and political education in new media era. China Land Urban Herald **07**, 122–124 (2021)
10. R. Chen, Current situation and suggestions for improving media literacy of college students in the era of new media—a case study of Guizhou normal university students. J. Journal. Res. **12**(13), 138–140 (2022)

The Teaching Design of Reduced-Order Differential Equations Can Be Explored Under Constructivist Theory

Fei Wang, Bingjie Li, and Hui Xu

Abstract Based on constructivist theory, this paper constructs the necessity of learning points by applying teaching cases to create scenarios. Through this design, students are stimulated to explore and solve practical problems, so that mathematical knowledge is no longer boring. Secondly, the scope of a problem that can be handled by the downgrade method is clarified through explicit application, so that mathematical applications are no longer ruleless. Then, through the construction of meaning, the constant commutation method of the first-order equation is inspired by the constant commutation method of the first-order equation based on the existing knowledge, and the mathematical method of solving the problem is provided. Finally, the extension section elaborates the limitations of the reduced-order method in solving higher-order differential equations and gives the extended solutions, so as to expand students' horizons and better understand the role and significance of the reduced-order method in solving differential equations.

Keyword Higher-order ordinary differential equation · Descending order · Transformation · General solution

1 The Analysis of Academic Conditions

First of all, before learning this part, students have learned the basic concepts of ordinary differential equations and the solution methods of four types of first-order differential equations, established a relevant knowledge system, laid the necessary

---

F. Wang (✉)
Department of Applied Mathematics and Military Cryptography, Air Force Engineering University, Xi'an, China
e-mail: wangfeishuxue@163.com

B. Li · H. Xu
Department of Applied Mathematics and Military, Air Force Engineering University, Xi'an, China

S. Patnaik and F. Paas (eds.), *Recent Trends in Educational Technology and Administration*, Learning and Analytics in Intelligent Systems 31,
https://doi.org/10.1007/978-3-031-29016-9_49

foundation for the learning of this content, and provided a favorable learning situation. This knowledge is mainly the use of lower-order thinking methods to solve higher-order differential equations, how to transition from the solution of first-order differential equations to the solution of higher-order differential equations, beginners have such a transformation difficulty when learning this content. Secondly, degradable differential equations have complex memory points, easy to confuse and abstract derivation processes, which pose great challenges to learners. Taking a step back, even if the types of downgradable equations are very clear, students still face the dilemma of using the downgrading method to solve higher-order differential equations in a complicated or unsolvable process, which is highlighted by why they should be downgraded, what kind of higher-order equations can be degraded, how to downgrade, and so on. In order to solve these outstanding factors that are detrimental to student learning, we design the classroom content through constructed ideas to achieve breakthroughs in classroom difficulties.

2 Teaching Objectives

2.1 *Syllabus Objectives*

According to the syllabus, the learning of this part of the content requires students to achieve the knowledge goals of understanding the nature of the downgrade method, mastering the types of differential equations that can be reduced, clarifying the transformations used by the downgrade method, mastering the downgrade process, and solving three types of degradable higher-order differential equations using the downgrading method. Under the requirements of such a knowledge goal, the core problem is the understanding of the essence of the downgrade method, which is also the key link in solving students' transition from solving first-order differential equations to solving higher-order differential equations. Then, by understanding the essence of the descending method, it requires mastering the basic transformations and solving the general solutions and special solutions of the three types of differential equations that can be reduced, and on this basis, students are further stimulated to explore the different requirements of different transformations for the treatment of the descending method, so as to achieve the cultivation of students' ability goals. Since the transformation used in the downgrade method is not unique, students are required to be able to flexibly use the transformation to solve problems. In order to stimulate the enthusiasm of students to learn actively, the teaching cases are used to explore the connection between practical problems and higher-order differential equations, and improve students' ability to analyze and solve problems.

2.2 *Emotional Goals*

The cultivation of emotional goals begins with the integration of the concept of "classroom thinking" into the learning of this classroom knowledge [1–3]. Trainees are subtly influenced by the mainstream values of society, which is embodied in the video case study of "torpedo chasing ships and sinking ships", so that students can experience the decisive role of a good commander in war [4, 5]. In this way, in the process of teaching and educating people, students are aroused to think about strategies and control of unseen things, so as to enhance students' decision-making ability and application ability [6–8]. Secondly, the cultivation of emotional goals highlights the students' attention to the learning of mathematical knowledge, mathematics in life is everywhere, and mathematical models can be established with the help of geometric relationships between variables to solve big problems in teaching cases [9, 10]. Finally, the emotional goal is landed, cultivating students' firm ideals and beliefs and stimulating students' cognition of science and technology leading the future according to the solution results listed in the case.

2.3 *Competency Objectives*

By understanding the essential requirements of the descending method, mastering the basic transformations, solving the general solutions and special solutions of three types of differential equations that can be reduced, and exploring the different requirements of different transformations for the treatment of the descending method. Secondly, because the transformations used in the downgrade method are not unique, students are required to flexibly use transformations to solve problems. Finally, the trainees' enthusiasm for active learning is stimulated, and the military cases are used to explore the connection between practical problems and high-order differential equations, so as to improve students' ability to analyze and solve problems.

3 Teaching Methods and Implementation Process

In the process of teaching implementation, the teaching method of this lesson is mainly task-driven and heuristic teaching method. Taking the teaching case "Torpedo Pursuit Ship Problem" as the precursor, a differential equation model is established, but the model cannot be solved with existing knowledge, and the "reduced order method" is proposed. We inspire thinking through the solution of first-order equations, getting several types of descending method ideas of higher-order differential equations. constitute the task-driven of this lesson, and inspire students to think about how to descend through demonstration, imitation, and classification, that is, the process of classification and interpretation, which is the focus of this lesson, and it is also a difficult point. Finally, the teaching case is solved by model classification and the teaching process as a whole is closed-loop.

4 Innovative Points in Classroom Teaching Theoretical Basis for Teaching Innovation

4.1 Theoretical Basis for Teaching Innovation

The theoretical basis of the teaching innovation in this lesson is constructivist learning theory, which includes four elements, scenario creation construction, collaborative interaction construction, conversational communication construction, and practical meaning construction. It is embodied in the first application case "Torpedo Pursuit Ship Problem" as the precursor to create a scenario, stimulate the curiosity of the trainees, and make the trainees intuitively understand the necessity of the knowledge points learned through the construction of the scenario. Secondly, the students are the center and the collaborative interaction is constructed in the form of rain classrooms and the enthusiasm and initiative of the students are fully utilized through collaborative construction. Again, the use of information teaching methods to construct conversational communication, through conversation construction to improve teaching efficiency. Finally, the problem is set as the driver, the meaning construction is solved to solve the difficult problem and the mathematical method of solving the problem is provided. Situational construction solves the problem of why to learn, meaning construction solves the problem of what to learn, and collaborative construction and conversation construction solve the problem of how to learn.

4.2 Layout and Refinement Innovation of Teaching Content

4.2.1 Scenario Creation and Mathematical Abstraction

Scenario creation is reflected in the actual problem of torpedo chasing ships and sinking ships with the video resource "The Problem of Torpedo Pursuit Ships" as the guide. According to the practical significance of the research question, the following assumptions are made. First of all, the direction of the torpedo is always aimed at the enemy ship and the speed is greater than the speed of the enemy ship. secondly, the torpedo trajectory is determined in advance to ensure that the enemy ship hits the enemy ship in the territorial waters of our territory. finally the torpedo hits the enemy ship during its voyage time when the attack is launched. Ask the question of when a torpedo attack can be launched, prompting students to think. This case is actually a strict mathematical abstraction of the torpedo trajectory equation and when the enemy ship is hit by a torpedo. When the enemy ship is set at a distance

a directly east of our ship, our ship fires guided torpedoes and the direction of the torpedo is always aimed at the enemy ship, and the enemy ship flees at maximum speed v_0 in the direction of due north. Set the torpedo speed to v_1 and $v_1 > v_0$, find the torpedo trajectory equation and ask when the enemy ship was hit by a torpedo.

4.2.2 Establishment of Mathematical Model

Figures 1 and 2 are abstracted into mathematical problems, and their corresponding mathematical models are as follows:

$$\begin{cases} xy'' = -\dfrac{v_0}{v_1}\sqrt{1+y'^2} \\ y|_{x=-a} = 0,\ y'|_{x=-a} = 0, \end{cases} \tag{1}$$

In order to explore the solution of the problem, the students are first guided to establish a suitable coordinate system, and then the mathematical model of the second-order differential equation is established by using the existing mathematical knowledge and geometric intuition with the help of hypothetical conditions, so as to lead to the research object of this lesson. By creating a scenario, a practical problem

Fig. 1 Torpedo chasing ships

Fig. 2 A torpedo hits a ship

that is generally of interest to the students is designed, which exercises the students' problem-solving ability, while improving the interest of mathematical knowledge and cultivating the modeling ability of the students to the greatest extent.

4.3 Problem Chain Drive Guidance

The main content of this lesson is to solve higher-order differential equations using the reduced-order method, focusing on the types of degradable second-order differential equations and the methods of solving them. The biggest confusion of objective analysis and judgment learning before the teaching activities are carried out mainly focuses on why to degrade, what kind of higher-order equations are degradable and how to reduce the order. Through the setting of the three questions, the enthusiasm and initiative of the students are fully mobilized, and the deep thinking is rooted in the entire teaching activities. At the same time, in view of the possible confusion of students, the principles of asking questions, analyzing problems and solving problems are gradually deepened in the setting of teaching content. Through the clear application, the scope of a problem that can be handled by the downgrade method is clarified, so that the mathematical application is no longer ruleless, so that the students understand why they learn and achieve the purpose of using it.

4.4 Innovation of Teaching Methods

In the process of teaching implementation, the teaching method of this lesson is mainly task-driven and heuristic pedagogy. The first is pre-class preparation, using the rain classroom to check the students' preparation in the form of multiple choice questions, and understand the dynamics of the students in real time. Then, taking the teaching case as the traction, using geometric intuition to establish the relationship between variables to obtain a mathematical model of the second-order differential equation, how to solve the model? We use the solution of first-order equations to inspire thinking to obtain several types of lower-order methods of higher-order differential equations, which constitute the task-driven of this lesson. Then we will talk about how to descend in the end, that is, the process of classifying and solving doubts, which is the focus of this lesson and the difficulty. Then use the rain classroom to check the learning effect of students in real time to consolidate the knowledge learned. Through demonstration imitation classification, students are inspired to think about solving three types of reducible differential equations. Finally, the demotion method is used to solve the teaching case, and according to the solution results of the model, the students in the new era are trained to establish the belief that they are called to come, can fight, and will win the battle, and the teaching process as a whole shows a closed-loop state. The construction of meaning is based on existing knowledge as a bridge, and the descending method of higher-order equations is inspired by the constant transmutation method of first-order equations. Collaborative sessions consolidate the knowledge learned in real time through rain class, pre-class preview, in-class inspection, and online exchange and discussion after class. Meaning construction solves the problem of what to learn, and collaborative conversation solves the problem of how to learn, and ultimately allows students to achieve learning.

5 The Teaching Contents Difficult to Break Through

Based on years of teaching experience and the results of the answers to the rain classroom preview questions, there are three difficulties in learning this part of the knowledge, which is also a difficulty in teaching.

5.1 Why Downgrade

Since there is no existing processing method to rely on in the previous learning content, it is difficult for students to think of solving the solution of higher-order differential equations using the descending order method. First of all, using the analogy method of thought, with the help of the constant variation method of the first-order linear non-homogeneous differential equation to solve the formula, from another angle through deformation to obtain the transformation $u = ye^{\int p(x)dx}$, The expressions for derivatives on both sides of the equation combined with first-order linear non-homogeneous differential equations are then calculated by integral $u = \int Q(x)e^{\int p(x)dx}dx + C$. Back to the original transformation, the general solution of the original equation can be obtained. Such a solution method inspires us that the essence of degradation is realized through transformation, that is, the general solution of the equation can be solved by using the idea of degradation. Inspire students to seek the general solution of the second order equation by using the thought method of the degradation method of the first order linear homogeneous differential equation.

5.2 What Kind of Higher-Order Equations Can Be Downgraded, and How to Downgrade

Although there is a thinking of descending order, it is necessary to look for the types of equations that can be reduced and the corresponding treatment methods. The easiest way to get started is to deal with special situations. Considering that the student has already understood through the reason why the order is downgraded, it is necessary to find a suitable transformation and the transformation must be included y'.

For this we choose the simplest transformation $u = y'$ trying it, thus the original equation

$$y'' = f(x, y, y') \tag{3}$$

It is deformed to $u' = f(x, y, u)$, at this point, the order of the equation is reduced to the first order to achieve the desired goal. Secondly, the special will be extended to the general to further solve the problem. Although the order of an equation is reduced from second order to first order, the equation $u' = f(x, y, u)$ Still can't be solved, Consider further that the original equation does not contain a dependent variable y situation, Thus a case of a second-order differential equation that can be downgraded was found $y'' = f(x, y')$. Analogous methods are also used, Consider not explicitly containing independent variables x, a second-order differential equation can be found $y'' = f(y, y')$ and only the independent variables are explicit x of second-order differential equations $y'' = f(x)$. Finally, through demonstration, imitation, and classification through multimedia visualization in the whole process, three types of differential equations and solution methods that can be reduced are found.

6 The Solution Method of the Second-Order Equation Can Be Reduced

Through the selection of example problems, students are shown that the transformation used in the equations that can be solved by the reduced order method is not unique, and then through such a phenomenon, the students' understanding of the non-uniqueness of the transformation is deepened, and the thinking of flexible selection of transformations is established to facilitate the solution of the equations. Secondly, through the example discussion, we find that if the equations that formally conform to the reduced order method cannot find a suitable transformation to solve their solutions or when the general solutions of many differential equations cannot be expressed as elementary functions, what should we do? In fact, in engineering applications, most of them only need to master the approximate solution, so we can use matlab to find its numerical solution, that is, the approximate solution, which will also be the problem that students will explore about the numerical solution of differential equations in the future, thus expanding the students' cognition and vision and deepening the classroom effect. The specific classroom activities are also carried out in the form of a chain of problems layer by layer, first of all, the problem can be solved by the reduced order method of the equation used in the transformation is the only one? To solve this problem, we have selected an example question. For example, the $y'' + y' - 2y = 0$ general solution. This is an implicit x second-order differential equation transformations $P = y'$ can be used to downgrade, But the solution process is very cumbersome. If a transform $u = y' + 2y$ is selected, then there is $u' = y'' + 2y' = (y'' + y' - 2y) + (y' + 2y)$, That is, the original equation is downgraded to a first-order differential equation $u' = u$, The general solution $u = C_1 e^x$ of the equation is obtained by separating the variables. Substitute the transformation $u = y' + 2y$ to obtain a first-order linear non-homogeneous differential equation $y' + 2y = C_1 e^x$, Thus the general solution of the original equation is $y = \frac{C_1}{3} e^x + C_2 e^{-2x}$. It is concluded that the way to achieve a reduced-order transformation is not unique and choosing different transformations will solve different processes, some troublesome, some simple. Through the setting of this problem, the students deepen their understanding of the uniqueness of the transformation, and the flexible selection of the transformation is easy to solve the equation.

Problem 2: How to solve the type of equation that conforms to the descending order, but cannot find a suitable transformation to find its general solution, or the general solution of the differential equation cannot be expressed as an elementary function?

$$\text{For an example,} \quad y'' - (1 - y^2)y' + y = 0 \tag{3}$$

It is a second-order differential equation with no apparent independent variable x, Falls within the scope of the equations we are discussing today, But its solution cannot be represented by elementary functions. How do we study such differential equations? In fact, in engineering applications, we only need to master its approximate solution, so we can use matlab to find its numerical solution, that is, the approximate solution, which will also be the problem that students will explore about the numerical solution of differential equations in the future. Here the selection $y(0) = 2$, $y'(0) = 0$ as an initial condition can be used matlab to make a numerical solution as shown in Fig. 3.

The numerical solution of a class of differential equations that cannot be solved by the reduced order method is expanded by the drawing tool, which complements the understanding method and is conducive to the exploration of students to expand the problem.

In addition to the above three types of second-order differential equations, how to solve for other types of second-order differential equations? How do you solve higher-order differential equations? We found that there are many problems worth studying, science is endless and the gradual expansion of problems arouses students' thinking and enhances their desire to explore problems.

7 Concluding Remarks

Reconstruct the teaching content, set up the situation, and take the problem as the driving force to deepen layer by layer, so as to realize the adherence to the student-centered classroom center, make the classroom teaching activities full of vitality, and let the abstract problems shine under the support of the information teaching model.

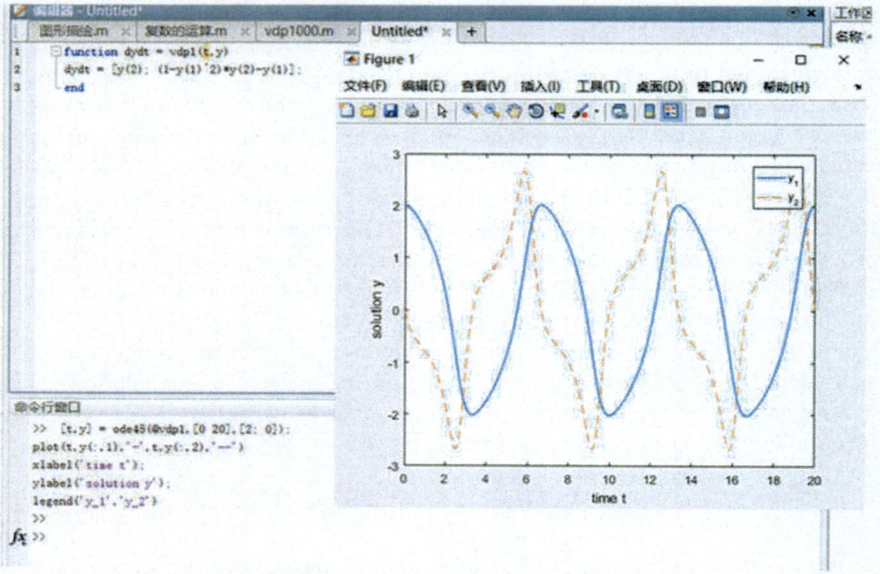

Fig. 3 Image of the solution of the equation

References

1. M. Ruibin, B. Li, Q. Xiwei, et al., Application and practice of interest induction in classroom teaching. Educ. Teach. Forum **11**(48), 123–124 (2017)
2. L. Ping, L. Hao, L. Deming, Y. Yi, Research and practice of interactive teaching methods based on cloud platform. Educ. Teach. Forum **11**(48), 118–122 (2017)
3. W. Guochen, Constructivist teaching theory and teaching model reform. J. Arm. Police College **12**(19), 48–50 (2003)
4. Z. Qiushi, University teaching method innovation and improvement of higher teaching quality. Yangtze River Series. Theoret. Res. (11), 244 (2017)
5. L. Qunxia, W. Xuan, Research and practice of case teaching method in curriculum teaching. Educ. Teach. Forum **11**(48), 157–158 (2017)
6. J. lianhe, On the application of heuristic education in teaching. Vocati. Educ. Tiandi (Xuezhou Weekly) **12**(36), 34–35 (2017)
7. Z. Song, D. Shuhua, Problems and path optimization of ideological and political construction of college curriculum. School Party Build. Ideol. Educ. **10**(19), 58–60 (2021)
8. W. Qian, D. Zhiyong, On the strategy of ideological and political construction of current college curriculum. Jiangsu Higher Educ. **5**(14), 94–98 (2021)
9. Z. Ye, W. Dejiang, Z. Huiling, Curriculum, teaching and educating people—construction and practice of ideological and political education in science and engineering disciplines and professional courses.Mech. Pract. (2), 214–218 (2020)
10. C. Shuli, The realistic predicament and coping mechanism of ideological and political construction of college curriculum in the perspective of collaborative education. Teach. Res. (3), 89–94 (2021)